T0238455

Lecture Notes in Computer Science 3256

Commenced Publication in 1973
Founding and Former Series Editors:
Gerhard Goos, Juris Hartmanis, and Jan van Leeuwen

Hartmut Ehrig   Gregor Engels
Francesco Parisi-Presicce
Grzegorz Rozenberg (Eds.)

Graph
Transformations

Second International Conference, ICGT 2004
Rome, Italy, September 28 – October 2, 2004
Proceedings

 Springer

Volume Editors

Hartmut Ehrig
Technical University Berlin
Department for Software Technology and Theoretical Informatics
Sekr. FR 6-1, Franklinstr. 28/29, 10587 Berlin, Germany
E-mail: ehrig@cs.tu-berlin.de

Gregor Engels
University of Paderborn
Faculty of Computer Science, Electrical Engineering, and Mathematics
Warburger Str. 100, 33098 Paderborn, Germany
E-mail: engels@upb.de

Francesco Parisi-Presicce
George Mason University
Department of Information and Software Engineering
4400 University Drive, Fairfax, VA 22030, USA
E-mail: fparisip@gmu.edu

Grzegorz Rozenberg
Leiden University
Leiden Institute of Advanced Computer Science (LIACS)
Niels Bohrweg 1, 2333 CA Leiden, The Netherlands
rozenber@liacs.nl

Library of Congress Control Number: 2004112162

CR Subject Classification (1998): E.1, G.2.2, D.2.4, F.1, F.2.2, F.3, F.4.2-3

ISSN 0302-9743
ISBN 3-540-23207-9 Springer Berlin Heidelberg New York

Springer is a part of Springer Science+Business Media

springeronline.com

© Springer-Verlag Berlin Heidelberg 2004
Printed in Germany

Typesetting: Camera-ready by author, data conversion by Olgun Computergrafik
Printed on acid-free paper   SPIN: 11325826   06/3142   5 4 3 2 1 0

Preface

ICGT 2004 was the 2nd International Conference on Graph Transformation, following the first one in Barcelona (2002), and a series of six international workshops on graph grammars with applications in computer science between 1978 and 1998. ICGT 2004 was held in Rome (Italy), Sept. 29–Oct. 1, 2004 under the auspices of the European Association for Theoretical Computer Science (EATCS), the European Association of Software Science and Technology (EASST), and the IFIP WG 1.3, Foundations of Systems Specification.

The scope of the conference concerned graphical structures of various kinds (like graphs, diagrams, visual sentences and others) that are useful when describing complex structures and systems in a direct and intuitive way. These structures are often augmented with formalisms that add to the static description a further dimension, allowing for the modelling of the evolution of systems via all kinds of transformations of such graphical structures. The field of graph transformation is concerned with the theory, applications, and implementation issues of such formalisms.

The theory is strongly related to areas such as graph theory and graph algorithms, formal language and parsing theory, the theory of concurrent and distributed systems, formal specification and verification, logic, and semantics. The application areas include all those fields of computer science, information processing, engineering, and the natural sciences where static and dynamic modelling using graphical structures and graph transformations, respectively, play important roles. In many of these areas tools based on graph transformation technology have been implemented and used.

The proceedings of ICGT 2004 consist of two parts. The first part comprises the contributions of the invited talks followed by the carefully reviewed and accepted 26 papers that were selected out of 58 submissions. The topics of the papers range over a wide spectrum, including graph theory and graph algorithms, theoretic and semantic aspects, modelling, applications in chemistry and biology, and tool issues. The second part contains two tutorial introductions to graph transformation and their relation to software and DNA computing, and short presentations of the satellite events of ICGT 2004.

We would like to thank the members of the program committee and the secondary reviewers for their enormous help in the selection process. We are also grateful to Reiko Heckel and Alexey Cherchago for their technical support in running the conference system and in editing the proceedings. Moreover, we would like to express our gratitude to the local organizers Paolo Bottoni (Chair), and Marta Simeoni who did a great job. Finally, we would like to acknowledge the always excellent cooperation with Springer, the publisher of the Lecture Notes in Computer Science.

July 2004

Gregor Engels, Hartmut Ehrig
Francesco Parisi-Presicce, Grzegorz Rozenberg

Program Committee

Michel Bauderon	Bordeaux (France)
Dorothea Blostein	Kingston (Ontario, Canada)
Andrea Corradini	Pisa (Italy)
Hartmut Ehrig	Berlin (Germany)
Gregor Engels (co-chair)	Paderborn (Germany)
Reiko Heckel	Paderborn (Germany)
Dirk Janssens	Antwerp (Belgium)
Hans-Jörg Kreowski	Bremen (Germany)
Barbara König	Stuttgart (Germany)
Bernd Meyer	Clayton (Victoria, Australia)
Ugo Montanari	Pisa (Italy)
Manfred Nagl	Aachen (Germany)
Fernando Orejas	Barcelona (Spain)
Francesco Parisi-Presicce (co-chair)	Rome (Italy) and Fairfax (Virginia, USA)
Mauro Pezzè	Milan (Italy)
John Pfaltz	Charlottesville (Virginia, USA)
Rinus Plasmeijer	Nijmegen (The Netherlands)
Detlef Plump	York (UK)
Leila Ribeiro	Porto Alegre (Brazil)
Grzegorz Rozenberg	Leiden (The Netherlands)
Andy Schürr	Darmstadt (Germany)
Gabriele Taentzer	Berlin (Germany)
Genny Tortora	Salerno (Italy)
Gabriel Valiente	Barcelona (Spain)

Secondary Referees

Alon Amsel	Olaf Chitil	Jan Hendrik Hausmann
Zena Ariola	Juan de Lara	Tobias Heindel
Thomas Baeck	Juergen Dingel	Dan Hirsch
Paolo Baldan	Carlotta Domeniconi	Berthold Hoffmann
Luciano Baresi	Claudia Ermel	Kathrin Hoffmann
Stefan Blom	Martin Erwig	Jon Howse
Achim Blumensath	Alexander Förster	Karsten Hölscher
Boris Böhlen	Giorgio Ghelli	Johannes Jakob
Tommaso Bolognesi	Stefania Gnesi	Renate Klempien-Hinrichs
Paolo Bottoni	Martin Grosse-Rhode	Peter Knirsch
Antonio Brogi	Roberto Grossi	Maciej Koutny
Roberto Bruni	Szilvia Gyapay	Vitali Kozioura
Alexey Cherchago	Annegret Habel	Sabine Kuske

Georgios Lajios
Marc Lohmann
Kim Marriott
Antoni Mazurkiewicz
Antoine Meyer
Mohamed Mosbah
George Paun
Lucia Pomello
Ulricke Prange

Arend Rensink
Davide Rossi
Jörg Schneider
Stefan Schwoon
Pawel Sobocinski
Volker Sorge
James Stewart
Sebastian Thöne
Emilio Tuosto

Niels van Eetvelde
Pieter van Gorp
Dániel Varró
Bernhard Westfechtel
Hendrik Voigt
Dobieslaw Wroblenski
Takashi Yokomori

Sponsoring Institutions

The European Association for Theoretical Computer Science (EATCS), the European Association of Software Science and Technology (EASST), the IFIP Working Group 1.3, Foundations of Systems Specification, Università di Roma "La Sapienza", Provincia di Roma, Comune di Roma.

Table of Contents

Invited Papers

Integration Technology

Chemistry and Biology

Graph Transformation Concepts

DPO Theory for High-Level Structures

Analysis and Testing

Graph Theory and Algorithms

Application Conditions and Logic

Transformation of Special Structures

Object-Orientation

Tutorials and Workshops

Improving Flow in Software Development Through Graphical Representations*

Margaret-Anne D. Storey

University of Victoria, British Columbia, Canada
mstorey@uvic.ca

Abstract. Software development is a challenging and time intensive task that requires much tool support to enhance software comprehension and collaborative work in software engineering. Many of the popular tools used in industry offer simple, yet highly effective, graphical aids to enhance programming tasks. In particular, tree views are frequently used to present features in the software and to facilitate navigation. General graph layouts, popular in many academic tools, are seen less frequently in industrial software development tools. Interactive graphs can allow a developer to visualize and manipulate non-structural relationships and abstractions in the software. In this presentation, I explore how graphical techniques developed in academia can improve "flow" for programmers using industrial development tools. The theory of "flow and optimal experiences" is used to offer rich explanations for the existence of many typical software tool features and to illuminate areas for potential improvements from graphical tool support.

---

* An extended version of this abstract is published in the IEEE proceedings of VL/HCC'04 (IEEE Symposium on Visual Languages and Human-Centric Computing), Rome, Italy, September 26-29, 2004.

H. Ehrig et al. (Eds.): ICGT 2004, LNCS 3256, p. 1, 2004.

A Perspective on Graphs
and Access Control Models

Ravi Sandhu

George Mason University and NSD Security
ISE Department, MS4A4
George Mason University
Fairfax, VA 22030, USA
sandhu@gmu.edu
http://www.list.gmu.edu

Abstract. There would seem to be a natural connection between graphs and information security. This is particularly so in the arena of access control and authorization. Research on applying graph theory to access control problems goes back almost three decades. Nevertheless it is yet to make its way into the mainstream of access control research and practice. Much of this prior research is based on first principles, although more recently there have been significant efforts to build upon existing graph theory results and approaches. This paper gives a perspective on some of the connections between graphs and their transformations and access control models, particularly with respect to the safety problem and dynamic role hierarchies.

1 Introduction

In concept there appears to be a strong potential for graphs and their transformations to be applied to information security problems. In practice, however, this potential largely remains to be realized. Applications of graph theory in the security domain go back almost three decades and there has been a steady trickle of papers exploring this potential. Nonetheless graph theory has yet to make its way into the mainstream of security research and practice. In part this may be due to the relative youth of the security discipline and the particular focus of the research community in the early years. Because of the versatility of graph representations and graph theory techniques perhaps it is only a matter of time before a strong and compelling connection is found.

Information security is a broad field and offers multiple avenues for application of graph theory. To pick just two examples, in recent years we have seen application of graph theory in penetration testing and vulnerability analysis [2, 7, 17, 20, 29] and in authentication metrics [21]. It is beyond the scope of this paper to consider the vast landscape of information security. Rather we will focus on the specific area of access control and authorization.

We begin with a brief review of access control and access control models, and then identify two specific problems of access control where graph theory has been

H. Ehrig et al. (Eds.): ICGT 2004, LNCS 3256, pp. 2–12, 2004.

employed in the past. These are the so-called safety problem and the problem of dynamic hierarchies. The rest of the paper explores past work in these two problem areas in some detail and concludes with a brief discussion of possible future research.

Access Control

Access control is concerned with the question of who can do what in a computer system. Clearly the same object (such as a file) may be accessible by different users in different ways. Some users may be able to read and write the file, others to just read it and still others who have no access to the file. Strictly speaking users do not manipulate files directly but rather do so via programs (such as a text editor or word processor). A program executing on behalf of a user is called a subject, so access control is concerned with enforcing authorized access of subjects to objects. This basic idea was introduced by Lampson in a classic paper [14] and continues to be the central abstraction of access control. Authorization in Lampson's access matrix model is determined by access rights (such as r for read and w for write) in the cells of an access matrix. An example of an access matrix is shown in figure 1. Here subject U can read and write file F but only read file G. Subject V can read and write file G but has no access to file F. A review of the essential concepts of access control is available in [25].

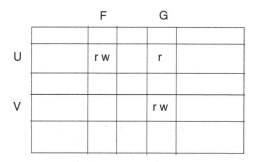

Fig. 1. Example of an Access Matrix.

The access matrix of figure 1 can be easily depicted as a directed graph with labelled edges as shown in figure 2. Thereby the intuitive feeling that there is a strong connection between graphs and access control. For convenience, we will henceforth talk of the access matrix and access graph as equivalent notions.

Access Control Models

A static access graph is not very interesting. Real computer systems are highly dynamic in that the access rights of subjects to objects change over time and new subjects and objects (and thereby new rights) are created and existing ones

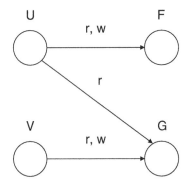

Fig. 2. Example of an Access Graph.

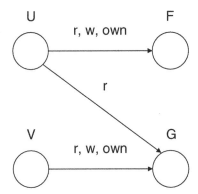

Fig. 3. Owner-Based Discretionary Access Control.

deleted. In terms of the access matrix this means that not only can contents of existing cells be changed but new rows and columns can be created and existing ones destroyed. In terms of the access graph, in addition to edge adding and deleting operations new nodes can be created and existing ones deleted.

An access control model specifies the operations by which the access graph can be changed. These operations are typically authorized by existing rights in the access graph itself. A common example of this is the "own" right shown in figure 3. The owner of a file has the own right for it and can add and delete rights for that file at the owner's free discretion. Thus subjects U and V control the rights of all subjects to files F and G respectively, i.e., U and V control the addition and deletion of edges labelled r or w terminating in F and G respectively.

The policy of owner-based discretionary access control is certainly reasonable but researchers quickly realized that there are many other policies of practical interest. For example, can the "own" right itself be granted? Some systems do not allow this. The creator of a file becomes its owner and remains its owner thereafter. Other systems allow ownership to be propagated from one subject to another. Some of these allow multiple simultaneous ownership, while other allow only one owner at a time. There does not appear to be any single policy that will be universally applicable. Hence the need for flexible access control models in this regard.

The Safety Problem

In a seminal paper Harrison, Ruzzo and Ullman [8] proposed a simple language for stating the rules by which an access graph can be changed. The resulting model is often called HRU. They then posed the safety problem as follows[1].

> Given an initial access graph and a fixed set of rules for making authorized changes to it, is it possible to reach an access graph in which subject X has α right to object Y (i.e., there is an edge labelled α from X to Y)?

It turns out that safety is undecidable in the HRU model. Surprisingly the quest to find useful models with efficiently decidable safety proved to be quite challenging. Although significant positive results have appeared over the years, a appropriate balance between safety and flexibility remains a challenge for access control models. The role of graph theory in progress on the safety problem is discussed in section 2.

Dynamic Hierarchies

Most practical access control systems go beyond the simple access graph we have discussed to provide a role (or group) construct. Thus subjects not only get the rights that they individually are granted but also acquire rights granted to roles (or groups) that they are a member of. For example, figure 4 shows an access graph in which subject U is a member of role G which in turn has the rights r and w for file F. Thereby U is authorized to read and write file F[2].

Roles are a powerful concept in aggregating permissions and simplifying their administration [5, 26]. Modern access control systems are typically role-based because of the power and flexibility of this approach. Roles are often organized into hierarchies as shown, for example, in figure 5. This is a Hasse diagram of a partial order where senior roles are shown towards the top and junior ones towards the bottom. The Supervising Engineer role inherits all permissions of its junior roles, thus this role can do everything that the junior roles can do plus more. Conversely, a user who is a member of a senior role is also considered to be a member of the junior roles. In other words permissions are inherited upwards in the hierarchy and membership is inherited downwards. In practice role hierarchies need to change and evolve over time. How to do this effectively is a challenging problem for role administration. The application of graph transformations in progress on this issue is discussed in section 3.

---

[1] The original HRU formulation is in terms of the access matrix but is easily restated as done here in terms of the access graph.

[2] This can be shown in the access graph by a "temporary" edge labelled r, w directed from U to F.

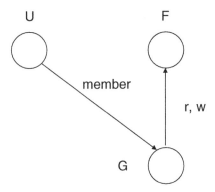

Fig. 4. Roles in Access Control.

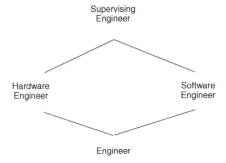

Fig. 5. An Example of Hierarchical Roles.

2   Graphs and the Safety Problem

The take-grant model of Lipton and Snyder [15] is among the earliest applications of graph theory to access control. This model was developed in reaction to the undecidable safety results of the HRU model. It takes its name from the two rights it introduces, t for take and g for grant. The depiction of these two rights in the access graph is shown in figure 6. The notation B/t∈dom(A) denotes the possession of the B/t capability in the A's domain, and is equivalent to stating that t∈[A,B] cell of the access matrix. Similarly for B/g∈dom(A). The take right in figure 6(a) enables any right that B has to be copied to A. That is any edge originating at B can be duplicated with the same label and termination node but originating at A. The grant right in figure 6(b) conversely enables any right that A has to be copied to B. That is any edge originating at A can be duplicated with the same label and termination node but originating at B.

Somewhat surprisingly it turns out that the flow of rights in the take-grant model is symmetric. This allows for efficient safety analysis in the model but severely limits its expressive power. The original formulation of the take-grant model depicted the take and grant rights in the access graph as shown in figure 6.

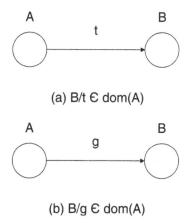

Fig. 6. Transport of Rights in the Take-Grant Model: The Original Access Graph View.

Lockman and Minsky [16] observed that a slightly different graph representation would demonstrate the symmetry of take-grant much more easily. They proposed to represent the ability for rights to flow from A to B by a directed edge from A to B labelled can-flow. The two situations of figure 6 are respectively shown in figure 7 in this modified representation. The focus of this representation is on the flow of rights rather than on the underlying right that enables the flow.

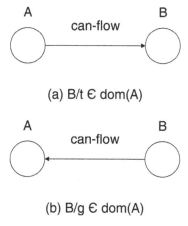

Fig. 7. Transport of Rights in the Take-Grant Model: The Modified Can-Flow View.

To complete description of the take-grant model we show the create operation shown in figure 8 using both styles of representation[3]. This diagram shows the

---

[3] The take-grant model also includes revoke and destroy operations. We omit their definition since they are not relevant here.

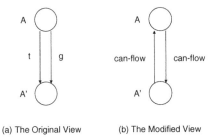

(a) The Original View   (b) The Modified View

Fig. 8. Creation in the Take-Grant Model.

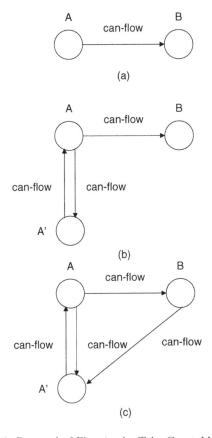

Fig. 9. Reversal of Flow in the Take-Grant Model.

result of A creating a new subject A'. A gets the A'/t and A'/g rights thus enabling can-flow in both directions.

With the modified can-flow representation symmetry of flow of rights in the take-grant model is easily demonstrated in figure 9. Figure 9(a) shows the initial situation with can-flow from A to B. This flow can be authorized by B/g∈dom(A) or by A/t∈dom(B). The specific authorization is not material. The first step is for A to create A' as shown in figure 9(b) with the resulting can-flow edges from A to A' and vice versa. These edges are authorized by A'/t∈dom(A) and A'/g∈dom(A). In particular, A'/g can be moved from dom(A) to dom(B) by virtue of the can-flow from A to B. This gives us the situation with can-flow from B to A' shown in figure 9(c). In conjunction with the can-flow from A' to A there is now a flow from B to A, thus reversing the original flow from A to B[4].

I came across these constructions during my doctoral research. I believe they demonstrate a fundamental truth. Graph representations are flexible and it is important to capture the correct properties in the edges and nodes of the graph. My subsequent work on the safety problem resulted in a number of models [1, 22, 24]. Some elements of graph theory are used in the analysis of these models but there is a probably a stronger tie with existing graph theory results. So there is potential for exploring a deeper connection here. Based on the discussion above much will depend upon a suitable representation of the access control problem in graph edges and nodes. Jaeger and Tidswell [6] have used graph notation to capture constraints and argue that this approach may lead to practical safety results. Nyanchama and Osborn have also discussed the representation of conflict of interest policies in their role-graph model [18]. Also recently Koch et al [9–12] have developed safety results directly based on the theory of graph transformations. Reconciliation of these results with the known safety results for access control models would be a step forward in understanding the insights that graphs and their transformations can provide in this domain.

3   Graphs and Dynamic Hierarchies

The use of role (or group) hierarchies in access control has a long history and considerable motivation for simplifying management of rights. The current commercial success of role-based access control products that support hierarchies is testimony to this fact. Mathematically a hierarchy is a partial order, that is a reflexive, transitive and anti-symmetric binary relation on roles. In this section we briefly look at two lines of research dealing with dynamic hierarchies for access control.

A particular kind of hierarchy called an ntree was introduced by this author [23]. The ntree has some very appealing properties for access control including the fact that it is a dimension 2 partial order, so it can be represented as the intersection of two linear orders. This allows us to label each node n with a pair of natural numbers $l(n)$ and $r(n)$, such that $u \leq v$ if and only if $l(u) \leq l(v)$ and $r(u) \leq r(v)$. The ntree also has a recursive definition based on refining an existing node into another ntree, with the base case being a forest of trees and inverted trees. There are efficient algorithms for recognizing whether or not a given hierarchy is an ntree. One of the open questions regarding ntrees is how

---

[4] Lockman and Minsky [16] went on to consider the grant-only and take-only models with the former having the symmetric flow property of take-grant but the latter allowing asymmetric flow.

to recognize hierarchies that are close to being ntrees and could benefit from a ntree representation augmented with some additional information. More generally research on hierarchies with special properties that are attractive for access control can be pursued.

The need for dynamic role hierarchies as part of the administration of role-based access control (RBAC) is motivated in the ARBAC97 model [27]. The fundamental question is how to allow localized evolution of a hierarchy without disrupting larger global relationships. The authors of ARBAC97 suggest the notion of an encapsulated range as the basis for determining a suitable unit for local modifications. Crampton and Loizou [3, 4] point out some problems with this notion and propose a mathematically better founded notion of administrative scope. Koch et al [13] discuss administrative scope in their graph-based approach to access control and provide an operational semantics for it. The issue of evolving hierarchies and reconciling pre-existing hierarchies is likely to grow in importance as enterprises deploy RBAC across multiple business units and business partners.

4   Conclusion

In this paper we have briefly explored the connection between graphs and access control models, focusing on the safety problem and on dynamic role hierarchies. There is a long history of attempts to apply graph theory to these problems. Much of the earlier work is based on first principles. In recent years we have seen a more direct application of graph theory results. There is strong potential in further exploring this connection.

The area of access control and authorization has had a resurgence of interest in recent years. Although the access matrix model has served as a reasonable foundation for access control research and practice it has become considerably dated. With the Internet explosion many new forms of access control are being deployed in various e-commerce scenarios. There is increasing realization that the foundations of access control need a deeper and richer model. A number of authors have proposed various extensions to traditional access control. Park and Sandhu [19, 28] recently proposed a unified model for next generation access control called usage control or UCON. Initial efforts to formalize this model have taken a logic-based approach [30]. It would be interesting to see how graphs and their transformations can be applied to UCON models. The framework of UCON is very rich so there is likely to be some aspect of UCON that can benefit from a graph-based formal foundation.

Acknowledgement

This material is based upon work supported by the National Science Foundation under Grant No. 0310776. Any opinions, findings and conclusions or recommendations expressed in this material are those of the author and do not necessarily reflect the views of the National Science Foundation (NSF).

References

1. Paul Ammann, Richard Lipton and Ravi Sandhu. The Expressive Power of Multi-Parent Creation in Monotonic Access Control Models. *Proc. IEEE Computer Security Foundations Workshop V*, Franconia, New Hampshire, June 1992, pages 148-156.
2. Paul Ammann, Duminda Wijesekera, and Saket Kaushik. Scalable, Graph-Based Network Vulnerability Analysis. *Proceedings CCS 2002: 9th ACM Conference on Computer and Communications Security*, pages 217-224, Washington, DC, November 2002.
3. Jason Crampton. Administrative Scope and Role Hierarchy Operations. *Proceedings of the Seventh ACM Symposium on Access Control Models and Technologies*, Monterey, California, 2002, pages 145-154.
4. Jason Crampton and George Loizou. Administrative scope: A Foundation for Role-Based Administrative Models. *ACM Trans. Inf. Syst. Secur.*, Volume 6, Number 2, pages 201-231, 2003.
5. David F. Ferraiolo, Ravi Sandhu, Serban Gavrila, D. Richard Kuhn and Ramaswamy Chandramouli. Proposed NIST Standard for Role-Based Access Control. *ACM Transactions on Information and System Security*, Volume 4, Number 3, August 2001, pages 224-274.
6. T. Jaeger and J. E. Tidswell. Practical safety in flexible access control models. *ACM Trans. on Info. and System Security*, 4(2), pages 158-190, 2001.
7. S. Jha, O. Sheyner and J. Wing. Two Formal Analyses of Attack Graphs. *Proceedings of the 15th IEEE Computer Security Foundations Workshop*, p.49-63, June 24-26, 2002.
8. Michael A. Harrison, Walter L. Ruzzo and Jeffrey D. Ullman. Protection in Operating Systems. *Commun. ACM*, 19:8, 461-471, 1976.
9. M. Koch, L.V. Mancini, and F. Parisi-Presicce. A Formal Model for Role-Based Access Control using Graph Transformation. In F.Cuppens, Y.Deswarte, D.Gollmann, and M.Waidner, editors, *Proc. of the 6th European Symposium on Research in Computer Security (ESORICS 2000)*, Lect. Notes in Comp. Sci. 1895, pages 122-139. Springer, 2000.
10. M. Koch, L.V. Mancini, and F. Parisi-Presicce. Decidability of Safety in Graph-Based Models of Access Control. In D.Gollmann, G.Karjoth, and M.Waidner, editors, *Proc. of the 7th European Symposium on Research in Computer Security (ESORICS 2002)*, Lect. Notes in Comp. Sci. 2502, pages 229-243. Springer, 2002.
11. M. Koch, L.V. Mancini, and F. Parisi-Presicce. A Graph Based Formalism for RBAC. *ACM Trans. on Info. and System Security*, 5(3):332-365, 2002.
12. M. Koch and F. Parisi-Presicce. Describing Policies with Graph Constraints and Rules. In A. Corradini, H. Ehrig, H.-J. Kreowski, and G. Rozenberg, editors, *Int. Conference on Graph Transformations*, Lect. Notes in Comp. Sci. 2505, pages 223-238. Springer, 2002.
13. M. Koch, L. V. Mancini and F. Parisi-Presicce. Administrative Scope in the Graph-Based Framework. *Proceedings of the ninth ACM Symposium on Access control Models and Technologies*, Yorktown Heights, New York, pages 97-104, 2004.
14. Lampson, B.W. Protection. *5th Princeton Symposium on Information Science and Systems*, pages 437-443, 1971. Reprinted in *ACM Operating Systems Review* 8(1):18-24, 1974.
15. R.J. Lipton and L. Snyder. A Linear Time Algorithm for Deciding Subject Security. *Journal of the ACM*, Volume 24, Number 3, pages 455-464, 1977.

16. Lockman, A. and Minsky, N. Unidirectional Transport of Rights and Take-Grant Control. *IEEE TSE*, Volume SE-8, Number 6, pages 597–604,1982.

17. J. P. McDermott. Attack Net Penetration Testing. *Proceedings of the 2000 workshop on New Security Paradigms*, Ballycotton, County Cork, Ireland, pages 15–21, 2000, ACM Press.

18. M. Nyanchama and S.L. Osborn. The Role Graph Model and Conflict of Interest. *ACM Trans. on Info. and System Security*, 1(2):3–33, 1999.

19. Jaehong Park and Ravi Sandhu. The UCON$_{ABC}$ Usage Control Model. *ACM Transactions on Information and System Security*, Volume 7, Number 1, February 2004, pages 128-174.

20. C. Phillips and L. Swiler. A graph-based system for network vulnerability analysis. *ACM New Security Paradigms Workshop*, pages 71–79, 1998.

21. Michael K. Reiter and Stuart G. Stubblebine. Authentication Metric Analysis and Design. *ACM Trans. Inf. Syst. Secur.*, Volume 2, Number 2, pages 138–158, 1999.

22. Ravi Sandhu. The Schematic Protection Model: Its Definition and Analysis for Acyclic Attenuating Schemes. *Journal of the ACM*, Volume 35, Number 2, April 1988, pages 404-432.

23. Ravi Sandhu. The NTree: A Two Dimension Partial Order for Protection Groups. *ACM Transactions on Computer Systems*, Volume 6, Number 2, May 1988, pages 197-222.

24. Ravi Sandhu. The Typed Access Matrix Model. *Proc. IEEE Symposium on Research in Security and Privacy*, Oakland, California, May 1992, pages 122-136.

25. Ravi Sandhu and Pierangela Samarati. Access Control: Principles and Practice. *IEEE Communications*, Volume 32, Number 9, September 1994, pages 40-48.

26. Ravi Sandhu, Edward Coyne, Hal Feinstein and Charles Youman,. Role-Based Access Control Models. *IEEE Computer*, Volume 29, Number 2, February 1996, pages 38-47.

27. Ravi Sandhu, Venkata Bhamidipati and Qamar Munawer. The ARBAC97 Model for Role-Based Administration of Roles. *ACM Transactions on Information and System Security*, Volume 2, Number 1, February 1999, pages 105-135.

28. Ravi Sandhu and Jaehong Park. Usage Control: A Vision for Next Generation Access Control. *Proc. Computer Network Security: Second International Workshop on Mathematical Methods, Models, and Architectures for Computer Network Security, MMM-ACNS*, Saint Petersburg, Russia, September 21-23, 2003, Springer-Verlag Lecture Notes in Computer Science 2776, pages 17-31.

29. Oleg Sheyner, Joshua Haines, Somesh Jha, Richard Lippmann and Jeannette M. Wing. Automated Generation and Analysis of Attack Graphs. *Proceedings of the 2002 IEEE Symposium on Security and Privacy*, p. 254-265, May 12-15, 2002.

30. Xinwen Zhang, Jaehong Park, Francesco Parisi-Presicce and Ravi Sandhu. A Logical Specification for Usage Control. *Proc. 9th ACM Symposium on Access Control Models and Technologies (SACMAT)*, New York, June 2-4, 2004, pages 1-10.

Transformation Language Design: A Metamodelling Foundation

Tony Clark, Andy Evans, Paul Sammut, and James Willans

Xactium Limited
andy.evans@xactium.com

1   Introduction

With the advent of the Model-Driven Architecture (MDA) [3] there is significant interest in the development and application of transformation languages. MDA recognises that systems typically consist of multiple models (possibly expressed in different modelling languages and at different levels of abstraction) that are precisely related. The relationship between these different models can be described by transformations (or mappings).

An important emerging standard for transformations is QVT (Queries, Views, Transformations). This standard, being developed by the Object Management Group (OMG), aims to provide a language for expressing transformations between models that are instances of the MOF (Meta Object Facility) [2] metamodel. The MOF is a standard language for expressing meta-data that is being used as the foundation for expressing language metamodels (models of languages). Because of the generic nature of MOF, it is also being used as the means of expressing the QVT language itself.

When the QVT process began (over two years ago), the task initially seemed quite straightforward. After all, many different transformation languages were already described in the literature, and it was felt that it would be straightforward to design such a language for MOF. Unfortunately, this has not been the case. Two key issues have made the task of designing such a language much harder. In the remainder of this paper we will examine these issues and propose a solution that is applicable across a wide variety of language definitions.

2   Design Issues

The first issue that impacts the design of a transformation language relates to transformation languages themselves. In practise, it turns out that there are many different flavours of transformation languages. Some of the choices of language features include:

- *Declarative vs. Imperative*: at what level of abstraction should transformations be expressed? Declarative languages enable transformations to be expressed in a more concise fashion, yet may suffer from being inefficient (or impossible) to implement.

H. Ehrig et al. (Eds.): ICGT 2004, LNCS 3256, pp. 13–21, 2004.

- *Compositionality*: whilst a transformation language should ideally be compositional, this is more readily achieved by the use of more declarative primitives, thus invoking the declarative vs imperative issue (above).
- *Patterns vs. Actions*: patterns are widely used as a declarative but executable abstraction for describing transformations (XSLT is a good example of this). Yet, should a transformation language be completely pattern based, or should a mixed language with imperative actions permitted for practicality?
- *Unidirectional vs. Bi-directional*: it is clear that there is a strong distinction between one-stop, unidirectional transformations, and bi-directional mappings that keep two models in sync. Should both be accommodated?

These, and many other choices make the decision process a difficult one for the designers of transformation languages. One approach to tackling the problem is to attempt to mudpack all of the different features into a single language. However, there is clearly a danger of producing an overly complex language. On the other hand, choosing a subset of the features will clearly omit use cases of the language that may be relevant to users of the language.

The second issue relates to MOF itself. During the QVT process, it has become more apparent than ever that current metamodelling practice is too weak. In particular, the standard approach to metamodelling, in which the main focus is on capturing the static properties of a language (i.e. the abstract syntax) does not enable two critical aspects of language design to be expressed: *semantics* (what the language does and means) and *concrete syntax* (how the language is represented). Thus, in order to describe these aspects, the design team must rely on informal textual descriptions or bespoke implementations. In the latter case, this often results in 'analysis paralysis', as there is insufficient information to validate the correctness of the design.

Clearly, in the context of an international standardisation process, this is not satisfactory. In particular, it will be difficult to ensure that implementations of the standard are conformant as there will be gaps in the definition that will be filled in by vendors in different ways.

3 The Way Forward

In order to fully address the needs of QVT and transformation language design in general, it is clear to us that two key changes are required. Firstly, the bar must be raised in the way in which we metamodel languages. Rather than just capturing abstract syntax, the metamodelling language must be rich enough to capture *all* aspects of a language, including concrete syntax, abstract syntax and semantics. This information should be sufficient to rapidly generate tools that implement the language and allow its properties to be fully explored and validated.

Secondly, it must be recognised that there is no single, all encompassing transformation language. Instead we must be prepared to embrace a diversity of languages, each with specific features. Furthermore, the standard must have the flexibility to accommodate this diversity in an interoperable manner, enabling different features to be mixed and matched as required.

At first, these proposals appear to be unconnected. Yet, they are in fact closely related. In practice, we have found that the richer the capability for expressing metamodels, the greater the flexibility and interoperability of the resulting language designs. This occurs because the metamodels capture cohesive language units that can be readily integrated within other languages.

4 XMF

We have constructed an approach (and associated tools) for language metamodelling that aims to realise these goals. This approach is based on what we call an eXecutable Metamodelling Framework (XMF). The basic philosophy behind XMF is that many different languages can be fully described via a metamodelling architecture that supports the following:

- A platform independent virtual machine for executing metamodels.
- A small, precise, executable metamodelling language that is bootstrapped independently of any implementation technology. This supports a generic parsing and diagramming language, a compiler and interpreter, and a collection of core executable MOF modelling primitives called XOCL (eXecutable OCL).
- A layered language definition architecture, in which increasingly richer languages and development technologies are defined in terms of more primitive languages via operational definitions of their semantics or via compilation to more primitive concepts.
- Support for the rapid deployment of metamodels into working tools. This involves linking executing metamodels with appropriate user-interface technology.

Using this architecture we have implemented many different modelling languages and development technologies for industrial clients. We have used exactly the same approach in the definition of transformation languages. Firstly, some core transformation language abstractions were implemented. These included a pattern matching language and synchronisation language. Two transformation languages were then defined on top of these. The first, XMap, provides a language for generative transformations based on pattern matching. XOCL is integrated in the language, thus enabling mixed declarative and imperative mappings. The second, XSync, supports the dynamic, bi-directional synchronisation of models, this time using XOCL as a means of writing the synchronisation rules.

In the following sections, we firstly give an example of one the languages, XMap, and then describe how the language is defined using a metamodel.

5 XMap Example

The example defines a mapping between two models: a simple model of state machines, and a simple model of C++. The simple state machines model is shown below:

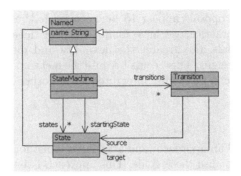

Both states and transitions are labelled with a name. A transition relates a source state to a target state.

The following model captures the basic features of C++. Note that the body of an operation is a string. However, if necessary the expressions and statements could also be modelled.

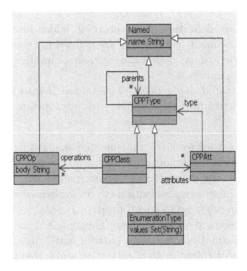

A mapping from the StateMachine model to the C++ model can now be written. It maps a StateMachine to a C++ class, where each state in the state machine is mapped to a value in an enumerated type called STATE. Each transition in the state machine is mapped to a C++ operation with the same name and a body, which changes the state attribute to the target of the transition.

The mapping can be modelled in XMap as shown below. The arrows represent mappings between elements of the two languages. The first mapping, SM2Class, maps a state machine to a C++ class. The second mapping, Transition2Op, maps a transition to an operation.

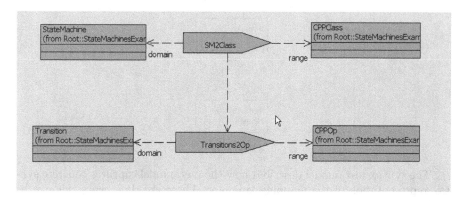

In order to describe the details of the mapping, XMap uses a textual mapping language based on pattern matching. As an example, the definition of the mapping between a transition and an operation is as follows:

```
context Transition2Op
  @Clause Transition2Op
    Transition
      [name = N,
       target = T]
    do
    CPPOp
      [name = N,
       body = B]
    where
      B = "state = " + T.name
end
```

A mapping consists of a collection of clauses, which are pattern matches between source and target objects. Whenever a source object is successfully matched to the input of the mapping, the resulting object in the do expression is generated. Variables can be used within clauses, and matched against values of slots in objects. Because XMap builds on XOCL, XOCL expressions can be used to capture complex relationships between variables.

In this example, whenever the mapping is given a Transition with a name equal to the variable N and a target equal to T, it will generate an instance of the class Operation, whose name is equal to N and whose body is equal to the variable B. The where clause is used to define values of variables, and it is used here to define the variable B to be concatenation of the text "state = " with the target state name. For instance, given a transition between the states "On" and "Off", the resulting operation will have the body "state = Off".

6   XMap Metamodel

The architecture of the XMap language metamodel is described in the figure below.

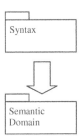

The syntax metamodel describes how the metamodel captures concrete syntax representation of the mapping language. There are three ways this can be achieved:

- A textual syntax can be defined for the language by constructing a grammar that states how a textual representation of models written in XMap can be parsed into an instance of the concepts represented in the syntax definition. This is achieved in XMF using XBNF: an extended BNF grammar language that provides information about how to turn textual elements into instance of XMF elements.
- A graphical syntax can be defined by defining a mapping from a model of the graphical syntax of the language to concepts in the syntax domain.
- A mixed approach can be used, in which both graphical and textual elements are defined. This is the approach taken with the XMap language.

6.1   Textual Syntax Metamodel

As an example a textual syntax metamodel, the following fragment of XBNF defines the rules for parsing a clause into an instance of a Clause class:

```
@Class Clause
 @Grammar extends OCL::OCL.grammar
    Clause ::= name = Name
           patterns = ClausePatterns 'do' body = Exp {
                      Clause(name, patterns, body) }.
    ClausePatterns ::= p = Pattern
           ps = (',' Pattern)* { Seq{p | ps } }.
    ClauseBindings :: 'where' Bindings | { Seq{}}.
    end
        ...
 end
```

The grammar extends the OCL grammar with the concept of a Clause, where a Clause is defined to be a name, followed by a collection of patterns, followed by a 'do' and a body, which can be an expression, and an optional collection of 'where' bindings. The result of matching any textual input of this form:

```
@Clause <name>
   <patterns>
   do
   <body>
   where
   <bindings>
end
```

is to create an instance of a Clause(), passing it the definitions of name, patterns and bindings. Of course, further XBNF definitions will be needed that define the grammar rules for Pattern, Binding, etc. These are omitted for brevity.

6.2   Diagram Syntax Metamodel

As an example of a metamodel of diagrammatical syntax, the following diagram describes a part of the syntax of XMap mapping diagrams. A mapping diagram extends a class diagram with MapNodes (uni-directional arrows).

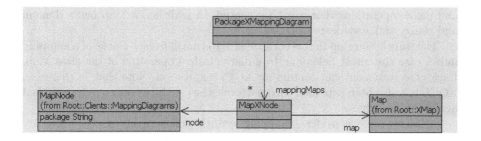

The relationship between a MapNode and a mapping (denoted here by the class Map) is kept constantly in step via the MapXNode mapping. This body of this mapping is written in XSync, thus synchronising the relevant aspect of the two elements. For instance, the name of the Map and the MapNode must always be kept in step.

6.3   Semantic Domain

The syntax of the language can be viewed as a syntactical sugar for concepts in the semantic domain. A semantic domain expresses the meaning of the concepts in terms of more primitive, but well-defined concepts. A semantics is thus defined for XMap via a translation from the syntactical representation of a mapping into a semantic domain model.

The semantic domain model for the XMap language is described by the following model:

Here, a Map denotes a mapping. It is a subclass of the class Class and therefore inherits all the properties of the class Class. It can therefore be instantiated

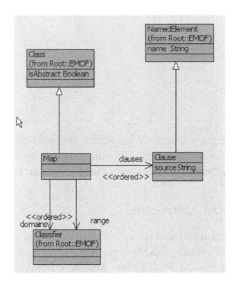

and define operations that can be executed. In addition, a Map has a domain and range, and a sequence of clauses.

The translation step that is performed is to translate each clause of a mapping into a case statement belonging to a distinguished operation of the class Map. Each case statement can contain an XOCL expression. Note that in this case, XOCL has also been extended with patterns, thus enabling values to be matched against other values.

As an example, consider a mapping with the following clauses:

```
@Clause Clause0
   Transition
     [name = ""]
   do
   CPPOp
     [name = "Empty"]
end

@Clause Clause1
   Transition
     [name = N]
   do
   CPPOp
     [name = N]
end
```

This would be translated into the following operation of the class Map:

```
@Operation invoke ():Element
   @Case of
```

```
Transition [name = N] do
  CPPOp [name = N]
end
Transition [name = [ "" ]] do
  CPPOp [name = "Empty"]
end
else self.error ("Mapping failed.")
  end
end
```

Running this operation will thus execute the appropriate case statements and perform the mapping.

The way in which the translation from syntax to semantic domain occurs is a matter of choice. At the diagram level, the desugaring is maintained by the synchronised mapping between the diagram and the semantic domain model. At the syntax level, a desugar() operation can be added to the grammar to tell it how to construct the appropriate case statement.

7   Conclusion

Our approach to defining transformation languages is to use rich metamodels to capture all aspects of their definition, including syntax and semantics. A key property is that definitions of existing languages and technologies (such as XOCL) can be merged in with the new language, creating richer, more expressive capabilities.

The result is a precise definition that is: platform independent (no reliance on external technology), transparent (the entire definition, including its semantics can be traced back through the metamodel architecture); extensible and interoperable (new features can be added by adding new language components), and executable (enabling the language to be tested and validated).

There has been much recent interest in the design of domain specific languages [1], and the approach described in this paper offers a scalable solution to the problem of how to generate new languages and tools that support those languages in a generic fashion.

In summary, our position is that a crucial step in the design of transformations languages must be the adoption of more complete and semantically rich approaches to metamodelling.

References

1. S. Cook. Domain-specific modeling and model driven architecture. *MDA Journal*, January 2004. http://www.bptrends.com/publicationfiles/01-04%20COL%20Dom%20Spec%20Modeling%20Frankel-Cook.pdf.
2. Object Management Group. Meta-object facility. http://www.omg.org/mof.
3. Object Management Group. Model driven architecture. http://www.omg.org/mda.

Rule Execution in Graph-Based Incremental Interactive Integration Tools

Simon M. Becker, Sebastian Lohmann, and Bernhard Westfechtel

Department of Computer Science III, RWTH Aachen University
Ahornstraße 55, D-52074 Aachen, Germany
{sbecker,slohmann,bernhard}@i3.informatik.rwth-aachen.de

Abstract. Development processes in engineering disciplines are inherently complex. Throughout the development process, different kinds of inter-dependent design documents are created which have to be kept consistent with each other. Graph transformations are well suited for modeling the operations provided for maintaining inter-document consistency. In this paper, we describe a novel approach to rule execution for graph-based integration tools operating interactively and incrementally. Rather than executing a rule in atomic way, we break rule execution up into multiple phases. In this way, the user of an integration tool may be informed about all potential rule applications and their mutual conflicts so that he may take a judicious decision how to proceed.

1 Introduction

Development processes in engineering disciplines are inherently complex. Throughout the development process, different kinds of inter-dependent *documents* are created which have to be kept consistent with each other. For example, in software engineering there are requirements definitions, software architectures, module bodies, etc. which describe a software system from different perspectives and at different levels of abstraction and granularity. Documents are connected by manifold dependencies and need to be kept consistent with each other. For example, the source code of a software system must match its high-level description in the software architecture.

Development processes may be viewed as multi-stage transformation processes from the initial problem statement to the final solution. Throughout the transformation process, many interacting decisions have to be performed. These decisions can be automated only to a limited extent; in many settings, human *interactions* are required. Moreover, transformation rarely proceeds stage-wise according to some waterfall model. Rather, *incremental* and iterative processes have been proposed, which require to propagate changes throughout a set of inter-dependent documents.

In such a setting, there is a need for incremental and interactive *integration tools* for supporting inter-document consistency maintenance. An integration tool has to manage *links* between parts of inter-dependent documents. These parts are called *increments* in the sequel. The tool assists the user in *browsing* (traversing the links in order to navigate between related increments in different documents), *consistency analysis* (concerning the relationships between the documents' contents), and *transformations* (of

H. Ehrig et al. (Eds.): ICGT 2004, LNCS 3256, pp. 22–38, 2004.

the increments contained in one document into corresponding increments of the related document).

Graphs and graph transformations have been used successfully for the specification and realization of integration tools [1, 2]. However, in the case of incremental and interactive integration tools specific requirements have to be met concerning the execution of integration rules. In this paper, we describe a novel approach to rule execution for graph-based integration tools operating incrementally and interactively. We have realized this approach, which is based on triple graph grammars [3, 4], in a research prototype called *IREEN*, an *I*ntegration *R*ule *E*valuation *EN*vinronment [5]. Rather than executing a rule in atomic way, IREEN breaks rule execution up into multiple phases. In this way, the user of an integration tool may be informed about all potential rule applications and their mutual conflicts so that (s)he may take a judicious decision how to proceed.

The rest of this paper is structured as follows: Section 2 presents a scenario which motivates our work by a practical example. Section 3 is devoted to the graph-based specification of integration tools. Section 4, the core part of this paper, presents our novel approach to rule execution. Section 5 discusses related work, and Section 6 presents a short conclusion.

2   Scenario

The research reported in this paper is carried out within the IMPROVE project [6], which is concerned with models and tools for design processes in *chemical engineering*. In this section, we present a small example which illustrates key features of incremental and interactive integration tools. This example is drawn from chemical engineering, but we could also have chosen an example from another engineering discipline (e.g., software engineering).

In chemical engineering, the *flow sheet* acts as a central document for describing the chemical process. The flow sheet is refined iteratively so that it eventually describes the chemical plant to be built. Simulations are performed in order to evaluate design alternatives. Simulation results are fed back to the flow sheet designer, who annotates the flow sheet with flow rates, temperatures, pressures, etc. Thus, information is propagated back and forth between flow sheets and *simulation models*. Although the flow sheet plays the role of a master document, it may also happen that a simulation model is created first and the flow sheet is derived from the simulation model (reverse engineering).

Unfortunately, the relationships between flow sheets and simulation models are not always straightforward. Different kinds of simulation models are created for different purposes. Often, simulation models have to be composed from pre-defined blocks which in general need not correspond 1:1 to structural elements of the flow sheet. Thus, maintaining consistency between flow sheets and simulation models is a demanding task requiring sophisticated tool support.

Figure 1 illustrates how an incremental integration tool assists in maintaining consistency between flow sheets and simulation models. In general, flow sheets and simulation models are created by different users at different times with the help of respective tools; an integration tool is used to establish mutual consistency on demand. In a cooper-

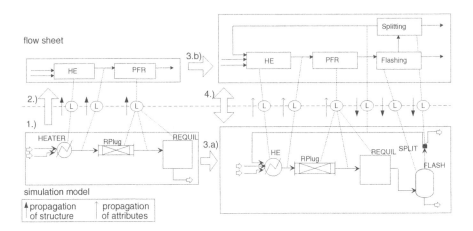

Fig. 1. Integration between flow sheet and simulation model.

ation with an industrial partner, we studied the coupling of COMOS [7], an environment for chemical engineering which in particular offers a flow sheet editor, and Aspen Plus [8], an environment for performing steady-state and dynamic simulations.

The chemical process taken as example produces ethanol from ethen and water. Flow sheet and simulation model are shown above and below the dashed line, respectively. The *integration document* for connecting them contains links which are drawn on the dashed line. The figure illustrates a design process consisting of four steps:

1. The simulation expert has already created a simulation model for a part of the chemical process (heating and reaction). The simulation model is composed of three blocks according to the capabilities of the respective simulation tool.
2. The simulation model is transformed into a flow sheet. This is achieved with the help of an integration tool. Multiple alternatives are available for this transformation. It turns out that the simplest one — a 1:1 transformation — does not result in an adequate flow sheet because the blocks do not correspond 1:1 to devices in the flow sheet. Rather, the user decides to group two blocks and their connecting stream into a single device (a plug flow reactor) in the flow sheet. The link between the PFR and the respective parts of the simulation model is established by firing a corresponding integration rule. In addition, another rule is available which just transforms the block called RPlug into a PFR. This 1:1 rule stands in conflict with the rule selected here. The integration tool presents conflicting rules to the user who may select the rule to be applied.
3. Steps 3a and 3b are carried out in parallel, using different tools. Using the simulation model created so far, a simulation is performed in the simulation tool. The simulation results comprise flow rates, temperatures, etc. In parallel, a flow sheet editor is used to extend the flow sheet with the chemical process steps that have not been specified so far (flashing and splitting).
4. Finally, the integration tool is used to synchronize the parallel work performed in the previous step. This involves information flow in both directions. First, the at-

tributes containing the simulation results are propagated from the simulation model back to the flow sheet[1]. Second, the extensions are propagated from the flow sheet to the simulation model. After these propagations have been performed, mutual consistency is re-established.

From this example, we may derive several features of the kinds of integration tools that we are addressing. Concerning the mode of operation, our focus lies on incremental integration tools rather than on tools which operate in a batch-wise fashion. Rather than transforming documents as a whole, incremental changes are propagated — in general in both directions — between inter-dependent documents. Often, the integration tool cannot operate automatically; rather, the user has to perform decisions interactively. In general, the user also maintains control on the time of activation, i.e., the integration tool is invoked to re-establish consistency whenever appropriate. Finally, it should be noted that integration tools do not merely support transformations. In addition, they are used for analyzing inter-document consistency or browsing along the links between inter-dependent documents.

3   Graph-Based Specification of Integration Tools

In complex scenarios as described in the previous section, an integration tool needs to maintain a data structure storing links between inter-dependent documents. This data structure has been called integration document. Altogether, there are three documents involved: the *source document*, the *target document*, and the *integration document*. Please note that the terms "source" and "target" denote distinct ends of the integration relationship between the documents, but it does not necessarily imply a unique direction of transformation (in fact, transformations are performed in both directions in our sample scenario).

All involved documents may be modeled as graphs, which are called *source graph*, *target graph*, and *correspondence graph*, respectively[2]. Moreover, the operations performed by the respective tools may be modeled by graph transformations. *Triple graph grammars* [3] were developed for the high-level specification of graph-based integration tools. The core idea behind triple graph grammars is to specify the relationships between source, target, and correspondence graphs by *triple rules*. A triple rule defines a coupling of three rules operating on source, target, and correspondence graph, respectively. By applying triple rules, we may modify coupled graphs synchronously, taking their mutual relationships into account.

An example of a triple rule is given in Figure 2 in PROGRES [10] syntax. The rule refers to the running example to be used throughout the rest of this paper, namely the creation of *connections* (appearing in both flow sheets and simulation models). In a flow sheet, a connection is used to relate structural elements such as *devices* and *streams*. An example of a device is a reactor, a stream is used to represent the flow of chemical substances between devices. In Figure 1, devices are represented as rectangles, streams

---

[1] For a description of the attribute propagation mechanism please refer to [9].

[2] If the tools operating on source and target document are not graph-based, the integration tool requires *wrappers* which establish corresponding graph views.

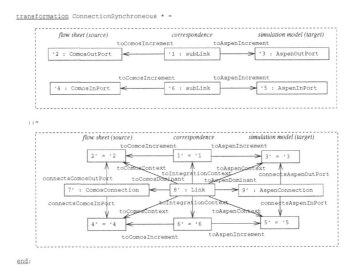

Fig. 2. Triple rule for a connection.

are shown as directed lines. Connections are not represented explicitly (rather, they may be derived from the layout), but they are part of the internal data model. Each device or stream has a set of ports; connections establish relationships between these ports.

The triple rule `ConnectionSynchronous` has a *left-hand side* (shown above the right-hand side) which spans all participating subgraphs: the source graph (representing the flow sheet) on the left, the correspondence graph in the middle, and the target graph (for the simulation model) on the right. The left-hand side is composed of port nodes in source and target graph, distinguishing between output ports and input ports[3]. Furthermore, it is required that the port nodes in both graphs correspond to each other. This requirement is expressed by the nodes of type `subLink` in the correspondence graph and their outgoing edges which point to nodes of the source and target graph, respectively. Port correspondences are established by other triple rules which transform the blocks the ports belong to, e.g. streams or devices. Correspondences between source and target patterns are represented by *links* and can be further structured by *sublinks*, e.g. to express port correspondences.

All elements of the left-hand side re-appear on the *right-hand side*. New nodes are created for the connections in source and target graph, respectively, as well as for the link between them in the correspondence graph. The connection nodes are embedded locally by edges to the respective port nodes. For the link node, three types of adjacent edges are distinguished. `toDominant` edges are used to connect the link to exactly one *dominant increment* in the source and target graph, respectively. In general, the source and target pattern related through the triple rule may consist of more than one increment in each participating graph. Then, there are additional edges to *normal increments* (not

---

[3] Only ports of different orientation may be connected.

needed in our running example)[4]. Finally, `toContext` edges point to nodes which are not themselves part of the transformation but are required as a context condition. These nodes are called *context increments*.

Figure 2 describes a *synchronous graph transformation*. As already explained earlier, we cannot assume in general that all participating documents may be modified synchronously. In case of asynchronous modifications, the triple rule shown above is not ready for use. However, we may derive *asynchronous rules* from the synchronous rule in the following ways:

- A *forward rule* assumes that the source graph has been extended, and extends the correspondence graph and the target graph accordingly. Thus, the forward rule derived from our sample rule would contain node 7 on the left-hand side.
- Analogously, a *backward rule* is used to describe a transformation in the reverse direction. In our example, node 9 would be part of the left-hand side.
- Finally, a *consistency analysis* rule is used when both documents have been modified in parallel. In our running example, this means that connections have been inserted into both the flow sheet and the simulation model and a link is created a posteriori. Thus, the consistency analysis rule would include nodes 7 and 9 on the left-hand side.

Unfortunately, even these rules are not ready for use in an integration tool as described in the previous section. In the case of non-deterministic transformations between inter-dependent documents, it is crucial that the user is made aware of conflicts between applicable rules. Thus, we have to consider all applicable rules and their mutual conflicts before selecting a rule for execution. To achieve this, we have to give up *atomic rule execution*, i.e., we have to decouple pattern matching from graph transformation.

4   Rule Execution

4.1   Overview

As explained in the previous section, an integration rule cannot be executed by means of a single graph transformation. To ensure the correct sequence of rule executions, to detect all conflicts between rule applications, and to allow the user to resolve conflicts, each integration rule is automatically translated to a set of graph transformations. These rule specific transformations are executed together with some generic ones following an *integration algorithm*. In this subsection, we will present the overall algorithm, while in the following subsections the phases of the algorithm are explained in detail, showing some of the rule specific and generic graph transformations involved. The simplified example in Figure 4 (to be explained later) is used to illustrate the algorithm.

While the algorithm in general supports the concurrent execution of forward, backward, and consistency analysis rules, we focus on forward transformations only, using

---

[4] The distinction between dominant and normal increments is not vital, but helpful for pragmatic reasons; see next section.

the forward transformation rule for a connection as running example. Some aspects of the algorithm are omitted, as the treatment of existing links that have become inconsistent due to modifications in the integrated documents.

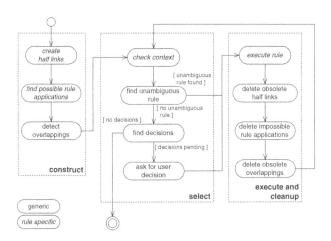

Fig. 3. Integration algorithm.

Figure 3 shows a UML activity diagram depicting the integration algorithm. To perform each activity, one ore more graph transformations are executed. Activities that require the execution of rule specific transformations are marked grey and italic. The overall algorithm is divided into three phases.

During the first phase (construct), all possible rule applications and conflicts between them are determined and stored in the graph. First, for each increment in the source document that has a type compatible to the dominant increment's type of any rule, a half link is created that references this increment. Then, for each half link the possible rule applications are determined. The last step of this phase is a generic transformation marking overlappings between possible rule applications.

In the next phase (select), for all rule applications the context is checked. If one rule application, whose context is present, is unambiguous, it is automatically selected for execution. Otherwise, the user is asked to select one rule among the rules with existing context. If there are no executable rules, the algorithm ends.

In the last phase (execute and cleanup), the selected rule is executed and some operations are performed to adapt the information that was collected in the construct phase to the new situation.

4.2   Construction Phase

In the construction phase, it is determined which rules can be possibly applied to which subgraphs in the source document. Conflicts between these rules are marked. This information is collected once in this phase and is updated later incrementally during the repeated executions of the other phases.

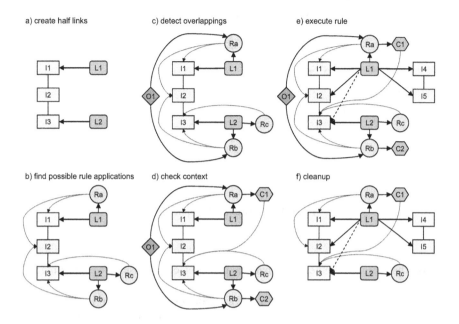

Fig. 4. Simplified example integration.

In the first step of the construction phase (create half links), for each increment, the type of which is the type of a *dominant increment* of at least one rule, a link is created that references only this increment (*half link*). Dominant increments are used as anchor for links and to group decisions for user interaction. Half links are the anchors for information about possible rule applications and are transformed to consistent links after one of the rules has been applied.

In the example, half links are created for the increments I1 and I3, named L1 and L2, respectively (c.f. Figure 4 a).

To achieve this, for each rule a PROGRES production is derived that matches an increment with the same type as the rule's dominant increment in its left-hand side, with the negative application condition that there is no half link attached to the increment, yet. Then on its right-hand side the half link node is created and connected to the increment with an edge. All these productions are executed repeatedly, until no more left-hand sides are matched, i.e., half links have been created for all possibly dominant increments.

The second step (find possible rule applications) determines the integration rules that are possibly applicable for each half link. A rule is *possibly applicable* for a given half link if the source document part of the left-hand side of the synchronous rule without the context increments is matched in the source document graph. The dominant increment of the rule has to be matched to the one belonging to the half link. For the possible applicability, context increments are not taken into account because missing context increments could be created later by the execution of other integration rules. For this

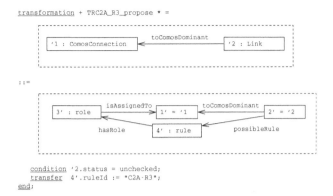

Fig. 5. Find possible rule applications.

reason, the context increments are matched in the selection phase before selecting a rule for execution.

Figure 5 shows the PROGRES transformation for the example rule. The left-hand side consists of the half link and the respective dominant increment only because all other increments of this rule are context increments. In general, all non-context increments and their connecting edges are part of the left-hand side. On the right-hand side, a rule node is created to identify the possible rule application (4′). This node carries the id of the rule and is connected to the half link. A role node is inserted to explicitly store the result of the pattern matching (3′). If there are more increments matched, role nodes can be distinguished by an id attribute. The asterisk (*) behind the production name tells PROGRES to apply this production for each possible matching of its left-hand side. When executed together with the corresponding productions for the other rules, as a result all possibly applicable rules are stored at each half link. Please note that if a rule is applicable for a half link with different matchings of its source increments, multiple rule nodes with the corresponding role nodes are added to the half link.

In the simplified example (Figure 4 b), three possible rule applications were found, e.g., Ra at the link L1 would transform the increments I1 and I2. Please note that the role nodes are omitted in the figure.

Each increment can be referenced by one link only as non-context increment. This leads to the fact that there can be *conflicts* between possible applications of integration rules. In the case of a conflict, the user has to choose one of the conflicting rules in the selection phase. There are two types of conflicts: First, there can be multiple rule nodes at one half link. These share at least the dominant increment, so only one of the corresponding rules can be executed. This is the case for link L2 in the example in Figure 4 c): Rb and Rc are conflicting. Second, an increment can be referenced by role nodes belonging to rule applications of different links. In the example, the increment I2 is referenced by Ra and Rb.

In the selection phase, for each link that is involved in a conflict all possible rule applications are presented to the user who has to resolve the conflict by selecting one. Thus, the conflicts of the first type are directly visible. Conflicts of the second type are

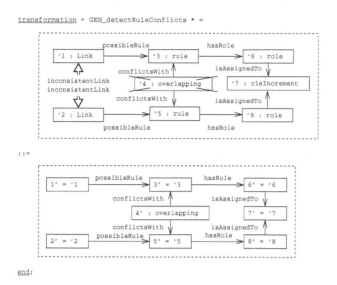

Fig. 6. Detect overlappings.

marked with cross references (hyperlinks) between the conflicting rule applications belonging to different links. To prepare the user interaction, conflicts of the second type are explicitly marked in the graph. This is done with the help of the generic PROGRES production in Figure 6. The pattern on the left-hand side describes an increment (' 7) that is referenced by two roles belonging to different rule nodes which belong to different links. The negative node ' 4 prevents multiple markings of the same conflict. On the right-hand side, an overlap node is inserted between the two rule nodes (O1 in the example). Again, this production is marked with an asterisk, so it is executed until all conflicts are detected. Besides detecting conflicts between different forward transformation rules, the depicted production also detects conflicts between forward, backward, and correspondency analysis rules generated from the same synchronous rule. As a result of that, it is not necessary to check whether the non-context increments of the right-hand side of the synchronous rule are already present in the target document when determining possible rule applications in the second step of this phase.

In the example in Figure 4 c), the overlap node O1 is created between Ra and Rb because they both reference I2. The conflict between Rb and Rc is not explicitly marked because it can be seen from the fact that they both belong to the same half link.

4.3   Selection Phase

The goal of the selection phase is to select one possible rule application for execution in the next phase. If there is a rule that can be executed without conflicts, the selection is performed automatically, otherwise the user is asked for his decision. Before a rule is selected, the contexts of all rules are checked because only a rule whose context has been found can be executed.

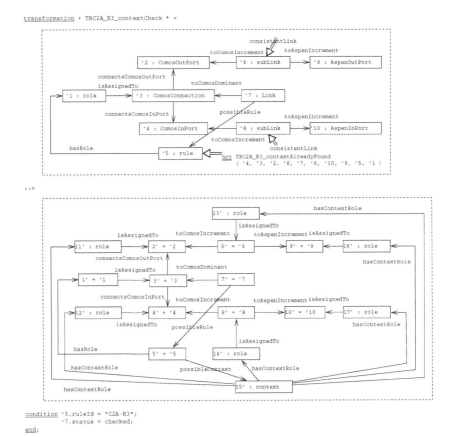

Fig. 7. Check context.

The context check is performed in the first step of this phase. The context is formed by all context elements from the synchronous rule. It may consist of increments of source and target documents and of links contained in the integration document.

In the example in Figure 4 d), the context for Ra consisting of increment I3 in the source document was found (C1). The context for Rb is empty (C2), the context for Rc is still missing.

Figure 7 shows the PROGRES production checking the context of the example integration rule. The left-hand side contains the half link (' 7), the non-context increments (here, only ' 3), the rule node (' 5), and the role nodes (' 1). The non-context increments and their roles are needed to embed the context and to prevent unwanted folding between context and non-context increments. For the example rule, the context consists of the two ports connected in the source document (' 2, ' 4), the related ports in the Aspen document (' 9, ' 10), and the relating sublinks (' 6, ' 8).

On the right-hand side, a new context node is created (' 15). It is connected to all nodes belonging to the context by role nodes (11', 12', 13', 14', 16', 17') and appropriate edges. If the matching of the context is ambiguous, multiple context nodes with their roles are created as the production is executed for all matches.

Because the selection phase is executed repeatedly, it has to be made sure that each context match (context node and role nodes) is added to the graph only once. The context match cannot be included directly as negative nodes on the left-hand side because edges between negative nodes are prohibited in PROGRES. Therefore, this is checked using an additional graph test which is called in the restriction on the rule node. The graph test is not presented here because it is rather similar to the right-hand side of this production.

The context is checked for all possible rule applications. To make sure, that the context belonging to the right rule is checked, the rule id is checked in the condition part of the productions. After the context of a possible rule application has been found, the rule can be applied.

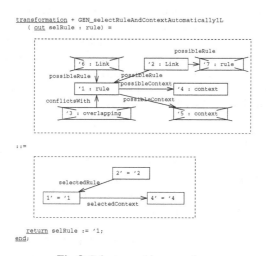

Fig. 8. Select unambiguous rule.

After the context has been checked for all possible rule applications, some rules can be applied, others still have to wait for their context. The next step of the algorithm (find unambiguous rule) tries to find a rule application that is not involved in any conflict. The conflicts have already been determined in the construction phase. Because any increment may be referenced by an arbitrary number of links as context, no new conflicts are induced by the context part of the integration rules. The generic PROGRES production in Figure 8 finds rule applications that are not part of a conflict. On the left-hand side a rule node is searched (′ 1) that has only one context node and is not related to any overlapping node. It has to be related to exactly one half link (′ 2) that does not have another rule node. For forward transformation rules, a rule node belongs to one link only, while nodes of consistency analysis rules are referenced by two half links. Therefore for consistency analysis rules, another production is used which is not shown here. A rule node is not selected for execution if there are conflicting rules, even if their context is still missing. As the context may be created later, the user has to decide whether to execute this rule and thereby making the execution of the other rules impossible.

If a match is found in the host graph, the rule node and the context node are selected for execution by substituting their referencing edges by `selectedRule` and `selectedContext` edges, respectively. The rule node is returned in the output parameter `selRule`. The corresponding rule can be applied in the execution phase.

In the example in Figure 4 d), no rule can be automatically selected for execution. The context of Rc is not yet available and Ra and Rb as well as Rb and Rc are conflicting.

If no rule could be selected automatically, the user has to decide which rule has to be executed. Therefore, in the next step (find decisions), all conflicts are collected and presented to the user. For each half link, all possible rule applications are presented. If a rule application conflicts with another rule of a different half link, this is presented as annotation at both half links. Rules that are not executable due to a missing context are included in this presentation but cannot be selected for execution. This information allows the user to select a rule manually, knowing which other rule applications will be made impossible by his decision. If there are no decisions left, the algorithm terminates. If there are still half links left at the end of the algorithm, the user has to perform the rest of the integration manually. If there are decisions, the result of the user interaction is stored in the graph (ask for user decision) and the selected rule is executed in the execution phase. In the example, the user selects rule Ra.

4.4   Execution Phase

The rule that was selected in the selection phase is executed in the execution phase. Afterwards, the information collected during the construction phase has to be updated.

In the example (Figure 4 e), the corresponding rule of the rule node Ra is executed. As a result, the increments I4 and I5 are created and references to all increments are added to the link L1.

Rule execution is performed by a rule specific PROGRES production, see Figure 9. The left-hand side of the production is nearly identical to the right-hand side of the context check production in Figure 7. The main difference is that the edge from the link (`'10`) to the rule node (`'7`) is now a `selectedRule` edge and the edge from the rule node to the context node (`'13`) is a `selectedContext` edge. The `possibleRule` and `possibleContext` edges are replaced when a rule together with a context is selected for execution either by the user or automatically.

On the right-hand side, the new increments in the target document are created and embedded by edges. In this case, the connection (`18'`) is inserted and connected to the two Aspen ports (`14'`, `15'`). The half link (`10'`) is extended to a full link, referencing all context and non-context increments in source and target document. The information about the applied rule and roles etc. is kept to be able to detect inconsistencies occurring later due to modifications in source and target documents.

The following steps of the algorithm are performed by generic productions that update the information about possible rule applications and conflicts. First, obsolete half links are deleted. A half link is obsolete if its dominant increment is referenced by another link as non-context increment. In the example this is not the case for any half link

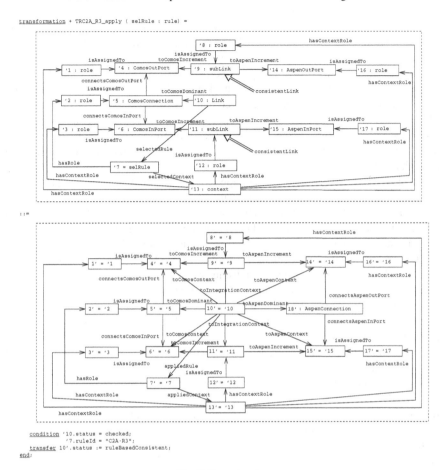

Fig. 9. Execute rule.

(Figure 4 e). Then, possible rule applications that are no longer possible are removed. In Figure 4 f), Rb is deleted because it depends on the availability of I2 which is now referenced by L1 as non-context increment. If there were alternative rule applications belonging to L1 they would be removed, as well. Last, obsolete overlappings have to be deleted. In the example, O1 is removed because Rb was deleted. Please note that the cleanup procedure may change depending on how detailed the integration process has to be documented.

5   Related Work

Our approach to incremental integration for development processes is based on the triple graph grammar approach introduced by Schürr [3] and early work at our department in the area of software engineering [11] during the IPSEN project [12]. We adapted the results to the domain of chemical engineering [9] and extended the original approach: now, we are dealing with the problem of *a-posteriori* integration, the rule definition formalism was modified [13] and the rule execution algorithm was further elaborated to support conflict detection (see Section 4).

Related areas of interest in computer science are (in-) *consistency checking* [14] and *model transformation*. Consistency checkers apply rules to detect inconsistencies between models which then can be resolved manually or by inconsistency repair rules. Model transformation deals with consistent translations between heterogeneous models. In the following a few projects of both areas are presented which are using graph transformations. Our approach contains aspects of both areas but is more closely related to model transformation.

In [15], a consistency management approach for different view points [16] of development processes is presented. The formalism of distributed graph transformations [17] is used to model view points and their interrelations, especially consistency checks and repair actions. To the best of our knowledge, this approach works incrementally but does not support detection of conflicting rules and user interaction.

The consistency management approach of Fujaba [18] supports inter-model consistency checks. The approach is based on triple graph grammars [3] as well. Comparable to our approach, different graph transformations are derived from each triple rule. User interaction is restricted to choosing the repair action for a detected inconsistency. Conflict detection between different inconsistency checking rules is supported only w.r.t. preventing endless loops if repair actions create new inconsistencies.

Model transformation recently gained increasing importance because of the model driven approaches for software development like the model driven architecture (MDA) [19]. In [20] and [21] some approaches are compared and requirements are proposed.

The PLCTools prototype [2] allows the translation between different specification formalisms for programmable controllers. The translation is inspired by the triple graph grammar approach [3] but is restricted to 1:n mappings. The rule base is conflict free so there is no need for conflict detection and user interaction. It can be extended by user defined rules which are restricted to be unambiguous 1:n mappings. Incrementality is not supported.

In the AToM project [1], modelling tools are generated from descriptions of their meta models. Transformations between different formalisms can be defined using graph grammars. The transformations do not work incrementally but support user interaction. Unlike in our approach, the control of the transformation is contained in the user-defined graph grammars.

The QVT Partner's proposal [22] to the QVT RFP of the OMG [23] is a relational approach based on the UML and very similar to the work of Kent [24]. While Kent is using OCL constraints to define detailed rules, the QVT Partners propose a graphical definition of patterns and operational transformation rules. Incrementality and user interaction are not supported.

BOTL [25] is a transformation language based on UML object diagrams. Comparable to graph transformations, BOTL rules consist of an object diagram on the left-hand side and another one on the right-hand side, both describing patterns. Unlike graph transformations, the former one is matched in the source document and the latter one is created in the target document. The transformation process is neither incremental nor interactive. There are no conflicts because of very restrictive constraints on the rules.

Transformations between documents are urgently needed (not only) in chemical engineering. They have to be incremental, interactive and bidirectional. Additionally, transformation rules are most likely ambiguous. There are a lot of transformation approaches and consistency checkers with repair actions that can be used for transformation as well, but none of them fulfills all of these requirements. Especially, the detection of conflicts between ambiguous rules is not supported. We address these requirements with the integration algorithm described in this contribution.

6   Conclusion

We have presented a novel approach to the execution of integration tools in incremental and interactive integration tools using graph transformations. Our approach was evaluated in an industrial cooperation with the German software company Innotec with a simplified prototype for the integration of flow sheets and simulation models implemented in C++. Experiments with the prototype showed that our approach considerably leverages the task of keeping dependent documents consistent to each other. Nevertheless, there is still the need for a lot of user interaction. Besides choosing among different possible rules, contradictory changes that have been made to the documents in parallel have to be resolved manually. The main ideas realized in the prototype have been incorporated into Innotec's product Comos PT and are well accepted by their customers. Of course, the rule execution formalism had to be simplified and the remaining complexity is hidden from the end user.

Acknowledgements

This work was in part funded by the CRC 476 IMPROVE of the Deutsche Forschungs-gemeinschaft (DFG). Furthermore, the authors gratefully acknowledge the fruitful cooperation with Innotec.

References

1. de Lara, J., Vangheluwe, H.: Computer aided multi-paradigm modelling to process petri-nets and statecharts. In: Proc. of 1st Int. Conf. on Graph Transformations (ICGT 2002). LNCS 2505, Springer (2002) 239–253
2. Baresi, L., Mauri, M., Pezzè, M.: PLCTools: Graph transformation meets PLC design. Electronic Notes in Theoretical Computer Science **72** (2002)
3. Schürr, A.: Specification of graph translators with triple graph grammars. In: Proc. of the 20th Intl. Workshop on Graph-Theoretic Concepts in Computer Science (WG 1994). LNCS 903, Herrsching, Germany, Springer (1995) 151–163
4. Becker, S.M., Westfechtel, B.: Incremental integration tools for chemical engineering: An industrial application of triple graph grammars. In: Proc. of the 29th Workshop on Graph-Theoretic Concepts in Computer Science (WG 2003). LNCS 2880, Springer (2003) 46–57

5. Lohmann, S.: Ausführung von Integrationsregeln mit einem Graphersetzungssystem. Master's thesis, RWTH Aachen University, Germany (2004)
6. Nagl, M., Marquardt, W.: SFB-476 IMPROVE: Informatische Unterstützung übergreifender Entwicklungsprozesse in der Verfahrenstechnik. In: Informatik '97: Informatik als Innovationsmotor. Informatik aktuell, Aachen, Germany, Springer (1997) 143–154
7. innotec GmbH: COMOS PT Documentation, http://www.innotec.de. (2003)
8. Aspen-Technology: Aspen Plus Documentation, http://www.aspentech.com. (2003)
9. Becker, S., Haase, T., Westfechtel, B., Wilhelms, J.: Integration tools supporting cooperative development processes in chemical engineering. In: Proc. of the 6th Biennial World Conf. on Integrated Design and Process Technology (IDPT-2002), Pasadena, California, USA, Society for Design and Process Science (2002) 10 pp.
10. Schürr, A., Winter, A., Zündorf, A.: The PROGRES approach: Language and environment. Volume 2. World Scientific (1999) 487–550
11. Lefering, M., Schürr, A.: Specification of integration tools. [12] 324–334
12. Nagl, M., ed.: Building Tightly-Integrated Software Development Environments: The IPSEN Approach. LNCS 1170. Springer, Berlin, Germany (1996)
13. Becker, S.M., Haase, T., Westfechtel, B.: Model-based a-posteriori integration of engineering tools for incremental development processes. Journal of Software and Systems Modeling (2004) to appear.
14. Spanoudakis, G., Zisman, A.: Inconsistency management in software engineering: Survey and open research issues. In: Handbook of Software Engineering and Knowledge Engineering. Volume 1. World Scientific (2001) 329–380
15. Enders, B.E., Heverhagen, T., Goedicke, M., Tröpfner, P., Tracht, R.: Towards an integration of different specification methods by using the viewpoint framework. Transactions of the SDPS 6 (2002) 1–23
16. Finkelstein, A., Kramer, J., Goedicke, M.: ViewPoint oriented software development. In: Intl. Workshop on Software Engineering and Its Applications. (1990) 374–384
17. Taentzer, G., Koch, M., Fischer, I., Volle, V.: Distributed graph transformation with application to visual design of distributed systems. In: Handbook on Graph Grammars and Computing by Graph Transformation: Concurrency, Parallelism, and Distribution. Volume 3. World Scientific (1999) 269–340
18. Wagner, R., Giese, H., Nickel, U.A.: A plug-in for flexible and incremental consistency mangement. In: Proc. of the Intl. Conf. on the Unified Modeling Language (UML 2003), San Francisco, California, USA, Springer (2003)
19. OMG Architecture Board ORMSC: Model driven architecture (MDA) (2001)
20. Gerber, A., Lawley, M., Raymond, K., Steel, J., Wood, A.: Transformation: The missing link of MDA. In: Proc. of 1st Intl. Conf. on Graph Transformations (ICGT 2002). LNCS 2505, Barcelona, Spain, Springer (2002) 90–105
21. Kent, S., Smith, R.: The Bidirectional Mapping Problem. Electronic Notes in Theoretical Computer Science 82 (2003)
22. Appukuttan, B.K., Clark, T., Reddy, S., Tratt, L., Venkatesh, R.: A model driven approach to model transformations. In: Proc. of the 2003 Model Driven Architecture: Foundations and Applications (MDAFA2003). CTIT Technical Report TR-CTIT-03-27, Univ. of Twente, The Netherlands (2003)
23. OMG: MOF 2.0 query / view / transformations, request for proposal (2002)
24. Akehurst, D., Kent, S., Patrascoiu, O.: A relational approach to defining and implementing transformations between metamodels. Journal on Software and Systems Modeling 2 (2003)
25. Braun, P., Marschall, F.: Transforming object oriented models with BOTL. Electronic Notes in Theoretical Computer Science 72 (2003)

Composition of Relations
in Enterprise Architecture Models

René van Buuren, Henk Jonkers, Maria-Eugenia Iacob, and Patrick Strating

Telematica Instituut, P.O. Box 589, 7500 AN Enschede, The Netherlands
{Rene.vanBuuren,Jonkers,Iacob,Strating}@telin.nl

Abstract. Enterprise architecture focuses on modelling different domains relevant for businesses or organisations. A major issue is how to
express and maintain the relations between different modelling domains.
Current architectural support focuses mainly on modelling techiques and
language for single domains. For enterprise architectures it is important
to have the flexibility to create cross domain models and views in which
inter-relations are made explicit. Therefore, a language for enterprise
architecture models should pay particular attention to the relations between domain models. In this paper we present a general approach to
derive an operator that allows for the composition of relations in architecture description languages. This general approach opens the door for
a number of interesting application areas, two of which are worked out
in more detail: the creation of more modelling flexibility, by allowing
to leave out certain details, and automated abstraction and complexity
reduction of models facilitating stakeholder-specific visualisations. For a
specific enterprise architecture modelling language, we explicitly derive
this composition operator. Because of the specific properties of this operator, the transitive closure of the metamodel of this language can be
determined.

1 Introduction

Many enterprises have only limited insight in the coherence between their business processes and ICT. An enterprise can be viewed as a complex 'system'
with multiple domains that may influence each other. In general, architectures
are used to describe components, relations and underlying design principles of a
system [8]. Constructing architectures for an enterprise may help to increase insight and overview required to successfully align the business and ICT. Although
the value of architecture has been recognised by many organisations, mostly separate architectures are constructed for various organisational domains, such as
business processes, applications, information and technical infrastructure. The
relations between these architectures often remain unspecified or implicit.

In contrast to architectural approaches for models within a domain (e.g., the
Unified Modelling Language, UML [2] for modelling applications or the technical infrastructure or the Business Process Modelling Notation BPMN [4] for
modelling business processes), enterprise architecture focuses on establishing a
coherent view of an enterprise. The term refers to a description of all the relevant

H. Ehrig et al. (Eds.): ICGT 2004, LNCS 3256, pp. 39–53, 2004.

elements that make up an enterprise and how those elements inter-relate. Models play an important role in all approaches to enterprise architecture. Models are well suited to express the inter-relations among the different elements of an enterprise and, especially if they can be visualised in different ways, they can help to alleviate the language barriers between the domains. Although a full-fledged enterprise architecture approach encompasses much more than just modelling, models play a central role in the support for enterprise architects that we develop within the ArchiMate project.

A similar movement towards integrated models can be recognised in the Model Driven Architecture (MDA) approach to software development [6]. MDA is a collection of standards of the Object Management Group (OMG) that raise the level of abstraction at which software solutions are specified. Typically, MDA results in software development tools that support specification of software in UML instead of in a programming language like Java. Recently, OMG has extended its focus to more business-oriented concepts and languages, to be developed within the MDA framework. These developments make MDA just as relevant for enterprise architecture as it is now for software development. The MDA trend reflects the growing awareness that it is important to take into account business considerations in software development decisions. Therefore, enterprise architectures form a natural starting point for automated software engineering.

In existing architectural description languages, the relations that are allowed between concepts are fixed: they are specified either informally or by means of a formal metamodel. Often, architects require the flexibility to create cross domain views or models. In practice such models are created but they often lack a formal and well-defined meaning. However, a need for a well-defined meaning becomes apparant in the context of maintainability and analyses performed on architectures, such e.g. impact of change analysis. Because relations play such a central role in enterprise architecture, we take a closer look at the properties of these inter-domain relations. In particular, this paper focuses on the question how indirect relations between concepts can be formally derived by defining how existing relations in models can be composed into new explicit relations.

This paper is organised as follows. Section 2 outlines our problem description using the ArchiMate language. Section 3 forms the core of this paper, in which we present a general approach for deriving a composition operator for architectural relations in any metamodel, and apply this approach to derive this operator for the metamodel of the ArchiMate language. In Section 4, two examples illustrate how the derived composition operator can be applied in modelling practice. In Section 5 we discuss related work. Finally, we draw our conclusions in Section 6.

2   Problem Description

As the basis for our approach, we use the core of the architecture description language of the ArchiMate project [9]. Figure 1 shows the metamodel with the core concepts of our language and the relations that they may have, expressed as a UML class diagram. The language can be used to model the business layer of an enterprise (e.g., the organisational structure and business processes), the

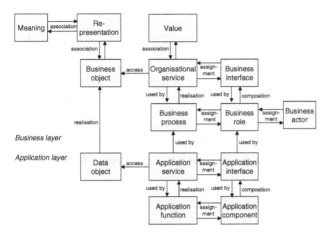

Fig. 1. Metamodel of the core of the enterprise architecture description language.

application layer (e.g., application components) and the relation between these layers. The UML classes in the metamodel represent the concepts, while the UML associations represent the possible relations that they may have. The roles in these associations indicate the relation type. Note that a relation between concepts in the metamodel defines a relation that is permitted between instances of these concepts. A model contains model elements, which are instances of the concepts in the metamodel, and actual relations between them.

Figure 2 shows a fragment from the ArchiMate metamodel, as well as a small model with instances of the concepts and relationships corresponding to this fragment. The relations represented by solid lines are not the only relations that exist: because of the coherent nature of the metamodel, all concepts are (explicitly or implicitly) linked to each other.

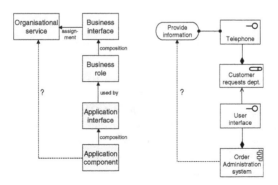

Fig. 2. Composition of relations in metamodel and model.

For example, there is a path between the application component (Order administration system) and the organisational service (Provide information), with three intermediary concepts. This means that these concepts are related to each other in some way, although it is not immediately clear what the nature of this relation is.

The direct relations as specified in the metamodel are well defined, but a rule for the composition of relations is not readily available. We will show in this paper how a composition operator for relations can be constructed based on the original metamodel.

The main question that we want to address is: *given a (meta)model with concepts (model elements) and relations between them, can we say something about the implicit relations between concepts (model elements) that are not captured in the (meta)model?*

In principle including these relations amounts to an extension of the original metamodel. Having extended the metamodel allows for the construction of a larger family of models that conform to the (extended) metamodel, i.e. it creates more modelling flexibility. This flexibility becomes apparent from the ability to define new models that conform to the extended metamomodel. Another issue relates to the composition of relations in existing models. Since there are an infinite number of potential models (conforming to a single metamodel), one has to show that models still conform to the (extended) metamodel after applying a composition of relations.

In the context of this paper, all of the relations have a direction, indicated by the arrows. The direction is needed for the definition of our composition operator. Certain relations (e.g., 'assignment' and 'association') can be used in either direction, which we denote by two separate directed relations.

Note that for clarity, we have omitted recent extensions to the language that cover, e.g., concepts for modelling the technical infrastructure. These extensions are not necessary to explain and illustrate the results described in this paper; however, the results do apply to these extensions as well. These could be described as structural relations, which model the coherence between different 'layers' and 'aspects' in enterprise architecture. Other relations, such as triggers and flows, as well as relations such as specialisation, are not considered.

3   Derivation of a Composition Operator

In this section we present a general approach to derive a composition operator for an architectural description language, and this approach is applied to the ArchiMate language. In this context, we assume that an architectural description language consists of a number of concepts that represent the 'components' in an architectural description and possible relations between these concepts. The abstract syntax of the language, i.e., the concepts and possible relations between them, is defined by the metamodel of the language.

3.1   Formalisation of Metamodel and Composition Operator

A metamodel (M) can be formally defined as a 3-tuple: $M = (C, T, R)$, where:

- C is a set of *concepts*.
- T is a set of *relation types*.
- $R \subseteq C \times C \times T$ is a set of *relations* between concepts. More precisely, an element $(c_1, c_2, t) \in R$ is called a 'relation' and it expresses the fact that a relation of type t exists from concepts c_1 to c_2.

Our relations types and relations correspond to relationships and links respectively in UML. Let us assume that there exists a composition operator for relation types, denoted by $\otimes$. Note that T only contains the relation types that currently occur in the metamodel. Since we cannot be certain beforehand that T is closed with respect to the composition operator, we define an extension of T, denoted by T^*, consisting of T and all relation types that may be the result of the composition of relation types in T^*. By definition, $T \subseteq T^*$. Now, we define the composition operator for relation types on the extended set of relation types: $\otimes : T^* \times T^* \mapsto T^*$.

With this composition operator for relation types, we can now define the set of 'two step' relations, R^2:

$$\forall_{c_1, c_2, c_3 \in C} \forall_{t_1, t_2 \in T} : (c_1, c_2, t_1) \in R \wedge (c_2, c_3, t_2) \in R \Rightarrow (c_1, c_3, t_1 \otimes t_2) \in R^2$$

Provided that the composition operator is associative, i.e., $\forall_{t_1, t_2, t_3 \in T^*} : (t_1 \otimes t_2) \otimes t_3 = t_1 \otimes (t_2 \otimes t_3)$ then for any $n > 2$, the set of 'n step' relations, R^n, can be defined recursively:

$$\forall_{c_1, c_{n-1}, c_n \in C} \forall_{t_1, t_2 \in T^*} : (c_1, c_{n-1}, t_1) \in R^{n-1} \wedge (c_{n-1}, c_n, t_2) \in R \Rightarrow (c_1, c_n, t_1 \otimes t_2) \in R^n$$

Because pair-wise composition in a chain of relations may be performed in an arbitrary order, associativity of $\otimes$ ensures that the resulting R^n is unambiguously defined, Now, the transitive closure, R^*, of the set of relations can be defined as:

$$R^* = \bigcup_{n=1}^{\infty} R^n$$

As an extension of this, we define the transitive closure of the metamodel as $M^* = (C, T^*, R^*)$.

Given the composition operator $\otimes$ for relation *types*, we can now also define an operator $\otimes_R : R^* \times R^* \mapsto R^*$ for *relations*. Consider $r_1, r_2 \in R^*$, with $r_1 = (c_1, c_2, t_1)$ and $r_2 = (c_2, c_3, t_2)$. First we define the projection operators $\pi_1, \pi_2,$ and π_3, where π_i takes the i-th element of each tuple in R^*. In case $\pi_1(r_2) = \pi_2(r_1)$ the operator $\otimes_R : R^* \times R^* \mapsto R^*$ is given by:

$$r_1 \otimes_R r_2 = (\pi_1(r_1), \pi_2(r_2), \pi_3(r_1) \otimes \pi_3(r_2))$$

If $\pi_1(r_2) \neq \pi_2(r_1)$, $\otimes_R(r_1, r_2)$ is not defined. It is easy to prove that the associativity of $\otimes_R$ results from the definition of $\otimes_R$ and the associativity of $\otimes$.

The transitive closure of the metamodel provides a specification of relations that may be drawn between concept instances in actual models. In this way, we can keep the metamodel relatively simple by only defining the direct relations between concepts, while the other relations that are allowed can be derived by means of the composition operator.

3.2   Extension to the Model Level

Thus far, we have derived a composition operator that applies to relations in a metamodel of a language. However, we would also like to be able to compose relations in *models* expressed in the language defined by a metamodel.

A model (A) can be formally defined as a 4-tuple: $A = (E, T^*, Q, F_c)$, where:

- E is a set of *model elements*
- T^* is a set of *relation types* (the same relation types that are used in the metamodel)
- $Q \subseteq E \times E \times T^*$ is a set of *relations* between model elements
- $F_c : E \mapsto C$ is a function that maps model elements to metamodel concepts

Definition 1. A model $A = (E, T^*, Q, F_c)$ *conforms to* a metamodel $M = (C, T^*, R)$ if and only if $\forall t \in T^* \forall e_1, e_2 \in E : (e_1, e_2, t) \in Q \Rightarrow (F_c(e_1), F_c(e_2), t) \in R$. We adopt the notation $A \gtrdot M$ for this.

Analogous to the metamodel (see Section 3.1) we can now define the transitive closure A^* of a model A. The set of 'two step' model relations, Q^2, is defined as:

$$\forall_{e_1, e_2, e_3 \in E} \forall_{t_1, t_2 \in T^*} : (e_1, e_2, t_1) \in Q \wedge (e_2, e_3, t_2) \in Q \Rightarrow (e_1, e_3, t_1 \otimes t_2) \in Q^2$$

Similarly, the sets Q^n of 'n step' model relations are determined, resulting in the transitive cosure of the set of model relations:

$$Q^* = \bigcup_{n=1}^{\infty} Q^n$$

As an extension of this, we define the transitive closure of the model as $A^* = (E, T^*, Q^*, F_c)$.

Next, we define a composition operator $\otimes_Q : Q^* \times Q^* \mapsto Q^*$ for *model relations*, defined in terms of the composition operator for relation types (again using the projection operators π_1, π_2 and π_3):

$$q_1 \otimes_Q q_2 = (\pi_1(q_1), \pi_2(q_2), \pi_3(q1) \otimes \pi_3(q_2))$$

in case $\pi_1(q_2) = \pi_2(q_1)$; otherwise, $q_1 \otimes_Q q_2$ is not defined.

Now we prove the following property, which says that applying the composition operator to any model that conforms to M^* always results in a new model that also conforms to M^*. This means that not only a 'normal' model, that conforms to M, can be extended, but also extensions to extended models conform to M^*.

Theorem 1. *Let $M = (C, T, R)$ be a metamodel for which an associative operator $\otimes$ for the composition of relation types exist and let $A = (E, T^*, Q, F_c)$ be a model. Let M^* and A^* be the transitive closures of the metamodel and model, respectively, with respect to the composition operator. Then: $A \gtrdot M^* \Rightarrow A^* \gtrdot M^*$.*

Proof. Consider any relation $(e, e', t) \in Q^*$ from A^*. According to the definition of Q^*, this means that $\exists n \geq 1 : (e, e', t) \in Q^n$. This further implies that there exist $e_1, \ldots, e_{n-1} \in E$ and $t_1, \ldots, t_n \in T^*$ such that:

$$(e, e', t) = (e, e_1, t_1) \otimes_Q (e_1, e_2, t_2) \otimes_Q \ldots \otimes_Q (e_{n-1}, e', t_n)$$

with $(e, e_1, t_1), (e_1, e_2, t_2), \ldots, (e_{n-1}, e', t_n) \in Q$ and $t = t_1 \otimes t_2 \otimes \ldots \otimes t_n$. Since $A \gtrapprox M^*$ it follows that

$$(F_c(e), F_c(e_1), t_1), (F_c(e_1), F_c(e_2), t_2), \ldots, (F_c(e_{n-1}), F_c(e'), t_n) \in R^*$$

The result of applying $\otimes_R$ to this chain of relations results in a new relations which belongs to R^*. According to the definition of $\otimes_R$, this relation is:

$$(F_c(e), F_c(e_1), t_1) \otimes_R (F_c(e_1), F_c(e_2), t_2) \otimes_R \ldots \otimes_R (F_c(e_{n-1}), F_c(e'), t_n) =$$
$$(F_c(e), F_c(e'), t_1 \otimes \ldots \otimes t_n)$$

This means that $\forall e, e' \in E \forall t \in T^* : (e, e', t) \in Q^* \Rightarrow (F_c(e), F_c(e'), t) \in R^*$, i.e., $A^* \gtrapprox M^*$.

In the remainder of this section, we show that such an associative composition operator $\otimes$ exists for the ArchiMate metamodel as described in Section 2, which means that the transitive closure of the metamodel can be defined.

3.3 Approach for Derivation of Composition Operator for ArchiMate

Figure 3 schematically shows the approach that we follow to derive a composition operator for relation types, which can be used to define the transitive closure of the metamodel. The approach consists of a constructive part and a reasoning part. As a starting point, we have the metamodel M as described in Section 2 (and formalised in Section 3.1). To determine the concepts and relations in the metamodel, we have used subjective knowledge about modelling phenomena from the 'real world'. We refer to this subjective knowledge as the 'architectural semantics' of the concepts.

The first constructive step in our approach is to determine all possible 'two step' relations between concepts, i.e., to determine R^2 (and M^2). For instance, there is a 'two step' relation between application component and business role, which is the composition of a 'composition relation' and a 'used by' relation. Having determined all the 'two step' relations, we derive the properties of the composition operator. However, it is important to realise that there is in general no formal, objective way to obtain the relation types for the 'two step' relations. In the same way as for the relations between concepts in the original metamodel, the most suitable relation type is determined based on the architectural semantics, i.e., on our knowledge of the 'real world'. In Section 3.4 we describe this constructive step in more detail. If we can show that the operator derived from the 'two-step' relations is associative, it can be used to derive the multiple-step relations and, ultimately, the transitive closure of the metamodel, M^*. Section 3.5 describes this formal step in more detail.

3.4 Derivation of 'Two Step' Relations for the ArchiMate Metamodel

The core of the ArchiMate metamodel, as shown in Figure 1, has a set C consisting of 14 concepts and the following set T of (directed) relation types:

$$T = \{\,association, realisation, used\ by, composition, assignment, aggregation, access\}$$

The set of relations is determined by all the arrows in the metamodel: for example, elements of R include ($application\ service, business\ behaviour, used\ by$) and ($business\ actor, business\ role, assignment$). All possible 'two step' relations can be determined by means of the square of an incidence matrix of the graph that represents the metamodel. Figure 4 shows an example. There is a 'two step' relation between the Application function and Business process, which is the composition of the realisation and the used by relations. In this case, based on our modelling experience (semantics), it makes sense to say that the Application function (which realises the Application service) is also used by the Business process; therefore, we state that the tuples ($application\ function, application\ service,$ $realisation$) and ($application\ service, business\ process, used\ by$) result in ($application\ function, business\ process, used\ by$). For this specific example $realisation$ $\otimes\ used\ by = used\ by$. Furthermore, by looking at all other combinations of two relations having the types $realisation$ and $used\ by$, respectively, we have observed that in all these cases a relation of the type $used\ by$ makes sense as the result of the $\otimes_R$ operator. This allows us to conclude that for the ArchiMate metamodel, in general, $realisation \otimes used\ by = used\ by$.

Note that we actually determine the relation between arbitrary instances of these concepts: only at the instance level, the relations have a meaning in reality. This supports the conjecture that it is also allowed to apply the composition operator at the model level, not only at the metamodel level. Section 3.2 elaborates on this.

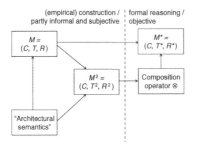

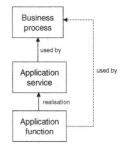

Fig. 3. Approach.   **Fig. 4.** Example of a 'two step' relation.

We determined all the individual 'two steps' relations in the metamodel. Not all the combinations of relation types occur in the metamodel M; in fact, we have determined $\otimes_R$ for $R \times R \subset R^* \times R^*$. Successive construction of the composed relations in R^2, R^3 etc. would ultimately yield R^* and the operator $\otimes_R$. We first concentrate on the composition operator for relation $types$, $\otimes$, because the operators $\otimes_R$ and $\otimes$ are strongly linked. In order to obtain a complete description of $\otimes$ we also consider the missing combinations of relations types. With the seven relation types that we have identified, this results in a table with $7 \times 7 = 49$ entries. A fragment of this table is shown below. The entries in italics are combinations that do not occur in the metamodel M.

The example of a 'two step' relation in figure 4 clarifies how to read the table. Since the relations in the ArchiMate metamodel have a direction, the relations described in the table correspond to this direction. From Figure 4, t_1 and t_2 correspond to the relations *realisation* and used by, respectively. the resulting composition relation is represented by the third column t_3, which in this example is a *used by* relation.

3.5 Definition of the Composition Operator for the ArchiMate Metamodel

The first thing that can be noticed from Table 1) is that for any pair of relation types t_1 and t_2, the composition $t_1 \otimes t_2$ always equals either t_1 or t_2; this implies that the composition operator is closed within T, and thus $T^* = T$.

Further observation discloses that a total order of the relation types can be recognised. A way to represent this total order is to define a 'weighing' function $W : T^* \mapsto \mathbb{N}$ (where $\mathbb{N}$ is the set of natural numbers), as shown in Table 2.

Table 1. Composed 'two step' relations types; the dots indicate that only a part of the table is presented.

t_1	t_2	t_3
...	...	...
Association	Access	Association
	Aggregation	Association
	Composition	Association
	Assignment	Association
	Association	Association
	Realisation	*Association*
	Used by	Association
Realisation	Access	Access
	Aggregation	Realisation
	Composition	Realisation
...	...	...

Table 2. Relations and their weights.

t	$W(t)$
Association	1
Access	2
Used by	3
Realisation	4
Assignment	5
Aggregation	6
Composition	7

By means of these weights the composition operator can be defined as follows:

$$W(t_1 \otimes t_2) = \min(W(t_1), W(T_2))$$

where 'min' is the traditional minimum operator for (integer) numbers. Informally speaking, we can say that the weight function determines the 'strength' of a relation; the composition of relations then always results in the 'weakest' relation. If we define

$$W_2 : T^* \times T^* \mapsto \mathbb{N} \times \mathbb{N}; W_2(t_1, t_2) = (W(t_1), W(t_2))$$

then the composition operator can be defined as

$$\otimes \equiv W^{-1} \circ \min \circ W_2$$

where $\circ$ denotes the traditional mathematical composition operator.

This operator is a formal representation of the composition table Table 1. For example, assume that t_1 is association. Because association has the lowest weight of all relation types, $t_1 \otimes t_2$ is always association, regardless of what t_2 is. Because the minimum operator is associative, our composition operator is also associative (we will omit the formal proof of this, which is rather straightforward when using the explicit formulation of the composition operator).

These properties of $\otimes$ fulfil the condition that is needed to be able to construct the transitive closure of our metamodel. We do not explicitly need to determine 'n step' relations, because these can be derived by applying the 'two step' composition repeatedly (in an arbitrary order).

3.6   Consequences for Metamodel Modifications and Extensions

Thus far, we have been successful in deriving an explicit composition operator (and thus the transitive closure, M^*) for the ArchiMate metamodel. This raises the question what happens to the composition operator if the metamodel is modified?

In this paper, we intentionally separated the formal approach to arrive at $\otimes$ (and thus M^*) from the actual determination of this operator for case of the ArchiMate metamodel. Both approaches remain valid for a modified version of the metamodel. However, changes might influence the outcome of the composition operator $\otimes$. For the ArchiMate metamodel, the composition operator corresponding to M^2 turned out to have the desired associativity property, which means that the transitive closure M^* can be determined immediately. Also, it was possible to express the operator in a closed form. For the composition operator in an arbitrary metamodel, this is not necessarily the case.

During the initial design of the ArchiMate language no explicit attention was paid to any desired formal properties of the metamodel itself. Emphasis was put on the applicability of the language. Therefore, the composition operator $\otimes$ is not enforced by design, but follows by uncovering the inherent properties of the metamodel. The composition operator has been derived by construction. This means that for any change or extension of the metamodel, parts of this construction have to be repeated. One might argue that conservation of the composition operator may become a design principle for metamodel extensions.

4   Example Applications

In this section we illustrate how composition of relations, as derived in this paper, can be used in architectural modelling practice. In Section 4.1 we show how the composition operator is used to extend the relations allowed by the metamodel, thus allowing for more flexible modelling. Section 4.2 illustrates the use of the composition operator as the basis for automatically generating model views and visualisations.

4.1 Modelling

Figure 5 shows a typical example of a high-level architectural model that we would like to express in our language. It shows that an end-user (modelled as a business actor) makes use of an application; this application makes use of two (application) services realised by supporting applications. One of the supporting applications, the 'customer database', accesses' a data object which represents the database content.

Formally, this model does not conform to the original ArchiMate metamodel as presented in Figure 1: e.g., in the metamodel, there is no 'used by' relation between the Application component (here end-user application) and the Business actor (here end-user) concepts, nor an 'access' relation between an Application component and a Data object. However, in the transitive closure of the ArchiMate metamodel, these relations do exist. In Figure 5, we specify how the relations are composed from the relations in the original metamodel. (Between parentheses, the intermediary concepts that have been omitted are shown.)

When we would strictly adhere to the original metamodel, we would always be forced to include all the intermediary concepts in our models. Because of the definition of the transitive closure, we can now also create a wide range of more abstract models like the one shown in Figure 5, without losing a precise definition of the meaning of the models.

4.2 Automated Abstraction and Visualisation

Integrated architectural descriptions may become very extensive and complex. For the presentation of architectural descriptions to specific stakeholders (many of which are not modelling experts), it may be useful to select only those aspects of a complex model that are relevant for their concerns, and visualise these aspects in a way that appeals to them.

For this purpose, the relevant information has to be extracted from the model. The notion of viewpoints, as defined in IEEE Standard 1471 [8], explicitly addresses this issue. Each stakeholder has its own concerns, which require a specific view on, and corresponding presentation of, the model. A viewpoint description addresses all of these issues. In most cases, this requires abstracting from certain details. The composition operator aids in (automated) abstraction of models.

Consider the following example model for an insurance company that describes the realisation of two organisational services offered to customers: 'Insurance selling' and 'Premium collection' (see Figure 6). The organisational services are realised by the business processes 'Take out insurance' and 'Collect premium', respectively. These business processes make use of application services realised by application components. The business process 'Collect premium' also

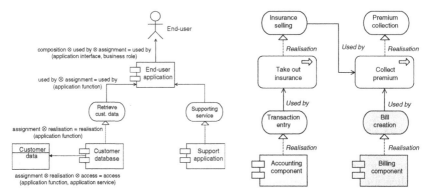

Fig. 5. Example of a model.   **Fig. 6.** Insurance example.

uses 'Insurance collection' as a supporting organisational service, to obtain the information needed to calculate the amount to be charged.

Assume that for a certain stakeholder it is relevant to know the mutual dependence of organisational services and business processes on application components. A landscape map [15] with business process and organisation services on the axes is a useful view for addressing these concerns. The construction of a landscape map is straightforward: an application component C covers a 'cell' (P, S) defined by business process P and organisational service S if and only if: (1) C is 'used by' P, and (2) P 'realises' S.

A useful intermediary step to produce such a landscape map is to derive the needed implicit relations between application components and business processes, by abstracting from the application services (and, for the 'used by' relation between the 'Accounting' component and the 'Collect premium' process, also from the intermediary business process and organisational service). When applying the composition operator on the model in Figure 6, these implicit relations can easily be derived, yielding the model in Figure 7. In this case, this results in the landscape map as shown in Figure 8.

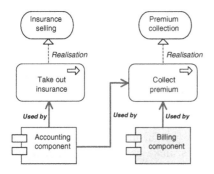

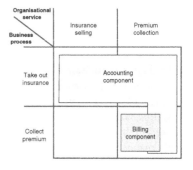

Fig. 7. Derived 'used by' relations.   **Fig. 8.** Landscape map.

Although the above example is fairly simple, the automated construction of landscape views can obviously be performed on arbitrary complex models. Moreover, the use of the composition operator for the abstraction from details in a model can be used as an intermediary step for a wide variety of visualisation and analysis techniques.

5   Related Work

The composition rule we present in this paper can, among other things, be applied for automated abstraction and complexity reduction of models for, e.g., the creation of stakeholder-oriented visualisations. In this section we refer to several papers that focus on such transformational processes and/or semantics that can be applied to diagrammatic specifications of models for automatic abstraction.

ArchiMate is ultimately a visual modelling language and we found several relevant approaches in the area of visual modelling languages. There is the growing interest of the visual language community in metamodelling approaches for the definition (in UML) of the semantics of visual languages [3]. Most of these approaches do not map the original model elements into a symbolic form, but rather into a graphical form (usually that of graphs composed of complex attributed graphical-type symbols – see [3]. Graph transformations are then defined and applied to this graphical from. Graph transformations and graph rewriting systems are used for instance in GenGEd [1] and Diagen [12, 10] to produce progressively more abstract and simple diagrams. In contrast, our approach provides support for a transformational process that applies to the symbolic form of the model. The result of this process is then visualised by means of a mapping to graphical constructs.

Another interesting approach that relates to our idea regarding the definition of an operator on the metamodel relation set is that of Holt [7]. Holt is concerned with architectures that can be characterised as directed graphs. In such architectures, the graph nodes represent entities (e.g. procedures, modules and subsystems), while the various types of relations between entities (e.g. 'calls', 'accesses', 'imports', 'includes', and 'uses') are mapped to the 'typed' or 'coloured' edges of the graph. Further on, Holt shows how a binary relational algebra (cf. Tarski [14]; Schmidt and Strohlein [13]) can be used to give constraints among the relations in architecture. In a completely different domain (geographical applications), a binary relational algebra [14, 11] is also referenced by Egenhofer [5] as a basis for the definition of a composition table of a number of eight different topological binary relations. The idea of building such a composition table is also present in our approach (see Table 1). We have explicitly formalised this composition table, in the form of a composition operator that can be used at both metamodel and model level.

6   Summary and Conclusions

In this paper a generic approach for deriving an operator for composition of relations in an architectural description language is presented, given the architectural semantics of the underlying metamodel. We applied this approach to a simplified version of the ArchiMate metamodel, resulting in a composition operator that can be expressed in closed form. This operator has a number of desirable properties: it is closed with respect to the set of relations and associative. The associativity allows for the construction of the transitive closure of the ArchiMate metamodel with respect to the relations.

We have shown in Section 3 how the composition operator $\otimes_R$ yields the transitive closure of a metamodel, M^*. This transitive closure extends the set of possible relations between concepts, thus providing more modelling flexibility. In Section 3.2, we have extended the result to existing models by deriving an operator $\otimes_Q$. We have proved that, given a model that conforms to M^*, the composition of relations in that model results in a model that conforms to M^*. Note that this includes models that conform to M. Figure 9 summarises this.

Fig. 9. Summary.

The composition operator can be brought to action in several relevant applications of architectural models. In this paper, we have presented two examples. First, it is now formally allowed to abstract from certain details by omitting intermediary concepts; thus, the operator creates more freedom in the construction of formal models. In practice most architects create formal models only for specific domains, and intuitively construct models that cross domains. With the approach suggested in this paper, it is possible to create cross domain models retaining a formal meaning, as shown in section 4.1. Second, we demonstrated that the composition rule can be successfully applied for the (automated) abstraction of complex models, which can form the basis of, e.g., stakeholder-specific visualisations.

We are currently exploring several other application of the composition rule. Among these, certain types of analyses appear obvious application areas. For instance, complexity reduction of the model may be used as 'preprocessing' to improve the efficiency of certain analysis algorithms. Also, quantifying relations, and showing how composition affects the quantitative attributes, may be used in certain types of quantitative analysis. This is a topic of current research.

Finally, we remark that the formal description or study of metamodel properties appears to be quite novel in the context of architecture description languages. In our view, metamodels might become much more expressive and powerful if at the design phase of the metamodel more attention would be paid to metamodel properties. The elegant but powerful property of the composition operator for relations has at least convinced us to use the preservation of this property as a design principle for changes to our metamodel.

Acknowledgement

This paper results from the ArchiMate project (http://archimate.telin.nl), a research initiative that aims to provide concepts and techniques to support enterprise architects in the visualisation, communication and analysis of integrated architectures. The ArchiMate consortium consists of ABN AMRO, Stichting Pensioenfonds ABP, the Dutch Tax and Customs Administration, Ordina, Telematica Instituut, Centrum voor Wiskunde en Informatica, Katholieke Universiteit Nijmegen, and the Leiden Institute of Advanced Computer Science.

References

1. R. Bardohl. GenGEd – A visual environment for visual languages. *Science of Computer Programming*, 44(2):181–203, 2002.
2. G. Booch, J. Rumbaugh, and I. Jacobson. *The Unified Modeling Language User Guide*. Addison-Wesley, 1999.
3. P. Bottoni. Dynamic aspects of visual modeling languages. *Electronic Notes in Theoretical Computer Science*, 82(7), 2003.
4. Business Process Management Initiative. Business process modeling notation. working draft (1.0), Aug. 2003.
5. M. Egenhofer. Deriving the composition on topological binary relations. *Journal of Visual Languages and Computing*, 5(2):133–149, 1994.
6. D. Frankel. *Model Driven Architecture : Applying MDA to Enterprise Computing*. John Wiley & Sons, 2003.
7. R. Holt. Binary relation algebra applied to software architecture. Technical Report 345, CSRI, University of Toronto, Canada, 1996.
8. IEEE Computer Society. IEEE standard 1471-2000: Recommended practice for architectural description of software-intensive systems, 2000.
9. H. Jonkers, R. van Buuren, et al. Towards a language for coherent enterprise architecture descriptions. In M. Steen and B. Bryant, editors, *Proceedings 7th IEEE International Enterprise Distributed Object Computing Conference (EDOC 2003)*, pages 28–37, Brisbane, Australia, 2003.
10. O. Köth, , and M. Minas. Abstraction in graph-transformation based diagram editors. *Electronic Notes in Theoretical Computer Science*, 50(3), 2001.
11. R. Maddux. Some algebras and algorithms for reasoning about time and space. Technical report, Department of Mathematics, Iowa State University, Ames, Iowa, 1990.
12. M. Minas. Concepts and realization of a diagram editor generator based on hypergraph transformation. *Science of Computer Programming*, 44:157–180, 2002.
13. G. Schmidt and T. Strohlein. *Relations and Graphs*. EATCS Monographs in Computer Science. Springer-Verlag, 1993.
14. A. Tarski. On the calculus of relations. *Journal of Symbolic Logic*, 6(3):73–89, 1941.
15. W. van der Sanden and B. Sturm. *Informatie Architectuur, de Infrastructurele Benadering*. Panfox, 1997.

Event-Driven Grammars: Towards the Integration of Meta-modelling and Graph Transformation

Esther Guerra and Juan de Lara

Escuela Politécnica Superior
Ingeniería Informática
Universidad Autónoma de Madrid
{Esther.Guerra_Sanchez,Juan.Lara}@ii.uam.es

Abstract. In this work we introduce *event-driven grammars*, a kind of graph grammars that are especially suited for visual modelling environments generated by meta-modelling. Rules in these grammars may be triggered by user actions (such as creating, editing or connecting elements) and in its turn may trigger other user-interface events. Its combination with (non-monotonic) triple graph grammars allows constructing and checking the consistency of the abstract syntax graph while the user is building the concrete syntax model. As an example of these concepts, we show the definition of a modelling environment for UML sequence diagrams, together with *event-driven grammars* for the construction of the abstract syntax representation and consistency checking.

1 Introduction

Traditionally, visual modelling tools have been generated from descriptions of the Visual Language (VL) given either in the form of a *graph grammar* [2] or as a *meta-model* [6]. In the former approach, one has to construct either a *creation* or a *parsing* grammar. The first kind of grammar gives rise to *syntax directed* environments, where each rule represents a possible user action (the user selects the rule to be applied). The second kind of grammars (for *parsing*) tries to reduce the model into an initial symbol in order to verify its correctness, and results in more *free editing* environments. Both kinds of grammars are indeed encodings of a *procedure* to check the validity of a model.

In the *meta-modelling* approach, the VL is defined by building a meta-model. This is a kind of type graph with multiplicities and other – possibly textual – constraints. Most of the times, the concrete syntax is given by assigning graphical appearances to both classes and relationships in the meta-model [6]. For example, in the AToM[3] tool, this is done by means of a special attribute that both classes and relationships have. In this approach the relationship between concrete (the appearances) and abstract syntax (the meta-model concepts) is one-to-one. The meta-modelling environment has to check that the model built by the user is a correct instance of the meta-model. This is done by finding a typing morphism between model and meta-model, and by checking the defined constraints on the model. In any case, whereas the graph-grammar approach is more *procedural*, the meta-modelling approach is more *declarative*.

In this paper we present a novel approach that combines the meta-modelling and the graph grammar approaches for VLs definition. To overcome the restriction of a

H. Ehrig et al. (Eds.): ICGT 2004, LNCS 3256, pp. 54–69, 2004.

one-to-one mapping between abstract and concrete syntaxes, we define separate meta-models for both kind of syntaxes. In a general case, both kinds of models can be very different. For example, in the definition of UML class diagrams [12], the meta-model defines concepts *Association* and *AssociationEnd* which are graphically represented together as a single line. In general, one can have abstract syntax concepts which are not represented at all, represented with a number of concrete syntax elements, and finally, concrete syntax elements without an abstract syntax representation are also possible. To maintain the correspondence between abstract and concrete syntax elements, we create a *correspondence meta-model* whose nodes have pairs of morphisms to elements of the concrete and abstract meta-models.

The concrete syntax part works in the same way as in the pure meta-modelling approach, but we define (non-monotonic) triple graph grammar rules [11] to build the abstract syntax model, and check the consistency of both kinds of models. The novelty is that we explicitly represent the user interface events in the concrete syntax part of the rules (creating, editing, connecting, moving, etc.) Events can be attached to the concrete syntax elements to which they are directed. In this way, rules may be triggered by user events, so we can use graph grammar rules in a free editing system. Additionally, we take advantage in the rules of the inheritance structure defined in the meta-model, and allow the definition of *abstract (triple) rules* [3]. These have abstract nodes (instances of abstract classes in the meta-model) in the LHS or RHS. These rules are equivalent to a number of *concrete rules* obtained from the valid substitutions of the abstract nodes by concrete ones (instances of the derived classes in the meta-model). We extend this concept to allow refinement of relationships.

As a proof-of-concept, we present a non-trivial example, in which we define the concrete and abstract syntax of sequence diagrams, define a grammar to maintain the consistency of both syntaxes, and define additional rules to check the consistency of the sequence diagram against existing class diagrams.

2 Meta-modelling in AToM^3

AToM^3 [6] is a meta-modelling tool that was developed in collaboration with Hans Vangheluwe from McGill University. The tool allows the definition of VLs by means of meta-modelling and model manipulation by means of graph transformation rules. The meta-modelling architecture is linear, and a *strict* meta-modelling approach is followed, where each element of the meta-modelling level n is an instance of exactly one element of the level $n + 1$ [1].

Figure 1 shows an example with three meta-modelling levels. The upper part shows a meta-metamodel for UML class diagrams, very similar to a subset of the core package of the UML 1.5 standard specification. The main difference is that *Associations* can also be refined, and that the types of attributes are specific AToM^3 types. Some of the concepts in this meta-metamodel are *Power types* [10], whose instances at the lower meta-level inherit from a common class. This is the case of *Class*, *Association* and *AssociationEnd*, whose instances inherit from *ASGNode* and *ASGConnection*. Classes *ATOM3AppearanceIcon*, *ATOM3AppearanceSegment* and *ATOM3AppearanceLink* are special types, which provide the graphical appearance of classes, association ends and

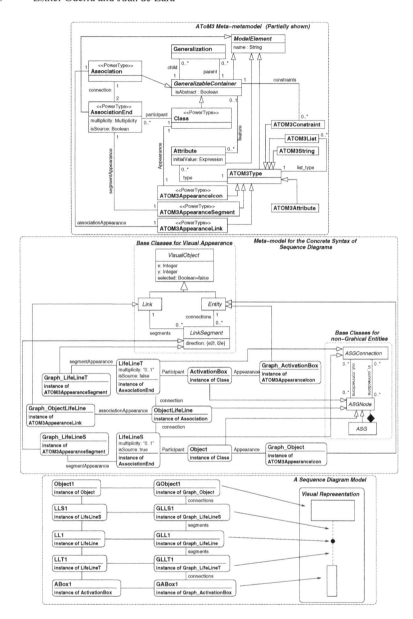

Fig. 1. Meta-modelling levels in AToM^3.

associations. They are also *Power types*, as their instances inherit from abstract classes *Entity*, *LinkSegments* and *Link*. The user can define the visual appearance of these instances with a graphical editor. Instances of *ATOM3AppearanceIcon* are icon-like, and they may include primitive forms such as circles, lines, text and show attribute values of the object associated with the instance through relationship *Appearance*. Instances of *ATOM3AppearanceLink* are similar to the previous one, but are associated with two *ATOM3AppearanceSegment* instances, which represent the incoming and outgoing segments to the link (which is itself drawn in the centre). Finally, the *ATOM3Attribute* class implements a special kind of attribute type, which is an instance of itself. In this way one can have arbitrary meta-modelling layers.

The second level in Figure 1 shows a part of the meta-model defined in Figure 4 (the lower part), but using an *abstract syntax* form (instead of the common graphical appearance of UML class diagrams that we have used in the upper meta-metamodel) where we indicate the elements of the upper meta-level from which they are instances. Only two classes are shown, *ActivationBox* and *Object*, together with the attributes for defining their appearances. In AToM^3, by default, the name of the appearance associated with a class or association begins with *"Graph_"* followed by the name of the class or association (that is, the *name* attribute defined in *ModelElement* is filled automatically). In the case of an *AssociationEnd* instance, it is similar, but followed by an *"S"* or *"T"*, depending if the end is source or target.

Finally, the lowest meta-level shows to the left (using an *abstract syntax* notation) a simple sequence diagram model. To the right, the same model is shown, using a visual representation, taking the graphical appearances designed for *Graph_Object*, *Graph_ActivationBox*, *Graph_LifeLine*, *Graph_LifeLineS* and *Graph_LifeLineT*. Note how the graphical forms are in a one-to-one correspondence with the non-graphical elements (*Object1*, *LL1*, *LLS1*, *LLT1* and *ABox1*). The non-graphical elements can be seen as the *abstract syntax* and the graphical ones as the *concrete syntax*. Nonetheless, as stated in the introduction, the one-to-one relationship is very restrictive. Therefore we propose building two separate meta-models, one for the concrete syntax representation (whose concepts are the graphical elements that the user draws on the screen) and another one for the abstract syntax. Both of them are related using a correspondence graph. The user builds the concrete syntax model, and a (triple, event-driven) graph grammar builds and checks the consistency of the abstract syntax model. These concepts are introduced in the following section.

3   Non-monotonic, Abstract Triple Graph Grammars

Triple Graph Grammars were introduced by Schürr [11] as a means to specify translators of data structures, check consistency, or propagate small changes of one data structure as incremental updates into another one. Triple graph grammar rules model the transformations of three separate graphs: source, target and correspondence graphs. The latter has morphisms from each node into source and target nodes. These concepts can be defined as follows(taken from [11]) [1]:

Definition 1 *(Graph Triple) Let* $CONC$, $ABST$ *and* $LINK$ *be three graphs and* $gs\colon LINK \to CONC$, $gt\colon LINK \to ABST$ *be two morphisms. The resulting graph triple is denoted as:* $CONC \xleftarrow{gs} LINK \xrightarrow{gt} ABST$.

---

[1] For space limitations, we have skipped all proofs referred to the constructions we introduce.

Morphisms gs and gt represent m-to-n relationships between $CONC$ and $ABST$ graphs via $LINK$ in the following way: $x \in CONC$ is related to $y \in ABST$ $\iff$ $\exists z \in LINK \mid x = gs(z)$ and $y = gt(z)$.

In [11] triple graph grammars were defined following the single pushout [7] (SPO) approach and were restricted to be monotonic (its LHS must be included in its RHS). In this way, only two morphisms were needed from the RHS of the $LINK$ graph to the RHS of the $CONC$ and $ABST$ graphs. Morphisms in LHS are defined thus as a restriction of the morphisms in RHS. Here we use the double pushout approach [7] (DPO) with negative application conditions (NAC) in rules and do not take the restriction of monotonicity. Hence, we have to define two morphisms from both LHS and RHS of the correspondence graph rule to the LHS and RHS of the $CONC$ and $ABST$ graphs.

Definition 2 *(Triple Rule) Let* $sp = (SL \xleftarrow{sl} SK \xrightarrow{sr} SR)$, $cp = (CL \xleftarrow{cl} CK \xrightarrow{cr} CR)$ *and* $tp = (TL \xleftarrow{tl} TK \xrightarrow{tr} TR)$ *be three rules.* $NAC = \{(NS \xleftarrow{nl} NC \xrightarrow{nr} NT, n)\}$ *is a set of tuples where the first component is a graph triple and n is a triple* $(n_S \colon SL \to NS, n_C \colon CL \to NC, n_T \colon TL \to NT)$ *of injective graph morphisms. Furthermore, let* $ls \colon CL \to SL$, $rs \colon CR \to SR$, $lt \colon CL \to TL$ *and* $rt \colon CR \to TR$ *be four graph morphisms, such that they coincide in the elements of CK as follows:* $\forall k_1 \in CK, \exists k_2 \in SK, ls(cl(k_1)) = sl(k_2) \wedge rs(cr(k_1)) = sr(k_2)$ [2] *(and analogously for the elements of TK). The resulting triple rule (see Figure 2) is defined as follows:* $p = (sp \xleftarrow{ls,rs} cp \xrightarrow{lt,rt} tp, NAC)$.

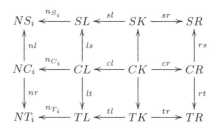

Fig. 2. A triple rule.

Figure 3 shows an example of two triple rules (where the dashed arrows depict morphisms ls, rs, lt and rt) with NACs, where only the additional elements to LHS and their context have been depicted. NACs have the usual meaning, if a match is found in the triple graph (which commutes with the LHS match and n), the rule cannot be applied. The kernel parts SK, CK and TK of the rules are not explicitly shown, but their elements have the same numbers in LHS and RHS. This is the notation that we use throughout the paper. For our purposes, we need to extend the previous definition of triple grammars to include attributes. This can be done in the way shown in [11].

---

[2] Which is equivalent to $\exists cs \colon CK \to SK$ such that $ls \circ cl = sl \circ cs$.

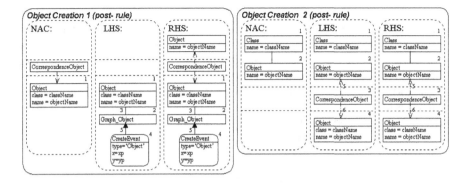

Fig. 3. An Example with two Triple Rules.

For the approach to be useful in meta-modelling environments, graphs must be consistent with a meta-model. We model this by defining *typing* morphisms between graphs and *type graphs*. We use the concept of *type graph with inheritance*[3] as defined in [3]:

Definition 3 *(Type Graph with Inheritance, taken from [3]) A type graph with inheritance is a triple (TG, I, A) of graphs TG and I sharing the same set of nodes N, and a set $A \subseteq N$, called abstract nodes. For each node n in I the* inheritance clan *is defined by $clan_I(n) = \{n' \in N \mid \exists \, path \; n' \xrightarrow{*} n \; in \; I\}$ where path of length 0 is included, i.e. $n \in clan_I(n)$.*

For the typing of a graph triple, we have to define meta-models for the $CONC$, $ABST$ and $LINK$ graphs. Additionally, as $LINK$ has morphisms to $CONC$ and $ABST$, we have to include information about the valid morphisms in the meta-model for the $LINK$ graph. Thus, we define a *meta-model triple* in the following way:

Definition 4 *(Meta-model triple) A meta-model triple is a triple of type graphs with inheritance, together with two morphisms (cs and ct) between nodes of one of the type graphs to the other two: $MMT = ((TG^{CONC}, I^{CONC}, A^{CONC}), (TG^{LINK}, I^{LINK}, A^{LINK}), (TG^{ABST}, I^{ABST}, A^{ABST}), cs, ct)$ where $cs \colon TG^{LINK} \to TG^{CONC}$ and $ct \colon TG^{LINK} \to TG^{ABST}$*

Figure 4 shows an example meta-model triple, which in the upper part (abstract syntax) depicts a slight variation of the UML 1.5 standard meta-model proposed by OMG for sequence diagrams. We have collapsed the triples (TG, I, A) into a unique graph, where the I graph is shown with hollow edges (following the usual UML notation) and the elements in A are shown in italics.

The lower meta-model in the figure declares the concrete appearance concepts and their relationships. The elements in this meta-model are in direct relationship with the graphical forms that will be used for graphical representation. As Figure 1 showed,

---

[3] In the following, we indistinctly use the terms "type graph" and "meta-model", although the latter may include additional constraints.

we allow the refinement of relationships, and this is shown with the usual notation for inheritance, but applied to relationships (arrows in the diagram). This is just a notation convenience, because each relationship (arrow) shown in Figure 4 is indeed an instance of class *Association* in the upper meta-model in Figure 1. In this way, the inheritance concept developed in [3] is immediately applicable to refinement of relationships.

The correspondence meta-model formalizes the kind of morphisms that are allowed from nodes of types *CorrespondenceMessage* and *CorrespondenceObject*. As it is defined, the declared morphism types in *cs* and *ct* are not *"inherited"* through I^{LINK} in the correspondence graph meta-model.

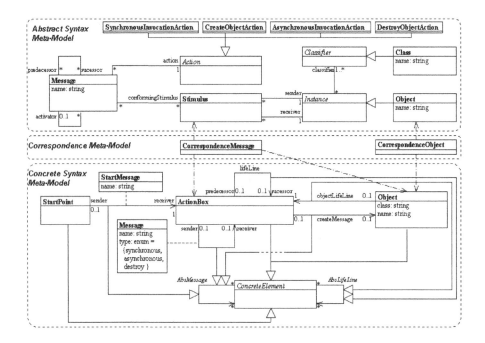

Fig. 4. An Example Meta-model triple.

Triple rules must be provided with typing morphisms to the meta-model triple. As in [3] we use the notion of *clan morphism* from graphs to type graphs with inheritance.

Definition 5 *(Clan Morphism, taken from [3]) Given a type graph with inheritance* (TG, I, A) *and graph* G, $type': G \rightarrow TG$ *is a* clan-morphism, *if for all* $e \in G_E$ $type'_N \circ s_G(e) \in clan_I(s_{TG} \circ type'_E(e))$ *and similar for* t_G.

We can define typed graph triples in a similar way as typed rules were defined in [3], but constraints regarding the morphisms of the correspondence graph should also be given. Additionally, we can define *abstract* triple rules by allowing the appearance of abstract nodes in LHS of each rule. If an abstract node appears in the RHS, then it must also appear in the LHS. An abstract rule is equivalent to a number of concrete rules where each abstract node is replaced by any concrete node in its inheritance clan. For the application of this concept here, first note that an *abstract* triple rule is equivalent to the combination of all its concrete subrules. Additionally, some of these combinations may be not valid, because of invalid morphisms between the resulting concrete rules of the correspondence graph and the source and target graphs.

Definition 6 *(Typed Graph Triple) A graph triple typed by a meta-model triple MMT $= ((TG^{CONC}, I^{CONC}, A^{CONC}), (TG^{LINK}, I^{LINK}, A^{LINK}), (TG^{ABST}, I^{ABST},$ $A^{ABST}), cs, ct)$ is depicted by $TRIG_{MMT} = (CONC \xleftarrow{gs} LINK \xrightarrow{gt} ABST,$ $type_C, type_L, type_A)$ where the last three components are typing clan morphisms from $CONC$, $LINK$ and $ABST$ to the first three components of MMT in which the following conditions hold: $\forall l \in LINK\ type_C(gs(l)) \in clan_{I^{CONC}}(cs(type_L(l)))$ and $type_A(gt(l)) \in clan_{I^{ABST}}(ct(type_L(l)))$*

If the image of any element of the triple graph belongs to some of the A sets, the typing is called abstract, otherwise it is called concrete.

Definition 7 *(Abstract Triple Rule) A triple rule typed by a meta-model triple MMT (defined as before) is depicted by $TRIP_{MMT} = (sp \xleftarrow{ls,rs} cp \xrightarrow{lt,rt} tp, NAC, type_{sp},$ $type_{cp}, type_{tp})$ where $type_{sp}$ is a triple of clan morphisms $(type_{sp}^L, type_{sp}^K$ and $type_{sp}^R)$ from SL, SK and SR $(sp = (SL \xleftarrow{sl} SK \xrightarrow{sr} SR))$ to TG^s (and similar for $type_{cp}$ and $type_{tp}$). Additionally, NACs are also typed as follows: $NAC = \{(NS \xleftarrow{nl}$ $NC \xrightarrow{nr} NT, n, type^N)\}$ is a set of tuples where the first two components are defined as in definition 2 and $type^N$ is a triple of clan morphisms $(type_S^N, type_C^N, type_T^N)$ from the graph triple to TG^s, TG^c and TG^t, which forms a typed graph triple with the first component (see definition 6).*

The following conditions hold for sp:

- *$type_{sp}^L \circ sl = type_{sp}^K = type_{sp}^R \circ sr$ (typing of preserved elements do not change).*
- *$type_{sp,N}^R(R'_{sp,N}) \cap A^s = \emptyset$, where $R'_{sp,N} := SR_N - sr_N(SK_N)$ (new nodes in RHS are not abstract)*
- *$type_S^N \circ n_S \leq type_{sp}^L$ for all $(N, n, type^N) \in NAC$ (where $\leq$ is the type refinement relationship [3]) (typing for NACs is finer than the corresponding elements in LHS)*

And analogously for cp and tp. As in previous definition, $\forall n \in CL, type_{sp}^L(ls(n)) \in clan_{I^{CONC}}(cs(type_{cp}^L(n)))$ and $type_{tp}^L(lt(n)) \in clan_{I^{ABST}}(ct(type_{cp}^L(n)))$ (and analogously for CK and CR)

Once we have defined the basic concepts regarding graph rules, next section presents event-driven grammars, which we use in combination with abstract triple rules in order to build the abstract syntax model associated with the concrete syntax. They are also useful for consistency checking, as we will see in section 5.

4   Event-Driven Grammars

In this section, we present *event-driven grammars*, as a means to formalize some of the user actions and their consequences when using a visual modelling tool. We have defined event-driven grammars to model the effects of editor operations in AToM^3 [6], although other tools could also be modelled. The actions a user can perform in AToM^3 are *creating*, *editing* and *deleting* an entity or a connection, and *connecting* and *disconnecting* two entities. All these events occur at the concrete syntax level.

The main idea of *event-driven* grammars is to make explicit these events in the models. Note how this is very different from the *syntax directed* approach, where graph grammar rules are defined for VL generation. In these environments the user chooses the rule to be applied. In our approach, the VLs are generated by means of metamodelling, and the user builds the model as in regular environments generated by metamodelling (*free-hand editing*). The events that the user generates may trigger the execution of some rules. In our approach, rules are triple rules and are used to build the abstract syntax model and to perform consistency checkings.

We have defined a set of rules (called *event-generator rules*, depicted as *evt* in Figure 5) that models the generation of events by the user. Another set of rules (called *action rules*, depicted as *sys-act* in Figure 5) models the actual action triggered by the event (creating, deleting entities, etc.), and finally, an additional set of rules (called *consume rules*, depicted as *del* in Figure 5) models the consumption of the events once the action has been performed. The VL designer can define his own rules to be executed after an event and before the execution of the *action rules* (depicted as *pre* in Figure 5), or after the *action rules* and before the *consume rules* (depicted as *post* in Figure 5)). These rules model pre- and post- actions respectively. In the pre-actions, rules can delete the produced events, if certain conditions are met. This is a means to specify pre-conditions for the event to take place. Additionally, in the post-actions, rules can delete the event and undo its actions, which is similar to a post-condition. The working scheme of an event-driven grammar is shown in Figure 5. All the sets of rules, (except the ones in *evt*, which just produce a user event) are executed as long as possible.

$$M_i \xrightarrow{evt} M_{evt} \xrightarrow{pre*} M_{evt-pre} \xrightarrow{sys-act*} M_{act} \xrightarrow{post*} M_{act-post} \xrightarrow{del*} M_f$$

Fig. 5. Application of an event driven grammar with user-defined rules.

In the example presented in this paper, models (M_i, M_{evt}, $M_{evt-pre}$, M_{act}, $M_{act-post}$ and M_f in Figure 5) are indeed typed graph triples. In this way, the set of rules *evt*, *sys* − *act* and *del* are applied to the $CONC$ graph, which represents the concrete syntax. In the example, rules in *pre* and *post* are abstract triple rules, used to propagate the changes due to the user-generated events to the abstract syntax model ($ABST$ graph).

Figure 6 shows the AToM^3 base classes for the concrete syntax. As stated before, all concrete syntax symbols inherit either from *Entity* (if it is an icon-like entity) or from *Link* (if it is an arrow-like entity). Both *Entity* and *Link* inherit from *VisualObject*, which has information about the object's location (x and y) and about if it is being dragged (*selected*). *Links* are connected to *Entities* via *Segments*; these can go either from *Entities* to *Links* (*e2l*) or the other way around (*l2e*).

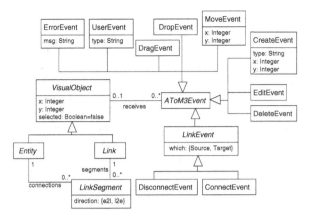

Fig. 6. AToM^3 base classes for concrete syntax objects and user events.

Some of the classes in Figure 6 model the events that can be generated by the user. All the events can be associated to a *VisualObject*. Some events have additional information, such as *CreateEvent*, which contains the type of the *VisualObject* to be created, and its position. The *MoveEvent* contains the position where the object has been moved. When connecting two *Entities*, two *ConnectEvent* objects are generated, one associated to the source and other one associated to the target. *ErrorEvent* signals an error associated with a certain object, AToM^3 presents the text of the error and highlights the associated object. Finally, the *UserEvent* class can be used to define new events.

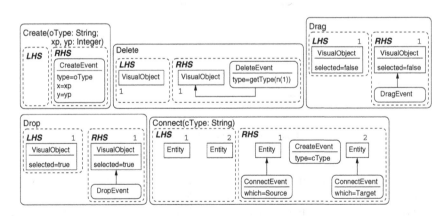

Fig. 7. Some of the *event-generator* rules.

Figure 7 shows some of the *event-generator* rules (depicted as *evt* in Figure 5), which model the generation of events by the user. The *Create* rule is triggered when the user clicks on the button to create a certain entity, and then on the canvas. The type of the object to be created is given by the button that the user clicks, and the x and y coordinates by the position of the cursor in the canvas. In AToM^3, a button is created for each non-abstract class in the meta-model. The *Delete* rule is triggered when the user deletes an object. The type of the object to be deleted is obtained by calling the *getType* function on node number one. This is a function which is available in Python (the implementation language of AToM^3) and returns the actual type of an object. Finally, the *Connect* rule is invoked when the user connects two *Entities*. In AToM^3 this is performed by clicking in the *connect* button and then on the source and the target entities. AToM^3 infers (with the meta-model information) the type of the subclass of *Link* that must be created in between. If several choices exist, then the user selects one of them. The type is then passed as a parameter of the rule, and the corresponding creation event is generated.

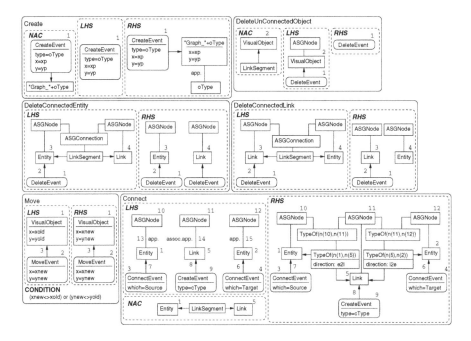

Fig. 8. Some of the *action* rules.

Figure 8 shows some of the rules that model the actual execution of the events (depicted as *sys-act* in Figure 5). The first rule models the actual creation of an instance (subclass of *ASGNode*, see Figure 1), together with its associated visual representation (whose type name is the same as the non-visual instance, but starting by *"Graph_"* and is a subclass of *Link*). The three following rules model the execution of a *delete* event. In the first case (*DeleteUnConnectedObject* rule), the object has no connections. In the second case (*DeleteConnectedEntity* rule), the icon-like object has connections, so a *delete* event is sent to the connected link, and the segment is erased. The third case (*DeleteConnectedLink* rule) models the deletion of a link (the "centre" of an arrow-like graphical form), which also deletes the associated segment. Please note that all the rules are executed as long as possible (see Figure 5). The *Move* rule simply modifies the position attributes of the object. Finally, the *Connect* rule models the connection of a link to two entities. Note, that rule *Connect* in Figure 7 generates a *CreateEvent* for the *link*, so rule *Create* in Figure 8 is executed first. The rule creates the *link* with the correct type. Next, rule *Connect* in Figure 8 can be applied, as classes *Entity* and *Link* are the base classes for all graphical objects. Note how the appropriate types for the segments in between links and entities are obtained (from the AToM[3] API) through function *TypeOf* which searches the information in the meta-model. Finally, a last set of rules (not shown in the paper) models the deletion of the events.

5   Example: Sequence Diagrams

As an example of the techniques explained before, we have built an environment to define UML sequence diagrams. By means of meta-modelling we define the abstract and concrete syntax of this kind of diagrams, as well as the correspondence relation between their elements (see meta-model triple in Figure 4). Starting from this triple meta-model, AToM[3] generates a tool where the user can build models according to that syntax. The user creates the diagrams at the concrete syntax level, therefore some automatic mechanism to generate the abstract syntax of the diagrams and support its mutual coherence has to be provided. With this aim we have built a set of event-driven rules triggered by user actions. Additionally, another set of triple rules check the consistency between the sequence diagram and existing class diagrams. Both set of rules are presented in the following subsections.

5.1   Abstract and Concrete Syntax of Sequence Diagrams

These rules manage the creation, edition and deletion of *Objects*, the creation, edition and deletion of *Messages*, and the creation and deletion of object *Life Lines*. The graphical actions that do not change the diagram abstract syntax (like creating an *Activation Box*) do not need the definition of extra event rules apart from the ones provided by AToM[3] (see Figures 7 and 8).

Rules for the creation, edition and deletion of *Objects* are the simplest of the set. These rules create, edit and delete *Objects* at the abstract syntax level (once the user generates the corresponding event at the concrete level). *Objects* at the abstract syntax are related to the concrete syntax *Objects* (which received the user event) through an element in the correspondence graph. Rules for creating objects (both *post*- actions, see Figure 5) are shown in figure 3. The rule on the left creates the object at the abstract syntax level, while the rule on the right connects (at the abstract syntax level) the object with its corresponding class. If the rule on the right cannot be applied, it means that the object class has not been created in any class diagram. This inconsistency is tolerated at this moment (we do not want to put many constraints in the way the user builds the different diagrams), but we have created a grammar to check and signal inconsistencies, including this one. The grammar is explained in the next subsection and can be executed at any moment in the modelling phase. For the deletion of an object (rules not shown in the paper), we ensure that it has no incoming or outgoing connection. This is done by a *pre-* condition rule (not shown in this paper) that erases the *delete* event on an object and presents a message if it has some connection.

The creation of a message is equivalent to connecting two elements belonging to the concrete syntax (*ConcreteElement*, see Figure 4) by means of a relationship of type *AbsMessage*. Obviously users cannot instantiate neither abstract entities nor abstract relationships, but only concrete ones. Therefore, at the user level the action to create messages includes three concrete cases: the connection of two *Activation Boxes* by means of a *Message* relation, the connection from an *Activation Box* to an *Object* by means of a *createMessage* relationship, and the connection from a *Start Point* to an *Activation Box* by means of a *startMessage* relation. The event rules for managing these three concrete cases are very similar except for the entities and relationships participating in the action. That is, we should have a first rule to create a *Message* relationship if its source and target are activation boxes; a second identical rule except for the relationship type (*createMessage*) and the target of the relationship (*Object*); and a third similar rule except for the relationship type (*StartMessage*) and source (*Start Point*). Since the three rules have the same structure, we use an abstract rule to reduce the grammar size. In Figure 9 we show the abstract rule compressing the first and third concrete rules mentioned above. We have used abstraction in many other rules, which highly reduces the total amount of rules. The rule in Figure 9 generates the abstract syntax of a new message created by the user, establishing a morphism between the concrete syntax of the new message (graphical appearance) and its respective abstract syntax. In this particular case the message concrete syntax is related to more than one abstract syntax entity: three abstract syntax entities (one *Message*, one *Stimulus* and one *Action*)) are graphically represented using a single symbol on the concrete syntax. On the other hand, the same event rule has to process the relationship between the newly created message and the rest of the model. In this way the successor, predecessor and activator messages of the created one have to be computed, as well as the objects sending and receiving the message. Additionally, we have to check if the new message activates in its turn another block of messages. We have broken down the creation event in a set of 6 user-defined events, each performing one step in the process. Thus the number of rules is reduced and the processing is easier.

Other rules (not shown in the paper) calculate the *predecessor* of a message. This is the previous one in the same processing block (the activation boxes corresponding with a method execution), or none if the message is the first one in the block. A total of 16 rules have been defined to manage the creation and edition of *objects* and *messages*. Some other rules, similar to the previous ones, manage the creation and deletion of *Life Lines*. The processing of the event (creation or deletion) triggers the execution of other user-defined events, simpler to process. Most of these events are the same as the ones generated by rule in Figure 9, therefore reutilization of rules has been possible. Due to space limitation, we do not show all the rules, which are 39 in total.

5.2 Consistency Checking

Triple rules can be used not only to maintain coherence between concrete and abstract syntax, but also to check consistency between different types of diagrams. The present

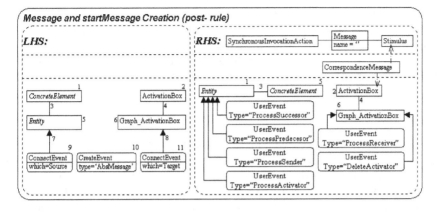

Fig. 9. Abstract rule for Creating *Messages* and *createMessages*.

work is part of a more general project with the aim to formalize the dynamic semantics of UML [8] by means of transformations into semantic domains (up to now Petri nets). Before translation, consistency checkings should be performed between the defined diagram (in this case a sequence diagram) and existing ones, such as class diagrams. Note how, while the user builds a sequence diagram, the previous rules add abstract syntax elements to a unique abstract syntax model. In this way, one has a unique abstract syntax model and possibly many concrete syntax models, one for each defined diagram (of any kind).

Using simple triple rules, we can perform consistency checkings between the sequence diagram and an existing abstract syntax model, generated by previously defined diagrams. For example, we may want to check that the class of the objects used in a sequence diagram has been defined in some of the existing class diagrams; if an object invokes a method of another object, the method should have been defined in its class, and there should be a navigable relationship between both object classes (see Figure 10), and that the invoked method is visible from the calling class.

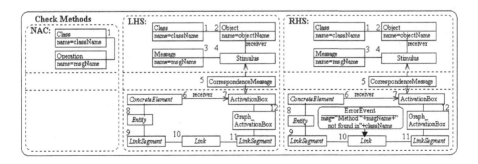

Fig. 10. One of the Rules for Consistency Checking.

We define *consistency triple rules* in such a way that their LHSs are conditions that are sought in the defined diagram (sequence diagrams in our case), possibly in the concrete and abstract parts. NACs are typically conditions to be sought in the existing abstract model with which we want to check consistency. If the rule is applied the rule's RHS sends an event of type *ErrorEvent* to some of the objects matched by the LHS.

6   Related Work

At a first glance, the present work may resemble the *syntax directed approach* for the definition of a VL. In this approach one defines a rule for each possible editing action, and the user builds the model by selecting the rules to be applied. Our approach is quite different, in the sense that we use a meta-model for the definition of the VL. The meta-model (which may include some constraints) provides all the information needed for the generation of the VL. The user builds the model by interacting with the user interface. In our approach we explicitly represent these events in the rules. Rules are triggered by the events, but the user may not be aware of this fact. In the examples, we have shown the combination of event-driven grammars with triple grammars to build the abstract syntax model and to perform consistency checks.

In the approach of [4], a restricted form of Statecharts was defined using a pure graph grammar approach (no meta-models). For this purpose, they used a *low level* (LLG, concrete syntax) and a *high level* (HLG, abstract syntax) representation. To verify the correctness, the LLG has to be transformed into an HLG (using a regular graph grammar), and a parsing grammar has to be defined for the latter. Other parsing approach based on constraint multiset grammars is the one of CIDER [9].

Other approaches for the definition of the VLs of the different UML diagrams, usually concentrate either on the concrete or the abstract syntax, but not on both. For example, in [5], graph transformation units are used to translate from sequence diagrams into collaboration diagrams. Note how, both kind of diagrams share the same abstract syntax, so in our case, a translation is not necessary, but we have to define triple rules to build the abstract syntax from the concrete one.

7   Conclusions

In this paper we have presented *event-driven grammars* in which user interface events are made explicit, and system actions in response to these events are modelled as graph grammar rules. Their combination with abstract triple rules and meta-modelling is an expressive means to describe the relationships between concrete and abstract syntax models (formally defined through meta-models). Rules can model pre- and post- conditions and actions for events to take place. Furthermore, we can use the information in the meta-models to define *abstract* rules, which are equivalent to a number of concrete ones, where nodes are replaced by each element in its inheritance clan. In this work, we have extended (in a straightforward way) the original work in [3] to allow refinement of relationships.

The applicability of these concepts has been shown by an example, in which we have defined a meta-model triple for the abstract and concrete syntax of sequence diagrams (according to the UML 1.5 specification). Additionally, we have presented some rules to check the consistency of sequence diagrams models with an existing abstract syntax model, generated by the previous definition of other diagrams.

Regarding future work, we want to derive validation techniques for triple, event-driven grammars. We also plan to use triple graph grammars to describe heuristics for the creation of UML diagrams. For example, if the user creates an object in a sequence diagram which belongs to a non-existing class, one option is to raise a consistency warning. Other possibility is to automatically derive the concrete syntax of a class diagram with the information of the abstract syntax (classes, methods, etc.) generated by the sequence diagram.

Acknowledgements

This work has been partially sponsored by the Spanish Ministry of Science and Technology (TIC2002-01948). The authors would like to thank the referees for their useful comments.

References

1. Atkinson, C., Kühne, T. 2002. *Rearchitecting the UML infrastructure.* ACM Transactions on Modeling and Computer Simulation, Vol 12(4), pp.: 290-321.
2. Bardohl, R. 2002. *A Visual Environment for Visual Languages.* Science of Computer Programming 44, pp.: 181-203. See also the GENGED home page: http://tfs.cs.tu-berlin.de/~genged/.
3. Bardohl, R., Ehrig, H., de Lara J., and Taentzer, G. 2004. *Integrating Meta Modelling with Graph Transformation for Efficient Visual Language Definition and Model Manipulation.* In proceedings of ETAPS/FASE'04, LNCS 2984, pp.: 214-228. Springer.
4. Bottoni, P., Taentzer, G., Schürr, A. 2000. *Efficient Parsing of Visual Languages based on Critical Pair Analysis and Contextual Layered Graph Transformation* Proc. of VL'2000, pp.: 59-60.
5. Cordes, B., Hölscher, Kreowski, H-J. 2003. *UML Interaction Diagrams: Correct Translation of Sequence Diagrams into Collaboration Diagrams.* Proc. of AGTIVE'03, pp.: 273-288.
6. de Lara, J., Vangheluwe, H. 2002. *AToM³: A Tool for Multi-Formalism Modelling and Meta-Modelling.* In ETAPS/FASE'02, LNCS 2306, pp.: 174 - 188. Springer-Verlag. See also the AToM³ home page at: http://atom3.cs.mcgill.ca
7. Ehrig, H., Engels, G., Kreowski, H.-J., Rozenberg, G. 1999. *Handbook of Graph Grammars and Computing by Graph Transformation.* (1). World Scientific.
8. Guerra, E., de Lara, J. 2003. *A Framework for the Verification of UML Models. Examples using Petri Nets.* Jornadas de Ingeniería del Software y Bases de Datos, JISBD. Alicante. Spain. pp.: 325-334.
9. Jansen, A.R, Marriott, K. and Meyer, B. 2003. *CIDER: A Component-Based Toolkit for Creating Smart Diagram Environments.* Proc. of the 9th Conference on Distributed and Multimedia Systems. pp.: 353-359. Knowledge Systems Institute.
10. Odell, J. 1994. *Power types.* Journal of Object Oriented Programming (May).
11. Schürr, A. 1994. *Specification of Graph Translators with Triple Graph Grammars.* In LNCS 903, pp.: 151-163. Springer.
12. UML specification at the OMG's home page: http://www.omg.org/UML.

Analysis of Metabolic Pathways
by Graph Transformation[*]

Francesc Rosselló[1] and Gabriel Valiente[2]

[1] Department of Mathematics and Computer Science,
Research Institute of Health Science (IUNICS), University of the Balearic Islands,
E-07122 Palma de Mallorca
[2] Department of Software, Technical University of Catalonia,
E-08034 Barcelona

Abstract. Biochemical pathways, such as metabolic, regulatory, and signal transduction pathways, constitute complex networks of functional and physical interactions between molecular species in the cell. They are represented in a natural way as graphs, with molecules as nodes and processes as arcs. In particular, metabolic pathways are represented as directed graphs, with the substrates, products, and enzymes as nodes and the chemical reactions catalyzed by the enzymes as arcs. In this paper, chemical reactions in a metabolic pathway are described by edge relabeling graph transformation rules, as explicit chemical reactions and also as implicit chemical reactions, in which the substrate chemical graph, together with a minimal set of edge relabeling operations, determines uniquely the product chemical graph. Further, the problem of constructing all pathways that can accomplish a given metabolic function of transforming a substrate chemical graph to a product chemical graph using a set of explicit chemical reactions, is stated as the problem of finding an appropriate set of sequences of chemical graph transformations from the substrate to the product, and the design of a graph transformation system for the analysis of metabolic pathways is described which is based on a database of explicit chemical reactions, a database of metabolic pathways, and a chemical graph transformation system.

1 Introduction

Biochemical pathways, such as metabolic, regulatory, and signal transduction pathways, constitute complex networks of functional and physical interactions between molecular species in the cell. They are represented in a natural way as graphs, with molecules as nodes and processes as arcs. In particular, metabolic pathways are represented as directed graphs, with the substrates, products, and enzymes as nodes and the chemical reactions catalyzed by the enzymes as arcs.

[*] This work has been partially supported by the Spanish CICYT, project MAVER-ISH (TIC2001-2476-C03-01), by the Spanish DGES and the EU program FEDER, project BFM2003-00771 ALBIOM, and by the Ministry of Education, Science, Sports and Culture of Japan through Grant-in-Aid for Scientific Research B-15300003 for visiting JAIST (Japan Advanced Institute of Science and Technology).

Chemical descriptions in a metabolic pathway can be made at different levels of resolution: a molecular descriptor uniquely identifies a molecule in a chemical database; a molecular formula indicates the number of each type of atom in a molecule; a constitutional formula or chemical graph also indicates which pairs of these atoms are bonded; and a structural formula also indicates those stereochemical distinctions that are required to uniquely identify a molecule.

In a detailed representation of metabolic pathways, at the level of the constitutional formula or the structural formula, structural change of chemical reactions can be modeled by superimposing the reactant and the product to match up the atoms and bonds that are unchanged in the transformation. A formalism called *imaginary transition structures* was introduced in [8–10] to model chemical reactions, where the chemical graphs representing the reactions' substrate and product are superimposed topologically, and the bonds are then distinguished and classified into three categories: *out-bonds* (bonds appearing only in the substrate molecules), *in-bonds* (bonds appearing only in the product molecules), and *par-bonds* (bonds appearing in both the substrate and the product molecules). Imaginary transition structures can be seen as double-pushout transformation rules [3] over chemical graphs: the left-hand side, context, and right-hand side are chemical graphs with set of labeled nodes corresponding to the atoms in its molecules; the left-hand side graph has edges representing out-bonds, the context graph has edges representing par-bonds, and the right-hand side graph has edges representing in-bonds. This is, essentially, the view of chemical reactions advocated in [1].

In this paper, chemical reactions in a metabolic pathway are described by edge relabeling graph transformation rules, as explicit chemical reactions and also as implicit chemical reactions, in which the substrate chemical graph, together with a minimal set of edge relabeling operations, determines uniquely the product chemical graph.

Further, the problem of constructing all pathways that can accomplish a given metabolic function of transforming a substrate chemical graph to a product chemical graph using a set of explicit chemical reactions, is stated as the problem of finding an appropriate set of sequences of chemical graph transformations from the substrate to the product, and the design of a graph transformation system for the analysis of metabolic pathways is described, which is based on a database of explicit chemical reactions, a database of metabolic pathways, and a chemical graph transformation system.

The rest of the paper is organized as follows. Chemical reactions are viewed in Section 2 as edge relabeling graph transformation rules, where both explicit and implicit chemical reactions are introduced. In Section 3, two related sets of axioms on the structure of metabolic pathways are reviewed and the graph transformation problem of analyzing a metabolic pathway is discussed. Feasible reaction pathway axioms model networks of chemical reactions that follow a series of accepted first principles and conditions, while combinatorially feasible reaction pathway axioms relax some of these conditions. Finally, some conclusions are outlined in Section 4.

2   Modeling Chemical Reactions

A chemical reaction is the change produced by two or more molecules acting upon each other. In a chemical reaction, *substrate* molecules are transformed into *product* molecules, often in the presence of a *catalyst*. For simplicity, we shall assume henceforth that the catalysts of a reaction are part of both its substrate and product.

Example 1. Consider, for example, the acidic hydrolysis of ethyl acetate, which is described by the following equation:

$$CH_3COOCH_2CH_3 + H_2O + HCl \longrightarrow CH_3COOH + CH_3CH_2OH + HCl$$

The substrate of the reaction, ethyl acetate ($CH_3COOCH_2CH_3$) and water (H_2O), is transformed into acetic acid (CH_3COOH) and ethanol (CH_3CH_2OH) in the presence of a catalyst, hydrochloric acid (HCl).

It is usual to describe molecules as graphs, with nodes representing the atoms, each one of them labeled by the name of the corresponding element, and edges representing the bonds, with a positive weight describing the order of the bond (1 for a single bond, 2 for a double bond, etc.). A set of molecules is consequently described by the disjoint union of the graphs representing them. For reasons that will be clear below, we shall allow the existence in these graphs of one or more edges labeled 0: they should be seen as non-existent. Let us call these graphs representing sets of molecules *chemical graphs*.

Definition 1. *A **chemical graph** is a weighted graph (V, E, μ), where (V, E) is an undirected graph (without multiple edges or self-loops) all whose nodes are labeled by means of chemical elements, and $\mu : E \to \mathbb{N}$ is a weight function. The **valence** of a node in a chemical graph is the total weight of the edges incident to it.*

Chemical reactions consist of breaking, forming and changing bonds in sets of molecules. Therefore, a chemical reaction can be represented by the transformation of the chemical graph representing the reaction's substrate into the chemical graph representing the product. This transformation will satisfy a set of specific conditions. First, the number and type of the atoms in the substrate and the product must be the same, and therefore the transformation must induce the identity on the set of labeled nodes. Besides, and for simplicity, we shall restrict ourselves in this paper to chemical reactions where each individual atom has the same valence in the substrate and in the product: from the point of view of graphs, this corresponds to ensure that the total weight of edges incident to each node remains constant after the transformation. In a more general setting we could simply impose the conservation of the total number of valence electrons, which would correspond to the conservation of the sum of all edges' weights.

A systematic study of organic chemical reactions was made in [8–10], where a formalism called *imaginary transition structures* was introduced to model chemical reactions. The imaginary transition structure of a given reaction is a structural formula in which, using our language, the unweighted chemical graphs representing the reactions' substrate and product are superimposed topologically, and the bonds are then distinguished and classified into three categories: *out-bonds* (bonds appearing only in the substrate molecules), *in-bonds* (bonds appearing only in the product molecules), and *par-bonds* (bonds appearing in both the substrate and the product molecules).

Example 2. The acidic hydrolysis of ethyl acetate from Example 1 can also be depicted as a transformation between the chemical graph to the left (representing the substrate) and the chemical graph to the right (representing the product) in the following diagram:

$$
\begin{array}{ccc}
O & CH_2 \longrightarrow CH_3 \\
\Vert & / \\
CH_3 \longrightarrow C \longrightarrow O \\
\\
H \longrightarrow O \qquad H \\
\diagdown \quad / \\
H \qquad Cl
\end{array}
\qquad \longrightarrow \qquad
\begin{array}{ccc}
O & CH_2 \longrightarrow CH_3 \\
\Vert & / \\
CH_3 \longrightarrow C \qquad O \\
/ \quad \diagdown \\
H \longrightarrow O \qquad H \\
\\
H \longrightarrow Cl
\end{array}
$$

The corresponding imaginary transition structure is shown in Fig. 1, where out-bonds are denoted by solid lines crossed by a bar, par-bonds are denoted by plain solid lines, and in-bonds are denoted by solid lines crossed by a small circle; for simplicity, we have assigned single nodes to groups of atoms like CH_3 and CH_2 that are not broken in the reaction.

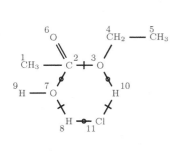

node	atom or group	neighborhood			
1	CH_3	2			
2	C	1	3	6	7
3	O	2	4	10	
4	CH_2	3	5		
5	CH_3	4			
6	O	2			
7	O	2	8	9	
8	H	7	11		
9	H	7			
10	H	3	11		
11	Cl	8	10		

Fig. 1. The imaginary transition structure of acidic hydrolysis of ethyl acetate.

An imaginary transition structure can be seen as a double-pushout transformation rule [3] over chemical graphs. The left-hand side, context, and right-hand side are chemical graphs with set of labeled nodes corresponding to the atoms in its molecules; the left-hand side graph has edges representing out-bonds and par-bonds, the context graph has edges representing par-bonds only, and the right-hand side graph has edges representing in-bonds and par-bonds. This is, essentially, the view of chemical reactions advocated in [1].

An imaginary transition structure can also be seen as a set of edge relabeling operations applied to a chemical graph. The substrate of the chemical reaction is represented by a chemical graph with set of labeled nodes corresponding to the atoms in its molecules and edges representing bonds that exist in the substrate or in the product of the reaction. The edges' weights are then assigned according to the definition of chemical graph given above: an edge is labeled 0 if the corresponding bond does not exist in the substrate (and then, by construction, it *must exist* in the product), and if the bond exists in the substrate, then it is weighted according to its order. The chemical reaction is then simply described by relabeling of the edges in this graph: a bond existing in some substrate molecule that breaks in the product molecules is relabeled by 0; a new bond appearing in some product molecule is assigned the label corresponding to its order; a bond that exists both in a substrate molecule and in a product molecule but they are of a different order, is relabeled according to its new order; and, finally, labels of bonds that are not modified at all by the chemical reaction are not modified either.

This description of a chemical reaction, based on Fujita's imaginary transition structures, motivates the introduction of the notion of an *explicit chemical reaction*.

Definition 2. *An **explicit chemical reaction** is a structure (V, E, σ, π), where (V, E, σ) and (V, E, π) are chemical graphs, called the **substrate** and the **product** chemical graphs respectively, satisfying the following conditions:*

(i) There is no $e \in E$ such that $\sigma(e) = \pi(e) = 0$.
(ii) For every $v \in V$, if $e_1, \ldots, e_k$ are the edges incident to it, then

$$\sigma(e_1) + \cdots + \sigma(e_k) = \pi(e_1) + \cdots + \pi(e_k) \geqslant 1.$$

Every imaginary transition structure, and hence every chemical reaction, can be represented by means of an explicit chemical reaction (V, E, σ, π) with (V, E, σ) and (V, E, π) being the chemical graphs describing its substrate and product.

Example 3. Consider again the acidic hydrolysis of ethyl acetate from Examples 1 and 2. The diagram in Fig. 2 represents the explicit chemical reaction describing it: the left-hand side graph depicts the graph (V, E), and the weight functions σ and π are given in the right-hand side table.

Application of an explicit chemical reaction to a given chemical graph, consists of relabeling the substrate by the product within the given chemical graph.

Definition 3. *A chemical graph (V, E, μ) is a **subgraph** of a chemical graph (V', E', μ') if $V \subseteq V'$, $E \subseteq E'$ and for all edges $e \in E$, $\mu(e) \leqslant \mu'(e)$. An explicit chemical reaction (V, E, σ, π) **can be applied** to a chemical graph (V', E', μ) if (V, E, σ) is a subgraph of (V', E', μ). In such a case, the **application** of (V, E, σ, π) to (V', E', μ) results in a chemical graph (V', E', μ'), where $\mu'(e) = \mu(e)$ for all edges $e \in E' \setminus E$ and $\mu'(e) = \pi(e)$ for all edges $e \in E$.*

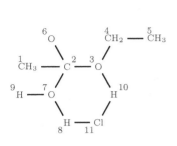

bond node–node	substrate weight	product weight
1–2	1	1
2–3	1	0
2–6	2	2
2–7	0	1
3–4	1	1
3–10	0	1
4–5	1	1
7–8	1	0
7–9	1	1
8–11	0	1
10–11	1	0

Fig. 2. The explicit chemical reaction of acidic hydrolysis of ethyl acetate.

Now, in general, it will not be necessary to provide the full substrate and product edge weight functions σ and π of an explicit chemical reaction, to describe the corresponding edge labeling transformation. Indeed, it is enough to specify the substrate chemical graph and a minimal set of edge relabeling operations which, when applied to a graph taking into account conditions (i) and (ii) in Definition 2, determine uniquely the product chemical graph. Since the undirected graph underlying the substrate chemical graph is finite, such a minimal set of edge relabeling operations will always exist, although it need not be unique.

Example 4. Consider one more time the acidic hydrolysis of ethyl acetate. As it can be seen in the table given in Fig. 2, this chemical reaction corresponds to six edge relabeling operations (see also Example 2). Now, it can be easily checked that each one of these edge relabeling operations, together with the conservation of the atoms' valences and the structure of the underlying undirected graph, entails the other five ones and hence it describes completely the chemical reaction.

For instance, assume we relabel from 0 to 1 the edge 8–11, joining a hydrogen in the water molecule and the chlorine in the hydrochloric acid. To preserve these atoms' valences, edges 7–8 and 10–11 must be relabeled to 0. And then to preserve the oxygen and hydrogen atoms' valences involved in these two last edges, the only possibility is to relabel edges 2–7 and 3–10 to 1. And finally, in order to preserve the valences of the carbon and oxygen atoms involved in these two edges, it can be checked that the only possibility is to relabel edge 2–3 to 0: any other relabeling modifies the valence of other atoms bound to the carbon or the hydrogen.

This leads us to the definition of an *implicit chemical reaction*.

Definition 4. *An **implicit chemical reaction** is a finite set* $\mathbf{R}$ *of edge relabeling operations of the form*

$$r = (A_{r,1}, A_{r,2}, w_{r,s}, w_{r,p}),$$

with $A_{r,1}$ *and* $A_{r,2}$ *atomic elements and* $w_{r,s}, w_{r,p} \in \mathbb{N}$ *such that* $w_{r,s} \neq w_{r,p}$.

*Such an implicit chemical reaction **can be applied** to a chemical graph* (V, E, σ) *when the following conditions are satisfied:*

(a) *For every* $r \in \mathbf{R}$ *there is one edge* e_r *in* E *whose nodes have labels* $A_{r,1}$ *and* $A_{r,2}$, *whose weight is* $w_{r,s}$, *and such that* $w_{r,p}$ *is less than or equal to the valence of both these nodes.*

(b) *There is one, and only one, chemical graph* (V, E, τ), *with the same nodes and edges as* (V, E, σ), *such that*
   - *all its nodes have the same valence as in* (V, E, σ);
   - $\tau(e_r) = w_{r,p}$ *for every* $r \in \mathbf{R}$;
   - *if* $\sigma(e) = 0$, *then* $\tau(e) \neq 0$.

The result of this application is, then, the chemical graph (V, E, τ), *and this application is said to **represent** the explicit chemical reaction* (V, E, σ, τ).

Notice that an implicit chemical reaction need not admit an application to a given chemical graph, even if it satisfies condition (a). And also that a given implicit chemical reaction may admit several applications to a given chemical graph, possibly giving different results. It could be interesting to study application and uniqueness conditions for implicit chemical reactions. But we shall not consider them here, as we are dealing with already existing chemical reactions, which we represent by means of an application to a suitably defined chemical graph of an implicit chemical reaction that has been determined beforehand.

3 Analyzing Metabolic Pathways

Metabolism can be regarded as a network of chemical reactions catalyzed by enzymes and connected via their substrates and products, and a metabolic pathway can be regarded as a coordinated sequence of chemical reactions [4]. The definition of a metabolic pathway is not exact, and most pathways constitute indeed highly intertwined cyclic networks. In a cell, a pathway's substrates are usually the products of another pathway, and there are junctions where pathways meet or cross [13]. For the purposes of this paper, we shall adopt the following definition.

Definition 5. *A **metabolic pathway** is a connected directed graph* (C, R), *where* C *is a set of chemical graphs and* $R \subset C \times C$ *is a set of explicit chemical reactions. The **substrate** of* (C, R) *is the set of chemical graphs* $S \subset C$ *such that for all* (V, E, π) *in* S, *there is no explicit chemical reaction of the form* (V, E, σ, π) *in* R. *The **product** of* (C, R) *is the set of chemical graphs* $P \subset C$ *such that for all* $(V, E, \sigma) \in P$, *there is no explicit chemical reaction of the form* (V, E, σ, π) *in* R.

The analysis of metabolic pathways is motivated by the rapidly increasing quantity of available information on metabolic pathways for different organisms. One of the most comprehensive sources of metabolic pathway data is [19]. There are also several databases on metabolic pathways, such as aMAZE [17], BRENDA [23], EcoCyc [14], KEGG [12], and WIT [21]. These databases contain hundreds of metabolic pathways and thousands of chemical reactions, and even the metabolic pathway for a small organism constitutes a large network. For instance, the proposed metabolic pathway for the bacterium *E. coli* consists of 436 compounds (substrates, products, and intermediate compounds) linked by 720 reactions [5].

One aspect of metabolic pathway analysis is *flux analysis*: the decomposition of a metabolic pathway into a complete set of nondecomposable steady state flux distributions. Two similar approaches to flux analysis are known which are based on the set of *elementary flux modes* [24] and on the set of *extreme pathways* [22]. In large metabolic networks, however, these approaches are hampered by the combinatorial explosion of possible routes: the maximal number of elementary flux modes in a metabolic pathway is exponential in the number of reactions, substrates, and products [15].

Another, complementary aspect of metabolic pathway analysis is *pathway synthesis*: the construction of all pathways that can accomplish a given metabolic function, which is: the transformation of a given set of substrates to a given set of products. Pathway synthesis belongs in pathway analysis, because it allows biologists and biochemists to contrast those metabolic pathways which exist in the cell for different organisms, against feasible metabolic pathways obtained by synthesis.

In pathway synthesis, much like in retrosynthetic analysis in organic chemistry [2, 26], the target chemical graph is subjected to a disconnection process, which corresponds to the reverse of a chemical reaction. As a result, the target chemical graph is transformed to a sequence of simpler chemical graphs in a stepwise manner, along a path that ultimately leads to simple chemical graphs. For a complex target chemical graph, some intermediate chemical graphs may undergo further retrosynthetic analysis. Thus, the repetition of this process eventually will result in a hierarchical synthesis tree for the target chemical graph.

In order to synthesize meaningful metabolic pathways, axioms on reaction pathways have been established in [6, 25], based on [7]. A first set of axioms, the *feasible reaction pathway* axioms, establish that (R1) every product is totally produced by reactions represented in the pathway; (R2) every substrate is totally consumed by reactions represented in the pathway; (R3) intermediate compounds are entirely produced by previous reactions and completely consumed by subsequent reactions; (R4) each reaction represented in the pathway is defined a priori; (R5) the network representing the pathway is acyclic; and (R6) at least one reaction represented in the pathway affects the activation of a substrate.

Among these axioms for feasible reaction pathways, (R4) follows from Definition 5. The remaining ones are enforced by the following definition.

Definition 6. *A metabolic pathway* (C, R) *is said to be a* **feasible reaction pathway** *if it satisfies the following axioms:*

- *For each product chemical graph of the form* (V, E, μ) *in* P*, there is a set of explicit chemical reactions* $R' \subseteq R$ *of the form* (V', E', σ', π') *such that* (V, E, μ) *is a subgraph of the union of the product chemical graphs* (V', E', π') *in* R'*.*
- *For each substrate chemical graph of the form* (V, E, μ) *in* S*, there is a set of explicit chemical reactions* $R' \subseteq R$ *of the form* (V', E', σ', π') *such that* (V, E, μ) *is a subgraph of the union of the substrate chemical graphs* (V', E', σ') *in* R'*.*
- *For each chemical graph of the form* (V, E, μ) *in* $C \setminus (S \cup P)$*, there is a set of explicit chemical reactions* $R' \subseteq R$ *of the form* (V', E', σ', π') *such that* (V, E, μ) *is a subgraph of the union of the product chemical graphs* (V', E', π') *in* R'*, and there is a set of explicit chemical reactions* $R'' \subseteq R$ *of the form* $(V'', E'', \sigma'', \pi'')$ *such that* (V, E, μ) *is a subgraph of the union of the substrate chemical graphs* (V'', E'', σ'') *in* R''*.*
- (C, R) *is acyclic.*
- *There is an explicit chemical reaction of the form* (V, E, σ, π) *in* R*, with* (V, E, σ) *a substrate chemical graph in* $S \subset C$*.*

A second set of axioms, the *combinatorially feasible reaction pathway* axioms, allow one to focus on the combinatorial properties of the network comprising the feasible reaction pathways, as the condition imposed by axiom (R5) is relaxed except for the cycles formed by the forward and reverse directions of a chemical reaction, and the condition imposed by axiom (R6) is discarded. These axioms establish that (T1) every product is represented in the network; (T2) every substrate is represented in the network; (T3) each reaction represented in the network is defined a priori; (T4) every compound represented in the network has at least one path leading to a product of the network; (T5) every compound represented in the network is a substrate or a product for at least one reaction represented in the network; (T6) a substrate of any reaction represented in the network is a substrate of the network if it is not a product of any reaction represented in the network; and (T7) each reaction represented in the network is either forward or backward, but not both.

Among these axioms for combinatorially feasible reaction pathways, (T3) and (T4) follow from Definition 5. The remaining ones are enforced by the following definition.

Definition 7. *Let* S' *and* P' *be fixed sets of chemical graphs. A metabolic pathway* (C, R) *is said to be a* **combinatorially feasible reaction pathway** *with respect to* S' *and* P' *if it satisfies the following axioms:*

- $P' \subseteq C$*.*
- $S' \subseteq C$*.*
- *For each chemical graph of the form* (V, E, μ) *in* C*, there is an explicit chemical reaction of the form* (V', E', σ', π') *in* R *such that either* (V, E, μ) *is a subgraph of* (V', E', σ')*, or* (V, E, μ) *is a subgraph of* (V', E', π')*.*

- *For each explicit chemical reaction of the form (V, E, σ, π) in R, (V, E, σ) is in the substrate S of (C, R) if there is no explicit chemical reaction of the form (V', E', σ', π') in R such that (V, E, σ) is a subgraph of (V', E', π').*
- *There are no two explicit chemical reactions of the form (V, E, σ, π) and (V, E, π, σ) in R.*

Now, in the analysis of metabolic pathways, the problem of constructing all pathways that can accomplish a given metabolic function can be stated as follows.

Problem 1 (Synthesis of metabolic pathways). Given a substrate chemical graph S', a product chemical graph P', and a set R' of explicit chemical reactions, find one or all feasible or combinatorially feasible metabolic pathways (C, R) with substrate $S = S'$, product $P = P'$, and set of explicit chemical reactions $R \subseteq R'$.

Since explicit chemical reactions are edge relabeling graph transformation rules, the metabolic pathway synthesis problem can be solved by graph transformation, as follows. Given a substrate chemical graph S', a product chemical graph P', and a set R' of explicit chemical reactions, a metabolic pathway (C, R) with substrate $S = S'$, product $P = P'$, and set of explicit chemical reactions $R \subseteq R'$ is given by a set of sequences of chemical graph transformations with substrate S and product P.

A graph transformation system for the analysis of metabolic pathways is being developed at the Technical University of Catalonia. The system is based on the following three main components:

- Database of explicit chemical reactions,
- Database of metabolic pathways, and
- Chemical graph transformation system.

The efficient implementation of the chemical graph transformation system relies on the CANON method for labeling a molecular structure with canonical labels [27–29], in which a molecular structure is treated as a graph with nodes (atoms) and edges (bonds), and each atom is given a unique numerical label on the basis of the topology of the molecular structure.

4   Conclusion

Chemical reactions in a metabolic pathway are described in this paper by edge relabeling graph transformation rules, both as explicit chemical reactions and as implicit chemical reactions, in which the substrate chemical graph, together with a minimal set of edge relabeling operations, determines uniquely the product chemical graph. On the basis of explicit chemical reactions, the problem of constructing all pathways that can accomplish a given metabolic function of transforming a substrate chemical graph to a product chemical graph using a set of explicit chemical reactions, is stated as the problem of finding an appropriate set of sequences of chemical graph transformations from the substrate to the product. The design of a graph transformation system for the analysis of metabolic pathways, based on a database of explicit chemical reactions, a database of metabolic pathways, and a chemical graph transformation system, is also described.

The formalism of chemical graphs and explicit chemical reactions is sufficient to describe most of the about 700 chemical reactions which have come to be recognized and referred to by name within the chemistry community [11, 16, 18, 20]. Future work includes extending this formalism to take compounds formed by ionic (instead of covalent) bonding, stereochemistry, and chirality into account, as well as modeling analysis problems upon more complex forms of biochemical networks, such as regulatory and signal transduction pathways, by graph transformation.

Acknowledgments

The authors would like to acknowledge the anonymous referees, and would also like to acknowledge with thanks detailed comments by Marta Cascante on a preliminary version of this paper.

References

1. Benkö, G., Flamm, C., Stadler, P.F.: A graph-based toy model of chemistry. Journal of Chemical Information and Computer Sciences **43** (2003) 1085–1093
2. Corey, E.J., Cheng, X.M.: The Logic of Chemical Synthesis. John Wiley & Sons, New York (1995)
3. Corradini, A., Montanari, U., Rossi, F., Ehrig, H., Heckel, R., Löwe, M.: Algebraic approaches to graph transformation. Part I: Basic concepts and double pushout approach. In Rozenberg, G., ed.: Handbook of Graph Grammars and Computing by Graph Transformation, Volume 1: Foundations. World Scientific (1997) 163–246
4. Deville, Y., Gilbert, D., van Helden, J., Wodak, S.J.: An overview of data models for the analysis of biochemical pathways. Briefings in Bioinformatics **4** (2003) 246–259
5. Edwards, J.S., Palsson, B.O.: The escherichia coli MG1655 in silico metabolic genotype: its definition, characteristics, and capabilities. Proc. Natural Academy of Science, USA **97** (2000) 5528–5533
6. Fan, L.T., Bertók, B., Friedler, F.: A graph-theoretic method to identify candidate mechanisms for deriving the rate law of a catalytic reaction. Computers and Chemistry **26** (2002) 265–292
7. Friedler, F., Tarján, K., Huang, Y.W., Fan, L.T.: Graph-theoretic approach to process synthesis: Axioms and theorems. Chemical Engineering Science **47** (1992) 1973–1988
8. Fujita, S.: Description of organic reactions based on imaginary transition structures. Part 1–5. Journal of Chemical Information and Computer Sciences **26** (1986) 205–242

9. Fujita, S.: Description of organic reactions based on imaginary transition structures. Part 6–9. Journal of Chemical Information and Computer Sciences **27** (1987) 99–120

10. Fujita, S.: Computer-Oriented Representation of Organic Reactions. Yoshioka Shoten, Kyoto (2001)

11. Hassner, A., Stumer, C.: Organic Syntheses based on Name Reactions. 2nd edn. Volume 22 of Tetrahedron Organic Chemistry Series. Pergamon Press, Oxford, England (2002)

12. Kanehisa, M., Goto, S.: KEGG: Kyoto encyclopedia of genes and genomes. Nucleic Acids Research **28** (2000) 27–30

13. Karp, P.D., Mavrovouniotis, M.L.: Representing, analyzing, and synthesizing biochemical pathways. IEEE Expert **9** (1994) 11–21

14. Karp, P.D., Riley, M., Saier, M., Paulsen, I.T., Collado-Vides, J., Paley, S.M., Pellegrini-Toole, A., Bonavides, C., Gama-Castro, S.: The EcoCyc database. Nucleic Acids Research **30** (2002) 56–58

15. Klamt, S., Stelling, J.: Combinatorial complexity of pathway analysis in metabolic networks. Molecular Biology Reports **29** (2002) 233–236

16. Laue, T., Plagens, A.: Named Organic Reactions. John Wiley & Sons, Chichester, England (1998)

17. Lemer, C., Antezana, E., Couche, F., Fays, F., Santolaria, X., Janky, R., Deville, Y., Richelle, J., Wodak, S.J.: The aMAZE LightBench: A web interface to a relational database of cellular processes. Nucleic Acids Research **32** (2004) 443–444

18. Li, J.J.: Name Reactions: A Collection of Detailed Reaction Mechanisms. 2nd edn. Springer-Verlag, Berlin (2003)

19. Michal, G., ed.: Biological Pathways: An Atlas of Biochemistry and Molecular Biology. John Wiley & Sons, New York (1999)

20. Mundy, B.P., Ellerd, M.G.: Name Reactions and Reagents in Organic Synthesis. John Wiley & Sons, Chichester, England (1988)

21. Overbeek, R., Larsen, N., Pusch, G.D., D'Souza, M., Selkov, E., Kyrpides, N., Fonstein, M., Maltsev, N., Selkov, E.: WIT: Integrated system for high-throughput genome sequence analysis and metabolic reconstruction. Nucleic Acids Research **28** (2000) 123–125

22. Schilling, C.H., Letscher, D., Palsson, B.O.: Theory for the systemic definition of metabolic pathways and their use in interpreting metabolic function from a pathway-oriented perspective. Journal of Theoretical Biology **203** (2000) 229–248

23. Schomburg, I., Chang, A., Schomburg, D.: BRENDA, enzyme data and metabolic information. Nucleic Acids Research **30** (2002) 47–49

24. Schuster, S., Fell, D.A., Dandekar, T.A.: A general definition of metabolic pathways useful for systematic organization and analysis of complex metabolic networks. Nature Biotechnology **18** (2000) 326–332

25. Seo, H., Lee, D.Y., Park, S., Fan, L.T., Shafie, S., Bertók, B., Friedler, F.: Graph-theoretical identification of pathways for biochemical reactions. Biotechnology Letters **23** (2001) 1551–1557

26. Serratosa, F., Xicart, J.: Organic Chemistry in Action: The Design of Organic Synthesis. 2nd edn. Number 51 in Studies in Organic Chemistry. Elsevier, Amsterdam (1996)

27. Weininger, D.: SMILES, a chemical language and information system. 1. Introduction to methodology and encoding rules. Journal of Chemical Information and Computer Sciences **28** (1988) 31–36

28. Weininger, D., Weininger, A., Weininger, J.L.: SMILES. 2. Algorithm for generation of unique SMILES notation. Journal of Chemical Information and Computer Sciences **29** (1989) 97–101
29. Weininger, D.: SMILES. 3. DEPICT. Graphical depiction of chemical structures. Journal of Chemical Information and Computer Sciences **30** (1990) 237–243

The Potential
of a Chemical Graph Transformation System

Maneesh K. Yadav[1], Brian P. Kelley[2], and Steven M. Silverman[1]

[1] The Scripps Research Institute, La Jolla CA 92037, USA
yadavm@scripps.edu
[2] Whitehead Institute for Biomedical Research, Cambridge MA 02142-1479, USA

Abstract. Chemical reactions can be represented as graph transformations. Fundamental concepts that relate organic chemistry to graph rewriting, and an introduction to the SMILES chemical graph specification language are presented. The utility of both deduction and unordered finite rewriting over chemical graphs and chemical graph transformations, is suggested. The authors hope that this paper will provide inspiration for researchers involved in graph transformation who might be interested in chemoinformatic applications.

1   Background

Few students of organic chemistry realize that there is a formal computational notion to the activity of writing down the products of chemical reactions. Graph transformation, which has appeared to meet broad interest in computer science, remains somewhat unnoticed in chemistry (in a formal sense) despite its obvious application. This paper will provide a broad outline of issues relating graph transformation to organic chemistry, in an attempt to provoke interest from computer scientists to the field of chemoinformatics.

Before we begin, an overview of organic chemistry is in order. We have deliberately made some omissions in our chemical explanations, in the interest of not obscuring the relevance of this paper to researchers involved in graph transformation. We refer the reader who is interested in learning some of the more subtle issues in graph-based representations of molecules (stereochemistry, tautomerism, aromaticity etc.) to any of the good textbooks[1], and your friendly local organic chemist.

Most physical things are made of *molecules*, collections of different types of atoms linked together via electronic orbitals. This paper will primarily concern itself with the construction of *organic* molecules. Such molecules contain carbon atoms in their structure, and are generally of biological origin or are synthesized from "salts of the earth" (carbon based compounds obtained from crude oil), via chemical *reactions*. A simple organic molecule is depicted Figure 1, and some common reactions are shown in Figure 2.

Molecules represented as graphs (where the typed "nodes" are atoms and typed "edges" represent bonds between atoms) have motivated solutions to

H. Ehrig et al. (Eds.): ICGT 2004, LNCS 3256, pp. 83–95, 2004.
© Springer-Verlag Berlin Heidelberg 2004

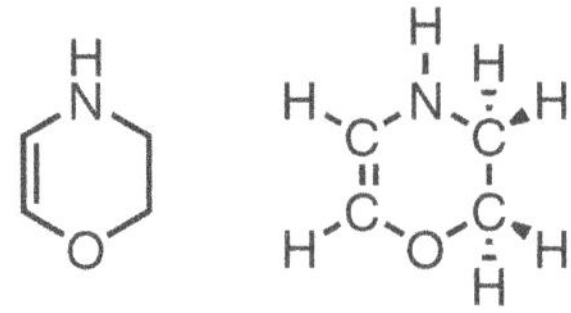

Fig. 1. Two depictions of the same molecule. Organic chemists tend to use the depiction on the left, omitting the labels for carbon (C) atoms and implicitly assuming the presence of hydrogen (H) atoms from the bond type (note the double bond). Wedged and and dashed edges represent bonds projecting out of and into the page, respectively. These figures primarily depict molecular *connectivity*, real molecules are embedded into distributions of three dimensional *conformations*. Atoms generally do not lie in one plane, and are tied to each other in a spring-like fashion by chemical bonds.

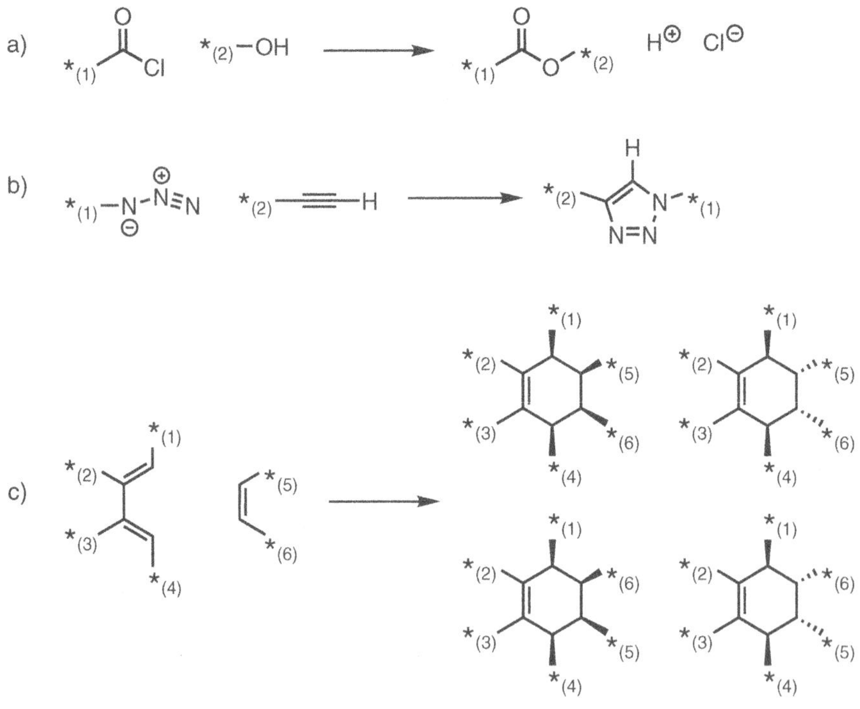

Fig. 2. Generalized depictions of three well known reactions a) esterfication b) acetylene-azide cycloaddition c) Diels-Alder cycloaddition. The molecules on the left hand side of the arrow are termed *reactants* and those on the right hand side, *products*. The asterisk denotes any molecular fragment. Depictions usually omit atoms that are not part of the main carbon skeleton, such as the H$^+$ (hydrogen cation) and Cl$^-$ (chlorine anion) in a). Note how some reactions can give multiple products, as in c).

problems in mathematics, under a field loosely called "mathematical chemistry". Chemical questions have lead to mathematical analogies which have inspired the investigation of topics such as constructive/analytic enumerations, graph canonicalization and (sub[2, 3])graph isomorphism[4]. These results provide a deeper understanding of molecular space and have lead to indispensable practical tools for machine aided storage, query and analysis of molecular structures.

Organic reactions can be represented as graph transformations, in fact, any chemical reaction can generally be represented by a sequence of applications of bond breaking and bond creation (atoms are generally conserved in *chemical transformations*). These bond manipulations physically correspond to sets of electrons re-arranging their quantum configurations to the most probable states. The precise order of these re-arrangements (which can occur on the order of femtoseconds) is the subject of reaction *mechanism* (see Figure 10), and is a fundamental physical concept that practicing chemists use to understand how certain products were achieved from a mixture of reactants[1]. From here on, we will refer to *chemical reactions* as the physical "territory" and *chemical graph transformations* as the "map"s which represent them.

Typically, multiple steps of chemical reactions are composed together (beginning with a set of starting materials) to produce *synthetic routes* to more complex molecules that have been targeted for construction, which are generally either

1. molecules which have been isolated from paucious natural sources, that have medicinal use and need to be synthetically produced on a large scale.
2. totally new molecules which have some predicted utility.

This paper will concern itself with the application of graph transformation towards both of these goals. An example synthetic route determined for an antibiotic is shown in Figure 3.

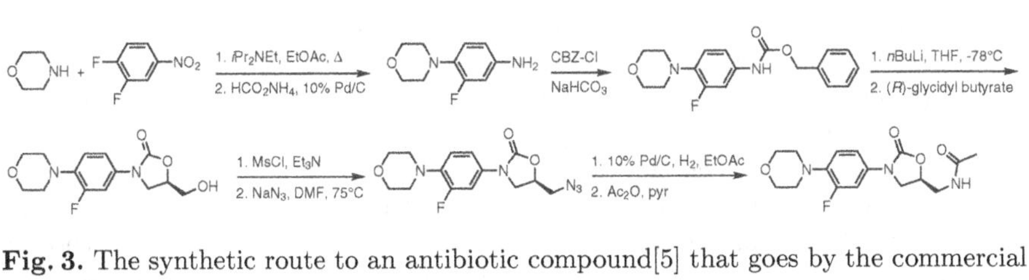

Fig. 3. The synthetic route to an antibiotic compound[5] that goes by the commercial name *Zyvox*.

The hitherto lack of theoretically founded graph rewriting applied to chemistry may be attributed to the shortage of practical underlying tools for its implementation, and its inherent intractability acquired from the requisite subgraph

---

[1] While chemists generally understand mechanism as the movements of electrons, the physical details of how this happens is remarkably complex, and the focus of many an investigator. We still cannot easily predict reactions based on first principles alone, which explains why new reactions are generally discovered by the experimentalist.

isomorphism step. It is important to note that despite theoretical intractability of subgraph isomorphism, constraints and heuristics can produce practically efficient implementations, demonstrated by the fact that chemical structure queries are routinely performed using subgraph isomorphism in many chemical database systems.

By considering molecules as expressions and the chemical transformations as rewriting rules, we propose the construction of a software system that would facilitate the implementation of automated deduction and exhaustive finite unordered rewriting over molecules and reactions. We outline the significance of such systems to organic chemistry in this paper. In the next section we describe a well known chemical graph specification language that such a system would likely employ.

2   The SMILES Graph Specification Language

Organic molecules are generally drawn in the canonical form as shown in this paper, a visual language which has evolved since the beginning of chemistry as the molecular nature of matter has become better understood.

A significant amount of historical[6, 7] effort has been spent on designing textual descriptions of molecular graphs which could be used for machine aided storage, query and analysis. The IUPAC naming system (which is actively refined as new molecules are discovered) is the standard naming system which chemists generally use in publications for unambiguous descriptions of molecules. IUPAC names are generally unwieldy when searching for molecules or even storing them in databases. The reader may recognize IUPAC names of organic molecules by looking on the ingredients label of many foodstuffs[2] and medicines.

The simple SMILES[8] language has emerged as the *de facto* standard for representing molecules in databases. While the grammar itself is rather straightforward, the generation of *canonical*[9] SMILES strings for molecules is of immense utility. The SMILES framework provides canonicalization, a convenient query-specification language (SMARTS) and representations for chemical reactions (SMIRKS).

The SMILES language contains nomenclature for describing atoms, bonds and for branches and closing loops. All atoms are denoted by their standard chemical symbols from the periodic table of elements. For example, carbon is 'C'. Bonds are denoted as '-' for single bonds, '=' for double bonds, '#' for triple bonds and '˜' for aromatic bonds. SMILES attempts to generate graph descriptions as small as possible so if a bond is not specified it is assumed to be either a single bond or an aromatic bond. Aromatic bonds are problematic in nature for most rewriting systems since they are bonds that resonate between single and double bonds and must be perceived by the SMILES parsing system. A simple SMILES example is shown in Figure 4.

---

[2] The IUPAC name for the artificial sweetener "aspartame", is *aspartyl-phenylalanine-L-methyl ester.*

Fig. 4. A simple four carbon chain (*butane*), specified by the SMILES string "CCCC".

The SMILES string "[CH3]-[CH2]-[CH2]-[CH3]" could also be used to describe the molecule in Figure 4, where hydrogen atoms are listed explicitly.

SMILES uses brackets to indicate isomers or non-standard configurations of atoms. To indicate closures, SMILES uses numbers after the atom designation. The SMILES string for a simple cyclic molecule is shown in Figure 5.

Fig. 5. A four carbon ring (*cyclobutane*), specified by the SMILES string "C1CCC1".

Finally, branches are denoted between parenthesis. The atom after the parenthesis is assumed to be attached to the immediately preceding atom (Figure 6).

Fig. 6. A branching example in *cyclobutanone*, specified by the SMILES string "C1CC(=O)C1".

Note that all of the following SMILES strings are valid for describing the molecule in Figure 6 (the last one being the canonical SMILES string):

"C1(=O)CCC1"

"C1C(=O)CC1"

"C1CC(=O)C1"

"C1CCC1=O"

"O=C1CCC1"

SMARTS is an extension of SMILES that basically allows one to query for sub-structures in SMILES strings. SMARTS look incredibly similar to SMILES, and most SMILES strings are also valid SMARTS. However, SMARTS adds AND and OR operations for graph matching. For instance, "[O,S]CCC" will match a set of four atoms that have an oxygen or a sulphur bonded to three carbons.

Finally, SMIRKS is a combination of SMILES and SMARTS used for specifying chemical graph transformations. Atom mapping is the main addition for
describing reactions, with the enumerated labeling of atoms used to describe the
embedding. A simple example, taken from the Daylight Inc.'s SMIRKS documentation, is esterfication of a carboxylic acid, shown in Figure 7.

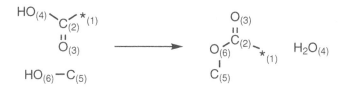

Fig. 7. A generalized form of the reaction a) in Figure 2, is described as
the following SMIRKS string, "[C:1][C:2](=[O:3])[O:4][H].[C:5][O:6][H] >>[C:1][C:2]
(=[O:3])[O:6][C:5].[O:4]" .

FROWNS is a free implementation of the SMILES system by one of the
authors[10]. The FROWNS implementation of SMIRKS, while incomplete, was
started mainly because of a disconnection between how SMILES and SMARTS
handle hydrogen atoms. Daylight has provided a wealth of of information on
the SMILES system[11], and continually improves the SMIRKS model, but the
authors feel that there may be room for some improvement by considering what
type of work has been done in the discipline of graph transformation and combining it with these current approaches in chemical representation.

3   Aspects of a Chemical Graph Transformation System

The authors imagine a software system that would be linked to a large database
of chemical structures (on the order of millions of compounds), reactions, and
validated synthetic routes where they can be manipulated in an algorithmic and
efficient fashion. Before the implementation on such a system were to begin, the
authors feel that careful attention must be paid to the following issues.

1. *Global Context Sensitivity.* The entire physical picture of a chemical transformation is not completely captured by only depicting the movements of
   bonds between atoms. Physical reactions have properties (yeild, rate, cost,
   etc.) which are dependant on the global context which the reaction is performed in. Almost all chemical reactions are run in liquid solvent (which
   itself can be a mixture of different molecules), the bulk properties of which
   (polarity, viscosity, specific heat, etc.) can greatly affect yield. Temprature,
   the presence of catalyst and the presence of radiation, are all aspects of a
   chemical reaction that are generally reported in any publication where an
   instance of that reaction is used (typically as one step in a multi-step synthesis). Any chemical graph transformation system must be able to associate
   such global physical descriptors with chemical graph tranformations.

2. *Local Context Senstivity.* Reactions are generally depicted in such a manner where the local molecular context is implied, or is generally understood by chemists (as in Figure 2). However, a chemical graph transformation system must be able to specify local preconditions under which a particular chemical graph transformation can be applied. For instance many potential reaction rules cannot be physically applied, due to the three dimensional conformation of reactive groups (Figure 8).

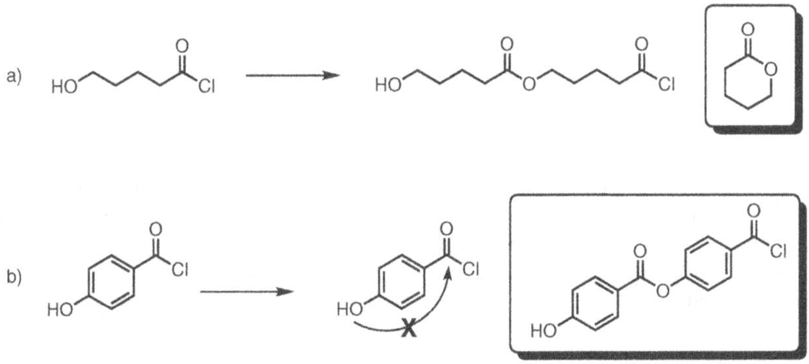

Fig. 8. Transformation a) in Figure 2 is applied once to two different molecules. In a) we would predict that the more abundant product (boxed) comes from the *intra*-molecular reaction, as the molecule will more "quickly" react internally with itself than with another molecule in solution. In b), the molecule is constrained conformationally so that only the *inter*-molecular product is formed (the graphs shown in reaction diagram actually depict physical multi-sets of molecules).

3. *Non-determinism.* Organic chemistry is in fact a game of numbers. By blindly applying chemical graph transformations we would find no chemical would be stable, as there are always routes that could operate on virtually any structure. Chemists take advantage of the fact that many potential reactions are separated by orders of magnitude in terms of rate (or *reactivity*). By applying local and global contexts in such a way that the chemist knows the desired reaction will occur orders of magnitude faster than any other potential reaction, she can isolate practical yields of intended product (Figure 9). Although a pure chemical graph transformation system will provide products "up to reactivity", the system should include provisions or hooks to physical simulation that will predict if certain reactions will (not) occur. In many cases the outcome of a particular reaction can be tuned to one product of a large potential set. A chemical graph transformation system needs to be able to backtrack to keep count of all possible outcomes of a reaction, to infer productive synthetic routes.

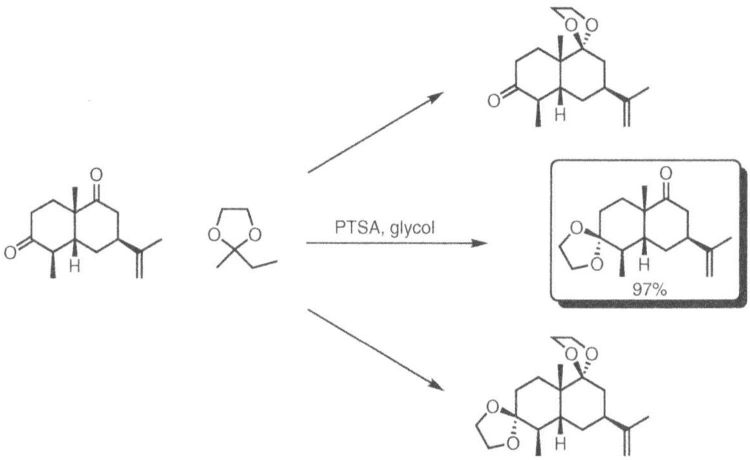

Fig. 9. An example[12] of a non-deterministic application of a chemical transformation. Either or both carbonyl groups (carbon double bonded to oxygen) can be transformed, but under the conditions of the reaction, only one (boxed product) is physically abundant. The reasons for this observed selectivity in reactivity, are due to both reaction conditions (written above the arrow) and the three dimensional conformation of the molecule.

4. *Mechanism.* Although we have not provided a background on chemical mechanism, we anticipate that a chemical graph transformation system would be more useful if reactions could be encoded and applied at different "levels of detail" (Figure 10). Students are often encouraged to not simply memorize organic transformations, but rather the mechanisms that they go through, which allows them to more easily infer the outcomes of particular reactions. In a similar sense, the ability to incorparate information about mechanism would greatly extend the utility of the system we propose.

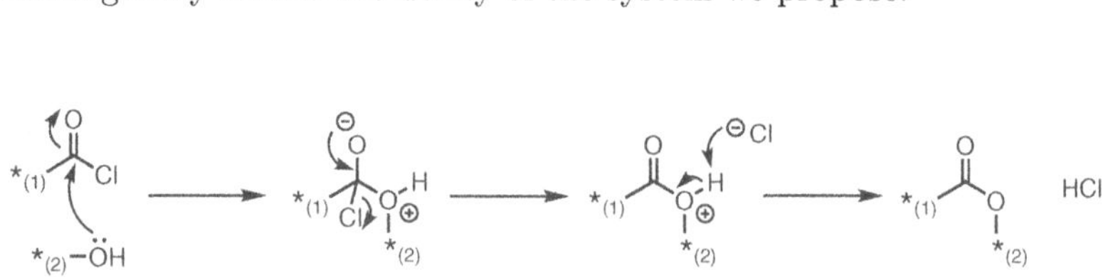

Fig. 10. A more detailed picture of reaction a) in Figure 2. The double dots above the oxygen (O) atom are an explicit representation of an electron pair. The curved arrows denote the movement of electron pairs.

4   Utility of a Chemical Graph Trasformation System

The references that are provided in this section demonstrate that there are many researchers who have implimented and applied graph rewriting over molecular graphs. The systems that are employed in these investigations are rather specialized for their particular application, and are not instances of a more general graph transformation framework. Despite the fact many day-to-day chemical questions are easily specified in terms of graph rewriting, such as: "What are all the potential mechanisms that could underly this reaction?", "What is the closest molecule I can make to my target, using only these reactions and materials?", "What potential degradation pathways can this molecule undergo?", such a framework does not exist.

Chemists require a language in which these questions can be specified as easily as they are thought of, and a system which can answer these queries efficiently. We hope that knowledge from the side of automated deduction, computer aided software engineering and visual specification languages can be applied to further this goal. We outline the general application areas to a greater depth below.

4.1 Computer Aided Organic Synthesis (CAOS)

An immense amount of knowledge in chemistry has been discovered along the way when chemists have attempted to construct molecules identical to those isolated from biological sources. We can consider these target structures as *chemical theorems* and the available set of starting materials and reactions which can be applied to them as *chemical axioms*. The computational process by which one tries to apply these axioms to physically construct (*chemically prove*) the target structures can be considered *chemical deduction*. Not only is it a computationally difficult task, but the physical constraints introduce uncertainty which requires that a large number of laborious experiments be done.

The computational complexity of finding a synthetic route to a target is equivalent to the semigroup word problem and is thus Turing undecidable[13]. By adding constraints such as bounding the number of applications and/or restricting the set of rewrite rules to only increase the size of the intermediates, the problem is equivalent to the tiling problem and is NP-complete[13]. These theoretical complexity bounds suggest why CAOS is difficult and not widely used, even modulo the difficulties airising from the physical unpredictability of chemical transformations.

The intuitive approach to chemical deduction, first formally outlined by E. J. Corey, is called *retrosynthetic analysis*[14]. The chemist recalls approximately 130 chemical transformations and attempts to mentally *disconnect* the target molecule into smaller pieces by applying the reversed transformation (the left and right hand side of the rule are switched, as in Figure 11), until he is left with starting materials that can be obtained from nature. A flexible chemical rewriting system should be able to reverse and apply rules, in addition to having an efficient mechanism by which preconditions can be specified and tested.

Current CAOS software systems (LHASA[15], Syngen, SYNCHEM, etc.) rely on expert-system based apporaches, and differ significantly in terms of the degree of interactivity, implemented constraints and supporting knowledge bases. It is important to note that none of these systems are considered widely used tools by practicing chemists[13]. While it is not our intent to be dismissive of these efforts

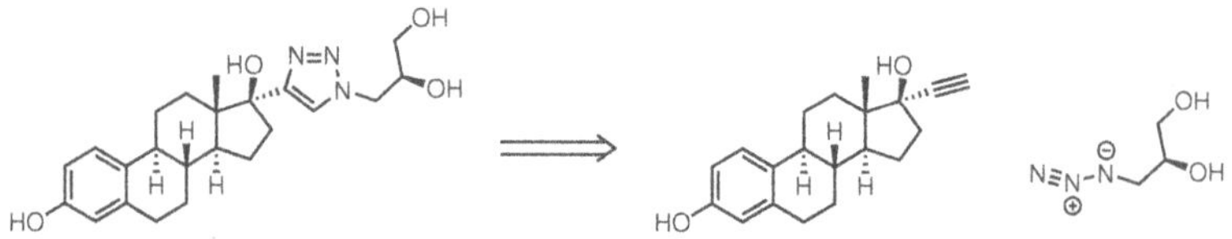

Fig. 11. A retrosynthetic transformation (denoted with the double lined arrow), using reaction b) from Figure 2. This is only a formal construction, the actual reaction happens in the forward sense[16].

(despite their impressive extensions, such as LHASA's LCOLI), these tools are not instances of a more general framework based on graph transformation and appear disjoint to general notions of deduction over graph grammars.

4.2   Exhaustive Finite Unordered Rewriting for Virtual Libraries

A common approach to the solution of computationally difficult problems is the strategy of "divide and conquer", by dividing a problem into easier sub-problems, and combining the resulting solutions to produce a solution of the original instance. Divide and conquer is an apt description of the underlying approach to search the physically intractable large space of small organic molecules (estimated to be between[17] 10^{62} and 10^{63} members) via combinatorial chemistry. By employing parallel synthetic routes and restricting the set of reactions and starting materials, chemists aim to quickly produce libraries of tens of thousands of different compounds that can be tested for some property. The reactions and starting materials should not only be practical and efficient, but provide libraries which maximize diversity, so as to sample as large a volume of molecular space as possible with the fewest number of compounds. By using a database of starting materials and encoding chemical transformations as graph rewrite rules, those rules can be applied in a exhaustive fashion for some practical number of steps[3] to produce a virtual library of compounds. Such a library could be subjected to *in silico* tests, to increase the efficiency with which chemists find new and useful compounds.

The commercial SMILES based LUCIA package from Sertanty Inc. promises an extremely general approach to virtual library creation, but the authors did not have access to this package at the time of writing.

4.3   Chemical Networks

Chemists typically perform reactions step by step with as few components as possible, generally to reduce cost and the possibility of any side products that

---

[3] Thermodynamics prevents any reaction from completely converting one molecule to another. Even as we compose ¿99% yield reactions together, the overall yield drops quickly as a function of the number of steps. The number of allowable steps in a large scale commercial synthesis varies with the value of the product, but a useful number to consider is ten.

may interfere in purifying the desired molecule. Pure substances are important for ease in characterization and reproducibility[4]. In general, it is difficult to predict all the possible outcomes when many different reactive species are put together in the same "soup".

One the fascinating aspects of biology is that many different and reactive small molecules productively co-exist in the same chemical soup, inside the living cell. Some suggest that it is the resulting complex network of interactions that distinguishes animate biological chemistry from everything else[18]; indeed, recent studies have espoused a "network-centric"[19–21] viewpoint of biology.

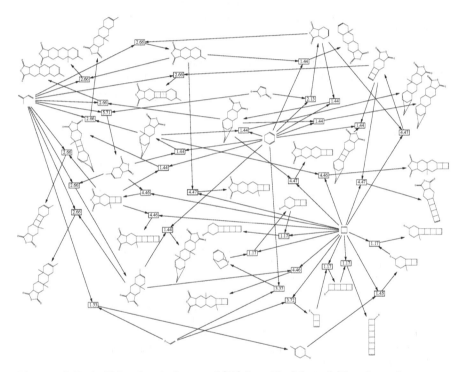

Fig. 12. A Diels-Alder chemical network[22] from Benkö et al. Boxed numbers represent the rate of the reaction between reactants (pointing in) to the product (pointed to).

A chemical graph transformation system would offer an algorithmic approach for enumerating all potential outcomes of "chemical networks", when a large number of potential reactions are applied iteratively to a set of starting materials. An interesting example[22] was reported where the Diels-Alder reaction (transformation c) in Figure 2) was iteratively applied for three cycles to an

---

[4] Consumers would be (rightfully) wary of purchasing medicine that was advertised as "80% pure".

initial set of five molecules. The resulting set of products is shown in Figure 12, taken from the referenced paper. Generalizing these systems to arbitrary reactivity would provide fertile ground for inferring outcomes of complicated reactive systems, and stochastic application could even be used for physical simulation.

5   Conclusion

We hope that this paper will act as a springboard for investigators to explore and apply techniques of graph transformation to chemistry. We restricted our suggestions to those related to reactivity, but there are obvious applications to the more visual aspects of chemistry, such as molecular modeling. An efficient implementation of a graph rewriting system would provide chemical investigators with an ability to more robustly determine retrosynthetic strategies, design synthetically accessible virtual libraries and simulate complicated chemical soups of reactive molecules. An even greater hope is that such physical analogies will provide reciprocal insight to the computer science of graph trasformation.

References

1. Clayden, J., Greeves, N., Warren, S., Wothers, P.: Organic Chemistry. Oxford University Press (2000)
2. Messmer, B.T., Bunke, H.: Subgraph isomorphism in polynomial time. Technical Report IAM 95-003, University of Bern, Institute of Computer Science and Applied Mathematics (1995)
3. Faulon, J.L.: Automorphism partitioning, and canonical labeling can be solved in polynomial- time for molecular graphs. Journal of Chemical Information and Computer Sciences **38** (1998) 432–444
4. Klin, M., Rücker, C., Rücker, G., Tinhofer, G.: Algebraic combinatorics in mathematical chemistry. Methods and algorithms. I. Permutation groups and coherent (cellular) algebras. Technical Report TUM M9510, Techn. Univ. München (1995)
5. Brickner, S.J., et al.: Synthesis and antibacterial activity of U-100592 and U-100766, two oxazolidinone antibacterial agents for the potential treatment of multidrug-resistant gram-positive bacterial infections. Journal of Medicinal Chemistry **39** (1996) 673–679
6. Garfield, E.: Are you ready for chemical linguistics? Chemical semantics? Chemical semiotics? Or, why WLN? Essays of an Information Scientist **1** (1972) 386–388 Paper available at:
   http://www.garfield.library.upenn.edu/essays/V1p386y1962-73.pdf.
7. Garfield, E.: Chemico-linguistics: Computer translation of chemical nomenclature. Nature **192** (1961) 192 Paper avilable at:
   http://www.garfield.library.upenn.edu/essays/v6p489y1983.pdf.
8. Weininger, D.: SMILES, a chemical language and information-system. 1. Introduction to methodlogy and encoding rules. Journal of Chemical Information and Computer Sciences **28** (1998) 31–36
9. Weininger, D., Weininger, A., Weininger, J.L.: SMILES 2: Algorithm for generation of unique SMILES notation. Journal of Chemical Information and Computer Sciences **29** (1989) 97–101

10. Kelley, B.P.: Graph canonicalization. Dr. Dobb's Journal **28** (2003) 66–69
11. Daylight Chemical Information Systems Inc.: (Daylight theory manual) Available at http://www.daylight.com/dayhtml/doc/theory/theory.toc.html.
12. Haaksna, A.A., Jansen, B.J.M., de Groot, A.: Lewis acid catalyzed Diels-Alder reactions of S-(+)-carvone with silyloxy dienes. Total synthesis of (+)-small alpha, greek-cyperone. Tetrahedron **48** (1992) 3121–3130
13. Smith, W.: Computational complexity of synthetic chemistry – basic facts. (1997) Paper available at http://citeseer.ist.psu.edu/192652.html.
14. Corey, E.J., Cheng, X.M.: The Logic of Chemical Synthesis. John Wiley and Sons (1995)
15. Johnson, A., Marshall, C.: Starting material oriented retrosynthetic analysis in the LHASA program. 2. Mapping the SM and target structures. Journal of Chemical Information and Computer Sciences **32** (1992) 418–425
16. Rostovtsev, V.V., Green, L.G., Fokin, V.V., Sharpless, K.B.: A stepwise huisgen cycloaddition process: Copper(I)-catalyzed regioselective "ligation" of azides and terminal alkynes. Angewandte Chemie International Edition **41** (2002) 2596–2599
17. Bohacek, R.S., McMartin, C., Guida, W.C.: The art and practice of structure-based drug design: A molecular modeling perspective. Medicinal Research Reviews **16** (1998) 3–50
18. Lee, D.H., Severin, K., Ghadiri, M.R.: Autocatalytic networks: The transition from molecular self-replication to molecular ecosystems. Current Opinion in Chemical Biology **1** (1997) 491–496
19. Jeong, H., Tombor, B., Albert, R., Oltvai, Z.N., Barabási, A.L.: The large-scale organization of metabolic networks. Nature **407** (2000) 651–654
20. Jeong, H., Mason, S., Barabási, A.L., Oltvai, N.Z.: Lethality and centrality in protein networks. Nature **411** (2001) 41–42
21. Ray, L.B., Jansy, B.R.: Life and the art of networks. Science **301** (2003) 1863
22. Benkö, G., Flamm, C., Stadler, P.F.: A graph-based toy model of chemistry. Journal of Chemical Information and Computer Sciences **43** (2003) 1085–1093
23. Blostein, D., Fahmy, H., Grbavec, A.: Practical use of graph rewriting. Technical Report 95-373, Queens University (1995)

Concepts for Specifying Complex Graph Transformation Systems

Boris Böhlen and Ulrike Ranger

RWTH Aachen University, Department of Computer Science III,
Ahornstrasse 55, 52074 Aachen, Germany
{boehlen,ranger}@cs.rwth-aachen.de

Abstract. As graph transformation systems evolve, they are used to solve complex problems and the resulting specifications tend to get larger. With specifications having more than one hundred pages they suffer from the same problems as large applications written in programming languages like C++ or Java do. Under the term programming in the large many different concepts have been developed to aid the solution of these problems. However, most graph transformation systems lack support for the problems.

With the introduction of UML-like packages, the PROGRES environment provides basic support for specifying in the large. Now that packages have been used in many different specifications, their shortcomings have become obvious. Additionally, our experience with large specifications showed that packages alone are not sufficient and therefore we have developed concepts for modularizing and coupling specifications. Our graph database provides the runtime-support required by these concepts.

1 Introduction

The idea of graph grammars arose in the late 1960s. With the help of graph grammars arbitrarily complex structures and relations can be represented [1]. Our department has been doing research in this area for many years. In 1989 we have started the development of the PROGRES language. PROGRES (PROgrammed Graph REwriting Systems) is a very high level programming language and an operational specification language for rapid prototyping [2]. The language is used to define, create, and manipulate graphs which are directed, typed, and attributed. The internal structure and the dynamic transformations of graphs are modeled in an intuitive way, for example the modification of complex structures can be specified visually.

To handle this voluminous language, a special environment emerged, which offers a comfortable way to make use of all the features. The PROGRES environment provides three different tools: First of all, the environment includes a syntax-controlled editor with an integrated analyzer, which annotates all violations against the static semantics of the language. Furthermore, an interpreter with a corresponding graph browser aids the user in debugging the specification by supporting the incremental execution of a specification and presenting the user the resulting graph structure in a graphical view. Finally, a compiler translates the specification into adequate and efficient C-code [3]. With this sophisticated and stable environment a user can easily specify complex graph

H. Ehrig et al. (Eds.): ICGT 2004, LNCS 3256, pp. 96–111, 2004.

transformation systems in a both textual and visual form. The graph schema and the actual graph are saved in the non-standard database GRAS (GRAph Storage). GRAS[4] has also been developed at our department and is a special database management system for storing graphs. As mentioned earlier, the PROGRES environment translates the specifications into efficient C-code, which can be loaded into the UPGRADE framework (Universal Platform for GRAph-based DEvelopment). UPGRADE[5] automatically creates a prototype from the given C-Code, which offers the user an adequate graphical interface. The layout of the generated prototype can be adapted to the user's needs. With the combination of PROGRES and UPGRADE several different tools have been developed, for example AHEAD[6], CHASID[7], and E-CARES[8].

As we have been specifying tools and experimenting with graph grammars for several years, the specifications have become more and more complex and huge. PROGRES provides a lot of useful concepts, but some specifications exceed the capabilities of the current PROGRES language. Additionally, the users need new features to structure large specifications in a clear manner. For example, decomposing a large specification into smaller parts is desirable. So, the user can work well-structured and can reuse parts of a certain specification in another one. To solve this problem, PROGRES has been extended with a package concept in [9]. As this concept has been used by some specificators, the shortcomings have become obvious. Furthermore, some users want to distribute the complex functionality of a tool over several prototypes which interact with each other. But the coupling of multiple prototypes is currently not supported by PROGRES. So we need new concepts in PROGRES for *specifying in the large*.

In this paper we will analyze the existing concepts in PROGRES for specifying in the large and revise them. Additionally, we present new concepts that will support this task. To demonstrate our approach, we will introduce AHEAD as a complex example implemented at our department in Section 2. In Section 3 we will discuss the problems of large specifications in PROGRES using the AHEAD example. Afterwards, we will point out approaches for the distribution of graph transformations that can be realized by extending PROGRES. In Section 5 we will present the Gras/GXL database management system and its role within the PROGRES system to support specifying in the large. At the end of this paper, we will give an overview of related work and summarize our results.

2   AHEAD: An Example for a Large PROGRES Specification

To illustrate the needs for specifying in the large, we will introduce a tool called AHEAD (Adaptable and Human-centered Environment for the mAnagement of Development Processes). As the development of products in disciplines such as mechanical, chemical, or software engineering is very complex and dynamic, AHEAD has been developed at our department. AHEAD constitutes a management system which supports the coordination of engineers through integrated management of products, activities, and resources. Therefore, AHEAD is composed of three combined partial models – CoMa, DYNAMITE, and RESMOD – which are all specified in PROGRES. Figure 1 shows a sample screenshot of AHEAD where the modeling of activities is demonstrated. Before we present these components in detail, we will explain some basic terms. A *process*

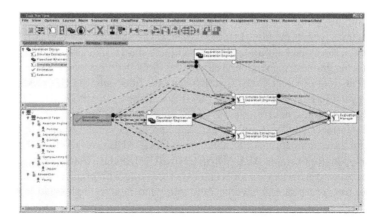

Fig. 1. Screenshot of the AHEAD system.

is an ordered set of working steps to achieve a certain result. The working steps are called *tasks*. The goal of a process consists in the creation of a certain *product*. In the case of software engineering the desired product is an extensive document. To achieve this objective, tasks consume input documents and produce output documents, which serve again as input for other tasks, like requirements definitions and design documents. For the execution of a task, *resources* are needed. We distinguish between human and non-human – for example computers – resources.

When developing for example a complex software system, *configuration management* is needed, which controls the evolution of documents during the long and dynamic process. The documents describe the results of activities, such as requirements definition, software architectures, and module implementations. Therefore CoMa (COnfiguration MAnagement)[10] has been developed which is used for the administration of these documents. The documents can be created within different tools. A configuration is composed of the documents and their dependencies. The documents and the configurations are subject to version control. Within CoMa composition hierarchies can be modeled, which determine the components a (sub-)system consists of. Consequently, the tool supports the engineers in maintaining system consistency. The model of CoMa is generic and can be adapted to a specific process.

To model dynamic development process, AHEAD uses DYNAMITE (DYNAMIc Task nEts)[11]. DYNAMITE is based on *dynamic task nets* and provides the seamless interleaving of planning, editing, analyzing, and execution of task nets. A task net consists of tasks which are connected through different relations. The execution sequence of tasks is determined by control flows. Data flows connect the input and output of tasks. Feedback flows present feedbacks in the development process, for example when a task has revealed to be incorrect. DYNAMITE offers also the possibility to define complex tasks which can consist of several other tasks. A task is composed of an interface and a realization. The interface describes the purpose and therefore defines all necessary information for the environment, for example inputs, outputs, pre- and postconditions. On the other hand, the realization determines how this task is being executed.

RESMOD (RESource management MODel)[12] manages the resources which are needed for the development process. All available *resources* are listed in RESMOD. Human and non-human resources are modeled in a uniform way. They have to be declared with their name, specific attributes, and their required resources. For example, a certain software system requires for the execution a computer which fulfills its requirements. A resource configuration is composed of a set of resources; one resource can be contained in several configurations. Additionally, RESMOD distinguishes between plan and actual resources. Plan resources support the manager in planning the resources for a project, for example he needs two programmers and three computers. In contrast to that, actual resources are concrete resources that can be mapped to the plan resources – like the two programmers Mrs. Smith and Mr. Scott. Furthermore, the manager can acquire special project resources which are only used in a certain project and are not employees in the company.

With CoMa, DYNAMITE, and RESMOD, AHEAD supports managers and engineers in the development process of a complex system. AHEAD can be used in various domains, because the models of the three parts are generic and can be easily adapted. Additionally, through the combinations of the three components all complex relations within a complicated development process can be described, which aids maintaining system consistency.

3   PROGRES and Specifying in the Large

As we have seen in Section 2, AHEAD is a complex management system which offers an extensive support for managers and engineers during the development process. AHEAD is modeled in PROGRES and because of all its features and the part models the resulting specification is very large and complicated. In the following we present the current state of PROGRES regarding the handling of huge specifications, which includes packages and modularization.

3.1   Package Structure

In [9] PROGRES was extended by a package structure, which is based on the package concept of the UML (Unified Modeling Language)[13]. The package structure can be used to define arbitrary nested packages within a specification. These packages allow the specification of *abstract data types* or *graph classes* such as binary trees.

A PROGRES package defines a *name space* for all the contained declarations, for example for the node classes and graph transformations. The declarations can be attributed with *visibilities*, like private, public, and protected similar to the regarding semantics of UML[14]. With the help of these attributes, the interface of a package is determined and so the access between two related packages is regulated. A package can import other packages, which enables the access on all public elements of the referred package. Additionally, specifiers may define inheritance hierarchies so that the inheriting package has access to the public and protected components of the superior package and can overwrite them. Packages may also be nested to allow successive refinement and structuring.

During the implementation of the package structure, it emerged that data abstraction contradicts visual programming. When the graph schema of a package is strictly hidden, the graph transformations of the importing package can not be specified visually by using the schema of the other package. Transformations of the importing package can only invoke methods in a textual way. On the other hand, when the graph schema is laid open, the importing package can change the graph schema managed by the imported package. This might lead to inconsistencies, for example if attributes are overwritten or modified. To solve the so-called *graph rewriting dilemma*[15], specifiers can utilize visibility attributes to restrict the access to the graph schema. In addition, *declaration markings* have been introduced which differ between read-only and read/write graph schema elements. Furthermore, three different types of *integrity constraints* can be used to guarantee consistency: constraints for pre- and post-conditions of graph transformations, schema constraints, and global constraints for general graph consistencies. If an integrity constraint has been violated, the PROGRES runtime environment ensures that a so-called *repair action* is executed. The repair action is part of the respecting constraint definition.

As packages have been used in many specifications, two more problems have become obvious. First, transitive inheritance of packages is not possible. For example, when package B inherits from package A and package C inherits from package B, then the schema and the transformations from A are not known in C. The problem is not a conceptual, but rather a technical one. Second, the schema view of PROGRES has to be adapted to the package structure. Even though these are minor problems – which do not limit the usability of packages – we will try to solve them.

3.2   Modularization

Although PROGRES provides a package structure, the PROGRES environment is not able to import packages from other specifications. Again, this is a technical limitation. Only packages within the same specification can be extended or imported. This limitation makes it on the one hand impossible to split large specifications into several parts. For example, the AHEAD meta-model presented in Section 2 consists of three parts for the tasks, resources, and products of a development process. Therefore, the separation of these parts into different specifications is desirable and leads to a clear structure and comfortable handling of the huge AHEAD specification. On the other hand, specifiers want to reuse parts of specifications, for example a model of a binary tree from a previous project. Thus, one of our goals is to eliminate this limitation by supporting the import of packages from other specifications.

To solve the mentioned problems, modularization has to be introduced into the PROGRES language. As a result, we can not only manage large specifications easier, but also structure complex problems better by distributing them over several graphs. Together with this enhancement we will have to provide support for graph-boundary crossing relations and graph hierarchies. While the support for boundary crossing relations can be partially hidden in the specification – the import of a specification automatically creates such a relation – we will have to provide new language elements for supporting graph hierarchies. However, we are still investigating if and how these concepts can be integrated into PROGRES using the existing package structure.

As this chapter has shown, specifiers have to be supported by means of packages and modularization. With inheritance, nesting, and the definition of graph classes, packages provide an adequate mechanism for handling large specifications easier. While this technique enables a clear structure of one specification, modularization offers the splitting of a specification into several parts. In this way individual components can be reused in other specifications.

4 Introducing Support for Distributed Specifications

An extension[16] of the AHEAD system supports distributed development processes. Two or more organizations use their own AHEAD instance, which are coupled at runtime. The basic idea of the extension is to propagate modifications of a task net to other AHEAD instances by events. This approach suffers from the problem that PROGRES does not support events on its own. However, the GRAS database system utilized by PROGRES prototypes uses events to indicate graph manipulations. Hence, the communication between the different AHEAD instances is realized by simply creating nodes representing events within the PROGRES specification. Then, GRAS creates appropriate events, which are transmitted to the other instances using the hand-coded communication server of the AHEAD prototype. In addition, the communication server is responsible for storing events, if an AHEAD instance is not online and thus can not receive events. A remote link manager, which is present in every prototype, receives events from the GRAS database, transmits them to the communication server, and invokes the appropriate graph transformation for an event vice versa. The transformation RI_ReleaseOutput, which releases a document after it has been consumed by a task, is an example for this approach:

```
transaction RI_ReleaseOutput ( Token : SEM_TOKEN ;
    dfTarget : SEM_TASK ; SessionID : SESSION) =
use eventNode : EVENT
  do
    IA_DF_ReleaseOutput ( Token, dfTarget, SessionID )
  & choose
      REM_MarkAsTransfered ( Token )
    & EVT_RaiseRelationEvent ( Token, dfTarget, EventTransfer,
          out eventNode )
    else
      EVT_RaiseRelationEvent ( Token, dfTarget,
          EventReleaseOutput, out eventNode )
    end
  end
end;
```

Because the whole coupling logic remains in the specification, we generate the remote link manager automatically from the specification. A naming convention determines which transformation has to be executed in response to a certain event. The implementation of the communication server is fairly generic and may be reused by other prototypes which apply the same coupling mechanism. However, up to now AHEAD is the only prototype using this mechanism. Due to the naming conventions the applicability of the concept in other specifications has to be analyzed.

The aforementioned approach has some drawbacks: Events have to be created and orphaned events must be removed, external programs have to be written to implement the communication, and implementation details clutter the specification. Obviously, adding appropriate language elements to PROGRES will solve these problems. Our first step is to propose language constructs for defining, raising, and intercepting events.

Before events can be raised, we have to define them. The following construct is used to specify an event class, which can have a super class. Each event class may define attributes which contain information that is transmitted together with the event.

```
EventClassDecl ::= "event" DeclEventId [ "is_a" ApplEventID ]
                        [ OptAttDeclList ]
                    "end" ";"
```

The attribute definition not only allows intrinsic attributes, but also derived and meta attributes. Attributes can be redefined in sub classes. The ability to use derived attributes is crucial for the coupling of different prototype instances. Certain attributes – especially node identifiers – do not make much sense in other instances. With derived attributes meaningful values can be computed for these attributes. For example, a node representing a task is not present in the other instances and thus has no meaning to them. But, the task's name – which can be computed by a derived attribute – can be used in the communication between the instances. After the definition, our specification can raise these events with the following statement:

```
EventRaiseStat ::= "raise" [CouplingMode]
                            ApplEventID OptActParList ";"
CouplingMode ::= "immediate" | "deferred" | "decoupled"
```

If intrinsic attributes are defined for the event, their concrete values must be set in the parameter list in the order of declaration. The derived attributes are evaluated directly after the values of the intrinsic attributes have been set. PROGRES executes each graph transformation in its own transaction. In the specification we can define how the execution of an event is coupled to the transaction. Hence, together with the raise statement one of three coupling modes can be defined. immediate raises the event immediately after the execution of the statement. The coupling mode decoupled raises the event only after the transaction has committed its changes successfully. Between these two modes is the default mode deferred, which raises the event just before the transaction is about to commit.

As a complementary operation, we have a construct to intercept events and perform a series of actions. The attributes transmitted with the event can be used as if they were local variables.

```
OnEventDecl ::= "on_event" ApplEventID "=" StatExpr "end" ";"
```

With these basic language constructs, the specificator can leave the event handling to the PROGRES runtime environment. Before, he had to manage the event handling on his own as shown in [16]. With the help of the event mechanism the graph transformation RI_ReleaseOutput is now realized like this:

```
transaction RI_ReleaseOutput( Token : SEM_TOKEN ;
   dfTarget : SEM_TASK ; SessionID : SESSION) =
use eventNode : EVENT
  do
    IA_DF_ReleaseOutput ( Token, dfTarget, SessionID )
  & choose
      REM_MarkAsTransfered ( Token )
    & raise EventTransfer(Token, dfTarget)
    else
      raise EventReleaseOutput(Token, dfTarget)
    end
  end
end;
```

Compared to the earlier version of the transformation, only little has changed. However, behind the scene we have far reaching changes. The specification gets simpler because the event handling is realized within the PROGRES environment and not the specification. Besides that, we can now explicitly define the actions which are executed when an event is intercepted. In the former specification there was a one-to-one match between the event's node type and the name of the transaction, which is executed for this event. Finally, the realization of the prototype gets easier, because we no longer need the implementation of the communication server and the remote link manager, which consist of 3000 lines of Java code and 1000 lines of XSL transformation scripts. All these components are now part of the PROGRES runtime environment.

Some of the problems described earlier can not be solved with this mechanism. Still, parts of the specification have to be modified. An adequate solution lies in raising events when a specific transaction is executed or certain graph modifications are made. Investigation of the AHEAD specification for distributed development processes showed that it is not that simple: For example, events may only be raised in very specific situations, thus conditions are necessary. Additionally, sometimes further transformations have to be executed before the event can be raised, as seen in the transformation RI_ReleaseOutput. There the transformation REM_MarkAsTransfered has to be executed to indicate that the token has been transfered. Another interesting extension to the event mechanism would be to raise events when a certain graph pattern has been created. Supporting this extension would be quite hard, because the graph pattern can be created by a series of transformations.

Of course, the new event concept for PROGRES is not limited to couple different specifications. The concept could also be used to realize reactive specifications which are executed in one prototype. We will first realize the aforementioned constructs and replace the event mechanism in the AHEAD specification. Afterwards, we will investigate which extensions should be realized to ease the specification of distributed and reactive systems. As long as PROGRES is not able to import other specifications, event definitions can not be shared and they have to be replicated in all participating specifications.

Even though PROGRES specifications can be coupled with the introduction of events, the specificator still has to worry about the "completeness" of the coupling. For example, he does not know if events are raised at all places in the specification

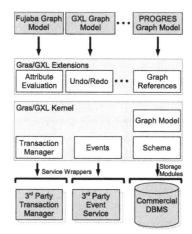

Fig. 2. System structure of the Gras/GXL database management system.

where they should. The problem is that the PROGRES environment can not provide any support for this. Therefore, we are investigating a second approach: the distribution of specifications can also be achieved by having a PROGRES specification which imports two other specifications and defines correspondences between them. Based on the correspondences, the coupling could be generated by the PROGRES environment automatically. But, we believe that this is not sufficient and the incorporation of other concepts – like the event mechanism described earlier – is still necessary. At the moment, we are studying approaches like triple graph grammars[17] and integrators[18, 19] to figure out how they fit into this vision.

5   Gras/GXL: A DBMS for Distributed Graph-Based Applications

Integrated development environments and visual language editors often use graphs or graph-like structures as data structures for their documents. Storing these graphs in a graph-oriented database management system has many advantages compared to storing them in main memory – data integrity, virtually unlimited graph size, etc. Thus, our department started the development on the graph-oriented database system GRAS in 1984. At that time, commercial database management systems did not provide features like incremental attribute evaluation or undo / redo of graph modifications which are especially used by PROGRES.

In the last few years PROGRES specifications and the graphs created by them got so large that certain limitations of GRAS hinder further development of both. At this time, the development of GRAS' successor Gras/GXL[20] has started. Unlike GRAS, Gras/GXL does not rely on its own storage management or transaction handling. Instead Gras/GXL uses third-party components whenever possible: commercial databases are used for storing graphs and transactional consistency is ensured by CORBA-compliant transaction managers.

Figure 2 illustrates the system structure of the Gras/GXL database management system. The Gras/GXL kernel defines the interfaces for the generic graph model and the graph schema. Storage modules tailored towards specific databases implement these two interfaces to store the graphs. Because of these storage modules Gras/GXL does not only support object-oriented and relational databases, the graphs can also be stored in main memory. At runtime these modules are plugged into the kernel. In addition, an abstraction layer to third-party components like transaction managers and event services is provided, to decouple the kernel from specific service implementations.

On top of the Gras/GXL kernel extensions can be implemented which provide services not available in the kernel. Examples for extensions are incremental evaluation of attributes, versioning, references to graph elements, etc. Extensions may be combined freely. But, some extensions may depend on others to realize their functionality, for example the undo/redo extension depends on the graph versioning extension. Another restriction for extensions is that they implement at least the interfaces of the generic graph model and the graph schema. The Gras/GXL graph model is richer than the ones used by most graph-based applications – for example n-ary relations are not supported by PROGRES. Thus, most applications will implement their own specific graph model on top of the Gras/GXL extensions to hide concepts they do not use. At the moment the specific graph model and its mapping to the Gras/GXL graph model have to be implemented manually, as we have already done for the PROGRES graph model. Currently, we are investigating how a specific graph model and its mapping can be generated from UML class diagrams.

5.1   Gras/GXL: Graph Model and Graph Schema

The UML class diagram for the Gras/GXL graph model is shown in Figure 3. A *graph pool* stores an arbitrary number of *graphs* which are identified by their *roles*. Graphs are a special kind of *graph element* – like *nodes*, *edges*, and *relations*. Each graph contains an arbitrary number of graph elements and other graphs. Graph elements can have an arbitrary number of attributes and meta-attributes. Because meta-attributes are not presented to the user, they are commonly used to store management information required by an extension, for example the cardinality of attributes for the PROGRES graph model. Relations and edges – which are just a shortcut for binary relations – connect graph elements stored in possibly different graph pools. As explained before, graphs are just ordinary graph elements in our graph model. Thus, they can be visited by edges and relations directly without using special graph elements. Edges and relations can be ordered. *Hierarchical graphs* are created either by a containment relationship or by graph-valued attributes. The containment relationship is used if a graph should be contained in another graph. Graph-valued attributes are used in all other situations, for example if a node should contain a graph. The use of graph-valued attributes together with the containment relationship allows us to create arbitrary hierarchical graphs and handle even complex situations uniformly – like hierarchies of graphs stored in different databases. As a result we get a clean and efficient realization of graph hierarchies.

The first version of the Gras/GXL graph model we presented in [20] supported references to graph elements. We have now dropped this support from the graph model, because it has a significant performance loss even if references are not used. If an ap-

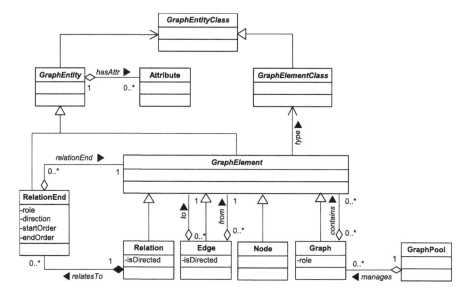

Fig. 3. Gras/GXL graph model.

plication requires references, they can be implemented on top of the graph model using our extension mechanisms. Thus, other applications no longer suffer from the inevitable performance loss.

The Gras/GXL graph model requires that every graph element has a type. Thus for each class in Figure 3 a corresponding graph schema class exists. Interesting in this paper are only two classes: GraphEntityClass and GraphElementClass. GraphEntityClass introduces attributes and meta-attributes and provides methods to declare, undeclare and enumerate them. GraphElementClass extends Graph-EntityClass and adds multiple inheritance and support of abstract classes. For all concrete graph elements a companion class exists, for example for a Node a Node-Class. Gras/GXL provides no means to support the structuring of a graph schema. Basically, two reasons speak against. First, a sophisticated structuring concept is likely to restrict the realization of a specific graph model. And second, basic support for structuring a schema would result in merely managing lists of classes, which clutters the interface and database support is unlikely to result in a performance gain. Thus, concepts for structuring a graph schema have to be implemented within the application graph model.

5.2   Gras/GXL: Support for Specifying in the Large

The graph-oriented database GRAS reduces the complexity of the PROGRES runtime environment, because some problems can be solved more easily. For example, static paths – which are essential for most specifications – are implemented using incrementally evaluated attributes; backtracking is realized by using the undo- and redo-operations of GRAS. The solutions developed and suggested in earlier sections demand for database support to ease their realization in the PROGRES runtime environment. Most of these required features are already present in Gras/GXL.

Generic Event Service: Events allow prototypes to exchange information about their state and modifications. The event service of Gras/GXL supports communication with arbitrary events and is not limited to the propagation of graph modifications. Events are typed and may carry any number of attributes. The event service compares any event against a set of rules which consists of an event pattern and an action. If the event matches an event pattern, the corresponding action is executed. Unlike its predecessor GRAS, Gras/GXL is aware of the transactional context and provides different modes (immediate, deferred, and decoupled) to couple the execution of the action to the transaction. Because an execution condition can be checked easily within an action, Gras/GXL only supports EA (event action) rules, and not ECA (event condition action) rules as introduced in [21].

Identification and Lookup of Prototypes: Gras/GXL provides a unique identifier for each graph and schema entity. The identifier consists not only of a simple number, but also contains the complete identification for the database which stores the entity by a so-called *data source URL.* Thus, graph elements stored in other graphs can be identified and accessed from other prototypes.

Structuring Complex Graphs: Gras/GXL provides several means to create graph hierarchies and to structure complex graphs, as explained earlier. If the graphs get too large, they can be distributed over several databases.

Graph-boundary Crossing Relations and References: Often, graphs are used to represent documents. Dependencies within and between them are expressed by edges or relations. When the different documents are distributed on many graphs and databases, boundary crossing relations are necessary to maintain these dependencies. In addition, these relations are used to couple prototypes. Fortunately, Gras/GXL supports boundary crossing edges. If one prototype has to access information managed by the other, for example to ensure document consistencies, distributed transactions come into play. Currently, we are investigating how the database can support this and how distributed transactions can be integrated into the PROGRES language. Instead of boundary crossing relations, references to graph elements can be used. However, the support of references is only supported through extensions.

Until now, the development for specifying in the large has just started and only a few requirements have been identified. The Gras/GXL database management system supports these requirements through its generic graph model, graph schema, and extensibility.

6   Related Work

In Section 3 an 4 we showed the problems caused by large specifications. These problems are not unique to the field of graph transformation systems. Modeling languages like the Unified Modeling Language [13] or programming languages like Ada[22] offer a couple of concepts to aid the solution of such problems. However, besides PROGRES only few graph transformation systems developed concepts for structuring huge specifications.

6.1   Graph Transformation Systems

With story diagrams Fujaba[23] introduced a new notation for graph rewriting rules, which are a combination of UML activity and collaboration diagrams. Fujaba is targeted towards the Java programming language and thus utilizes Java packages to structure specifications. However, Java packages do not support package extension by inheritance or similar concepts. PROGRES packages are used to model the similarity of graph classes. In contrast, Java packages can only structure a software system – or specification in the case of Fujaba – but not more. A plug-in developed at the Technical University of Darmstadt aims to bring MOF 2.0[24] support to Fujaba. At the moment, Fujaba is based on UML 1. Full support for packages – like package merge or inheritance – seems to be out of the scope of the plug-in. We intend to fill the resulting gap with the package concepts developed for PROGRES and implement a Fujaba plug-in for this.

Another example of a graph transformation system is GRACE (GRAph and rule CEntered specification language)[25]. Various researchers from different German universities work on the development of GRACE. GRACE is *approach independent* due to the consideration of graph transformations as a uniform framework. Many different and competing graph transformation approaches are available in the literature[1], so the specifier can choose the type of graphs, rule, and rule applications according to his taste. The main syntactic entities in GRACE are *transformation units*, which are composed of rules, an initial and a terminal graph class expression, a control condition, and import components. With their help, binary relations on graphs can be modeled. To handle large specifications, the transformation units can be structured into *modules*, which can again be imported by other modules. Additionally, a *concurrent semantics* of transformation units is defined in GRACE which provides the simultaneous execution of imported transformation units and rules to a graph. Therefore, the graph is divided into several graphs; a boundary graph contains all overlapping parts. The boundary graph can only be read and must not be changed. Furthermore, *distributed graph transformations*[26] can be realized through distributed transformation units, which consist of local transformation units that are connected by interface units. In [27] a different approach for the realization of hierarchically distributed graph transformation is presented. Within a *network graph*, the so-called network nodes constitute *local systems* which are again modeled as graphs. They are connected through relations in the network graph which can specify consistency constraints. All transformations within the distributed system are described with *hierarchical distributed graph* (HD-graph) productions. The productions consist of a network production and, if also the local systems should be transformed, local productions. In future, we will analyze the mentioned concepts and investigate if they can be adapted to PROGRES.

Other examples for graph transformation systems are AGG[28] and DiaGen[29]. Until now, both systems do not provide any modularization concepts.

6.2   Unified Modeling Language 2.0

The Unified Modeling Language is a general purpose language for modeling artifacts of systems. As these models can get very large, UML provides packages to make these artifacts manageable. Most UML elements can be contained in a package and can be accessed by its full qualified name. The visibility of an element limits its accessibility from other packages. For the sake of simplicity we will ignore the effect of an elements' visibility in the following description of UML packages.

Elements from one package can be *imported* into another. The import relationship makes the element directly visible in the importing package and the element can be accessed by its full qualified name. Another relationship between packages is the package *merge*. The UML 2.0 Infrastructure[30] distinguishes between two different kinds of package merge semantics: *define* and *extend*. For *extend*, elements from the merged packages with the same kind and name are merged into a single element using specialization and redefinition, respectively. *Extend* can be seen as a short-hand for explicitly defining the appropriate specializations and generalizations. In contrast, for a package merge of kind *define* the contents of the merged package are copied into the merging package using deep copy, where applicable.

The PROGRES packages, their relation to UML packages and other modularization concepts, are discussed in [14]. Currently, we are investigating the influences of the recent developments of UML 2.0 on PROGRES packages.

7   Conclusion

Although PROGRES is a very complex environment which has been extended in several dissertations, PROGRES lacks additional extensions like modularization. As the demonstration of the AHEAD specification showed, PROGRES demands for a comfortable handling of large specifications. This includes on the one hand an elaborate package structure which allows the modeling of graph classes. On the other hand, PROGRES lacks a modularization concept, so that parts of a specification can be reused in other ones. Furthermore, distributed graph transformations are necessary, which can be realized through an event mechanism. Therefore, we have introduced new language elements for defining and handling events. Gras/GXL provides the foundation for realizing the specification in the large and distribution concepts within the PROGRES runtime environment.

We will analyze the mentioned approaches and investigate which concepts can be adapted to PROGRES. As PROGRES is a very large and complicated environment, the planned extensions to this environment must be well considered and coherent to all other implemented concepts. After the realization of the new concepts, their influence on specification styles has to be examined. As several researchers at our department use PROGRES to specify tools from various domains, the applicability can be locally tested. Afterwards, we will introduce adequate mechanisms to FUJABA, because FUJABA lacks support for distributed graph transformations.

References

1. Rozenberg, G., et al., eds.: Handbook on Graph Grammars and Computing by Graph Transformation. Volume 1–3. World Scientific, Singapore (1997)
2. Schürr, A.: Operationales Spezifizieren mit programmierten Graphersetzungssystemen. Dissertation, RWTH Aachen (1991)

3. Zündorf, A.: Eine Entwicklungsumgebung für PROgrammierte GRaphErsetzungsSysteme. Dissertation, RWTH Aachen (1995)

4. Kiesel, N., Schürr, A., Westfechtel, B.: GRAS, a graph-oriented (software) engineering database system. Information Systems **20** (1995) 21–51

5. Böhlen, B., Jäger, D., Schleicher, A., Westfechtel, B.: UPGRADE: A framework for building graph-based interactive tools. In Mens, T., Schürr, A., Taentzer, G., eds.: Graph-Based Tools (GraBaTs 2002). Volume 72 of Electronical Notes in Theoretical Computer Science., Barcelona, Spain, Elsevier Science Publishers (2002)

6. Jäger, D., Schleicher, A., Westfechtel, B.: AHEAD: A graph-based system for modeling and managing development processes. [31] 325–339

7. Gatzemeier, F.: CHASID - A Semantic-Oriented Authoring Environment. Dissertation, RWTH Aachen (to appear in 2004)

8. Marburger, A., Westfechtel, B.: Graph-based reengineering of telecommunication systems. [32] 270–285

9. Schürr, A., Winter, A.J.: Uml packages for programmed graph rewriting systems. [33] 396–409

10. Westfechtel, B.: Using programmed graph rewriting for the formal specification of a configuration management system. [34] 164–179

11. Heimann, P., Joeris, G., Krapp, C.A., Westfechtel, B.: DYNAMITE: Dynamic task nets for software process management. In: Proceedings of the 18th International Conference on Software Engineering (ICSE'96), Berlin, Germany, IEEE Computer Society Press, Los Alamitos, CA, USA (1996) 331–341

12. Krapp, C.A., Krüppel, S., Schleicher, A., Westfechtel, B.: Graph-based models for managing development processes, resources, and products. [33] 455–474

13. Rumbaugh, J., Jacobson, I., Booch, G.: The Unified Modelling Language Reference Manual. Object Technology Series. Addison Wesley, Reading, MA, USA (1999)

14. Schürr, A., Winter, A.J.: UML Packages for PROgrammed Graph REwriting Systems. [33]

15. Winter, A.J.: Visuelles Programmieren mit Graphtransformationen. Dissertation, RWTH Aachen (2000)

16. Heller, M., Jäger, D.: Graph-based tools for distributed cooperation in dynamic development processes. [35] 352–368

17. Schürr, A.: Specification of graph translators with triple graph grammars. [34] 151–163

18. Becker, S.M., Haase, T., Westfechtel, B.: Model-based a-posteriori integration of engineering tools for incremental development processes. Journal of Software and Systems Modeling (2004) to appear.

19. Enders, B.E., Heverhagen, T., Goedicke, M., Tröpfner, P., Tracht, R.: Towards an integration of different specification methods by using the viewpoint framework. Transactions of the SDPS **6** (2002) 1–23

20. Böhlen, B.: Specific graph models and their mappings to a common model. [35] 45–60

21. Hsu, M.C., Ladin, R., McCarthy, D.: An execution model for active DB management systems. In Beeri, C., Dayal, U., Schmidt, J.W., eds.: Proceedings of the 3rd International Conference on Data and Knowledge Bases – Improving Usability and Responsiveness, Jerusalem, Israel, Morgan Kaufmann, San Francisco, CA, USA (1988) 171–179

22. ISO/IEC 8652:1995: Annotated Ada Reference Manual. Intermetrics, Inc. (1995)

23. Fischer, T., Niere, J., Torunski, L., Zündorf, A.: Story diagrams: A new graph rewrite language based on the Unified Modelling Language and Java. [33]

24. Object Managment Group, Needham, MA, USA: MetaObject Facility (MOF) Specification, Version 2.0. (2004). URL http://www.omg.org/uml

25. Kreowski, H.J., Busatto, G., Kuske, S.: GRACE as a unifying approach to graph-transformation-based specification. In Ehrig, H., Ermel, C., Padberg, J., eds.: UNIGRA 2001: Uniform Approaches to Graphical Process Specification Techniques. Volume 44 of Electronical Notes in Theoretical Computer Science., Genova, Italy, Elsevier Science Publishers (2001)

26. Knirsch, P., Kuske, S.: Distributed graph transformation units. [32] 207–222

27. Taentzer, G.: Hierarchically distributed graph transformation. In Cuny, J., Ehrig, H., Engels, G., Rozenberg, G., eds.: Proceedings 5th International Workshop on Graph Grammars and Their Application to Computer Science. Volume 1073 of Lecture Notes in Computer Science., Williamsburg, VA, USA, Springer-Verlag, Heidelberg (1995) 304–320

28. Taentzer, G.: AGG: A tool enviroment for algebraic graph transformation. [31]

29. Minas, M.: Bootstrapping visual components of the DiaGen specification tool with DiaGen. [35] 398–412

30. Object Managment Group, Needham, MA, USA: UML 2.0 Infrastructure Specification. (2003). URL http://www.omg.org//uml

31. Nagl, M., Schürr, A., Münch, M., eds.: Proceedings International Workshop on Applications of Graph Transformation with Industrial Relevance (AGTIVE'99). Volume 1779 of Lecture Notes in Computer Science. Kerkrade, The Netherlands, Springer-Verlag, Heidelberg (2000)

32. Corradini, A., Ehrig, H., Kreowski, H.J., Rozenberg, G., eds.: Proceedings 1st International Conference on Graph Transformation (ICGT'02). Volume 2505 of Lecture Notes in Computer Science. Barcelona, Spain, Springer-Verlag, Heidelberg (2002)

33. Ehrig, H., Kreowski, G.E.H.J., Rozenberg, G., eds.: Proceedings 6th International Workshop on Theory and Application of Graph Transformation (TAGT'98). Volume 1764 of Lecture Notes in Computer Science. Paderborn, Germany, Springer-Verlag, Heidelberg (1999)

34. Mayr, E., Schmidt, G., Tinhofer, G., eds.: Proceedings WG '94 20th International Workshop on Graph-Theoretic Concepts in Computer Science. Volume 903 of Lecture Notes in Computer Science. Herrsching, Germany, Springer-Verlag, Heidelberg (1995)

35. Pfaltz, J.L., Nagl, M., Böhlen, B., eds.: Proceedings 2nd International Workshop on Applications of Graph Transformation with Industrial Relevance (AGTIVE'03). Volume 3062 of Lecture Notes in Computer Science. Charlottesville, VA, USA, Springer-Verlag, Heidelberg (2004)

Typing of Graph Transformation Units*

Renate Klempien-Hinrichs, Hans-Jörg Kreowski, and Sabine Kuske

University of Bremen, Department of Computer Science
P.O.Box 33 04 40, 28334 Bremen, Germany
{rena,kreo,kuske}@informatik.uni-bremen.de

Abstract. The concept of graph transformation units in its original sense is a structuring principle for graph transformation systems which allows the interleaving of rule applications with calls of imported units in a controlled way. The semantics of a graph transformation unit is a binary relation on an underlying type of graphs. In order to get a flexible typing mechanism for transformation units and a high degree of parallelism this paper introduces typed graph transformation units that transform k-tuples of typed input graphs into l-tuples of typed output graphs in a controlled and structured way. The transformation of the typed graph tuples is performed with actions that apply graph transformation rules and imported typed units simultaneously to the graphs of a tuple. The transformation process is controlled with control conditions and with graph tuple class expressions. The new concept of typed graph transformation units is illustrated with examples from the area of string parsing with finite automata.

1 Introduction

The area of graph transformation brings together the concepts of rules and graphs with various methods from the theory of formal languages and from the theory of concurrency, and with a spectrum of applications, see the three volumes of the Handbook of Graph Grammars and Computing by Graph Transformation as an overview [15,5,7]. The key of rule-based graph transformation is the derivation of graphs from graphs by applications of rules. In this way, a set of rules specifies a binary relation of graphs with the first component as input and the second one as output. If *graph* names the class of graphs $\mathcal{G}$, the type of such a specified relation is *graph* $\rightarrow$ *graph* where each graph is a potential input and output. To get a more flexible typing, one can employ graph schemata or graph class expressions X that specify subclasses $\mathcal{G}(X)$ of the given class of graphs. This allows one typings of the form $I \rightarrow T$ restricting the derivations to those that start in initial graphs from $\mathcal{G}(I)$ and end in terminal graphs from $\mathcal{G}(T)$. Alternatively, one may require that all graphs involved in derivations stem from

---

* Research partially supported by the EC Research Training Network SegraVis (Syntactic and Semantic Integration of Visual Modeling Techniques) and the Collaborative Research Centre 637 (Autonomous Cooperating Logistic Processes: A Paradigm Shift and Its Limitations) funded by the German Research Foundation (DFG).

H. Ehrig et al. (Eds.): ICGT 2004, LNCS 3256, pp. 112–127, 2004.

$\mathcal{G}(X)$ for some expression X (cf. the use of graph schemata in PROGRES [17]). Another form of typing in the area of graph transformation can be found in the notion of pair grammars and triple grammars where a pair resp. a triple of graphs is derived in parallel by applying rules in all components simultaneously (see, e.g., [14, 16]).

In this paper, we propose a new, more general typing concept for graph transformation that offers the parallel processing of arbitrary tuples of graphs. Moreover, some components can be selected as input components and others as output components such that relations of the type $I_1 \times \cdots \times I_k \rightarrow T_1 \times \cdots \times T_l$ can be specified. The new typing concept is integrated into the structuring concept of graph transformation units (see, e.g., [1, 11, 12]).

The concept of graph transformation units in its original sense is a structuring principle for graph transformation systems which allows the interleaving of rule applications with calls of imported units in a controlled way. The semantics of a graph transformation unit is a binary relation on an underlying type of graphs that transforms initial graphs into terminal ones. In order to get a flexible typing mechanism for transformation units and a high degree of parallelism this paper introduces typed graph transformation units that transform k-tuples of typed input graphs into l-tuples of typed output graphs in a controlled and structured way. The transformation of the typed graph tuples is performed with actions that apply graph transformation rules and imported typed units simultaneously to the graphs of a tuple. The transformation process is controlled with control conditions and with graph tuple class expressions. The new concept of typed graph transformation units is illustrated with examples from the area of string parsing with finite automata.

2   Typed Graph Transformation

Graph transformation in general transforms graphs into graphs by applying rules, i.e. in every transformation step a single graph is transformed with a graph transformation rule. In typed graph transformation this operation is extended to tuples of graphs. This means that in every transformation step a tuple of graphs is transformed with a tuple of rules. The graphs, the rules, and the ways the rules have to be applied are taken from a so-called base type which consists of a tuple of rule bases. A rule base is composed of graphs, rules, and a rule application operator.

2.1   Rule Bases

A rule base $B = (\mathcal{G}, \mathcal{R}, \Longrightarrow)$ consists of a type of graphs $\mathcal{G}$, a type of rules $\mathcal{R}$, and a rule application operator $\Longrightarrow$. In the following the components $\mathcal{G}$, $\mathcal{R}$, and $\Longrightarrow$ of a rule base B are also denoted by $\mathcal{G}_B$, $\mathcal{R}_B$, and $\Longrightarrow_B$, respectively.

Examples for graph types are labelled directed graphs, graphs with a structured labelling (e.g. *typed graphs* in the sense of [3]), hypergraphs, trees, forests, finite automata, Petri nets, etc. The choice of graphs depends on the kind of applications one has in mind and is a matter of taste.

In this paper, we explicitly consider directed, edge-labelled graphs with individual, possibly multiple edges. A *graph* is a construct $G = (V, E, s, t, l)$ where V is a set of *vertices*, E is a set of *edges*, $s, t \colon E \to V$ are two mappings assigning to each edge $e \in E$ a *source* $s(e)$ and a *target* $t(e)$, and $l \colon E \to \Sigma$ is a mapping labelling each edge in a given label alphabet Σ.

For instance the graph

consists of seven nodes and six directed edges. It is a string graph which represents the string *aaba*. The beginning of the string is indicated with the *begin*-edge pointing to the source of the leftmost *a*-edge. Analogously, there is an *end*-edge originating from the end of the string, i.e. from the target of the rightmost *a*-edge.

Another instance of a graph is the following deterministic finite state graph where the edges labelled with a and b represent transitions, and the sources and targets of the transitions represent states. The start state is indicated with a *start*-edge and every final state with a *final*-edge. Moreover there is an edge labelled with *current* pointing to the current state of the deterministic finite state graph.

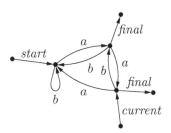

To be able to transform the graphs in $\mathcal{G}$, rules are applied to the graphs yielding graphs again. Hence, each rule $r \in \mathcal{R}$ defines a binary relation $\underset{r}{\Longrightarrow} \subseteq \mathcal{G} \times \mathcal{G}$ on graphs. If $G \underset{r}{\Longrightarrow} H$, one says that G *directly derives* H by applying r. There are many possibilities to choose rules and their applications. Types of rules may vary from the more restrictive ones, like edge replacement [4] or node replacement [8], to the more general ones, like double-pushout rules [2], single-pushout rules [6], or PROGRES rules [17].

In this paper, we concentrate on a simplified notion of double-pushout rules, i.e. every rule is a triple $r = (L, K, R)$ where L and R are graphs (the *left-* and *right-hand side* of r, respectively) and K is a set of nodes shared by L and R. In a graphical representation of r, L and R are drawn as usual, with numbers uniquely identifying the nodes in K. Its application means to replace an occurrence of L with R such that the common part K is kept. In particular, we will use rules that add or delete a node together with an edge and/or that redirect an edge.

A rule $r = (L, K, R)$ can be applied to some graph G directly deriving the graph H if H can be constructed up to isomorphism (i.e. up to the renaming of nodes and edges) in the following way.

1. Find an isomorphic copy of L in G, i.e. a subgraph that coincides with L up to the naming of nodes and edges.
2. Remove all nodes and edges of this copy except the nodes corresponding to K, provided that the remainder is a graph (which holds if the removal of a node is accompanied by the removal of all its incident edges).
3. Add R by merging K with its corresponding copy.

For abbreviating sets of rules, we use also variables instead of concrete labels. For every instantiation of a variable with a label we get a rule as described above. For example, the following rule $read(x)$ has as left-hand side a graph consisting of an x-edge and a $begin$-edge. The right-hand side consists of the target of a new $begin$-edge pointing from the source of the old $begin$-edge to the target of the x-edge. The common part of the rule $read(x)$ consists of the source of the $begin$-edge and the target of the x-edge.

If the variable x is instantiated with a, the resulting rule $read(a)$ can be applied to the above string graph. Its application deletes the $begin$-edge and the leftmost a-edge together with its source. It adds a new $begin$-edge pointing from the source of the old $begin$-edge to the target of the a-edge. The resulting string graph represents the string aba.

The following rule $go(x)$ redirects a $current$-labelled edge from the source of some x-labelled edge to the target of this edge.

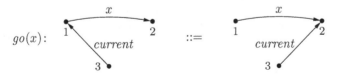

If x is instantiated with a, its application to the above deterministic finite state graph results in the same deterministic finite state graph except that the current state is changed to the start state.

2.2   Graph Tuple Transformation

As the iterated application of rules transforms graphs into graphs yielding an input-output relation, the natural type declaration of a graph transformation in a rule base $B = (\mathcal{G}, \mathcal{R}, \Longrightarrow)$ is $B \colon \mathcal{G} \to \mathcal{G}$. But in many applications one would like to have a typing that allows one to consider several inputs and maybe even several outputs, or at least an output of a type different from all inputs. Moreover, one may want to be able to transform subtypes of the types of input and output graphs. In order to reach such an extra flexibility in the typing of graph transformations we introduce in this section the transformation of tuples of typed graphs, which is the most basic operation of the typed graph transformation units presented in Section 4.

Graph tuple transformation over a base type is an extension of ordinary rule application in the sense that graphs of different types can be transformed in parallel. For example to check whether some string can be recognized by a deterministic finite automaton, one can transform three graphs in parallel: The first graph is a string graph representing the string to be recognized, the second graph is a deterministic finite state graph the current state of which is the start state, and the third graph represents the boolean value *false*. To recognize the string one applies a sequence of typed rule applications which consume the string graph while the corresponding transitions of the deterministic finite state graph are traversed. If after reading the whole string the current state is a final state, the third graph is transformed into a graph representing *true*. This example will be explicitly modelled in Section 4.

In graph tuple transformation, tuples of rules are applied to tuples of graphs. A tuple of rules may also contain the symbol $-$ in some components where no change is desired. The graphs and the rules are taken from a *base type*, which is a tuple of rule bases $BT = (B_1, \ldots, B_n)$. Let $(G_1, \ldots, G_n)$ and $(H_1, \ldots, H_n)$ be graph tuples over BT, i.e. $G_i, H_i \in \mathcal{G}_{B_i}$ for $i = 1, \ldots, n$. Let $a = (a_1, \ldots, a_n)$ with $a_i \in \mathcal{R}_{B_i}$ or $a_i = -$ for $i = 1, \ldots, n$. Then $(G_1, \ldots, G_n) \xrightarrow{a} (H_1, \ldots, H_n)$ if for $i = 1, \ldots, n$, $G_i \underset{a_i}{\Longrightarrow} H_i$ if $a_i \in \mathcal{R}_{B_i}$ and $G_i = H_i$ if $a_i = -$. In the following we call a a *basic action of BT*.[1] For a set ACT of basic actions of BT, $\xrightarrow[ACT]{}$ denotes the union $\bigcup_{a \in ACT} \xrightarrow{a}$, and $\xrightarrow[ACT]{*}$ its reflexive and transitive closure.

For example, let I be some finite alphabet and let $B_{string}^{basic} = (string, \{read(x) \mid x \in I\}, \Longrightarrow)$ and $B_{dfsg}^{basic} = (dfsg, \{go(x) \mid x \in I\}, \Longrightarrow)$ be two rule bases such that *string* consists of all string graphs over I and *dfsg* consists of all deterministic finite state graphs over I. Let G_1 be the string graph representing *aaba* and let G_2 be the above deterministic finite state graph. Then $(G_1, G_2) \xrightarrow{a} (H_1, H_2)$ for the basic action $a = (read(a), go(a))$ of base type $(B_{string}^{basic}, B_{dfsg}^{basic})$ if H_1 represents *aba* and H_2 is obtained from G_2 by redirecting the *current*-edge to the start state.

Let ACT be the set of all basic actions of $BT = (B_1, \ldots, B_n)$. Then obviously the following holds: $(G_1, \ldots, G_n) \xrightarrow[ACT]{*} (H_1, \ldots, H_n)$ if and only if $G_i \underset{\mathcal{R}_{B_i}}{\overset{*}{\Longrightarrow}} H_i$ for $i = 1, \ldots, n$. This means that the transformation of graph tuples via a sequence of basic actions is equivalent to the transformation of tuples of typed graphs where every component is transformed independently with a sequence of direct derivations of the corresponding type.

In [9] the transformation of tuples of typed graphs is generalized in the sense that the transformations are performed by a product of transformation units instead of tuples of rules. In every transformation step of a product of transformation units $tu_1, \ldots, tu_n$, one can apply rules as well as imported transformation units of tu_i to the ith graph in the current graph tuple ($i = 1, \ldots, n$). Hence, in

---

[1] More sophisticated actions with more expressive power will be introduced further on.

every transformation step the graphs of the current graph tuple are transformed in parallel. Such a transformation step in a product unit is called an action. The nondeterminism of actions is restricted by the control conditions and the graph class expressions of the units $tu_1, \ldots, tu_n$. Moreover one can specify control conditions on the level of actions. In order to get a flexible kind of typing, i.e. to declare a sequence of input components and a sequence of output components independently, the embedding and projection of graph products is introduced. For the same reasons, similar operations will be introduced for typed transformation units in this paper. The most striking difference of the product of transformation units in [9] and the typed graph transformation units presented here is the import component. Typed units can import other typed units whereas a product of transformation units is composed of transformation units in the original sense, i.e. it does not use other typed units.

3   Restricting the Nondeterminism

Application of rule tuples is highly nondeterministic in general. For many applications of graph transformation it is meaningful to restrict the number of possible ways to proceed with a transformation process. Hence, in order to employ typed transformation units meaningfully, they are equipped with graph tuple class expressions and control conditions to restrict the number of possible sequences of transformation steps.

3.1   Graph Tuple Class Expressions

The aim of graph tuple class expressions is to restrict the class of graph tuples to which certain transformation steps may be applied, or to filter out a subclass of all the graph tuples that can be obtained from a transformation process. Typically, a graph tuple class expression may be some logic formula describing a tuple of graph properties like connectivity, or acyclicity, or the occurrence or absence of certain labels. In this sense, every graph tuple class expression e over a base type $BT = (B_1, \ldots, B_n)$ specifies a set $SEM(e) \subseteq \mathcal{G}_{B_1} \times \cdots \times \mathcal{G}_{B_n}$ of graph tuples in BT.

In many cases such a graph tuple class expression will be a tuple $e = (e_1, \ldots, e_n)$ where the ith item e_i restricts the graph class $\mathcal{G}_{B_i}$ of the rule base B_i, i.e. $SEM_{B_i}(e_i) \subseteq \mathcal{G}_{B_i}$ for $i = 1, \ldots, n$. Consequently, the semantics of e is $SEM_{B_1}(e_1) \times \cdots \times SEM_{B_n}(e_n)$. Hence, each item e_i is a graph class expression as defined for transformation units without explicit typing.

The graph tuple class expressions used in this paper are also tuples of graph class expressions. A simple example of a graph class expression is all which specifies for any rule base B the graph type of B, i.e. $SEM_B(all) = \mathcal{G}_B$. Consequently, the graph tuple class expression $(e_1, \ldots, e_n)$ with $e_i = all$ for $i = 1, \ldots, n$ does not restrict the graph types of the rule bases, i.e. $SEM((e_1, \ldots, e_n)) = \mathcal{G}_{B_1} \times \cdots \times \mathcal{G}_{B_n}$. Another example of a graph class expression over a rule base B is a set of graphs in $\mathcal{G}_B$. The semantics of a set $e \subseteq \mathcal{G}_B$ is e itself. In particular we will use the set $initialized$ consisting of all deterministic finite state graphs the current state of which is the start state.

3.2   Control Conditions

A control condition is an expression that determines, for example, the order in which transformation steps may be applied to graph tuples. Semantically, it relates tuples of start graphs with tuples of graphs that result from an admitted transformation process. In this sense, every control condition C over a base type BT specifies a binary relation $SEM(C)$ on the set of graph tuples in BT. More precisely, for a base type $BT = (B_1, \ldots, B_n)$ $SEM(C)$ is a subset of $(\mathcal{G}_{B_1} \times \cdots \times \mathcal{G}_{B_n})^2$.

As control condition we use in particular actions, sequential composition, union, and iteration of control conditions, as well as the expression *as-long-as-possible* (abbreviated with the symbol !). An action prescribes which rules or imported typed units should be applied to a graph tuple, i.e. an action is a control condition that allows one to synchronize different transformation steps. The basic actions of the previous section are examples of actions. Roughly speaking, an action over a base type $BT = (B_1, \ldots, B_n)$ is a tuple $act = (a_1, \ldots, a_n)$ that specifies an n, n-relation $SEM(act) \subseteq (\mathcal{G}_{B_1} \times \cdots \times \mathcal{G}_{B_n})^2$. Actions will be explained in detail in Section 4.

In particular, an action act is a control condition that specifies the relation $SEM(act)$. For control conditions C, C_1, and C_2 the expression $C_1; C_2$ specifies the sequential composition of both sematic relations, $C_1 | C_2$ specifies the union, and C^* specifies the reflexive and transitive closure, i.e. $SEM(C_1; C_2) = SEM(C_1) \circ SEM(C_2)$, $SEM(C_1 | C_2) = SEM(C_1) \cup SEM(C_2)$, and $SEM(C^*) = SEM(C)^*$. Moreover, for a control condition C the expression $C!$ requires to apply C as long as possible, i.e. $SEM(C)$ consists of all pairs $(G, H) \in SEM(C)^*$ such that there is no H' with $(H, H') \in SEM(C)$. In the following the control condition $C_1 | \cdots | C_n$ will also be denoted by $\{C_1, \ldots, C_n\}$.

For example, let C_1, C_2, and C_3 be control conditions that specify n, n-relations on graphs of different types. Then the expression $C_1!; C_2^*; (C_3 | C_1)$ prescribes to apply first C_1 as long as possible, then C_2 arbitrarily often, and finally C_3 or C_1 exactly once.

4   Typed Graph Transformation Units

Typed transformation units provide a means to structure the transformation process from a sequence of typed input graphs to a sequence of typed output graphs. More precisely, a typed graph transformation unit transforms k-tuples of graphs into l-tuples of graphs such that the graphs in the k-tuples as well as the graphs in the l-tuples may be of different types. Hence, a typed transformation unit specifies a k, l-relation on typed graphs. Internally a typed transformation unit transforms n-tuples of typed graphs into n-tuples of typed graphs, i.e. it specifies internally an n, n-relation on typed graphs. The transformation of the n-tuples is performed according to a base type which is specified in the declaration part of the unit. The k, l-relation is obtained from the n, n-relation by embedding k input graphs into n initial graphs and by projecting n terminal graphs onto l output graphs. The embedding and the projection are also given in the declaration part of a typed unit.

4.1   Syntax of Typed Graph Transformation Units

Base types, graph tuple class expressions, and control conditions form the ingredients of typed graph transformation units. Moreover, the structuring of the transformation process is achieved by an import component, i.e. every typed unit may import a set of other typed units. The transformations offered by an imported typed unit can be used in the transformation process of the importing typed unit.

The basic operation of a typed transformation unit is the application of an action, which is a transformation step from one graph tuple into another where every component of the tuple is modified either by means of a rule application, or is set to some output graph of some imported typed unit, or remains unchanged. Since action application is nondeterministic in general, a transformation unit contains a control condition that may regulate the graph tuple transformation process. Moreover, a typed unit contains an initial graph tuple class expression and a terminal graph tuple class expression. The former specifies all possible graph tuples a transformation may start with and the latter specifies all graph tuples a transformation may end with. Hence, every transformation of an n-tuple of typed graphs with action sequences has to take into account the control condition of the typed unit as well as the initial and terminal graph tuple class expressions.

A tuple of sets of typed rules, a set of imported typed units, a control condition, an initial graph tuple class expression, and a terminal graph tuple class expression form the body of a typed transformation unit. All components in the body must be consistent with the base type of the unit.

Formally, let $BT = (B_1, \ldots, B_n)$ be a base type. A *typed graph transformation unit tgtu with base type BT* is a pair (*decl*, *body*) where *decl* is the *declaration part* of *tgtu* and *body* is the *body* of *tgtu*. The declaration part is of the form $in \rightarrow out$ on BT where $in \colon [k] \rightarrow [n]$ and $out \colon [l] \rightarrow [n]$ are mappings with $k, l \in \mathbb{N}$.[2] The body of *tgtu* is a system $body = (I, U, R, C, T)$ where I and T are graph tuple class expressions over BT, U is a set of imported typed graph transformation units, R is a tuple of rule sets $(R_1, \ldots, R_n)$ such that $R_i \subseteq \mathcal{R}_{B_i}$ for $i = 1, \ldots, n$, and C is a control condition over BT. The numbers k and l of *tgtu* are also denoted by k_{tgtu} and l_{tgtu}. Moreover, the ith input type $\mathcal{G}_{B_{in(i)}}$ of *tgtu* is also denoted by $intype_{tgtu}(i)$ for $i = 1, \ldots, k$ and the jth output type $\mathcal{G}_{B_{out(j)}}$ by $outtype_{tgtu}(j)$ for $j = 1, \ldots, l$.

To simplify technicalities, we assume in this first approach that the import structure is acyclic (for a study of cyclic imports of transformation units with a single input and output type see [13]). Initially, one builds typed units of level 0 with empty import. Then typed units of level 1 are those that import only typed units of level 0, and typed units of level $n + 1$ import only typed units of level 0 to level n, but at least one from level n.

---

[2] For a natural number $n \in \mathbb{N}$, $[n]$ denotes the set $\{1, \ldots, n\}$.

4.2   Examples for Typed Graph Transformation Units

Example 1. The base type of the following example of a typed transformation unit is the tuple $(B_{string}, B_{dfsg}, B_{bool})$. The rule base B_{string} is $(string, \{read(x) \mid x \in I\} \cup \{is\text{-}empty\}, \Longrightarrow)$, where the rule *is-empty* checks whether the graph to which it is applied represents the empty string. It has equal left- and right-hand sides consisting of a node to which a *begin*- and from which an *end*-edge are pointing.

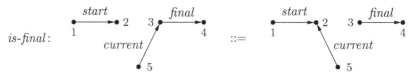

The rule base B_{dfsg} is $(dfsg, \{go(x) \mid x \in I\} \cup \{is\text{-}final\}, \Longrightarrow)$. The rule *is-final* checks whether the current state of a deterministic finite state graph is a final state, resetting it to the start state in that case, and can be depicted as follows.

The rule base B_{bool} contains the graph type *bool* which consists of the two graphs *TRUE* and *FALSE*, where *TRUE* represents the value *true* and *FALSE* the value *false*. Both graphs consist of a single node with a loop that is labelled *true* and *false*, respectively:

$$TRUE \; = \; \bullet\!\!\circlearrowleft true \qquad\qquad FALSE \; = \; \bullet\!\!\circlearrowleft false$$

The rule type of B_{bool} consists of the four rules

where *set-to-true* changes a *false*-loop into a *true*-loop, *set-to-false* does the same the other way round, *is-true* checks whether a graph of type *bool* is equal to *TRUE*, and *is-false* checks the same for *FALSE*.

Now we can define the typed unit *recognize* shown in Figure 1. It has as input graphs a string graph and a deterministic finite state graph and as output graph a boolean value. The mapping *in* of the declaration part of *recognize* is defined by $in: [2] \to [3]$ with $in(1) = 1$ and $in(2) = 2$. We use the more intuitive tuple notation $(string, dfsg, -)$ for this. The mapping *out* is denoted by $(-, -, bool)$ which means that $out: [1] \to [3]$ is defined by $out(1) = 3$. Hence, $intype_{recognize}(1) = string$, $intype_{recognize}(2) = dfsg$, and $outtype_{recognize}(1) = bool$.

The initial graph tuple class expression is $(string, initialized, FALSE)$, i.e. it admits all tuples $(G_1, G_2, G_3) \in string \times dfsg \times bool$ where the *current*-edge of G_2

recognize

decl:	$(string, dfsg, -) \rightarrow (-, -, bool)$ on $(B_{string}, B_{dfsg}, B_{bool})$
initial:	$(string, initialized, FALSE)$
rules:	$(\mathcal{R}_{B_{string}}, \mathcal{R}_{B_{dfsg}}, \{set\text{-}to\text{-}true\})$
cond:	$a_1!; a_2!$ where
	$a_1 = \{(read(x), go(x), -) \mid x \in I\}$ and
	$a_2 = (is\text{-}empty, is\text{-}final, set\text{-}to\text{-}true)$
terminal:	$(string, dfsg, bool)$

Fig. 1. A typed unit with empty import.

points to the start state and G_3 is equal to *FALSE*. The rules are restricted to the tuple $(\mathcal{R}_{B_{string}}, \mathcal{R}_{B_{dfsg}}, \{set\text{-}to\text{-}true\})$, i.e. just one rule from B_{bool} is admitted. The control condition requires to apply first the action a_1 as long as possible and then the action a_2 as long as possible, where a_1 applies $read(x)$ to the first component of the current graph tuple and $go(x)$ to the second component (for any $x \in I$). The action a_2 sets the third component to *TRUE* if the current string is empty, the current state of the state graph is a final state, and the third component is equal to *FALSE*. Note that a_2 can be applied at most once because of $set\text{-}to\text{-}true$, and only in the case where a_1 cannot be applied anymore because of $is\text{-}empty$. The terminal graph tuple class expression does not restrict the graph types of the base type, i.e. it is equal to $(string, dfsg, bool)$. The unit *recognize* does not import other typed units.

Example 2. The unit *recognize-intersection* shown in Figure 2 is an example of a typed unit with a non-empty import component. It has as input graphs a string graph and two deterministic finite state graphs. The output graph represents again a boolean value. The base type of *recognize-intersection* is the six-tuple $(B_{string}, B_{dfsg}, B_{dfsg}, B_{bool}, B_{bool}, B_{bool})$. The mapping *in* of the declaration part requires to take a string graph from the first rule base of the base type, one deterministic finite state graph from the second and one from the third rule base as input graphs. The mapping *out* requires to take a graph from the last rule base as output graph.

recognize-intersection

decl:	$(string, dfsg, dfsg, -, -, -) \rightarrow (-, -, -, -, -, bool)$ on
	$(B_{string}, B_{dfsg}, B_{dfsg}, B_{bool}, B_{bool}, B_{bool})$
initial:	$(string, dfsg, dfsg, bool, bool, FALSE)$
uses:	*recognize*
rules:	$(\emptyset, \emptyset, \emptyset, \{is\text{-}true\}, \{is\text{-}true\}, \{set\text{-}to\text{-}true\})$
cond:	$a_1; a_2!$ where
	$a_1 = (-, -, -, recognize(1, 2), recognize(1, 3), -)$ and
	$a_2 = (-, -, -, is\text{-}true, is\text{-}true, set\text{-}to\text{-}true)$
terminal:	$(string, dfsg, dfsg, bool, bool, bool)$

Fig. 2. A typed unit with imported units combined in an action.

The unit *recognize-intersection* imports the above unit *recognize* and has as local rules *is-true* and *set-to-true* where *is-true* can be applied to the fourth and the fifth component of the current graph tuples and *set-to-true* to the sixth component. The control condition requires the following.

1. Apply *recognize* to the first and the second component and write the result into the fourth component and
2. apply *recognize* to the first and the third component and write the result into the fifth component.
3. If then possible apply the rule *is-true* to the fourth and the fifth component and the rule *set-to-true* to the sixth component.

This means that in the first point *recognize* is applied to the input string graph and the first one of the input deterministic finite state graphs. In the second point *recognize* must be applied to the input string graph and to the second deterministic finite state graph. These two transformations can be performed in parallel within one and the same action denoted by the tuple $(-, -, -, recognize(1, 2), recognize(1, 3), -)$. (The precise semantics of this action will be given in the next subsection where actions and their semantics are introduced formally.) The rule application performed in the third point corresponds to applying the basic action $(-, -, -, is\text{-}true, is\text{-}true, set\text{-}to\text{-}true)$ as long as possible. Since the initial graph tuple class expression requires that the sixth graph represent *false*, this means at most one application due to *set-to-true*. The terminal graph tuple class expression admits all graph tuples of the base type.

Example 3. Let I be the alphabet consisting of the symbols a, b, let L, L_a, L_b be regular languages, and let $subst: I \to \mathcal{P}(I^*)$ be a substitution with $subst(a) = L_a$ and $subst(b) = L_b$. The aim of the following example is to model the recognition of the substitution language $subst(L) = \{subst(w) \mid w \in L\}$ based on a description of L, L_a, L_b by deterministic finite automata. (The model can of course be extended to arbitrarily large alphabets.)

First, consider the typed unit *reduce* shown in Figure 3. It takes a string graph and a deterministic finite state graph as input, requiring through the initial component that the state graph be in its start state. It then reduces the string graph by arbitrarily often applying actions of the form $(read(x), go(x))$, i.e. by consuming an arbitrarily large prefix of the string and changing states

reduce	
decl:	$(string, dfsg) \to (string, -)$ on (B_{string}, B_{dfsg})
initial:	$(string, initialized)$
rules:	$(\mathcal{R}_{B_{string}}, \mathcal{R}_{B_{dfsg}})$
cond:	$a_1{}^*; a_2$ where
	$a_1 = \{(read(x), go(x)) \mid x \in I\}$ and
	$a_2 = (-, is\text{-}final)$
terminal:	$(string, dfsg)$

Fig. 3. A typed unit that returns a modified input graph as output.

accordingly in the state graph, and returns the residue of the string graph as output, but only if the consumed prefix is recognized by the state graph, i.e. only if the action $(-, is\text{-}final)$ is applied exactly once.

$recognize\text{-}substitution$
decl:	$(string, dfsg, dfsg, dfsg, -) \rightarrow (-, -, -, -, bool)$ on
	$(B_{string}, B_{dfsg}, B_{dfsg}, B_{dfsg}, B_{bool})$
initial:	$(string, initialized, initialized, initialized, FALSE)$
uses:	$reduce$
rules:	$(\{is\text{-}empty\}, \mathcal{R}_{B_{dfsg}}, \emptyset, \emptyset, \{set\text{-}to\text{-}true\})$
cond:	$(a_1|a_2)^*; a_3$ where
	$a_1 = \{(reduce(1,3), go(a), -, -, -),$
	$a_2 = \{(reduce(1,4), go(b), -, -, -),$ and
	$a_3 = (is\text{-}empty, is\text{-}final, -, -, set\text{-}to\text{-}true)$
terminal:	$(string, dfsg, dfsg, bool, bool, bool)$

Fig. 4. A typed unit with imported units combined in an action.

The typed unit $recognize\text{-}substitution$ shown in Figure 4 makes use of $reduce$ in order to decide whether an input string graph is in the substitution language given as further input by three deterministic finite state graphs A, A_a, A_b that define L, L_a, L_b, in that order. Initially, the state graphs must once again be in their respective start states and the value in the output component is $false$. The idea is to guess, symbol by symbol, a string $w \in L$ such that the input string is in $subst(w)$. If the next symbol is guessed to be a, the action $(reduce(1,3), go(a), -, -, -)$ is applied that runs A_a to delete a prefix belonging to L_a from the input string $(reduce(1,3))$ and simultaneously executes the next state transition for a in A $(go(a))$. The action $(reduce(1,4), go(b), -, -, -)$ works analogously for the symbol b. Thus, $recognize\text{-}substitution$ is an example of a typed unit that combines an imported unit $(reduce)$ and a rule $(go(x))$ in an action. Finally, a mandatory application of the action $(is\text{-}empty, is\text{-}final, -, -, set\text{-}to\text{-}true)$ produces the output value $true$, but only if the input string is completely consumed and A is in some final state.

It may be noted that even though the finite state graphs are deterministic, there are two sources of nondeterminism in this model: The symbols of the supposed string $w \in L$ must be guessed as well as a prefix of the input string for each such symbol. Consequently, the model admits only tuples with output $TRUE$ in its semantics.

4.3 Semantics of Typed Graph Transformation Units

Typed transformation units transform initial graph tuples to terminal graph tuples by applying a sequence of actions so that the control condition is satisfied. Moreover, the mappings in and out of the declaration part prescribe for every such transformation the input and output graph tuples of the unit. Hence, the semantics of a typed transformation unit can be defined as a k, l-relation between input and output graphs.

Let $tgtu = (in \rightarrow out\; on\; BT, (I, U, R, C, T))$ be a typed transformation unit with $BT = (B_1, \ldots, B_n)$, $in\colon [k] \rightarrow [n]$, $out\colon [l] \rightarrow [n]$, and $R = (R_1, \ldots, R_n)$. If $U = \emptyset$, $tgtu$ transforms internally a tuple $G \in \mathcal{G}_{B_1} \times \cdots \times \mathcal{G}_{B_n}$ into a tuple $H \in \mathcal{G}_{B_1} \times \cdots \times \mathcal{G}_{B_n}$ if and only if

1. G is an initial graph tuple and H is a terminal graph tuple, i.e. $(G, H) \in SEM(I) \times SEM(T)$;
2. H is obtained from G via a sequence of basic actions over $(R_1, \ldots, R_n)$, i.e.
   $$G \xrightarrow[ACT(tgtu)]{*} H$$ where $ACT(tgtu)$ is the set of all basic actions $a = (a_1, \ldots, a_n)$
   of BT such that for $j = 1, \ldots, n$, $a_i \in R_i$ if $a_i \neq -$, and
3. the pair (G, H) is allowed by the control condition, i.e. $(G, H) \in SEM(C)$.

If the transformation unit $tgtu$ has a non-empty import, the imported units can also be applied in a transformation from G to H. This requires that we extend the notion of basic actions so that calls of imported typed units are allowed, leading to the notion of (general) actions.

Formally, an *action of* $tgtu$ is a tuple $a = (a_1, \ldots, a_n)$ such that for $i = 1, \ldots, n$ we have $a_i \in R_i$, or $a_i = -$, or a_i is of the form $(u, input, output)$ where $u \in U$, $input\colon [k_u] \rightarrow [n]$ with $\mathcal{G}_{B_{input(j)}} \subseteq intype_u(j)$ for $j = 1, \ldots, k_u$, and $output \in [l_u]$ with $outtype_u(output) \subseteq \mathcal{G}_{B_i}$. In the latter case, we denote a_i by $u(input(1), \ldots, input(k_u))(output)$, and shorter by $u(input(1), \ldots, input(k_u))$ if u has a unique output, i.e. $l_u = 1 = output$.

The application of an action $a = (a_1, \ldots, a_n)$ to a current graph tuple of n typed graphs works as follows: As for typed rule application, if a_i is a rule of R_i, it is applied to the ith graph. If a_i is equal to $-$, the ith graph remains unchanged. The new aspect is the third case where a_i is of the form $(u, input, output)$. In this case, the mapping $input\colon [k_u] \rightarrow [n]$ determines which graphs of the current tuple of typed graphs should be chosen as input for the imported unit u. The output $output \in [l_u]$ specifies which component of the computed output graph tuple of u should be assigned to the ith component of the graph tuple obtained from applying the typed unit u to the input graphs selected by $input$.

For example the action $(-, -, -, recognize(1, 2), recognize(1, 3), -)$ of the typed unit *recognize-intersection* has as semantics every pair $((G_1, \ldots, G_6), (H_1, \ldots, H_6))$ such that $G_i = H_i$ for $i \in \{1, 2, 3, 6\}$, H_4 is the output of *recognize* applied to (G_1, G_2), and H_5 is the output of *recognize* applied to (G_1, G_3).

Formally, assume that every imported typed unit u of $tgtu$ defines a semantic relation

$$SEM(u) \subseteq (intype_u(1) \times \cdots \times intype_u(k_u)) \times (outtype_u(1) \times \cdots \times outtype_u(l_u)).$$

Then every pair $((G_1, \ldots, G_n), (H_1, \ldots, H_n))$ of graph tuples over BT is in the *semantics of an action* $a = (a_1, \ldots, a_n)$ of $tgtu$ if for $i = 1, \ldots, n$:

- $G_i \underset{a_i}{\Longrightarrow} H_i$ if $a_i \in R_i$,
- $G_i = H_i$ if $a_i = -$, and

- $H_i = H'_{output}$ if $a_i = (u, input, output)$ and $((G_{input(1)}, \ldots, G_{input(k_u)}),$
  $(H'_1, \ldots, H'_{l_u})) \in SEM(u)$.

The set of all actions of $tgtu$ is denoted by $ACT(tgtu)$ and the semantics of an action $a \in ACT(tgtu)$ by $SEM(a)$.

Now we can define the semantics of $tgtu$ as follows. Every pair $((G_1, \ldots, G_k),$ $(H_1, \ldots, H_l))$ is in $SEM(tgtu)$ if there is a pair $(\bar{G}, \bar{H})$ with $\bar{G} = (\bar{G}_1, \ldots, \bar{G}_n),$ $\bar{H} = (\bar{H}_1, \ldots, \bar{H}_n)$ such that the following holds.

- $(G_1, \ldots, G_k) = (\bar{G}_{in(1)}, \ldots, \bar{G}_{in(k)})$,
- $(H_1, \ldots, H_l) = (\bar{H}_{out(1)}, \ldots, \bar{H}_{out(l)})$,
- $(\bar{G}, \bar{H}) \in (SEM(I) \times SEM(T)) \cap SEM(C)$,
- $(\bar{G}, \bar{H}) \in (\bigcup_{a \in ACT(tgtu)} SEM(a))^*$.

For example, the semantics of the typed unit *recognize* consists of all pairs of the form $((G_1, G_2), (H))$ where G_1 is a string graph, G_2 is a deterministic finite state graph with its start state as current state, and $H = TRUE$ if G_1 is recognized by G_2; otherwise $H = FALSE$. The semantics of the typed unit *recognize-intersection* consists of every pair $((G_1, G_2, G_3), (H))$ where G_1 is a string graph, G_2 and G_3 are deterministic finite state graphs with their respective start state as current state, and $H = TRUE$ if G_1 is recognized by G_2 and G_3; otherwise $H = FALSE$. The semantics of the typed unit *reduce* contains all pairs $((G_1, G_2), (G_3))$ where G_1 and G_3 are string graphs and G_2 is a deterministic finite state graph with its start state as current state such that G_3 represents some suffix of the string represented by G_1 and G_2 recognizes the corresponding "prefix" of G_1. The semantics of *recognize-substitution* contains all pairs $((G_1, G_2, G_3, G_4), (TRUE))$ where G_1 represents a string in the substitution language $subst(L)$, G_2 recognizes the language L, and G_3 and G_4 recognize the languages $subst(a)$ and $subst(b)$, respectively.

5   Conclusion

In this paper, we have introduced the new concept of typed graph transformation units, which is helpful to specify structured and parallel graph transformations with a flexible typing. To this aim a typed transformation unit contains an import component which consists of a set of other typed transformation units. The semantic relations offered by the imported typed units are used by the importing unit. The nondeterminism inherent to rule-based graph transformation can be reduced with control conditions and graph tuple class expressions.

Typed transformation units are a generalization of transformation units [10] in the following aspects. (1) Whereas a transformation unit specifies a binary relation on a single graph type, a typed transformation unit specifies a k, l-relation of graphs of different types. (2) The transformation process in transformation units is basically sequential whereas in typed transformation units typed graphs are transformed simultaneously. Moreover, as described in Section 2.2 typed transformation units generalize the concept of product units [9] that also specify k, l-relations of typed graphs. With product units, however, the possibilities of structuring (and modelling) are more restrictive in the sense that only rules and transformation units can be applied to graph tuples but no imported typed transformation unit.

Further investigation of typed transformation units may concern the following aspects. (1) We used graph-transformational versions of the truth values, but one may like to combine graph types directly with arbitrary abstract data types, i.e. without previously modelling the abstract data types as graphs. (2) In the presented definition, we consider acyclic import structures. Their generalization to networks of typed transformation units with an arbitrary import structure is an interesting task. (3) In the presented approach the graphs of the tuples do not share common parts. Hence, one could consider graph tuple transformation where some relations (like morphisms) can be explicitly specified between the different graphs of a tuple. (4) Apart from generalizing the concept of typed transformation units, a comparison with similar concepts such as pair grammars [14] and triple grammars [16] is needed. (5) Finally, case studies of typed units should also be worked out that allow to get experience with the usefulness of the concept for the modelling of (data-processing) systems and systems from other application areas.

Acknowledgement

We are grateful to the referees for their valuable remarks.

References

1. Marc Andries, Gregor Engels, Annegret Habel, Berthold Hoffmann, Hans-Jörg Kreowski, Sabine Kuske, Detlef Plump, Andy Schürr, and Gabriele Taentzer. Graph transformation for specification and programming. *Science of Computer Programming*, 34(1):1–54, 1999.
2. Andrea Corradini, Hartmut Ehrig, Reiko Heckel, Michael Löwe, Ugo Montanari, and Francesca Rossi. Algebraic approaches to graph transformation part I: Basic concepts and double pushout approach. In Rozenberg [15], pages 163–245.
3. Andrea Corradini, Ugo Montanari, and Francesca Rossi. Graph processes. *Fundamenta Informaticae*, 26(3,4):241–265, 1996.
4. Frank Drewes, Annegret Habel, and Hans-Jörg Kreowski. Hyperedge replacement graph grammars. In Rozenberg [15], pages 95–162.
5. Hartmut Ehrig, Gregor Engels, Hans-Jörg Kreowski, and Grzegorz Rozenberg, editors. *Handbook of Graph Grammars and Computing by Graph Transformation, Vol. 2: Applications, Languages and Tools*. World Scientific, Singapore, 1999.
6. Hartmut Ehrig, Reiko Heckel, Martin Korff, Michael Löwe, Leila Ribeiro, Annika Wagner, and Andrea Corradini. Algebraic approaches to graph transformation II: Single pushout approach and comparison with double pushout approach. In Rozenberg [15], pages 247–312.
7. Hartmut Ehrig, Hans-Jörg Kreowski, Ugo Montanari, and Grzegorz Rozenberg, editors. *Handbook of Graph Grammars and Computing by Graph Transformation, Vol. 3: Concurrency, Parallelism, and Distribution*. World Scientific, Singapore, 1999.

8. Joost Engelfriet and Grzegorz Rozenberg. Node replacement graph grammars. In Rozenberg [15], pages 1–94.

9. Renate Klempien-Hinrichs, Hans-Jörg Kreowski, and Sabine Kuske. Rule-based transformation of graphs and the product type. In Patrick van Bommel, editor, *Handbook on Transformation of Knowledge, Information, and Data*. To appear.

10. Hans-Jörg Kreowski and Sabine Kuske. On the interleaving semantics of transformation units — a step into GRACE. In Janice E. Cuny, Hartmut Ehrig, Gregor Engels, and Grzegorz Rozenberg, editors, *Proc. Graph Grammars and Their Application to Computer Science*, volume 1073 of *Lecture Notes in Computer Science*, pages 89–108, 1996.

11. Hans-Jörg Kreowski and Sabine Kuske. Graph transformation units and modules. In Ehrig, Engels, Kreowski, and Rozenberg [5], pages 607–638.

12. Hans-Jörg Kreowski and Sabine Kuske. Graph transformation units with interleaving semantics. *Formal Aspects of Computing*, 11(6):690–723, 1999.

13. Hans-Jörg Kreowski, Sabine Kuske, and Andy Schürr. Nested graph transformation units. *International Journal on Software Engineering and Knowledge Engineering*, 7(4):479–502, 1997.

14. Terrence W. Pratt. Pair grammars, graph languages and string-to-graph translations. *Journal of Computer and System Sciences*, 5:560–595, 1971.

15. Grzegorz Rozenberg, editor. *Handbook of Graph Grammars and Computing by Graph Transformation, Vol. 1: Foundations*. World Scientific, Singapore, 1997.

16. Andy Schürr. Specification of graph translators with triple graph grammars. In G. Tinnhofer, editor, *Proc. WG'94 20th Int. Worhshop on Graph-Theoretic Concepts in Computer Science*, volume 903 of *Lecture Notes in Computer Science*, pages 151–163, 1994.

17. Andy Schürr. Programmed graph replacement systems. In Rozenberg [15], pages 479–546.

Towards Graph Programs for Graph Algorithms

Detlef Plump and Sandra Steinert

Department of Computer Science, The University of York
York YO10 5DD, UK
{det,sandra}@cs.york.ac.uk

Abstract. Graph programs as introduced by Habel and Plump [8] provide a simple yet computationally complete language for computing functions and relations on graphs. We extend this language such that numerical computations on labels can be conveniently expressed. Rather than resorting to some kind of attributed graph transformation, we introduce conditional rule schemata which are instantiated to (conditional) double-pushout rules over ordinary graphs. A guiding principle in our language extension is syntactic and semantic simplicity. As a case study for the use of extended graph programs, we present and analyse two versions of Dijkstra's shortest path algorithm. The first program consists of just three rule schemata and is easily proved to be correct but can be exponential in the number of rule applications. The second program is a refinement of the first which is essentially deterministic and uses at most a quadratic number of rule applications.

1 Introduction

The graph transformation language introduced by Habel and Plump in [8] and later simplified in [7] consists of just three programming constructs: nondeterministic application of a set of rules (in the double-pushout approach) either in one step or as long as possible, and sequential composition. The language has a simple formal semantics and is both computationally complete and minimal [7]. These properties are attractive for formal reasoning on programs, but the price for simplicity is a lack of programming comfort.

This paper is the first step in developing the language of [7] to a programming language GP (for *graph programs*) that is usable in practice. The goal is to design – and ultimately implement – a semantics-based language that allows high-level problem solving by graph transformation. We believe that such a language will be amenable to formal reasoning if programs can be mapped to a core language with a simple formal semantics. Also, graphs and graph transformations naturally lend themselves to visualisation which will facilitate the understanding of programs.

The language of [7] has no built-in data types so that, for example, numerical computations on labels must be encoded in a clumsy way. We therefore extend graph programs such that operations on labels are performed in a predefined algebra. Syntactically, programs are based on rule schemata labelled with terms over the algebra, which prior to their application are instantiated to ordinary

H. Ehrig et al. (Eds.): ICGT 2004, LNCS 3256, pp. 128–143, 2004.
© Springer-Verlag Berlin Heidelberg 2004

double-pushout rules. In this way we can rely on the well-researched double-pushout approach to graph transformation [2, 6] and avoid resorting to some kind of attributed graph transformation. We also introduce conditional rule schemata which are rule schemata equipped with a Boolean term over operation symbols and a special **edge** predicate. This allows to control rule schema applications by comparing values of labels and checking the (non-)existence of edges.

To find out what constructs should be added to the language of [7] to make GP practical, we intend to carry out various case studies. Graph algorithms are a natural choice for the field of such a study because the problem domain need not be encoded and there exists a comprehensive literature on graph algorithms. In Section 7 we present and analyse two graph programs for Dijkstra's shortest path algorithm. The first program contains just three rule schemata but can be inefficient, while the second program is closer to Dijkstra's original algorithm and needs at most a quadratic number of rule applications. We prove the correctness of the first program and the quadratic complexity of the second program to demonstrate how one can formally reason on graph programs.

In general, we want to keep the syntax and semantics of GP as simple a possible while simultaneously providing sufficient programming comfort. Of course there is a trade-off between these aims; for example, we found it necessary to introduce a while loop in order to efficiently code Dijkstra's algorithm in the second program.

2   Preliminaries

A *signature* $\Sigma = (S, OP)$ consists of a set S of *sorts* and a family $OP = (OP_{\bar{s},s})_{\bar{s} \in S^*, s \in S}$ of *operation symbols*. A family $X = (X_s)_{s \in S}$ of *variables* consists of sets X_s that are pairwise disjoint and disjoint with OP. The sets $T_{OP,s}(X)$ of *terms* of sort s are defined by $x, c \in T_{OP,s}(X)$ for all $x \in X_s$ and all $c \in OP_{\lambda,s}$, and $op(t_1, \ldots, t_n) \in T_{OP,s}(X)$ for all $op \in OP_{s_1 \ldots s_n, s}$ and all $t_1 \in T_{OP,s_1}(X), \ldots, t_n \in T_{OP,s_n}(X)$. The set of all terms over Σ and X is denoted by $T_\Sigma(X)$.

A *Σ-algebra* A consists of a family of nonempty sets $(A_s)_{s \in S}$, elements $c_A \in A_s$ for all $c \in OP_{\lambda,s}$, and functions $op_A \colon A_{s_1} \times \ldots \times A_{s_n} \to A_s$ for all $op \in OP_{s_1 \ldots s_n, s}$.

An *assignment* $\alpha \colon X \to A$ is a family of mappings $(\alpha_s \colon X_s \to A_s)_{s \in S}$. The extension $\hat{\alpha} \colon T_\Sigma(X) \to A$ of α is defined by $\hat{\alpha}(x) = \alpha(x)$ and $\hat{\alpha}(c) = c_A$ for all variables x and all constant symbols c, and $\hat{\alpha}(op(t_1, \ldots, t_n)) = op_A(\hat{\alpha}(t_1), \ldots, \hat{\alpha}(t_n))$ for all $op(t_1, \ldots, t_n) \in T_\Sigma(X)$. If t is a variable-free term, then $\hat{\alpha}(t)$ is denoted by t_A.

A *label alphabet* is a pair $\mathcal{C} = (\mathcal{C}_V, \mathcal{C}_E)$, where $\mathcal{C}_V$ is a set of *node labels* and $\mathcal{C}_E$ is a set of *edge labels*. A *partially labelled graph* over $\mathcal{C}$ is a system $G = (V_G, E_G, s_G, t_G, l_{G,V}, l_{G,E})$, where V_G and E_G are finite sets of *nodes* and *edges*, $s_G, t_G \colon E_G \to V_G$ are *source* and *target* functions for edges, $l_{G,V} \colon V_G \to \mathcal{C}_V$ is the partial node labelling function and $l_{G,E} \colon E_G \to \mathcal{C}_E$ is the partial edge labelling function[1]. A graph is *totally labelled* if $l_{G,V}$ and $l_{G,E}$ are total functions. We write $\mathcal{G}(\mathcal{C})$ for the set of partially labelled graphs, and $\mathcal{G}^t(\mathcal{C})$ for the set of totally labelled graphs over $\mathcal{C}$.

A *premorphism* $g: G \rightarrow H$ between two graphs G and H consists of two source and target preserving functions $g_V: V_G \rightarrow V_H$ and $g_E: E_G \rightarrow E_H$, that is, $s_H \circ g_E = g_V \circ s_G$ and $t_H \circ g_E = g_V \circ t_G$. If g also preserves labels in the sense that $l_H(g(n)) = l_G(n)$ for all n in $\text{Dom}(l_{G,V})$ and $\text{Dom}(l_{G,E})$, then it is a *graph morphism*. Moreover, g is *injective* if g_V and g_E are injective, and it is an *inclusion* if $g(n) = n$ for all nodes and edges n in G.

Assumption 1 We assume a signature $\Sigma = (S, OP)$ such that $\texttt{Bool} \in S$, $OP_{\lambda,\texttt{Bool}} = \{\texttt{true}, \texttt{false}\}$, $OP_{\texttt{Bool},\texttt{Bool}} = \{\neg\}$ and $OP_{\texttt{Bool Bool},\texttt{Bool}} = \{\wedge, \vee, \rightarrow , \leftrightarrow\}$. The signature is interpreted in a fixed Σ-algebra A such that $A_{\texttt{Bool}} = \{\texttt{tt}, \texttt{ff}\}$, $\texttt{true}_A = \texttt{tt}$, $\texttt{false}_A = \texttt{ff}$ and $\neg_A, \wedge_A, \vee_A, \rightarrow_A, \leftrightarrow_A$ are the usual Boolean operations. We also assume a family of variables $X = (X_s)_{s \in S}$ and that S contains two distinguished sorts s_V and s_E for nodes and edges. The label alphabets $\mathcal{C}_T$ and $\mathcal{C}_A$ are defined by

$$\mathcal{C}_T = (T_{OP,s_V}(X), T_{OP,s_E}(X)) \quad \text{and} \quad \mathcal{C}_A = (A_{s_V}, A_{s_E}).$$

3 Rules and Rule Schemata

We recall the definition of double-pushout rules with relabelling given in [9], before introducing rule schemata over $\mathcal{G}(\mathcal{C}_T)$.

Definition 1 (Rule). A *rule* $r = (L \leftarrow K \rightarrow R)$ consists of two graph morphisms $K \rightarrow L$ and $b: K \rightarrow R$ over $\mathcal{G}(\mathcal{C}_A)$ such that $K \rightarrow L$ is an inclusion and

(1) for all $n \in L$, $l_L(n) = \perp$ implies $n \in K$ and $l_R(b(n)) = \perp$, and
(2) for all $n \in R$, $l_R(n) = \perp$ implies $l_L(n') = \perp$ for exactly one $n' \in b^{-1}(n)$.

The rule r is *injective* if $b: K \rightarrow R$ is injective. All rules in the graph programs for Dijkstra's algorithm in Section 7 will be injective, but in general we want to allow non-injective rules.

Definition 2 (Direct derivation). Let G and H be graphs in $\mathcal{G}^t(\mathcal{C}_A)$ and $r = (L \leftarrow K \rightarrow R)$ a rule. A *direct derivation* from G to H by r consists of two natural pushouts[2] as in Figure 1, where $L \rightarrow G$ is injective.

We write $G \Rightarrow_{r,g} H$ or just $G \Rightarrow_r H$ if there exists a direct derivation as in Definition 2. If $\mathcal{R}$ is a set of rules, then $G \Rightarrow_{\mathcal{R}} H$ means that there is some r in $\mathcal{R}$ such that $G \Rightarrow_r H$. Figure 2 shows an example of a rule where we assume $A_{s_V} = A_{s_E} = \mathbb{R}$. (In pictures like this, numbers next to the nodes are used to represent graph morphisms.)

---

[1] Given a partial function $f: A \rightarrow B$, the set $\text{Dom}(f) = \{x \in A \mid f(x) \text{ is defined}\}$ is the *domain* of f. We write $f(x) = \perp$ if $f(x)$ is undefined.
[2] A pushout is *natural* if it is also a pullback. See [9] for the construction of natural pushouts over partially labelled graphs.

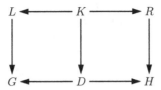

Fig. 1. A direct derivation.

Definition 3 (Match). Given a rule $r = (L \leftarrow K \rightarrow R)$ and a graph G in $\mathcal{G}^t(\mathcal{C}_A)$, an injective graph morphism $g: L \rightarrow G$ is a *match* for r if it satisfies the *dangling condition*: no node in $g(L) - g(K)$ is incident to an edge in $G - g(L)$.

In [9] it is shown that, given r and an injective morphism $g: L \rightarrow G$, there exists a direct derivation as in Figure 1 if and only if g is a match for r. Moreover, in this case D and H are determined uniquely up to isomorphism.

Definition 4 (Rule schema). If $K \rightarrow L$ and $K \rightarrow R$ are graph morphisms over $\mathcal{G}(\mathcal{C}_T)$ satisfying the conditions of Definition 1, then $r = (L \leftarrow K \rightarrow R)$ is a *rule schema*.

An example of a rule schema is shown in Figure 3, where x, y and z are variables of sort **Real**.

Fig. 2. A rule.

Fig. 3. A rule schema.

Rule schemata are instantiated by evaluating their terms according to some assignment $\alpha: X \rightarrow A$.

Definition 5 (Instances of graphs and rule schemata). Given a graph G over $\mathcal{C}_T$ and an assignment $\alpha: X \rightarrow A$, the *instance* G^α of G is the graph over $\mathcal{C}_A$ obtained from G by replacing the labelling functions l_G with $\hat{\alpha} \circ l_G$. The instance of a rule schema $r = (L \leftarrow K \rightarrow R)$ is the rule $r^\alpha = (L^\alpha \leftarrow K^\alpha \rightarrow R^\alpha)$.

For example, the rule in Figure 2 is an instance of the rule schema in Figure 3; the associated assignment α satisfies $\alpha(\mathbf{x}) = 1$, $\alpha(\mathbf{y}) = 2$ and $\alpha(\mathbf{z}) = 4$. Note that a rule schema may have infinitely many instances if A contains infinite base sets.

Given graphs G and H in $\mathcal{G}^t(\mathcal{C}_A)$ and a rule schema r, we write $G \Rightarrow_r H$ if there is an assignment α such that $G \Rightarrow_{r^\alpha} H$. For a set $\mathcal{R}$ of rule schemata, $G \Rightarrow_{\mathcal{R}} H$ means that there is some r in $\mathcal{R}$ such that $G \Rightarrow_r H$.

4   Conditional Rules and Conditional Rule Schemata

We introduce *conditional* rule schemata which allow to control the application of a rule schema by comparing values of terms in the left-hand side of the schema. This concept will be crucial to express graph algorithms conveniently.

Analogously to the instantiation of rule schemata to rules, conditional rule schemata will be instantiated to conditional rules. We define a conditional rule as a rule together with a set of admissible matches.

Definition 6 (Conditional rule). A *conditional rule* $q = (r, M)$ consists of a rule $r = (L \leftarrow K \rightarrow R)$ and a set M of graph morphisms such that $M \subseteq \{g \colon L \rightarrow G \mid G \in \mathcal{G}^t(\mathcal{C}_A) \text{ and } g \text{ is a match for } r\}$.

Intuitively, M is a predicate on the matches of r in totally labelled graphs. Given a conditional rule $q = (r, M)$ and graphs G and H in $\mathcal{G}^t(\mathcal{C}_A)$, we write $G \Rightarrow_q H$ if there is a morphism g in M such that $G \Rightarrow_{r,g} H$.

Our concept of a conditional rule is similar to that of [5] where rules are equipped with two sets of morphisms (representing positive and negative application conditions, respectively). Because [5] is based on the so-called single-pushout approach, admissible morphisms need not satisfy the dangling condition.

Conditional rules as defined above are a semantic concept in that the set M of admissible matches will usually be infinite. To represent conditional rules in the syntax of a programming language, we introduce conditional rule schemata which consist of a rule schema and a Boolean term. This term may contain any operation symbols of the predefined signature Σ and, in addition, a special binary predicate **edge** on the nodes of the left-hand side of the rule schema.

Definition 7 (Conditional sule schema). Given a rule schema $(L \leftarrow K \rightarrow R)$, extend the signature Σ to $\Sigma^L = (S^L, OP^L)$ by $S^L = S \cup \{\text{Node}\}$, $OP^L_{\lambda,\text{Node}} = V_L$, $OP^L_{\text{NodeNode,Bool}} = \{\text{edge}\}$, $OP^L_{w,s} = OP_{w,s}$ if $w \in S^*$ and $s \in S$, and $OP^L_{w,s} = \emptyset$ otherwise. Then a term c in $T_{OP^L,\text{Bool}}(X)$ is a *condition* and $\langle (L \leftarrow K \rightarrow R), c \rangle$ is a *conditional rule schema*.

A conditional rule schema is also written as $(L \leftarrow K \rightarrow R)$ **where** c. In pictures, a rule or rule schema $(L \leftarrow K \rightarrow R)$ is often given in the form $L \Rightarrow R$. In this case we assume that K consists of the numbered nodes of L and that these nodes are unlabelled in K. For example, Figure 4 shows a conditional rule schema that is applicable to a graph G only if x, y and z are instantiated such that $\alpha(x) + \alpha(y) < \alpha(z)$ and if there is no edge in G from the image of node 2 to the image of node 1.

Conditional rule schemata are instantiated by instantiating the rule schema according to some assignment α and by evaluating the condition by an extension of α which takes into account the meaning of the **edge** predicate.

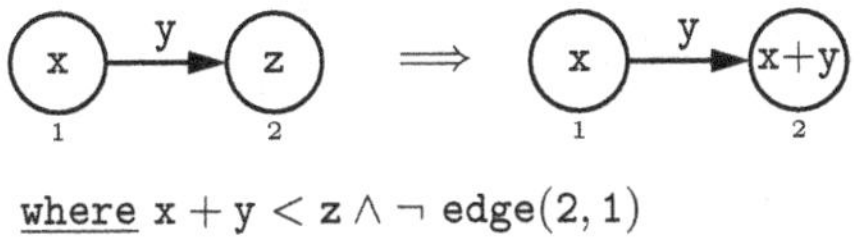

where $x + y < z \wedge \neg\ \text{edge}(2, 1)$

Fig. 4. A conditional rule schema.

Definition 8 (Instance of a conditional rule schema). Given a conditional rule schema $r = \langle (L \leftarrow K \rightarrow R), c \rangle$, an assignment $\alpha\colon X \rightarrow A$ and a graph morphism $g\colon L^\alpha \rightarrow G$ with $G \in \mathcal{G}^t(\mathcal{C}_A)$, define the extension $\alpha_g\colon T_{\Sigma^L}(X) \rightarrow A$ as follows:

(1) $\alpha_g(x) = \alpha(x)$ and $\alpha_g(c) = c_A$ for all variables x and all constants c in Σ.[3]

(2) $\alpha_g(\text{edge}(v, w)) = \begin{cases} \text{tt} & \text{if there is an edge in } G \text{ from } g(v) \text{ to } g(w), \\ \text{ff} & \text{otherwise.} \end{cases}$

(3) $\alpha_g(op(t_1, \ldots, t_n)) = op_A(\alpha_g(t_1), \ldots, \alpha_g(t_n))$
   for all $op(t_1, \ldots, t_n) \in T_{OP^L, S^L}(X)$ with $op \in OP$.

Then the *instance* r^α of r is the conditional rule $\langle (L^\alpha \leftarrow K^\alpha \rightarrow R^\alpha), M \rangle$ where $M = \{g\colon L^\alpha \rightarrow G \mid G \in \mathcal{G}^t(\mathcal{C}_A), \ g \text{ is a match and } \alpha_g(c) = \text{tt}\}$.

Given graphs G and H in $\mathcal{G}^t(\mathcal{C}_A)$ and a conditional rule schema $q = r$ **where** c, we write $G \Rightarrow_q H$ if there is an assignment $\alpha\colon X \rightarrow A$ and a graph morphism g such that $G \Rightarrow_{r^\alpha, g} H$ and $\alpha_g(c) = \text{tt}$.

Operationally, the application of a conditional rule schema $(L \leftarrow K \rightarrow R)$ **where** c to a graph G in $\mathcal{G}^t(\mathcal{C}_A)$ amounts to the following steps:

1. Find an injective premorphism $g\colon L \rightarrow G$ satisfying the dangling condition.
2. Find an assignment $\alpha\colon X \rightarrow A$ such that for all n in $\text{Dom}(l_L)$, $\hat{\alpha}(l_L(n)) = l_G(g(n))$.
3. Check whether $\alpha_g(c) = \text{tt}$.
4. Construct for $(L^\alpha \leftarrow K^\alpha \rightarrow R^\alpha)$ and g the natural pushouts of Definition 2 (according to [9]).

5   Deterministic Conditional Rule Schemata

For an implementation of a programming language based on rule schemata it is prohibitive to enumerate all instances of a rule schema $r = (L \leftarrow K \rightarrow R)$ in order to find an instance that turns a given premorphism $g\colon L \rightarrow G$ into a graph morphism. This is because r may have infinitely many instances. Even if one restricts attention to instances r^α where α evaluates the terms in L to labels of corresponding nodes and edges in G, there may be infinitely many instances left. For example, consider the conditional rule schema in Figure 5 and an associated premorphism $g\colon L \rightarrow G$. Whereas the values $\alpha(\text{k})$ and $\alpha(\text{z})$ are uniquely determined by g, there are infinitely many choices for $\alpha(\text{x}), \alpha(\text{y})$ and

---

[3] Note that α_g is undefined for all constants in $OP^L_{\lambda, \text{Node}}$.

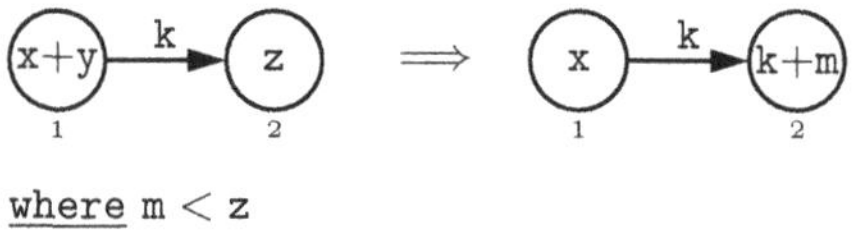

where m < z

Fig. 5. A conditional rule schema that is not deterministic.

$\alpha(\mathtt{m})$ if nodes are labelled with integers, say. We therefore introduce a subclass of (conditional) rule schemata which are instantiated by premorphisms in at most one way.

A term t in $T_\Sigma(X)$ is *simple* if it is a variable or does not contain any variables. We denote by $\mathrm{Var}(t)$ and $\mathrm{Var}(G)$ the sets of variables occuring in a term t or graph G.

Definition 9 (Deterministic conditional rule schema). A rule schema $(L \leftarrow K \rightarrow R)$ is *deterministic*, if

(1) all labels in L are simple terms, and
(2) $\mathrm{Var}(R) \subseteq \mathrm{Var}(L)$.

A conditional rule schema $\langle r, c \rangle$ with $r = (L \leftarrow K \rightarrow R)$ is deterministic if r is deterministic and $\mathrm{Var}(c) \subseteq \mathrm{Var}(L)$.

For example, the conditional rule schema in Figure 4 is deterministic.

Proposition 1. Let $r = \langle (L \leftarrow K \rightarrow R), c \rangle$ be a deterministic conditional rule schema and $g\colon L \rightarrow G$ a premorphism with $G \in \mathcal{G}^t(\mathcal{C}_A)$. Then there is at most one instance r' of r such that g is a match for r'.

Proof. Let r^α and r^β be instances of r such that g is a match for both. By Definition 5 and Definition 8, we have $r^\alpha = r^\beta$ if $\hat{\alpha}(t) = \hat{\beta}(t)$ for all terms t in L and R, and $\alpha_g(c) = \beta_g(c)$ (note that every term in K occurs also in L). Therefore it suffices to show that $\alpha(x) = \beta(x)$ for each variable x in $\mathrm{Var}(L) \cup \mathrm{Var}(R) \cup \mathrm{Var}(c)$. Since r is deterministic, we have $x \in \mathrm{Var}(L)$. Hence there is a node or an edge in L that is labelled with a term containing x. Without loss of generality let v be a node such that $x \in \mathrm{Var}(l_{L,V}(v))$. Because all terms in L are simple, $x = l_{L,V}(v)$. Thus, by Definition 5, $\alpha(x) = \hat{\alpha}(x) = \hat{\alpha}(l_{L,V}(v)) = l_{G,V}(g_V(v)) = \hat{\beta}(l_{L,V}(v)) = \hat{\beta}(x) = \beta(x)$. $\qquad\square$

Proposition 1 ensures that premorphisms cannot "instantiate" deterministic (conditional) rule schemata in more than one way. The next proposition gives a necessary and sufficient condition for such an instantiation to take place. The condition makes precise how to find an assignment α as required in the second step of the description of rule-schema application, given at the end of Section 4.

Proposition 2. Let $g\colon L \rightarrow G$ be a premorphism where $L \in \mathcal{G}(\mathcal{C}_T)$ is labelled with simple terms and $G \in \mathcal{G}^t(\mathcal{C}_A)$. Then there is an assignment $\alpha\colon X \rightarrow A$ such that g is a graph morphism from L^α to G, if and only if for all nodes and edges n, n' in L,

(1) $l_G(g(n)) = t_A$ if $l_L(n)$ is a variable-free term t, and
(2) $l_G(g(n)) = l_G(g(n'))$ if $l_L(n) = l_L(n') \in X$.

Proof. Suppose first that g is a graph morphism from L^α to G. If n is labelled with a variable-free term t in L, then n's label in L^α is $\hat{\alpha}(t) = t_A$. Since g is label-preserving, $g(n)$ is labelled with t_A, too. Moreover, if n and n' are labelled with the same variable x in L, then both are labelled with $\alpha(x)$ in L^α. Hence $l_G(g(n)) = \alpha(x) = l_G(g(n'))$.

Conversely, suppose that conditions (1) and (2) are satisfied. For every sort s in S, let d_s be a fixed element in A_s. Then, by (2),

$$\alpha(x) = \begin{cases} l_G(g(n)) & \text{if there is a node or edge } n \text{ with } l_L(n) = x, \\ d_s & \text{otherwise, where } x \in X_s \end{cases}$$

defines an assignment $\alpha \colon X \to A$. Consider any node or edge n in L^α. If $l_L(n)$ is variable-free, then (1) gives $l_G(g(n)) = t_A = \hat{\alpha}(t) = l_{L^\alpha}(n)$. Otherwise $l_L(n)$ is a variable x, and hence by definition of α, $l_G(g(n)) = \alpha(x) = l_{L^\alpha}(n)$. Thus $g \colon L^\alpha \to G$ is label-preserving. $\square$

6   Graph Programs

We extend the language of [8, 7] by replacing rules with deterministic conditional rule schemata and adding a while-loop.

Definition 10 (Syntax of programs). *Programs* are defined as follows:

(1) For every finite set R of deterministic conditional rule schemata, R and $R{\downarrow}$ are programs.
(2) For every graph B in $\mathcal{G}(\mathcal{C}_T)$ and program P, **while** B **do** P **end** is a program.
(3) If P and Q are programs, then $P; Q$ is a program.

A finite set of conditional rule schemata is called an *elementary* program. Our syntax is ambiguous because a program $P_1; P_2; P_3$ can be parsed as both $(P_1; P_2); P_3$ and $P_1; (P_2; P_3)$. This is irrelevant however as the semantics of sequential composition will be relation composition which is associative.

Next we define a relational semantics for programs. Given a binary relation $\phi \subseteq A \times B$ between two sets A and B, the *domain* of ϕ is the set $\text{Dom}(\phi) = \{a \in A \mid a \, \phi \, b \text{ for some } b \in B\}$. If $A = B$ we write ϕ^* for the reflexive-transitive closure of ϕ. The composition of two relations ϕ and ϱ on A is the relation $\phi \circ \varrho = \{\langle a, c \rangle \mid a \, \phi \, b \text{ and } b \, \varrho \, c \text{ for some } b\}$. Given a graph B in $\mathcal{G}(\mathcal{C}_T)$, let $B? = \{(B \leftarrow B \to B)\}$ with $B \to B$ being the identity morphism on B.

Definition 11 (Semantics of programs). The *semantics* of a program P is a binary relation $[\![P]\!]$ on $\mathcal{G}^t(\mathcal{C}_A)^4$ which is inductively defined as follows:

---

[4] Strictly speaking, the graphs in $\mathcal{G}^t(\mathcal{C}_A)$ should be considered as *abstract graphs*, that is, as isomorphism classes of graphs. For simplicity we stick to ordinary graphs and consider them as representatives for isomorphism classes; see [8, 7] for a precise account.

(1) For every elementary program R, $[\![R]\!] = \Rightarrow_R$.

(2) $[\![R\!\downarrow]\!] = \{\langle G, H\rangle \mid G \Rightarrow_R^* H \text{ and } H \notin \mathrm{Dom}(\Rightarrow_R)\}$.

(3) $[\![\text{\underline{while} } B \text{ \underline{do} } P \text{ \underline{end}}]\!] = \{\langle G, H\rangle \in [\![B?; P]\!]^* \mid H \notin \mathrm{Dom}([\![B?]\!])\}$.

(4) $[\![P; Q]\!] = [\![P]\!] \circ [\![Q]\!]$.

By clause (3), the operational interpretation of $\text{\underline{while} } B \text{ \underline{do} } P \text{ \underline{end}}$ is that P is executed as long as B occurs as a subgraph. In particular, the loop has no effect on a graph G not containing B: in this case we have G $[\![\text{\underline{while} } B \text{ \underline{do} } P \text{ \underline{end}}]\!]$ H if and only if $G = H$. Note also that if G contains B but P fails on input G either because a set of rules in P is not applicable or because P does not terminate, then the whole loop fails in the sense that there is no graph H such that G $[\![\text{\underline{while} } B \text{ \underline{do} } P \text{ \underline{end}}]\!]$ H.

Consider now subsets $\mathcal{G}_1$ and $\mathcal{G}_2$ of $\mathcal{G}^t(\mathcal{C}_A)$ and a relation $\phi \subseteq \mathcal{G}_1 \times \mathcal{G}_2$. We say that a program P *computes* ϕ if $\phi = [\![P]\!] \cap (\mathcal{G}_1 \times \mathcal{G}_2)$, that is, if ϕ coincides with the semantics of P restricted to $\mathcal{G}_1$ and $\mathcal{G}_2$. This includes the case of partial functions $\phi: \mathcal{G}_1 \to \mathcal{G}_2$, which are just special relations.

7   Dijkstra's Shortest Path Algorithm

The so-called single-source shortest path algorithm by Dijkstra [1, 11] computes the distances between a given start node and all other nodes in a graph whose edges are labelled with nonnegative numbers. Given a graph G and nodes v and w, a *path* from v to w is a sequence $e_1, \ldots, e_n$ of edges such that $s_G(e_1) = v$, $t_G(e_n) = w$ and $t_G(e_i) = s_G(e_{i+1})$ for $i = 1, \ldots, n-1$. The *distance* of such a path is the sum of its edge labels. A *shortest path* between two nodes is a path of minimal distance.

Dijkstra's algorithm stores the distance from the start node to a node v in a variable $d(v)$. Initially, the start node gets the value 0 and every other node gets the value ∞. Nodes for which the shortest distance has been computed are added to a set S, which is empty in the beginning. In each step of the algorithm, first a node w from $V_G - S$ is added to S, where $d(w)$ is minimal. Then for each edge e outgoing from w, $d(t_G(e))$ is changed to $\min(d(t_G(e)), d(w) + l_{G,E}(e))$.

7.1   A Simple Graph Program for Dijkstra's Algorithm

Before giving our graph programs, we specify the signature Σ and the algebra A of Assumption 1. The programs will store calculated distances as node labels, so we need some numerical type for both edge and node labels. Let Real be a sort in Σ, $s_V = s_E = \text{Real}$, and let $\mathbb{R}^+$ be the set of nonnegative real numbers. We assume the following operation symbols in Σ:[5] $OP_{\lambda,\text{Real}} = \mathbb{R}^+ \cup \{\infty, *, \square\}$, $OP_{\text{RealReal,Bool}} = \{<\}$ and $OP_{\text{RealReal,Real}} = \{+\}$. The algebra A is given by $A_{\text{Real}} = \mathbb{R}^+ \cup \{\infty, *, \square\}$, $c_A = c$ for all $c \in OP_{\lambda,\text{Real}}$, $x <_A y = \text{tt}$ if and only if

---

[5] Note that all numbers in $\mathbb{R}^+$ are used as constant symbols. The representation of numbers in an implementation of our programming language is beyond the scope of this paper.

$(x, y \in \mathbb{R}^+$ and $x < y)$ or $(x \neq \infty$ and $y = \infty)$, $x +_A y = x + y$ if $x, y \in \mathbb{R}^+$ and $x +_A y = \infty$ otherwise.

Our first program for Dijkstra's algorithm, Simple Dijkstra, is given in Figure 6. We assume that the program is started from a graph in $\mathcal{G}^t(\mathcal{C}_A)$ whose edges are labelled with nonnegative numbers and whose start node is marked by a unique loop labelled with $*$. The rule schema S Prepare relabels every node of the input graph with ∞, S Start deletes the unique loop and relabels the start node with 0, and S Reduce changes a stored distance whenever a shorter path has been found.

Simple Dijkstra = S Prepare ↓; S Start; S Reduce ↓

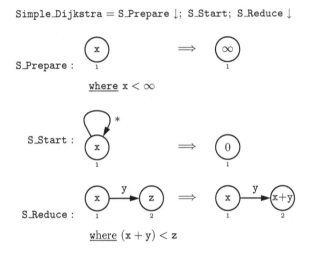

Fig. 6. The program Simple Dijkstra.

Proposition 3 (Correctness of Simple Dijkstra). *Let G be a graph in $\mathcal{G}^t(\mathcal{C}_A)$ containing a unique loop e, where $l_{G,E}(e) = *$ and $l_{G,E}(e') \in \mathbb{R}^+$ for all other edges e'. When started from G, Simple Dijkstra terminates and produces a unique graph H which is obtained from G by removing e and labelling each node v with the shortest distance from $s_G(e)$ to v.*

Proof. Termination of Simple Dijkstra follows from the fact that every application of S Prepare reduces the number of nodes not labelled with ∞, and that every application of S Reduce reduces the sum of all node labels in a graph.

Let now H be a graph such that $G \, \llbracket \text{Simple Dijkstra} \rrbracket \, H$. Since there are no rule schemata for adding or deleting nodes, and S Start is the only rule schema that alters G's edges, it is clear that H can be obtained from G by removing the loop e and relabelling the nodes. Thus, H is uniquely determined if each node v is labelled with the shortest distance from $s_G(e)$ to v. To show the latter, we need the following invariance property.

Claim. Let G [S_Prepare$\downarrow$; S_Start] $H_0 \Rightarrow^*_{\text{S\_Reduce}} H'$. Then for each node v in H', either $l_{H',V}(v) = \infty$ or $l_{H',V}(v)$ is the distance of a path from $s_G(e)$ to v.

Proof. The proposition holds for H_0, because $s_G(e)$ is labelled with 0 and every other node is labelled with ∞. Moreover, it is easy to see that every application of S_Reduce preserves the claimed property. $\square$

Suppose now that there is a node v in H such that $l_{H,V}(v)$ is not the shortest distance from $s_G(e)$ to v. We distinguish two cases.

Case 1: $v = s_G(e)$. Since v is labelled with 0 after the application of S_Start, and $l_{H,V}(v) \neq 0$, there must be an application of S_Reduce that changes $v's$ label to a negative number. But this contradicts the above claim.

Case 2: $v \neq s_G(e)$. By the above claim, there is a path from $s_G(e)$ to v (as otherwise $l_{H,V}(v) \neq \infty$). Let $e_1, \ldots, e_n$ be a shortest path from $s_G(e)$ to v. Let $v_0 = s_G(e)$ and $v_i = t_H(e_i)$ for $i = 1, \ldots, n$. By Case 1, $l_{H,V}(v_0) = 0$. Hence, there is some k, $1 \leq k \leq n$, such that $l_{H,V}(v_k)$ is not the shortest distance from v_0 to v_k and for $i = 0, \ldots, k-1$, $l_{H,V}(v_i)$ is the shortest distance from v_0 to v_i. Now since $e_1, \ldots, e_n$ is a shortest path to v_n it follows that $e_1, \ldots, e_k$ is a shortest path to v_k and that $e_1, \ldots, e_{k-1}$ is a shortest path to v_{k-1}. So the shortest distance from v_0 to v_k is $\sum_{i=1}^{k-1} l_{H,E}(e_i) + l_{H,E}(e_k) = l_{H,V}(v_{k-1}) + l_{H,E}(e_k)$. As this sum is smaller than $l_{H,V}(v_k)$, S_Reduce is applicable to e_k. But this contradicts the fact that $H \notin \text{Dom}(\Rightarrow_{\text{S\_Reduce}})$. $\square$

The correctness of Simple_Dijkstra was easy to show, however the program can be expensive in the number of applications of the rule schema S_Reduce. For example, the right-hand derivation sequence in Figure 7 contains 48 applications of S_Reduce and represents the worst-case program run for the given input graph of 5 nodes. In contrast, Dijkstra's algorithms (as sketched at the beginning of this section) changes distances only 10 times when applied to the same graph. Although Simple_Dijkstra needs only 4 applications of S_Reduce in the best case, there is no guarantee that it does not choose the worst case. We therefore refine Simple_Dijkstra by modelling more closely the original algorithm.

7.2   A Refined Program

The program Dijkstra of Figure 8 uses a **while**-loop to repeatedly select a node of minimal distance and to update the distances of the target nodes of the outgoing edges of that node. Nodes that have not yet been selected are marked by a $\square$-labelled loop. Removing the $*$-labelled loop from a node by Next corresponds to adding that node to the set S of the original algorithm. Note that Dijkstra is essentially deterministic: Min $\downarrow$ always determines a node of minimal distance among all nodes marked with loops, and Reduce is applied only to edges outgoing from this node.

The left-hand derivation sequence of Figure 7 is a worst-case run of Dijkstra, containing 26 rule-schema applications. Among these are only 10 applications of Reduce, which correspond to the 10 distance changes done by the original algorithm. The next proposition establishes the worst-case complexity of Dijkstra

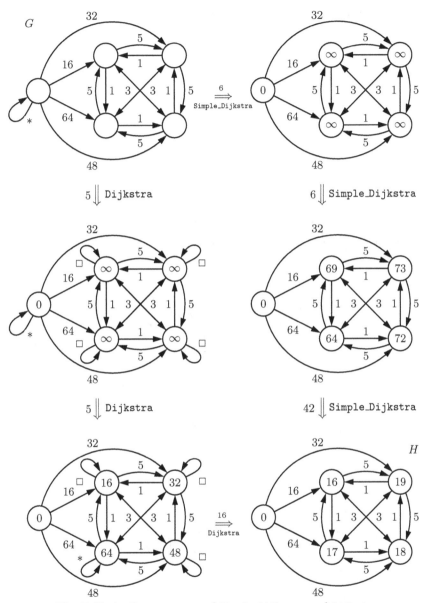

Fig. 7. Derivation sequences of Simple_Dijkstra and Dijkstra.

in terms of the number of rule-schema applications, where we assume that input graphs satisfy the precondition of Proposition 3.

Proposition 4 (Complexity of Dijkstra). *When started from a graph containing n nodes and e edges, Dijkstra terminates after $O(n^2 + e)$ rule-schema applications.*

Dijkstra = Prepare ↓; Start; while B do Min ↓; Reduce ↓; Next end; CleanUp

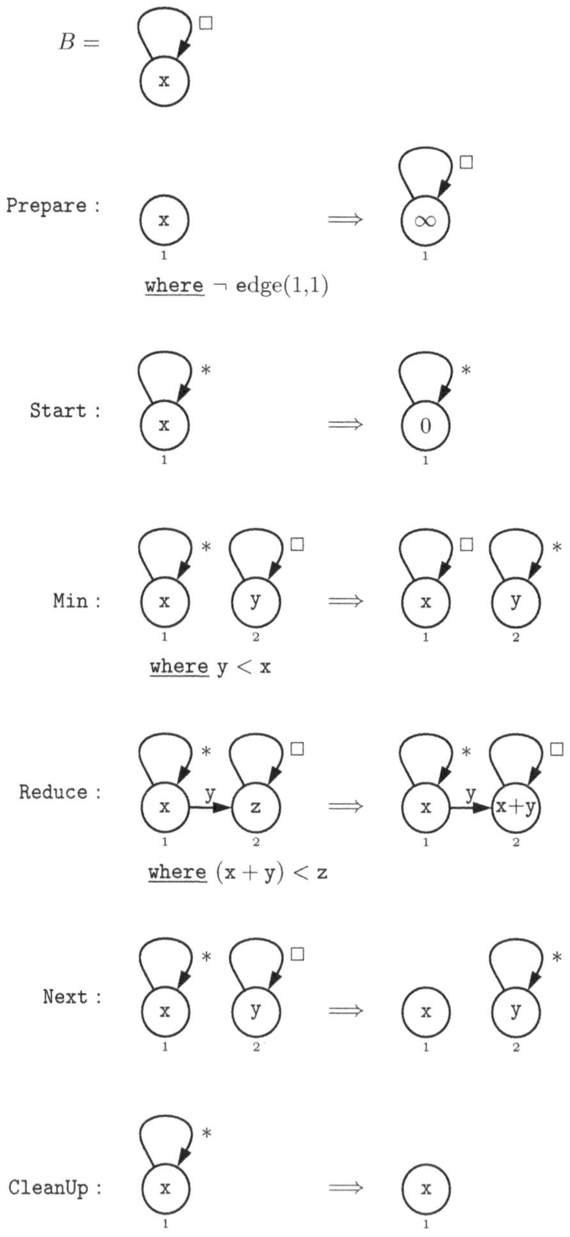
B =
x
Prepare :
x
∞
where ¬ edge(1,1)
Start :
x
0
Min :
x
y
where y < x
Reduce :
x
y
z
x+y
where (x + y) < z
Next :
CleanUp :

Fig. 8. The program Dijkstra.

Proof. The initialisation phase **Prepare** ↓; **Start** uses n rule-schema applications. The body of the <u>while</u>-loop is executed $(n - 1)$-times because initially there are $n - 1$ loops labelled with □, and each execution of the body reduces this number by one. So the overall number of **Next**-applications is $n - 1$, too. Each execution of **Min** ↓ takes at most $n - 1$ steps because there is only one *-labelled loop. Hence, there are at most $(n - 1)^2$ applications of **Min** overall. The total number of **Reduce**-applications is at most e since **Reduce** cannot be applied twice to the same edge. This is because **Reduce** is applied only to edges outgoing from the *-marked node, and the * mark is removed by **Next**. Thus, a bound for the overall number of rule-schema applications is $n + (n - 1) + (n - 1)^2 + e$, which is in $O(n^2 + e)$.   □

Note that if we forbid parallel edges in input graphs, then e is bounded by n^2 and hence the complexity of **Dijkstra** is $O(n^2)$.

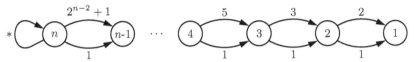

Fig. 9. A worst-case input for **Simple_Dijkstra**.

The quadratic complexity of **Dijkstra** means a drastic improvement on the running time of **Simple_Dijkstra** which may be exponential. More precisely, one can show that for every $n \geq 2$ there is a graph with n nodes and $2(n - 1)$ edges such that there is a run of **Simple_Dijkstra** in which the rule schema **S_Reduce** is applied $\sum_{k=1}^{n-1} 2^k$ times. Such a graph is shown in Figure 9. (The running time of **Dijkstra** for this graph is actually linear.)

8   Related Work

A guiding principle in our ongoing design of the graph programming language GP is syntactic and semantic simplicity, which distinguishes GP from the complex PROGRES language [15]. It remains to be seen how much we have to compromise this principle to enable practical programming in application areas. Our approach also differs from a language such as AGG [4] in that we insist on a formal semantics. We want GP to be semantics-based since we consider the ability to formally reason on programs as a key feature.

The rule schemata intoduced in this paper are not the only way to extend graph transformation with calculations on labels. An alternative is to use one of the approaches to attributed graph transformation that have been proposed in the literature. The recent papers [10, 3], for example, merge graphs and algebras so that attributed graphs are usually infinite. We rather prefer to work with finite graphs in which "attributes" are ordinary labels.

Our method of working with rule schemata and their instances is close to Schied's approach to double-pushout transformations on graphs labelled with algebra elements [14]. (A single-pushout version of this approach is outlined in [13].) Roughly, his double-pushout diagrams can be decomposed into our diagrams with rule schema instantiations on top of them. A major difference between the present paper and [14] is that our rules can relabel and merge items whereas the rules in [14] are label preserving and injective. Schied also introduces conditions for rules, in the form of propositional formulas over term equations, but he does not consider built-in predicates on the graph structure such as our **edge** predicate.

Surprisingly, there seems to be hardly any work on studying graph algorithms in the framework of graph transformation languages. We are only aware of a case study on Floyd's all-pairs shortest path algorithm in Kreowski and Kuske's paper [12]. The paper presents a program for Floyd's algorithm and proves its correctness as well as a cubic bound for the number of rule applications. (The program consists of rules with parameters, similar to our rule schemata, but [12] does not give a general formalism for such rules.)

9 Conclusion

As pointed out in the Introduction, this paper is only the first step in extending the language of [7] to a graph programming language GP. We have introduced graph programs over rule schemata to incorporate numerical data and other basic data types. Rule schemata can have Boolean application conditions which may contain built-in predicates on the graph structure. We have identified deterministic conditional rule schemata as a class of schemata that admit a reasonable implementation in that their applicability and the graphs resulting from applications are uniquely determined by premorphisms from left-hand sides into graphs. As a case study for extended graph programs, we have given two programs for Dijkstra's shortest path algorithm and have analysed their correctness and complexity.

In future work, more case studies on graph algorithms and in other areas will be pursued to find out what additional programming constructs are needed to make GP a practical language. We hope that new constructs can be mapped to a small core of GP – possibly the language used in this paper – to keep the semantics comprehensible and to facilitate formal reasoning on programs, static program analysis, program transformation, etc. And, of course, GP should eventually be implemented so that its practical usefulness can be proved.

References

1. T. Cormen, C. Leiserson, and R. Rivest. *Introduction to Algorithms*. The MIT Press and McGraw-Hill, 2000.
2. A. Corradini, U. Montanari, F. Rossi, H. Ehrig, R. Heckel, and M. Löwe. Algebraic approaches to graph transformation — Part I: Basic concepts and double pushout approach. In G. Rozenberg, editor, *Handbook of Graph Grammars and Computing by Graph Transformation*, volume 1, chapter 3, pages 163–245. World Scientific, 1997.

3. H. Ehrig, U. Prange, and G. Taentzer. Fundamental theory for typed attributed graph transformation. In *Proc. International Conference on Graph Transformation (ICGT 2004)*, Lecture Notes in Computer Science. Springer-Verlag, 2004. This volume.

4. C. Ermel, M. Rudolf, and G. Taentzer. The AGG approach: Language and environment. In H. Ehrig, G. Engels, H.-J. Kreowski, and G. Rozenberg, editors, *Handbook of Graph Grammars and Computing by Graph Transformation*, volume 2, chapter 14, pages 551–603. World Scientific, 1999.

5. A. Habel, R. Heckel, and G. Taentzer. Graph grammars with negative application conditions. *Fundamenta Informaticae*, 26(3/4):287–313, 1996.

6. A. Habel, J. Müller, and D. Plump. Double-pushout graph transformation revisited. *Mathematical Structures in Computer Science*, 11(5):637–688, 2001.

7. A. Habel and D. Plump. A minimal and complete programming language for graph transformation. In preparation.

8. A. Habel and D. Plump. Computational completeness of programming languages based on graph transformation. In *Proc. Foundations of Software Science and Computation Structures (FOSSACS '01)*, volume 2030 of *Lecture Notes in Computer Science*, pages 230–245. Springer-Verlag, 2001.

9. A. Habel and D. Plump. Relabelling in graph transformation. In *Proc. International Conference on Graph Transformation (ICGT 2002)*, volume 2505 of *Lecture Notes in Computer Science*, pages 135–147. Springer-Verlag, 2002.

10. R. Heckel, J. Küster, and G. Taentzer. Confluence of typed attributed graph transformation systems. In *Proc. International Conference on Graph Transformation (ICGT 02)*, volume 2505 of *Lecture Notes in Computer Science*, pages 161–176. Springer-Verlag, 2002.

11. D. Jungnickel. *Graphs, Networks and Algorithms*. Springer-Verlag, 2002.

12. H.-J. Kreowski and S. Kuske. Graph transformation units and modules. In H. Ehrig, G. Engels, H.-J. Kreowski, and G. Rozenberg, editors, *Handbook of Graph Grammars and Computing by Graph Transformation*, volume 2, chapter 15, pages 607–638. World Scientific, 1999.

13. M. Löwe, M. Korff, and A. Wagner. An algebraic framework for the transformation of attributed graphs. In R. Sleep, R. Plasmeijer, and M. van Eekelen, editors, *Term Graph Rewriting: Theory and Practice*, pages 185–199. John Wiley, 1993.

14. G. Schied. *Über Graphgrammatiken, eine Spezifikationsmethode für Programmiersprachen und Verteilte Regelsysteme*. Doctoral dissertation, Universität Erlangen-Nürnberg, 1992. Volume 25(2) of *Arbeitsberichte des Instituts für Informatik* (in German).

15. A. Schürr, A. Winter, and A. Zündorf. The PROGRES approach: Language and environment. In H. Ehrig, G. Engels, H.-J. Kreowski, and G. Rozenberg, editors, *Handbook of Graph Grammars and Computing by Graph Transformation*, volume 2, chapter 13, pages 487–550. World Scientific, 1999.

Adhesive High-Level Replacement Categories and Systems

Hartmut Ehrig[1], Annegret Habel[2], Julia Padberg[1], and Ulrike Prange[1]

[1] Technical University of Berlin, Germany
{ehrig,padberg,ullip}@cs.tu-berlin.de
[2] Carl v. Ossietzky University of Oldenburg, Germany
annegret.habel@informatik.uni-oldenburg.de

Abstract. Adhesive high-level replacement (HLR) categories and systems are introduced as a new categorical framework for graph transformation in a broad sense, which combines the well-known concept of HLR systems with the new concept of adhesive categories introduced by Lack and Sobociński.

In this paper we show that most of the HLR properties, which had been introduced ad hoc to generalize some basic results from the category of graphs to high-level structures, are valid already in adhesive HLR categories. As a main new result in a categorical framework we show the Critical Pair Lemma for local confluence of transformations. Moreover we present a new version of embeddings and extensions for transformations in our framework of adhesive HLR systems.

1 Introduction

High-level replacement systems have been introduced in [1] to generalize the well-known double pushout approach from graphs [2] to various kinds of high-level structures, including also algebraic specifications and Petri nets. In order to generalize basic results, like the local Church-Rosser, parallelism and concurrency theorem, several different conditions have been introduced in [1], called HLR conditions. The theory of HLR systems has been applied to a large number of example categories, where all the HLR conditions have been verified explicitly. Unfortunately, however, these conditions have some kind of ad hoc character, because they are just a collection of all the properties which are used in the categorical proofs of the basic results. Up to now it has not been analyzed how far these HLR properties are independent from each other or are consequences of a more general principle.

This problem concerning the ad hoc character of the HLR conditions has been solved recently by Lack and Sobociński in [3] by introducing the notion of adhesive categories. They have shown that the concept of "van Kampen squares", short VK squares, known from topology [4], can be considered as such a general principle. Roughly spoken a VK square is a pushout square which is stable under pullbacks. The key idea of adhesive categories is the requirement that pushouts along monomorphisms are VK squares. This property is valid not

H. Ehrig et al. (Eds.): ICGT 2004, LNCS 3256, pp. 144–160, 2004.

only in the categories **Sets** and **Graph**, but also in several varieties of graphs, which have been used in the theory of graph grammars and graph transformation [5] up to now. On the other hand Lack and Sobociński were able to show in [3] that most of the ad hoc HLR conditions required in [1] can be shown for adhesive categories. Together with the results in [1] this implies that the basic results for the theory of graph transformation mentioned above are valid in adhesive categories, where only the Parallelism Theorem requires in addition the existence of binary coproducts.

Unfortunately the concept of adhesive categories incorporates an important restriction, which rules out several interesting application categories. The HLR framework in [1] is based on a distinguished class M of morphisms, which is restricted to the class of all monomorphisms in adhesive categories. This restriction rules out the category (**SPEC**, M) of all algebraic specifications with class M of all strict injective specification morphisms (see [1]) and several other integrated specification techniques like algebraic high-level nets [6, 7] and different kinds of attributed graphs [8, 9], which are important in the area of graph transformation and HLR systems.

In this paper we combine the advantages of HLR and of adhesive categories by introducing the new concept of "adhesive HLR categories". Roughly spoken an adhesive HLR category is an adhesive category with a suitable subclass M of monomorphisms, which is closed under pushouts and pullbacks. As main results of this paper we are able to show that adhesive HLR categories are closed under product, slice, coslice and functor category constructions and that most of the important HLR properties of [1] are valid. These results are generalizations of corresponding results in [3], where we remove the restrictions, that M is the class of all monomorphisms and that adhesive categories in [3] are required to have all pullbacks instead of pullbacks along M-morphisms only.

In sections 2 - 4 of this paper we review and recover the basic results for HLR systems in [1] and adhesive grammars in [3] in the framework of adhesive HLR categories and systems. Moreover, we present in section 5 a new version of the results for embedding and extension of transformations [2, 10, 11]. This is the basis to show in section 6 another main result of this paper: For the first time we present a categorical version of the Critical Pair Lemma for local confluence of transformations, discussed for hypergraphs in [12] and attributed graphs in [9], in our new framework of adhesive HLR systems.

For lack of space we only give proof ideas for some of our results in this paper. For a more detailed version we refer to our technical report [13].

2   Review of Van Kampen Squares and Adhesive Categories

In this section we review adhesive categories as introduced by Lack and Sobociński in [3].

The basic notion of adhesive categories is that of a so called van Kampen square. The intuitive idea of a van Kampen square is that of a pushout which is stable under pullbacks and vice versa pushout preservation implies pullback stability. The name van Kampen derives from the relationship between these squares and the Van Kampen Theorem in topology [4].

Definition 1 (van Kampen square). *A pushout (1) is a van Kampen (VK) square, if for any commutative cube (2) with (1) in the bottom and back faces being pullbacks holds: the top is pushout ⇔ the front faces are pullbacks.*

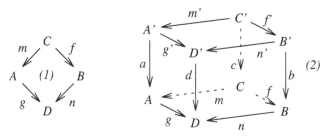

In the definition of adhesive categories only those VK squares are considered, where m is a monomorphism. In this case the square is called a pushout along a monomorphism. The first interesting property of VK squares in [3] shows that in this case also n is a monomorphism and the square is also a pullback.

Definition 2 (adhesive category). *A category C is an adhesive category, if*

1. *C has pushouts along monomorphisms, i.e. pushouts, where at least one of the given morphisms is monomorphism,*
2. *C has pullbacks,*
3. *pushouts along monomorphisms are VK squares.*

The most basic example of an adhesive category is the category **Sets** of sets. Moreover it is shown in [3] that adhesive categories are closed under product, slice, coslice and functor category construction. This implies immediately that also the category **Graphs** of graphs $G = (E \overset{s,t}{\rightrightarrows} V)$, and also several variants like typed graphs, labelled graphs and hypergraphs are adhesive categories. This is a first indication that adhesive categories are suitable for graph transformation. Counterexamples for adhesive categories are **Pos** (partially ordered sets), **Top** (topological spaces), **Gpd** (groupoids) and **Cat** (categories), where pushouts along monomorphisms fail to be VK squares (see [3]).

The main reason why adhesive categories are important for the theory of graph transformation and its generalization to high-level replacement systems (see [1]) is the fact that most of the HLR conditions required in [1] are shown to be valid already in adhesive categories (see [3]). This implies that basic results like the Local Church-Rosser Theorem and the Concurrency Theorem (see [1]) are valid already in the framework of adhesive categories, while the Parallelism Theorem needs in addition the existence of binary coproducts.

The main advantage of adhesive categories compared with HLR categories in [1] is the fact that the requirements for adhesive categories are much more smooth than the variety of different HLR conditions in [1], which have been stated "ad hoc" as needed in the categorical proofs of the corresponding results mentioned above.

On the other hand HLR categories in [1] are based on a class M of morphisms, which is restricted to the class of all monomorphisms in adhesive categories. This rules out several interesting examples. In order to avoid this problem we combine the two concepts leading to the notion of adhesive HLR categories in the next section.

3   Adhesive HLR Categories

As motivated in the previous section we will combine the concepts of adhesive categories [3] and HLR categories [1] leading to the new concept of adhesive HLR categories in this section. Most of the results presented in this section are generalizations of results for adhesive categories in [3], but we present new interesting examples which are not instantiations of adhesive categories.

The main difference of adhesive HLR categories compared with adhesive categories is the fact that we consider a suitable subclass M of monomorphisms instead of the class of all monomorphisms. Moreover we require only pullbacks along M-morphisms and not for general morphisms.

Definition 3 (adhesive HLR category (C, M)). *A category C with a morphism class M is called adhesive HLR category, if*

1. *M is a class of monomorphisms closed under isomorphisms and closed under composition ($f : A \to B \in M$, $g : B \to C \in M \Rightarrow g \circ f \in M$) and decomposition ($g \circ f \in M$, $g \in M \Rightarrow f \in M$),*
2. *C has pushouts and pullbacks along M-morphisms and M-morphisms are closed under pushouts and pullbacks,*
3. *pushouts in C along M-morphisms are VK squares.*

Remark 1. Most of the results in this paper can also be formulated under slightly weaker assumptions, where the existence of pullbacks is required only if both given morphisms are in M and pushouts along M-morphisms are required to be M-VK squares only, i.e. only for the case $f \in M$ or $a, b, d \in M$. This weaker version is called "weak adhesive HLR category". But presently we have no interesting example of this weak case that is not also an adhesive HLR category.

Example 1. 1. All examples of adhesive categories are adhesive HLR categories for the class M of all monomorphisms. As shown in [3] this includes the category **Sets** of sets, **Graphs** of graphs and several variants of graphs like typed, labelled and hypergraphs discussed above. Moreover this includes the category **PT-Net** of place transition nets considered in [1].
2. The category (**Spec**, M_1) of algebraic specifications with class M_1 of all monomorphisms is not adhesive, because pushouts along monomorphisms are not necessarily pullbacks. But (**Spec**, M_2) with class M_2 of all strict injective specification morphisms is an HLR2 category in the sense of [1] and also an adhesive HLR category.

For similar reasons the category **AHL-Net** of algebraic high-level nets (see [7]) has to be considered with strict injective specification morphisms concerning the specification part of the net morphism.

3. An important new example is the category (**AGraphs$_{\text{ATG}}$**, M) of typed attributed graphs with type graph ATG and class M of all injective morphisms with isomorphisms on the data part. In our paper [14] we explicitly show that this is an adhesive HLR category satisfying all additional HLR properties considered later in this paper.

The first important result shows that adhesive HLR categories are closed under product, slice, coslice and functor category construction. This allows to construct new examples from given ones.

Theorem 1 (construction of adhesive HLR categories). *Adhesive HLR categories can be constructed as follows:*

- *If (C, M_1) and (D, M_2) are adhesive HLR, then ($C \times D$, $M_1 \times M_2$) is adhesive HLR.*
- *If (C, M) is adhesive HLR, then so are the slice category ($C \backslash C$, $M \cap C \backslash C$) and the coslice category ($C \backslash C$, $M \cap C \backslash C$) for any object C in C.*
- *If (C, M) is adhesive HLR, then every functor category ([X, C], M-functor transformations) is adhesive HLR.*

Remark 2. An M-functor transformation is a natural transformation $t : F \to G$ where all morphisms $t(X) : F(X) \to G(X)$ are in M.

Proof idea. In the case of product and functor categories the properties of adhesive HLR categories can be shown componentwise. For slice and coslice categories some standard constructions for pushouts and pullbacks can be used to show the properties.   □

The second important result shows that most of the HLR conditions stated in [1, 7] are already valid in adhesive HLR categories.

Theorem 2 (HLR properties of adhesive HLR categories). *Given an adhesive HLR category (C, M), the following HLR conditions are satisfied.*

1. *Pushouts along M-morphisms are pullbacks.*
2. *Pushout-pullback decomposition: Given the following diagram with $l, w \in M$, (1) + (2) pushout and (2) pullback. Then (1) and (2) are pushouts and also pullbacks.*
3. *Cube pushout-pullback property: Given the following commutative cube (3), where all morphisms in top and bottom are in M, the top is pullback and the front faces are pushouts. Then we have: the bottom is pullback $\Leftrightarrow$ the back faces of the cube are pushouts.*

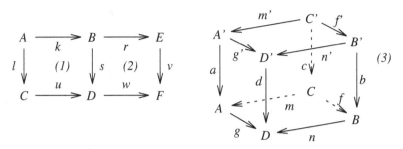

4. *Uniqueness of pushout complements for M-morphisms: Given*
$k : A \rightarrow B \in M$ *and* $s : B \rightarrow D$ *then there is up to isomorphism at*
most one C *with* $l : A \rightarrow C$ *and* $u : C \rightarrow D$ *such that diagram (1) is a*
pushout.

Proof idea. These properties are shown for adhesive categories with class M of
all monomorphisms in [3]. The proofs can be reformulated for any subclass M
of monomorphisms as required for adhesive HLR categories.   □

Remark 3. The HLR conditions stated above together with the existence of binary coproducts compatible with M (see Thm. 3) correspond roughly to the
HLR conditions in [7] resp. HLR2 and two of the HLR2* conditions in [1]. The
HLR2 condition of [1] stating that M is closed under isomorphisms is not needed
in our context.

4   Adhesive HLR Systems

In this section we use the concept of adhesive HLR categories introduced in the
previous sections to present the basic notions and results of adhesive HLR systems in analogy to HLR systems in [1]. The Local Church-Rosser Theorem and
the Parallelism Theorem are shown to be valid in [1] for HLR1 categories, and
the Concurrency Theorem for HLR2 categories, where the existence of binary
coproducts is only needed for the Parallelism Theorem. Using the properties of
adhesive HLR categories in the previous section we can immediatly conclude
that the Local Church-Rosser Theorem and the Concurrency Theorem are valid
in adhesive HLR categories and the Parallelism Theorem in adhesive HLR categories with binary coproducts.

Definition 4 (adhesive HLR system). *An adhesive HLR system*
$AS = (C, M, S, P)$ *consists of an adhesive HLR category* (C, M), *a start object*
S *and a set of productions* P, *where*

1. *a production* $p = L \xleftarrow{l} K \xrightarrow{r} R$ *consists of objects* L, K *and* R *called*
   left-hand side, gluing object and right-hand side respectively, and morphisms
   $l : K \rightarrow L$, $r : K \rightarrow R$ *with* $l, r \in M$,
2. *a direct transformation* $G \xRightarrow{p,m} H$ *via a production* p *and a morphism*
   $m : L \rightarrow G$, *called match, is given by the following diagram, called DPO-*
   diagram, where (1) and (2) are pushouts,

$$
\begin{array}{ccccc}
L & \xleftarrow{\quad l \quad} & K & \xrightarrow{\quad r \quad} & R \\
\Big\downarrow m & (1) & \Big\downarrow k & (2) & \Big\downarrow n \\
G & \xleftarrow{\quad f \quad} & D & \xrightarrow{\quad g \quad} & H
\end{array}
$$

3. a transformation is a sequence $G_0 \Rightarrow G_1 \Rightarrow \ldots \Rightarrow G_n$ of direct transformations, written $G_0 \stackrel{*}{\Rightarrow} G_n$,
4. the language $L(AS)$ consists of all objects G in $\mathbf{C}$ derivable from the start object S by a transformation, i.e. $L(AS) = \{G \mid S \stackrel{*}{\Rightarrow} G\}$.

Remark 4.   1. An adhesive HLR system is on the one hand an HLR system in the sense of [1], where in [1] we have in addition a distinguished class T of terminal objects, and on the other hand an adhesive grammar in the sense of [3], provided that the class M is the class of all monomorphisms.
   2. A direct transformation $G \stackrel{p,m}{\Longrightarrow} H$ is uniquely determined up to isomorphism by the production p and the match m, because due to Thm. 2 It. 4 pushout complements along M-morphisms in adhesive HLR categories are unique up to isomorphism.
   3. All the examples for HLR1 and HLR2 systems considered in [1] and all systems over adhesive HLR categories considered in Ex. 1 are adhesive HLR systems, which includes especially the classical graph transformation approach in [2].

The following basic results are shown for HLR2 categories in [1] and they are rephrased for adhesive categories in [3]. According to Thm. 2 they are also valid for adhesive HLR systems. A more detailed version of these results is presented in [13].

Theorem 3 (Local Church-Rosser, Parallelism and Concurrency Theorem). *The Local Church-Rosser Theorems I and II, the Parallelism Theorem and the Concurrency Theorem as stated in [1] are valid for all adhesive HLR systems $AS = (\mathbf{C}, M, S, P)$. Only for the Parallelism Theorem we have to require in addition that $(\mathbf{C}, M)$ has binary coproducts which are compatible with M, i.e. $m_1, m_2 \in M$ implies $m_1 + m_2 \in M$.*

Proof. Follows from [1] and Thm. 2.   □

5   Embedding and Extension of Adhesive HLR Transformations

In this section we present a categorical version of the Embedding Theorem for graph transformation (see [2]) using the concept of initial pushouts first introduced in [10]. The embedding theorem is not only important for the theory of graph transformation, but also for the component framework for system modelling introduced in [15]. In [11] it is shown how to verify the extension properties used in the generic component concept of [15] in the framework of HLR systems.

The Embedding Theorem and the Extension Theorem presented for adhesive HLR systems in this section combine the results for both areas and will also be used in the next section to show the Local Confluence Theorem. The key notion is the concept of initial pushouts, which formalizes the construction of boundary and context in [2]. The important new property going beyond [10] is the fact that initial pushouts are closed under double pushouts under certain conditions. As in Sec. 4 we assume also in this section that we have an adhesive HLR system.

We start with the definition of an extension diagram in the sense of [15, 11] which means that a transformation t is extended to a transformation t' via an extension morphism.

Definition 5 (extension diagram). *An extension diagram is a diagram (1)*

$$
\begin{array}{ccc}
G_0 & \overset{t}{\underset{*}{\Longrightarrow}} & G_n \\
{\scriptstyle k_0}\downarrow & (1) & \downarrow{\scriptstyle k_n} \\
G_0' & \underset{t'}{\overset{*}{\Longrightarrow}} & G_n'
\end{array}
$$

where $k_0 : G_0 \to G_0'$ is a morphism, called extension morphism, and $t : G_0 \overset{*}{\Rightarrow} G_n$ and $t' : G_0' \overset{*}{\Rightarrow} G_n'$ are transformations via the same productions $(p_0, ..., p_{n-1})$ and matches $(m_0, ..., m_{n-1})$ resp. $(k_0 \circ m_0, ..., k_{n-1} \circ m_{n-1})$ defined by the following DPOs.

$$
\begin{array}{ccccc}
p_i: & L_i & \longleftarrow K_i \longrightarrow & R_i & \\
& {\scriptstyle m_i}\downarrow & \downarrow \qquad \downarrow & & (i = 0, ..., n\text{-}1) \\
& G_i & \longleftarrow D_i \longrightarrow & G_{i+1} & \\
& {\scriptstyle k_i}\downarrow & \downarrow \qquad \downarrow{\scriptstyle k_{i+1}} & & \\
& G_i' & \longleftarrow D_i' \longrightarrow & G_{i+1}' &
\end{array}
$$

Remark 5. 1. The extension diagram (1) is completely determined (up to isomorphism) by $t : G_0 \overset{*}{\Rightarrow} G_n$ and $k_0 : G_0 \to G_0'$ (using the uniqueness of pushout complements).

2. Extension diagrams are closed under horizontal and vertical composition (using corresponding composition properties of pushouts).

The main problem is now to determine under which condition a transformation $t : G_0 \overset{*}{\Rightarrow} G_n$ and an extension morphism $k_0 : G_0 \to G_0'$ lead to an extension diagram. The key notion is that of an initial pushout, which will be required for the extension morphism k_0 in the consistency condition below.

Definition 6 (initial pushout, boundary and context). *Given $f : A \to A'$ a morphism $b : B \to A$ with $b \in M$ is called boundary over f if there is a pushout complement such that (1) is an initial pushout over f. Initiality of (1) over f means, that for every pushout (2) with $b' \in M$ there exist unique morphisms $b^* : B \to D$ and $c^* : C \to E$ with $b^*, c^* \in M$ such that $b' \circ b^* = b$, $c' \circ c^* = c$ and (3) is pushout. Then B is called boundary object and C context w.r.t. $f : A \to A'$.*

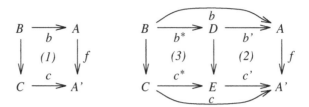

Remark 6. In the classical case of graph transformations [2] the boundary B of a graph morphism $f : A \to A'$ consists of all nodes in $b \in A$ such that $f(b)$ is adjacent to an edge in $A' \backslash f(A)$. These nodes are necessary to glue A to the context graph $C = A' \backslash f(A) \cup f(b(B))$ in order to obtain A' as gluing of A and C via B in the initial pushout (1).

As pointed out in the introduction of this section the closure of initial pushouts under double pushouts is an important technical lemma.

Lemma 1 (closure property of initial pushouts). *Let M' be a class of morphisms closed under pushouts and pullbacks along M-morphisms. Moreover we assume to have initial pushouts over M'-morphisms. Then initial pushouts over M'-morphisms are closed under double pushouts. That means given an initial pushout (1) over $h_0 \in M'$ and a double pushout diagram (2) with $d_0, d_1 \in M$, then (3) and (4) are initial pushouts over $d \in M'$ respectively $h_1 \in M'$ for the unique $b : B \to D$ with $d_0 \circ b = b_0$ obtained by initiality of (1).*

$$B \xrightarrow{b_0} G_0 \qquad G_0 \xleftarrow{d_0} D \xrightarrow{d_1} G_1$$
$$\downarrow (1) \quad \downarrow h_0 \qquad h_0 \downarrow \quad \downarrow d \quad \downarrow h_1 (2)$$
$$C \to G_0' \qquad G_0' \xleftarrow{} D' \xrightarrow{} G_1'$$

$$B \xrightarrow{} D \qquad B \xrightarrow{d_1 \circ b} G_1$$
$$\downarrow b \quad (3) \quad \downarrow d \qquad \downarrow \quad (4) \quad \downarrow h_1$$
$$C \to D' \qquad C \xrightarrow{} G_1'$$

Proof idea. This can be shown shown stepwise for pushouts in the opposite and in the same direction by using the properties of M and M'. The complete proof can be found in [13].   □

The following consistency condition for a transformation $t : G_0 \overset{*}{\Rightarrow} G_n$ and an extension morphism $k_0 : G_0 \to G_0'$ means intuitively that the boundary B of k_0 is preserved by t. In order to formulate this property we use the notion of a derived span $der(t) = G_0 \leftarrow D \to G_n$ of the transformation t, which connects the first and the last object.

Definition 7 (derived span and consistency). *The derived span of a direct transformation $G \overset{p,n}{\Longrightarrow} H$ as shown in Def. 4 is the span $G \leftarrow D \to H$. The derived span $der(t) = (G_0 \overset{d_0}{\leftarrow} D \overset{d_n}{\to} G_n)$ of a transformation $t : G_0 \overset{*}{\Rightarrow} G_n$ is the* *composition via pullbacks of the spans of the corresponding direct transforma-tions.*

A morphism $k_0 : G_0 \to G'_0$ is called consistent w.r.t. a transformation $t : G_0 \overset{*}{\Rightarrow} G_n$ with derived span $der(t) = (G_0 \overset{d_0}{\leftarrow} D \overset{d_n}{\to} G_n)$ if there exist an initial pushout (1) over k_0 and a morphism $b \in M$ with $d_0 \circ b = b_0$.

$$
\begin{array}{ccccccc}
 & \overset{b}{} & & & & & \\
B & \Rightarrow & G_0 & \Longleftarrow & D & \longrightarrow & G_n \\
\downarrow b_0 & & \downarrow k_0 & d_0 & & d_n & \\
\; (1) & & & & & & \\
C & \longrightarrow & G_0' & & & &
\end{array}
$$

Remark 7. 1. The morphisms of the span $G \leftarrow D \to H$ are in M because M is closed under pushouts. This implies that the compositions of these spans exist and are M-morphisms, because pullbacks along M-morphisms exist and M-morphisms are closed under pullbacks in adhesive HLR categories.
  2. The consistency condition in [2], called JOIN condition, requires a suitable family $b_i : B \to D_i$ of morphisms from the boundary B to the context graphs D_i of the direct transformations. In fact, our consistency condition is equivalent to the existence of a corresponding family $(b_i)_{i=0,\ldots,n-1}$.
  3. For the definition of consistency and for Thm. 4 below – but not for Thm. 5 – it would be sufficient to require the existence of a pushout over k_0 instead of an initial one. Moreover we need only conditions 1 and 2 of Def. 3.

Now we are able to prove the Embedding and the Extension Theorem which show that consistency is sufficient and also necessary for the construction of extension diagrams. Moreover, we obtain a direct construction of the extension $k_n : G_n \to G'_n$ in the extension diagram.

Theorem 4 (Embedding Theorem). *Given a transformation $t : G_0 \overset{*}{\Rightarrow} G_n$ and a morphism $k_0 : G_0 \to G'_0$ which is consistent w.r.t. t, then there is an extension diagram for t and k_0 (see (1) in Def. 5).*

Proof idea ($n = 2$). We construct pullback (0) leading to the derived span $G_0 \leftarrow D_0 \leftarrow D \to D_1 \to G_2$ of the transformation $t : G_0 \Rightarrow^* G_2$. Given k_0 consistent w.r.t. t we have initial pushout (2) over k_0 and $b : B \to D$.

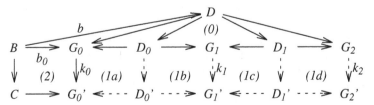

This leads to M-morphisms $B \to D_0$ and $B \to D_1$ such that first D'_0 can be constructed as pushout object of $B \to D_0$ and $B \to C$ leading by decomposition to pushout (1a) and by construction to pushout (1b). Then D'_1 can be constructed as pushout of $B \to D_1$ and $B \to C$ leading by decomposition to pushout (1c)

and by construction to pushout (1d). The given transformation $t : G_0 \overset{*}{\Rightarrow} G_2$ together with the pushouts (1a) - (1d) constitutes the required extension diagram. □

Theorem 5 (Extension Theorem). *Given a transformation* $t : G_0 \overset{*}{\Rightarrow} G_n$ *with derived span* $der(t) = (G_0 \overset{d_0}{\leftarrow} D \overset{d_n}{\rightarrow} G_n)$ *and an extension diagram (1)*

$$
\begin{array}{ccc}
B \overset{b_0}{\longrightarrow} & G_0 \overset{t}{\underset{}{\Longrightarrow}} & G_n \\
\downarrow \quad (2) & \downarrow k_0 \ (1) & \downarrow k_n \\
C \longrightarrow & G_0' \overset{*}{\underset{t'}{\Longrightarrow}} & G_n'
\end{array}
$$

with initial pushout (2) over $k_0 \in M'$ *for some class* M' *closed under pushouts and pullbacks along* M*-morphisms and initial pushouts over* M'*-morphisms, then we have*

1. k_0 *consistent w.r.t.* $t : G_0 \overset{*}{\Rightarrow} G_n$ *with morphism* $b : B \to D$,
2. *a direct transformation* $G_0' \Rightarrow G_n'$ *via* $der(t)$ *and match* k_0 *given by pushouts (3) and (4) with* $d, k_n \in M'$,
3. *initial pushouts (5) and (6) over* d *resp.* k_n.

$$
\begin{array}{ccccc}
G_0 \overset{d_0}{\longleftarrow} D \overset{d_n}{\longrightarrow} G_n & \quad & B \longrightarrow D & \quad & B \overset{d_n \circ b}{\longrightarrow} G_n \\
k_0 \downarrow \ (3) \quad \downarrow d \ (4) \quad \downarrow k_n & \quad & \downarrow \ (5) \ \downarrow d & \quad & \downarrow \ (6) \quad \downarrow k_n \\
G_0' \longleftarrow D' \longrightarrow G_n' & \quad & C \longrightarrow D' & \quad & C \longrightarrow G_n'
\end{array}
$$

Remark 8. The extension theorem shows

1. Consistency of k_0 w.r.t. t is necessary for the existence of the extension diagram.
2. The extension diagram (1) can be represented by a direct transformation with match k_0 and comatch k_n.
3. The extension $k_n : G_n \to G_n'$ can be constructed by a pushout (6) of G_n and context C along the boundary B with $d_n \circ b : B \to G_n$.

Proof idea $(n = 2)$. Given t and k_0 with initial pushout (2) and the extension diagram given by pushouts (1a) - (1d) in proof of Thm. 4, where D is pullback in (0). Initiality of (2) and pushout (1a) lead to $b_0^* : B \to D_0$ and by Lem. 1 to an initial pushout over k_1. This new initiality and pushout (1c) leads to $b_1 : B \to D_1$. The morphisms b_0^* and b_1 lead to an induced $b : B \to D$ - using the pullback properties of (0) - which allows to show consistency of k_0 w.r.t. t. This consistency immediately implies the pushout complement D' in (3) and pushout (4) of d and d_n. Finally, the double pushout (3), (4) implies by Lem. 1 initial pushouts (5) and (6) from (2). □

6   Critical Pairs and Local Confluence of Adhesive HLR Systems

Critical pairs and local confluence have been studied for hypergraph transformations in [12] and for typed attributed graph transformation in [9]. In this section we present a categorical version in adhesive HLR categories. As additional requirements we only need an E'-M' pair factorization for cospans of morphisms - in analogy to the well-known epi-mono-factorization of morphisms - and initial pushouts over M'-morphisms. These assumptions are stated where necessary. Otherwise we only assume to have an adhesive HLR system.

It is well-known that local confluence and termination imply confluence. But we only analyze local confluence in this paper and no termination nor general confluence.

Definition 8 (confluence, local confluence). *A pair of transformations* $H_1 \overset{*}{\Leftarrow} G \overset{*}{\Rightarrow} H_2$ *is confluent if there are transformations* $H_1 \overset{*}{\Rightarrow} X$ *and* $H_2 \overset{*}{\Rightarrow} X$. *An adhesive HLR system is locally confluent, if this property holds for each pair of direct transformations, it is confluent, if it holds for all pairs of transformations.*

In order to define and construct critical pairs we introduce the notion of E'-M' pair factorization.

Definition 9 (E'-M' pair factorization). *An adhesive HLR category has E'-M' pair factorization, if M' is a class of morphisms closed under pushouts and pullbacks along M-morphisms and E' a class of morphism pairs with same codomain and we have for each pair of morphisms $f_1 : A_1 \rightarrow C$, $f_2 : A_2 \rightarrow C$ that there is an object K and morphisms $e_1 : A_1 \rightarrow K$, $e_2 : A_2 \rightarrow K$, $m : K \rightarrow C$ with $(e_1, e_2) \in E'$, $m \in M'$ such that $m \circ e_1 = f_1$ and $m \circ e_2 = f_2$.*

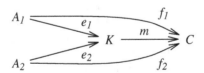

Remark 9. It is sufficient to require this property for matches $f_i = m_i : L_i \rightarrow G$ (i = 1, 2). The closure properties of M' are needed in Lem. 2 and Thm. 6.

The intuitive idea of morphism pairs $(e_1, e_2) \in E'$ in most example categories is that the pair is jointly surjective resp. jointly epimorphic. This can be achieved in categories C with binary coproducts and E_0-M_0 factorization of morphisms, where $E_0 \subseteq Epis$ and $M_0 \subseteq Monos$. Given $A_1 \overset{f_1}{\rightarrow} C \overset{f_2}{\leftarrow} A_2$ we simply take an E_0-M_0 factorization $f = m \circ e$ of the induced morphism $f : A_1 + A_2 \rightarrow C$ and define $e_1 = e \circ i_1$ and $e_2 = e \circ i_2$, where i_1, i_2 are the coproduct injections. If the category has no binary coproducts, or the construction above is not always adequate - as in the case of typed attributed graph transformation - we may have another alternative to obtain an E'-M' pair factorization. In [14] an explicit E'-M' pair factorization for typed attributed graph transformation is provided, where M'-morphisms are not necessarily injective on the data type part and hence $M' \not\subseteq M$.

The main idea to prove local confluence is to show local confluence explicitly only for critical pairs based on the notion of parallel independence (see [1]).

Definition 10 (critical pair). *Given an E'-M' pair factorization, a critical pair is a pair of non-parallel independent direct transformations $P_1 \overset{p_1,o_1}{\Longleftarrow} K \overset{p_2,o_2}{\Longrightarrow} P_2$ such that $(o_1,o_2) \in E'$ for the corresponding matches o_1 and o_2.*

The first step towards local confluence is to show completeness of critical pairs.

Lemma 2 (completeness of critical pairs). *Consider an adhesive HLR system with E'-M' pair factorization and $M' \subseteq M$. For each pair of non-parallel independent direct transformations $H_1 \overset{p_1,m_1}{\Longleftarrow} G \overset{p_2,m_2}{\Longrightarrow} H_2$ there is a critical pair $P_1 \overset{p_1,o_1}{\Longleftarrow} K \overset{p_2,o_2}{\Longrightarrow} P_2$ with extension diagrams (1) and (2) and $m \in M'$.*

$$
\begin{array}{ccccc}
P_1 & \Longleftarrow & K & \Longrightarrow & P_2 \\
\downarrow & (1) & \downarrow m & (2) & \downarrow \\
H_1 & \Longleftarrow & G & \Longrightarrow & H_2
\end{array}
$$

Remark 10. If $M' \not\subseteq M$ we have to require in addition that the pushout-pullback decomposition (Thm. 2 It. 2) holds also with $l \in M$ and $w \in M'$ in diagram (1)+(2) of Thm. 2.

Proof. With the E'-M' pair factorization for m_1 and m_2 we get an object K and morphisms $m : K \to G \in M'$, $o_1 : L_1 \to K$ and $o_2 : L_2 \to K$ with $(o_1,o_2) \in E'$ such that $m_1 = m \circ o_1$ and $m_2 = m \circ o_2$. We can now build the following extension diagram. First we construct the pullback over q_1 and m, derive the morphism t_1 and by applying Thm. 2 It. 2 both squares are pushouts. In the case $M' \not\subseteq M$ of Rem. 10 we have the pushout-pullback decomposition because $l_1 \in M$ and $m \in M'$. With Def. 3 we can build the pushout over r_1 and t_1, derive the morphism z_1 and with pushout decomposition this square is a pushout. The same construction is applied for the second transformation.

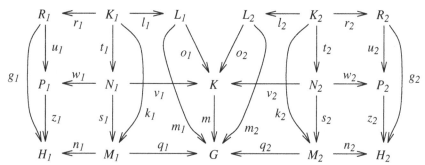

$P_1 \overset{p_1,o_1}{\Longleftarrow} K \overset{p_2,o_2}{\Longrightarrow} P_2$ are non-parallel independent. Otherwise there are morphisms $i : L_1 \to N_2$ and $j : L_2 \to N_1$ with $v_2 \circ i = o_1$ and $v_1 \circ j = o_2$. Then $q_2 \circ s_2 \circ i = m \circ v_2 \circ i = m \circ o_1 = m_1$ and $q_1 \circ s_1 \circ j = m \circ v_1 \circ j = m \circ o_2 = m_2$, that means $H_1 \overset{p_1,m_1}{\Longleftarrow} G \overset{p_2,m_2}{\Longrightarrow} H_2$ □
are parallel independent, contradiction.

From [12] in the case of hypergraph transformation it is known already that confluence of critical pairs is not sufficient to show local confluence in general. In fact, we need a slightly stronger property, called strict confluence.

Definition 11 (strict confluence of critical pairs). *A critical pair* $P_1 \overset{p_1,o_1}{\Longleftarrow} K \overset{p_2,o_2}{\Longrightarrow} P_2$ *is called strictly confluent, if we have*

1. *confluence: the critical pair is confluent, i.e. there are transformations* $P_1 \overset{*}{\Rightarrow} K'$, $P_2 \overset{*}{\Rightarrow} K'$ *with derived spans* $der(P_i \overset{*}{\Rightarrow} K') = P_i \overset{v_{i+2}}{\Longleftarrow} N_{i+2} \overset{w_{i+2}}{\to} K'$ *for* $i = 1, 2$.
2. *strictness: Let* $der(K \overset{p_i,o_i}{\Longrightarrow} P_i) = K \overset{v_i}{\leftarrow} N_i \overset{w_i}{\to} P_i$ $(i = 1, 2)$ *and* N_5, N_6 *and* N *pullback objects of pullbacks (1), (2) resp. (3) then there are morphisms* z_5 *and* z_6 *such that (4), (5) and (6) commute.*

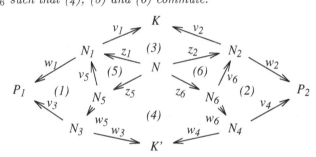

Remark 11. The strictness condition is a combination of corresponding conditions stated in [12] and [9]. More precisely, commutativity of (4) is required in [12] and that of (5) and (6) in [9]. In [12] however, commutativity of (5) and (6) seems to be a consequence of inclusion properties. The intuitive idea of strictness is that the common part N, which is preserved by each transformation of the critical pair, is also preserved by the transformations $P_1 \overset{*}{\Rightarrow} K'$ and $P_2 \overset{*}{\Rightarrow} K'$ and mapped by the same morphism $N \to K'$.

Finally our last main result states that strict confluence of all critical pairs implies local confluence. This result is also known as Critical Pair Lemma (see [12, 9]).

Theorem 6 (Local Confluence Theorem - Critical Pair Lemma). *An adhesive HLR system with* E'-M' *pair factorization,* $M' \subseteq M$ *and initial pushouts over* M'-*morphisms is locally confluent, if all its critical pairs are strictly confluent.*

Remark 12. See Rem. 10 for the case $M' \not\subseteq M$. In the proof we need that M is closed under decomposition (see Def. 3 It. 1) in order to show $b_3 \in M$.

Proof. Given a pair of direct transformations $H_1 \overset{p_1,m_2}{\Leftarrow} G \overset{p_2,m_2}{\Rightarrow} H_2$ we have to show the existence of transformations $t'_1 : H_1 \overset{*}{\Rightarrow} G'$ and $t'_2 : H_2 \overset{*}{\Rightarrow} G'$. If the given pair is parallel independent this follows from the local Church-Rosser theorem. If the given pair is not parallel independent Lem. 2 implies the existence of a critical pair $P_1 \overset{p_1,o_1}{\Leftarrow} K \overset{p_2,o_2}{\Rightarrow} P_2$ with extension diagrams (7) and (8) and $m \in M'$. By assumption this critical pair is strictly confluent leading to transformations $t_1 : P_1 \overset{*}{\Rightarrow} K'$, $t_2 : P_2 \overset{*}{\Rightarrow} K'$ and the diagram in Def. 11.

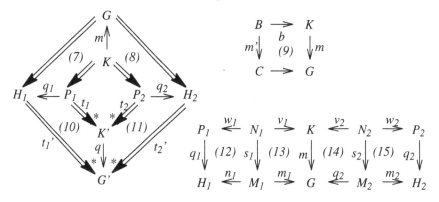

Now let (9) be an initial pushout over $m \in M'$ and consider the double pushouts (12), (13) and (14), (15) corresponding to extension diagrams (7) and (8) respectively.

Initiality of (9) applied to pushout (13) leads to unique $b_1, c_1 \in M$ such that (16) and (17) commute and (18) is pushout. By Lem. 1 (18) is initial pushout over s_1. Dually we obtain $b_2, c_2 \in M$ with $v_2 \circ b_2 = b$. Using pullback property of (3) in Def. 11 we obtain a unique $b_3 : B \to N$ with $z_1 \circ b_3 = b_1$ and $z_2 \circ b_3 = b_2$. Moreover $b_1, z_1 \in M$ implies $b_3 \in M$ by decomposition property of M. In order to show consistency of q_1 w.r.t. t_1 we have to construct $b'_3 \in M$ such that (19) commutes where (18)+(12) is initial pushout over q_1 by Lem. 1. In fact $b'_3 = w_5 \circ z_5 \circ b_3 \in M$ makes (19) commutative using (5) in Def. 11.

Dually q_2 is consistent w.r.t. t_2 using $b'_4 = w_6 \circ z_6 \circ b_3 \in M$ and (6) in Def. 11. By the Embedding Theorem we obtain extension diagrams (10) and (11), where the morphism $q : K' \to G'$ is the same in both cases. This equality can be shown using part 3 of the Extension Theorem, where q is determined by an initial pushout of $m' : B \to C$ and $w_3 \circ b'_3 : B \to K'$ in the first and $w_4 \circ b'_4 : B \to K'$ in the second case and we have $w_3 \circ b'_3 = w_4 \circ b'_4$ using commutativity of (4) in Def. 11. □

7   Conclusion

In this paper we have introduced adhesive HLR categories and systems combining the framework of adhesive categories in [3] and of HLR systems in [1]. We claim that this new framework is most important for different theories of graphs and graphical structures in computer science, which are mainly based on pushout constructions. As shown in this paper this includes first of all the double pushout approach in the theory of graph transformation and HLR systems [2, 1, 5], where important new results have been presented in this framework which are already applied to typed attributed graph transformation in [14]. Constraints and application conditions for DPO-transformations of adhesive HLR systems are considered already in [16]. On the other hand pushouts have also been used in semantics in order to derive well-behaved labeled transition systems by Leifer and Milner in [17], by Sassone and Sobociński in [18] and by König and Ehrig in [19]. We agree with [3] that the role of adhesive categories - and even more adhesive HLR categories - for this kind of applications is most likely to become comparable to the role of cartesian closed categories for simply typed lambda calculi as pointed out by Lambek and Scott in [20].

References

1. Ehrig, H., Habel, A., Kreowski, H.J., Parisi-Presicce, F.: Parallelism and Concurrency in High-Level Replacement Systems. Math. Struct. in Comp. Science **1** (1991) 361–404
2. Ehrig, H.: Introduction to the Algebraic Theory of Graph Grammars (A Survey). In: Graph Grammars and their Application to Computer Science and Biology. Volume 73 of LNCS. Springer (1979) 1–69
3. Lack, S., Sobociński, P.: Adhesive Categories. In: Proc. of FOSSACS '04. Volume 2987 of LNCS., Springer (2004) 273–288
4. Brown, R., Janelidze, G.: Van Kampen Theorems for Categories of Covering Morphisms in Lextensive Categories. Journal of Pure and Applied Algebra **119** (1997) 255–263
5. Corradini, A., Ehrig, H., Heckel, R., Löwe, M., Montanari, U., Rossi, F.: Algebraic Approaches to Graph Transformation I : Basic Concepts and Double Pushout Approach. In Rozenberg, G., ed.: Handbook of Graph Grammars and Computing by Graph Transformation, Volume 1: Foundations. World Scientific (1997)
6. Padberg, J., Ehrig, H., Ribeiro, L.: Algebraic High-Level Net Transformation Systems. In: Mathematical Structures in Computer Science Vol 2. (1995) 217–256
7. Ehrig, H., Gajewsky, M., Parisi-Presicce, F.: High-Level Replacement Systems with Applications to Algebraic Specifications and Petri Nets. In Rozenberg, G., Montanari, U., Ehrig, H., Kreowski, H.J., eds.: Handbook of Graph Grammars and Computing by Graph Transformations, Volume 3: Concurrency, Parallelism, and Distribution. World Scientific (1999) 341–400
8. Löwe, M., Korff, M., Wagner, A.: An Algebraic Framework for the Transformation of Attributed Graphs. In: Term Graph Rewriting: Theory and Practice. John Wiley and Sons Ltd. (1993) 185–199
9. Heckel, R., Küster, J., Taentzer, G.: Confluence of Typed Attributed Graph Transformation with Constraints. In: Proc. ICGT 2002. Volume 2505 of LNCS., Springer (2002) 161–176

10. Padberg, J., Taentzer, G.: Embedding of Derivations in High-Level Replacement Systems. Technical Report 1993/9, TU Berlin (1993)
11. Ehrig, H., Orejas, F., Braatz, B., Klein, M., Piirainen, M.: A Component Framework for System Modeling Based on High-Level Replacement Systems. In: Software and Systems Modeling 3(2), Springer (2004) 114–135
12. Plump, D.: Hypergraph Rewriting: Critical Pairs and Undecidability of Confluence. In Sleep, M., Plasmeijer, M., van Eekelen, M., eds.: Term Graph Rewriting: Theory and Practice. John Wiley & Sons Ltd (1993) 201–213
13. Ehrig, H., Habel, A., Padberg, J., Prange, U.: Adhesive High-Level Replacement Categories and Systems: Long Version. Technical Report TU Berlin. (2004)
14. Ehrig, H., Prange, U., Taentzer, G.: Fundamental Theory for Typed Attributed Graph Transformation. In: Proc. ICGT 2004. LNCS, Springer (2004) (this volume).
15. Ehrig, H., Orejas, F., Braatz, B., Klein, M., Piirainen, M.: A Generic Component Concept for System Modeling. In: Proc. FASE 2002. Volume 2306 of LNCS., Springer (2002) 33–48
16. Ehrig, H., Ehrig, K., Habel, A., Pennemann, K.: Constraints and Application Conditions: From Graphs to High-Level Structures. In: Proc. ICGT 2004. LNCS, Springer (2004) (this volume).
17. Leifer, J., Milner, R.: Deriving Bisimulation Congruences for Reactive Systems. In: Proc. CONCUR 2000. Volume 1877 of LNCS., Springer (2000) 243–258
18. Sassone, V., Sobociński, P.: Deriving Bisimulation Congruences: 2-Categories vs Precategories. In: Proc. FOSSACS 2003. Volume 2620 of LNCS., Springer (2003) 409–424
19. Ehrig, H., König, B.: Deriving bisimulation congruences in the DPO approach to graph rewriting. In: Proc. FOSSACS 2004. Volume 2987 of LNCS., Springer (2004) 151 – 166
20. Lambek, J., Scott, P.: Introduction to Higher Order Categorical Logic. Cambridge University Press (1986)

Fundamental Theory
for Typed Attributed Graph Transformation

Hartmut Ehrig, Ulrike Prange, and Gabriele Taentzer

Technical University of Berlin, Germany
{ehrig,ullip,gabi}@cs.tu-berlin.de

Abstract. The concept of typed attributed graph transformation is most significant for modeling and meta modeling in software engineering and visual languages, but up to now there is no adequate theory for this important branch of graph transformation. In this paper we give a new formalization of typed attributed graphs, which allows node and edge attribution. The first main result shows that the corresponding category is isomorphic to the category of algebras over a specific kind of attributed graph structure signature. This allows to prove the second main result showing that the category of typed attributed graphs is an instance of "adhesive HLR categories". This new concept combines adhesive categories introduced by Lack and Sobociński with the well-known approach of high-level replacement (HLR) systems using a new simplified version of HLR conditions. As a consequence we obtain a rigorous approach to typed attributed graph transformation providing as fundamental results the Local Church-Rosser, Parallelism, Concurrency, Embedding and Extension Theorem and a Local Confluence Theorem known as Critical Pair Lemma in the literature.

1 Introduction

The algebraic theory of graph transformation based on labeled graphs and the double-pushout approach has already a long tradition (see [1]) with various applications (see [2,3]). Within the last decade graph transformation has been used as a modeling technique in software engineering and as a meta-language to specify and implement visual modeling techniques like the UML. Especially for these applications it is important to use not only labeled graphs as considered in the classical approach [1], but also typed and attributed graphs. In fact, there are already several different concepts for typed and attributed graph transformation in the literature (see e.g. [4–6]). However, there is no adequate theory for this important branch of graph transformation up to now. The key idea in [5] is to model an attributed graph with node attribution as a pair $AG = (G, A)$ of a graph G and a data type algebra A. In this paper we use this idea to model attributed graphs with node and edge attribution, where G is now a new kind of graph, called E-graph, which allows also edges from edges to attribute nodes. This new kind of attributed graphs combined with the concept of typing leads to a category **AGraphs$_{ATG}$** of attributed graphs typed over an attributed type

H. Ehrig et al. (Eds.): ICGT 2004, LNCS 3256, pp. 161–177, 2004.

graph ATG. This category seems to be an adequate formal model not only for various applications in software engineering and visual languages, but also for the internal representation of attributed graphs in our graph transformation tool AGG [7].

The main purpose of this paper is to provide the basic concepts and results of graph transformation known in the classical case [1] for this new kind of typed attributed graphs. The straightforward way would be to extend the classical theory in [1] step by step first to attributed graphs and then to typed attributed graphs. In this paper we propose the more elegant solution to obtain the theory of typed attributed graph transformation as an instantiation of the corresponding categorical theory developed in [8]. In [8] we have proposed the new concept of "adhesive HLR categories and systems", which combines the concept of "adhesive categories" presented by Lack and Sobociński in [9] with the concept of high-level replacement systems, short HLR systems, introduced in [10]. In [8] we have shown that not only the Local Church-Rosser, Parallelism and Concurrency Theorem - presented already in [10] for HLR systems -, but also several other results known from the classical theory [1, 9] are valid for adhesive HLR systems satisfying some additional HLR properties.

For this purpose we have to show that the category **AGraphs$_{ATG}$** of typed attributed graphs is an adhesive HLR category in this sense. In Thm. 1 we show that the category **AGraphs$_{ATG}$** is isomorphic to a category of algebras over a suitable signature $AGSIG(ATG)$, which is uniquely defined by the attributed type graph ATG. In fact, it is much easier to verify the categorical properties required for adhesive HLR categories for the category of algebras **AGSIG(ATG)-Alg** and to show the isomorphism between **AGSIG(ATG)-Alg** and **AGraphs$_{ATG}$**, than to show the categorical properties directly for the category **AGraphs$_{ATG}$**. In Thm. 2 we show that **AGSIG(ATG)-Alg** and hence also **AGraphs$_{ATG}$** is an adhesive HLR category. In fact, we show this result for the category **AGSIG-Alg**, where $AGSIG$ is a more general kind of attributed graph structure signature in the sense of [4, 11, 12]. Combining the main results of this paper with those of [8] we are able to show that the following basic results shown in Thm. 3 - 5 are valid for typed attributed graph transformation:

1. Local Church-Rosser, Parallelism and Concurrency Theorem,
2. Embedding and Extension Theorem,
3. Local Confluence Theorem (Critical Pair Lemma).

Throughout the paper we use a running example from the area of model transformation to illustrate the main concepts and results. We selected a small set of model elements, basic for all kinds of object-oriented models. It describes the abstract syntax, i.e. the structure of method signatures. These structures are naturally represented by node and edge attributed graphs where node attributes store e.g. names, while edge attributes are useful to keep e.g. the order of parameters belonging to one method. Attributed graph transformation is used to specify simple refactorings on this model part such as adding a parameter, exchanging two parameters, etc. Usually such refactorings are not always independent of each other. Within this paper we analyse the given refactoring rules concerning potential conflicts and report them as critical pairs.

Node and edge attributed graphs build the basic structures in the graph transformation environment AGG [7]. The attribution is done by Java objects and expressions. We use AGG to implement our running example and to compute all its critical pairs. In GenGED [13], a visual environment for the definition of visual languages, the internal structures are $AGSIG$-algebras for attributed graph structure signatures $AGSIG$ discussed above.

This paper is organized as follows. In section 2 we introduce node and edge attributed graphs and typing and present our first main result. Typed attributed graphs in the framework of adhesive HLR categories are discussed in section 3 together with our second main result. This allows to present the theory of typed attributed graph transformation in section 4 as an instance of the general theory in [8]. Finally we discuss related work and future perspectives in section 5.

For lack of space we can only present short proof ideas in this paper and refer to our technical report [14] for more detail.

2   Node and Edge Attributed Graphs and Typing

In this section we present our new notion of node and edge attributed graphs, which generalizes the concept of node attributed graphs in [5], where node attributes are modelled by edges from graph nodes to data nodes. The new concept is based on graphs, called E-graphs, which allows also edges from graph edges to data nodes in order to model edge attributes. An attributed graph $AG = (G, D)$ consists of an E-graph G and a data type D, where parts of the data of D are also vertices in G. This leads to the category **AGraphs** of attributed graphs and **AGraphs$_{ATG}$** of typed attributed graphs over an attributed type graph ATG. The main result in this section shows that **AGraphs$_{ATG}$** is isomorphic to a category **AGSIG(ATG)-Alg** of algebras over a suitable signature $AGSIG(ATG)$, which is in one-to-one correspondence with ATG.

In our notion of E-graphs we distinguish between two kinds of vertices, called graph and data vertices, and three different kinds of edges, according to the different roles they play for the representation and implementation of attributed graphs.

Definition 1 (E-graph). *An E-graph* $G = (V_1, V_2, E_1, E_2, E_3, (source_i, target_i)_{i=1,2,3})$ *consists of sets*

- V_1 *and* V_2 *called graph resp. data nodes,*
- E_1, E_2, E_3 *called graph, node attribute and edge attribute edges respectively,*

and source and target functions

- $source_1 : E_1 \rightarrow V_1$, $source_2 : E_2 \rightarrow V_1$, $source_3 : E_3 \rightarrow E_1$,
- $target_1 : E_1 \rightarrow V_1$, $target_2 : E_2 \rightarrow V_2$, $target_3 : E_3 \rightarrow V_2$.

An E-graph morphism $f : G_1 \rightarrow G_2$ *is a tuple* $(f_{V_1}, f_{V_2}, f_{E_1}, f_{E_2}, f_{E_3})$ *with* $f_{V_i} : G_{1,V_i} \rightarrow G_{2,V_i}$ *and* $f_{E_j} : G_{1,E_j} \rightarrow G_{2,E_j}$ *for* $i = 1, 2$, $j = 1, 2, 3$ *such that* f *commutes with all source and target functions.*
E-graphs combined with E-graph morphisms form the category **EGraphs**.

The following notions of attributed and typed attributed graphs are in the spirit of (node) attributed graphs of [5], where graphs are replaced by E-graphs in order to allow node and edge attribution. A data signature *DSIG* is an ordinary algebraic signature (see [15]).

Definition 2 (attributed graph). *Consider a data signature* $DSIG = (S_D, OP_D)$ *with attribute value sorts* $S'_D \subseteq S_D$. *An attributed graph* $AG = (G, D)$ *consists of an E-graph* G *together with a DSIG-algebra* D *such that* $\dot{\bigcup}_{s \in S'_D} D_s = G_{V_2}$.
An attributed graph morphism is a pair $f = (f_G, f_A)$ *with an E-graph morphism* f_G *and an algebra homomorphism* f_A *such that (1) is a pullback for all* $s \in S'_D$.

$$
\begin{array}{ccc}
D_{1.s} & \xrightarrow{f_{A,s}} & D_{2.s} \\
\Big\uparrow & (1) & \Big\uparrow \\
G_{1.V_2} & \xrightarrow{f_{G,V_2}} & G_{2.V_2}
\end{array}
$$

Attributed graphs and attributed graph morphisms form the category **AGraphs**.

Remark 1. The pullback property for the graph morphism is required for Thm. 1, otherwise the categories in this theorem would not be isomorphic.

Definition 3 (typed attributed graph). *An attributed type graph is an attributed graph* $ATG = (TG, Z)$ *where* Z *is the final DSIG-algebra.*
A typed attributed graph (AG, t) *over* ATG *consists of an attributed graph* AG *together with an attributed graph morphism* $t : AG \rightarrow ATG$. *A typed attributed graph morphism* $f : (AG_1, t_1) \rightarrow (AG_2, t_2)$ *is an attributed graph morphism* $f : AG_1 \rightarrow AG_2$ *such that* $t_2 \circ f = t_1$.
Typed attributed graphs over ATG *and typed attributed graph morphisms form the category* **AGraphs**$_{\mathbf{ATG}}$. *The class of all attributed type graphs* ATG *is denoted by* **ATG-Graphs**.

Example 1 (typed attributed graphs). Given suitable signatures *CHAR*, *STRING* and *NAT*, the data signature *DSIG* is defined by
$DSIG = CHAR + STRING + NAT +$
sorts: ParameterDirectionKind
opns: in, out, inout, return: $\rightarrow$ ParameterDirectionKind
and the set of all data sorts used for attribution is $S'_D = \{$String, Nat, Parameter-DirectionKind$\}$. Fig. 1 shows an attributed type graph $ATG = (TG, Z)$ for method signatures. It is an attributed graph where each data element is named after its corresponding sort, because the final *DSIG*-algebra Z has sorts $Z_s = \{s\}$ for all $s \in S_D$. Note that TG is an E-graph with edge attribute edge "order" from "parameter" to "Nat". An attributed graph AG typed over ATG is given in Fig. 2, where only those algebra elements are shown explicitly which are used

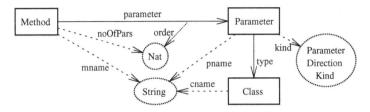

Fig. 1.

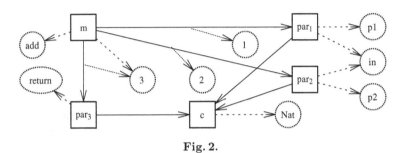

Fig. 2.

for attribution. The graph AG is typed over ATG by the attributed graph morphism $t : AG \to ATG$ defined on vertices by $t(\text{m}) = \text{Method}$, $t(\text{par}_1) = t(\text{par}_2)$ $= t(\text{par}_3) = \text{Parameter}$, $t(\text{c}) = \text{Class}$, $t(1) = t(2) = t(3) = \text{Nat}$, $t(\text{return}) = t(\text{in})$ $= \text{ParameterDirectionKind}$ and $t(\text{p1}) = t(\text{p2}) = t(\text{add}) = t(\text{Nat}) = \text{String}$. In AGG, a typed attributed graph like the one in Fig. 2 is depicted in a more compact notation like the graph in Fig. 3. Each node and edge inscription has two compartments. The upper compartment contains the type of a graph element, while the lower one holds its attributes. The attributes are ordered in a list, just for convenience. Nodes and edges are not explicitly named. While the formal concept of an attributed graph allows partial attribution in the sense that there is no edge from a graph node or edge to a data node, this is not possible in AGG. Thus, parameter par_3 has to be named by an empty string. Furthermore, the formal concept allows several outgoing attribute edges from one graph node or edge which is also not possible in AGG.   □

The category **AGraphs$_{\text{ATG}}$** introduced above is the basis for our theory of typed attributed graph transformation in this paper. In order to prove properties for **AGraphs$_{\text{ATG}}$**, however, it is easier to represent **AGraphs$_{\text{ATG}}$** as a category **AGSIG(ATG)-Alg** of classical algebras (see [15]) over a suitable signature $AGSIG(ATG)$. For this purpose we introduce the notion of general respectively well-structured attributed graph structure signatures $AGSIG$ where the well-structured case corresponds to attributed graph signatures in the LKW-approach [4]. The signature $AGSIG(ATG)$ becomes a special case of a well-structured $AGSIG$.

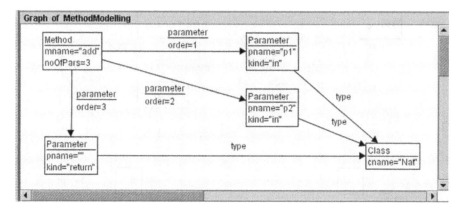

Fig. 3.

Definition 4 (attributed graph structure signature). *A graph structure signature* $GSIG = (S_G, OP_G)$ *is an algebraic signature with unary operations* $op : s \to s'$ *in* OP_G *only. An attributed graph structure signature* $AGSIG = (GSIG, DSIG)$ *consists of a graph structure signature* $GSIG$ *and a data signature* $DSIG = (S_D, OP_D)$ *with attribute value sorts* $S'_D \subseteq S_D$ *such that* $S'_D = S_D \cap S_G$ *and* $OP_D \cap OP_G = \emptyset$.
AGSIG is called well-structured if for each $op : s \to s'$ *in* OP_G *we have* $s \notin S_D$. *The category of all AGSIG-algebras and AGSIG-homomorphisms (see [15]) is denoted by* **AGSIG-Alg**.

Theorem 1 (Characterization of AGraphs$_{\mathbf{ATG}}$). *For each attributed type graph ATG there is a well-structured attributed graph structure signature* $AGSIG(ATG)$ *such that* **AGraphs$_{\mathbf{ATG}}$** *is isomorphic to the category* **AGSIG(ATG)-Alg** *of* $AGSIG(ATG)$*-algebras:*
AGraphs$_{\mathbf{ATG}}$ $\cong$ **AGSIG(ATG)-Alg**.

Construction. Given $ATG = (TG, Z)$ with final $DSIG$-algebra Z we have $TG_{V_2} = \dot{\bigcup}_{s \in S'_D} Z_s = S'_D$ and define $AGSIG(ATG) = (GSIG = (S_G, OP_G),$ $DSIG)$ with $S_G = S_V \dot{\cup} S_E$ and $S_V = TG_{V_1} \dot{\cup} TG_{V_2}$, $S_E = TG_{E_1} \dot{\cup} TG_{E_2} \dot{\cup}$ TG_{E_3} and $OP_G = \dot{\bigcup}_{e \in S_E} OP_e$ with $OP_e = \{src_e, tar_e\}$ defined by

- $src_e : e \to v(e)$ for $e \in TG_{E_1}$ with $v(e) = source_1^{TG}(e) \in TG_{V_1}$,
- $tar_e : e \to v'(e)$ for $e \in TG_{E_1}$ with $v'(e) = target_1^{TG}(e) \in TG_{V_1}$,
- src_e, tar_e for $e \in TG_{E_2}$ and $e \in TG_{E_3}$ defined analogously.

Proof idea. Based on the construction above we are able to construct a functor $F : \boldsymbol{AGraphs}_{\mathbf{ATG}} \to \boldsymbol{AGSIG(ATG)\text{-}Alg}$ and a corresponding inverse functor F^{-1}. □

3   Typed Attributed Graphs in the Framework of Adhesive HLR Categories

As pointed out in the introduction we are not going to develop the theory of typed attributed graph transformation directly. But we will show that it can be obtained as an instantiation of the theory of adhesive HLR systems, where this new concept (see [8]) is a combination of adhesive categories and grammars (see [9]) and HLR systems introduced in [10]. For this purpose we present in this section the general concept of adhesive HLR categories and we show that **AGSIG-Alg, AGSIG(ATG)-Alg** and especially the category **AGraphs$_{ATG}$** of typed attributed graphs are adhesive HLR categories for a suitable class M of morphisms. Moreover our categories satisfy some additional HLR conditions, which are required in the general theory of adhesive HLR systems (see [8]). This allows to apply the corresponding results to typed attributed graph transformation systems, which will be done in the next section.

We start with the new concept of adhesive HLR categories introduced in [8] in more detail.

Definition 5 (adhesive HLR category). *A category C with a morphism class M is called adhesive HLR category, if*

1. *M is a class of monomorphisms closed under isomorphisms and closed under composition ($f : A \to B \in M$, $g : B \to C \in M \Rightarrow g \circ f \in M$) and decomposition ($g \circ f \in M$, $g \in M \Rightarrow f \in M$),*
2. *C has pushouts and pullbacks along M-morphisms, i.e. if one of the given morphisms is in M, then also the opposite one is in M, and M-morphisms are closed under pushouts and pullbacks,*
3. *pushouts in C along M-morphisms are van Kampen (VK) squares, where a pushout (1) is called VK square, if for any commutative cube (2) with (1) in the bottom and pullbacks in the back faces we have: the top is pushout $\Leftrightarrow$ the front faces are pullbacks.*

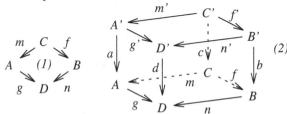

Important examples of adhesive HLR categories are the category (**Sets**, M_{inj}) of sets with class M_{inj} of all injective functions, the category (**Graph**, M_{inj}) of graphs with class M_{inj} of injective graph morphisms and different kinds of labelled and typed graphs (see [8]). Moreover all HLR1 and HLR2 categories presented in [10] are adhesive HLR categories. In the following we will show that also our categories **AGSIG-Alg, AGSIG(ATG)-Alg** and **AGraphs$_{ATG}$** presented in the previous section are adhesive HLR categories for the class M of all injective morphisms with isomorphic data type part, which is used for typed attributed graph transformation systems in the next section.

Definition 6 (class M for typed attributed graph transformation). *The distinguished class M is defined by $f \in M$ if*

1. *f_G is injective, f_A is isomorphism for $f = (f_G, f_A)$ in* **AGraphs$_{ATG}$** *and $AG = (G, A)$,*
2. *f_{GSIG} is injective, f_{DSIG} is isomorphism for f in* **AGSIG-Alg** *or* **AGSIG(ATG)-Alg** *and $AGSIG = (GSIG, DSIG)$.*

Remark 2. The corresponding categories (**AGraphs$_{ATG}$**, M), (**AGSIG-Alg**, M) and (**AGSIG(ATG)-Alg**, M) are adhesive HLR categories (see Thm. 2). For simplicity we use the same notation M in all three cases. For practical applications we assume that f_A and f_{DSIG} are identities.

This class M of morphisms is on the one hand the adequate class to define productions of typed attributed graph transformation systems (see Def. 7), on the other hand it allows to construct pushouts along M-morphisms componentwise. This is essential to verify the properties of adhesive HLR categories.

Lemma 1 (properties of pushouts and pullbacks in (AGSIG-Alg, M)).

1. *Given $m : C \to A$ and $f : C \to B$ with $m \in M$ then there is a pushout (1) in* **AGSIG-Alg** *with $n \in M$.*

$$
\begin{array}{ccc}
 & C & \\
m \swarrow & & \searrow f \\
A & (1) & B \\
g \searrow & & \swarrow n \\
 & D &
\end{array}
$$

   Moreover given (1) commutative with $m \in M$ then (1) is a pushout in **AGSIG-Alg** *iff (1) is a componentwise pushout in* **Sets**. *If $m \in M$ then also $n \in M$.*
2. *Given $g : A \to D$ and $n : B \to D$ then there is a pullback (1) in* **AGSIG-Alg**. *Moreover given (1) commutative then (1) is a pullback in* **AGSIG-Alg** *iff (1) is a componentwise pullback in* **Sets**. *If $n \in M$ then also $m \in M$.*

Proof. (see [14])

Remark 3. Since $AGSIG(ATG)$ is a special case of $AGSIG$, the lemma is also true for **AGSIG(ATG)-Alg**. It is well-known that **AGSIG-Alg** - as a category of algebras - has pushouts even if $m \notin M$, but in general such pushouts cannot be constructed componentwise.

Theorem 2 (adhesive HLR categories). *The category (**AGraphs$_{ATG}$**, M) of typed attributed graphs and also (**AGSIG-Alg**, M) and (**AGSIG(ATG)-Alg**, M) are adhesive HLR categories.*

Proof. It is sufficient to prove the properties for (**AGSIG-Alg**, M), because (**AGSIG(ATG)-Alg**, M) is a special case of (**AGSIG-Alg**, M) and (**AGraphs**$_\mathbf{ATG}$, M) $\cong$ **AGSIG(ATG)-Alg** by Thm. 1.

1. The class M given in Def. 6 is a subclass of monomorphisms, because monomorphisms in **AGSIG-Alg** are exactly the injective homomorphisms, and it is closed under composition and decomposition.
2. (**AGSIG-Alg**, M) has pushouts and pullbacks along M-morphisms and M-morphisms are closed under pushouts and pullbacks due to Lem. 1.
3. Pushouts along M-morphisms in **AGSIG-Alg** are VK squares because pushouts and pullbacks are constructed componentwise in **Sets** (see Lem. 1) and for (**Sets**, M_{inj}) pushouts along monomorphisms are VK squares as shown in [9].

4   Theory of Typed Attributed Graph Transformation

After the preparations in the previous sections we are now ready to present the basic notions and results for typed attributed graph transformation systems. In fact, we obtain all the results presented for adhesive HLR systems in [8] in our case, because we are able to show the corresponding HLR conditions for (**AGraphs**$_\mathbf{ATG}$, M). This category (**AGraphs**$_\mathbf{ATG}$, M) is fixed now for this section.

Definition 7 (typed attributed graph transformation system). *A typed attributed graph transformation system* $GTS = (DSIG, ATG, S, P)$ *based on (* **AGraphs**$_\mathbf{ATG}$, M *) consists of a data type signature* $DSIG$, *an attributed type graph* ATG, *a typed attributed graph* S, *called start graph, and a set* P *of productions, where*

1. *a production* $p = (L \xleftarrow{l} K \xrightarrow{r} R)$ *consists of typed attributed graphs* L, K *and* R *attributed over the term algebra* $T_{DSIG}(X)$ *with variables* X, *called left hand side* L, *gluing object* K *and right hand side* R *respectively, and morphisms* $l, r \in M$, *i.e.* l *and* r *are injective and isomorphisms on the data type* $T_{DSIG}(X)$,
2. *a direct transformation* $G \xRightarrow{p,m} H$ *via a production* p *and a morphism* $m : L \to G$, *called match, is given by the following diagram, called double pushout diagram, where (1) and (2) are pushouts in* **AGraphs**$_\mathbf{ATG}$,

$$
\begin{array}{ccccc}
L & \xleftarrow{\ l\ } & K & \xrightarrow{\ r\ } & R \\
{\scriptstyle m}\downarrow & (1) & \downarrow & (2) & \downarrow \\
G & \xleftarrow{\quad} & D & \xrightarrow{\quad} & H
\end{array}
$$

3. *a typed attributed graph transformation, short transformation, is a sequence* $G_0 \Rightarrow G_1 \Rightarrow ... \Rightarrow G_n$ *of direct transformations, written* $G_0 \overset{*}{\Rightarrow} G_n$,
4. *the language* $L(GTS)$ *is defined by* $L(GTS) = \{G \mid S \overset{*}{\Rightarrow} G\}$.

Remark 4. A typed attributed graph transformation system is an adhesive HLR system in the sense of [8] based on the adhesive HLR category ($\mathbf{AGraphs_{ATG}}$, M).

Example 2 (typed attributed graph transformation system). In the following, we start to define our typed attributed graph transformation system *MethodModelling* by giving the productions. All graphs occuring are attributed by term algebra $T_{DSIG}(X)$ with $DSIG$ being the data signature presented in Ex. 1 and $X = \bigcup_{s \in S'_D} X_s$, i.e. $X = X_{String} \cup X_{int} \cup X_{ParameterDirectionKind}$ with $X_{String} = \{m, p, ptype, P1, P2\}$, $X_{ParameterDirectionKind} = \{k\}$ and $X_{int} = \{n, x, y\}$. We present the productions in the form of AGG productions where we have the possibility to define a subset of variables of X as input parameters. That means a partial match is fixed by the user before the proper matching procedure starts. Each production is given by its name followed by the left and the right-hand side as well as a partial mapping from left to right given by numbers. From this partial mapping the gluing graph K can be deduced being the domain of the mapping. Parameters are m, p, k, ptype, x and y. We use a graph notation similarly to Fig. 3.

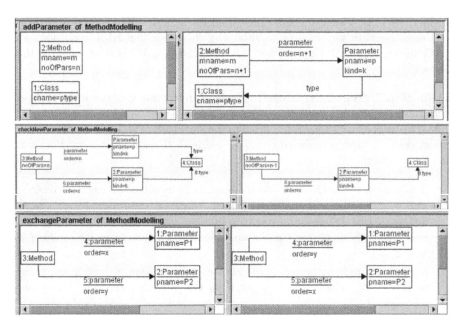

AGG productions are restricted concerning the attribution of the left-hand sides. To avoid the computation of most general unifiers of two general terms, nodes and edges of left-hand sides are allowed to be attributed only by constants and variables. This restriction is not a real one, since attribute conditions may be used. A term in the left-hand side is equivalent to a new variable and a new attribute condition stating the equivalence of the term and this new variable.

Productions $addMethod$ and $addClass$ with empty left-hand side and a single method respectively class on the right-hand side are not shown. Together with production $addParameter$ they are necessary to build up method signatures. A new parameter is inserted as last one in the parameter list. Production $checkNewParameter$ checks if a newly inserted parameter is already in the list. In this case it is removed. Production $exchangeParameter$ is useful for changing the order of parameters in the list.

The start graph S is empty, i.e. $S = \emptyset$, while data signature $DSIG$ and type graph T have already been given in Ex. 1. Summarizing, the typed attributed graph transformation system is given by $MethodModelling = (DSIG, ATG, S, P)$ with $P = \{addMethod, addClass, addParameter, exchangeParameter, checkNewParameter\}$. $\qquad\square$

In the following we show that the basic results known in the classical theory of graph transformation in [1] and in the theory of HLR systems in [10] are also valid for typed attributed graph transformation systems.

The Local Church-Rosser Theorem states that direct transformations $G \overset{p_1,m_1}{\Longrightarrow} H_1$ and $G \overset{p_2,m_2}{\Longrightarrow} H_2$ can be extended by direct transformations $H_1 \overset{p_2,m_2'}{\Longrightarrow} X$ and $H_2 \overset{p_1,m_1'}{\Longrightarrow} X$ leading to the same X, provided that they are parallel independent. Parallel independence means that the matches m_1 and m_2 overlap only in common gluing items, i.e. $m_1(L_1) \cap m_2(L_2) \subseteq m_1(l_1(K_1)) \cap m_2(l_2(K_2))$.

The Parallelism Theorem states that in the case of parallel independence we can apply the parallel production $p_1 + p_2 = (L_1 + L_2 \overset{l_1+l_2}{\Longleftarrow} K_1 + K_2 \overset{r_1+r_2}{\Longrightarrow} R_1 + R_2)$ in one step $G \overset{p_1+p_2,m}{\Longrightarrow} X$ from G to X. Vice versa each such direct parallel derivation can be sequentialized in any order leading to two sequential independent sequences $G \overset{p_1,m_1}{\Longrightarrow} H_1 \overset{p_2,m_2'}{\Longrightarrow} X$ and $G \overset{p_2,m_2}{\Longrightarrow} H_2 \overset{p_1,m_1'}{\Longrightarrow} X$.

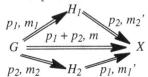

The case of general sequences, which may be sequentially dependend, is handled by the Concurrency Theorem. Roughly spoken, for each sequence $G \overset{p_1,m_1}{\Longrightarrow} H_1 \overset{p_2,m_2'}{\Longrightarrow} X$ there is a production $p_1 * p_2$, called concurrent production, which allows to construct a direct transformation $G \overset{p_1*p_2}{\Longrightarrow} X$ and vice versa, leading, however, only to one sequentialization.

Theorem 3 (Local Church-Rosser, Parallelism and Concurrency Theorem). *The Local Church-Rosser Theorems I and II, the Parallelism Theorem and the Concurrency Theorem as stated in [10] are valid for each graph transformation system based on (* **AGraphs**$_{\mathbf{ATG}}$*, M).*

Proof idea. The Local Church-Rosser, Parallelism and Concurrency Theorem are verified for HLR2 categories in [10] and they are shown for adhesive HLR systems in [8], where only the Parallelism Theorem requires in addition the existence of binary coproducts compatible with M. Compatibility with M means f, $g \in M$ implies $f + g \in M$. In Thm. 2 we have shown that $(\mathbf{AGraphs_{ATG}}, M)$ is an adhesive HLR category. Binary coproducts compatible with M can be constructed already in $\mathbf{AGSIG\text{-}Alg}$ with well-structured $AGSIG$ and transfered to $\mathbf{AGraphs_{ATG}}$ by Thm. 1. In $(\mathbf{AGSIG\text{-}Alg}, M)$ binary coproducts can be constructed separately for the $DSIG$-part and componentwise in $\mathbf{Sets}$ for all sorts $s \in S_G \backslash S_D$, which implies compatibility with M. If $AGSIG$ is not well-structured we still have binary coproducts in $\mathbf{AGSIG\text{-}Alg}$ - as in any category of algebras - but they may not be compatible with M.   □

The next basic result in the classical theory of graph transformation systems is the Embedding Theorem (see [1]) in the framework of adhesive HLR systems. The main idea of the Embedding Theorem according to [1] is to show under which conditions a transformation $t : G_0 \overset{*}{\Rightarrow} G_n$ can be extended to a transformation $t' : G_0' \overset{*}{\Rightarrow} G_n'$ for a given "embedding" morphism $k_0 : G_0 \to G_0'$. In the case of typed attributed graph transformation we consider the following class M' of "graph part embeddings": M' consists of all morphisms k_0 where the E-graph part of k_0 is injective except of data nodes (see Def. 1 - 3). This means, that the algebra part of k_0 is not restricted to be injective.

Similar to the graph case it is also possible in the case of typed attributed graphs to construct a boundary B and context C leading to a pushout (1) over k_0 in Fig. 4 with $b_0 \in M$, i.e. G_0' is the gluing of g_0 and context C along the boundary B. This boundary-context pushout (1) over k_0 turns out to be an initial pushout over k_0 in the sense of [8].

Now the morphism $k_0 \in M'$ is called consistent with respect to the transformation $t : G_0 \overset{*}{\Rightarrow} G_n$, if the boundary B is "preserved" by t leading to a morphism $b_n : B \to G_n \in M$ in Fig. 4. (For a formal notion of consistency see [8].)

The following Embedding and Extension Theorem shows that consistency is necessary and sufficient in order to extend $t : G_0 \overset{*}{\Rightarrow} G_n$ for $k_0 : G_0 \to G_0'$ to $t' : G_0' \overset{*}{\Rightarrow} G_n'$.

Theorem 4 (Embedding and Extension Theorem). *Let GTS be a typed attributed graph transformation system based on $(\mathbf{AGraphs_{ATG}}, M)$ and M' the class of all graph part embeddings defined above. Given a transformation $t : G_0 \overset{*}{\Rightarrow} G_n$ and a morphism $k_0 : G_0 \to G_0' \in M'$ with boundary-context pushout (1) over k_0 we have: The transformation t can be extended to a transformation $t' : G_0' \overset{*}{\Rightarrow} G_n'$ with morphism $k_n : G_n \to G_n' \in M'$ leading to diagram (2), called extension diagram, and a boundary-context pushout (3) over k_n if and only if the morphism k_0 is consistent with respect to t.*

Proof idea. This theorem follows from the Embedding and Extension Theorems in [8] shown for adhesive HLR systems over an adhesive HLR category $(\mathbf{C}, M)$. It requires initial pushouts over M'-morphisms for some class M', which is closed under pushouts and pullbacks along M-morphisms. By Thm. 2 we know that $(\mathbf{AGraphs_{ATG}}, M)$ is an adhesive HLR category. In addition it can be shown

$$B \xrightarrow{b_0} G_0 \overset{t}{\underset{*}{\Longrightarrow}} G_n \xleftarrow{b_n} B$$

$$\downarrow c \ (1) \quad \downarrow k_0 \ (2) \quad \downarrow k_n \ (3) \quad \downarrow c$$

$$C \longrightarrow G_0' \underset{t'}{\overset{*}{\Longrightarrow}} G_n' \longleftarrow C$$

Fig. 4.

that for each $k_0 \in M'$, where M' is the class of all graph part embeddings, there is a boundary-context pushout (1) over k_0, which is already an initial pushout over k_0 in the sense of [8]. Moreover it can be shown that M' is closed under pushouts and pullbacks along M-morphisms. □

The Embedding and Extension Theorems are used in [8] to show the Local Confluence Theorem, also known as critical pair lemma, in the framework of adhesive HLR systems, where in addition to initial pushouts also the existence of an E'-M' pair factorization is used.

Definition 8 (E'-M' pair factorization). *Given a class E' of morphism pairs (e_1, e_2) with the same codomain and M' the class of all graph part embeddings defined above. We say that a typed attributed graph transformation system based on ($\mathbf{AGraphs_{ATG}}$, M) has E'-M' pair factorization, if for each pair of matches $f_1 : L_1 \to G$, $f_2 : L_2 \to G$ there is a pair $e_1 : L_1 \to K$, $e_2 : L_2 \to K$ with $(e_1, e_2) \in E'$ and a morphism $m : K \to G$ with $m \in M'$ such that $m \circ e_1 = f_1$ and $m \circ e_2 = f_2$.*

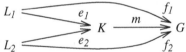

Remark 5. For simplicity we have fixed M' to be the class of all graph part embeddings, which implies that M' is closed under pushouts and pullbacks along M-morphisms as required for E'-M' pair factorization in [8] with general class M'.

Example 3. 1. Let E' be the class of jointly surjective morphisms in $\mathbf{AGraphs_{ATG}}$ with same codomain. Given f_1 and f_2 we obtain an induced morphism $f_{12} : L_1 + L_2 \to G$ with coproduct injections $i_1 : L_1 \to L_1 + L_2$ and $i_2 : L_2 \to L_1 + L_2$. Now let $f_{12} = m \circ e$ an epi-mono factorization of f_{12} leading to $e_1 = e \circ i_1$ and $e_2 = e \circ i_2$ with $(e_1, e_2) \in E'$. In this case $m : K \to G$ is injective and the data type part of K is a quotient term algebra $T_\Sigma(X_1 + X_2)|_\equiv$, where $T_\Sigma(X_1)$ and $T_\Sigma(X_2)$ are the algebras of L_1 and L_2 respectively. This corresponds to one possible choice of congruence $\equiv$ considered in [5].

2. In order to obtain a minimal number of critical pairs it is essential to consider also the case, where $\equiv$ is the trivial congruence with $T_\Sigma(X) \cong T_\Sigma(X)|_\equiv$. In fact, a most general unifier construction $\sigma_n : X \to T_\Sigma(X)$ considered in [5] leads to a different E'-M' pair factorization of f_1, f_2 with $(e_1, e_2) \in E'$,

$m \in M'$, $e_1 : L_1 \to K$, $e_2 : L_2 \to K$ and $m : K \to G$, where (e_1, e_2) is jointly surjective for non-data nodes, the data type part of K is $T_\Sigma(X)$ and m is injective for non-data nodes, where non-data nodes are all nodes and edges, which are not data nodes (see Def. 1).

Dependent on the choice of an E'-M' pair factorization we are now able to define critical pairs and strict confluence.

Definition 9 (critical pair and strict confluence). *Given an E'-M' pair factorization, a critical pair is a pair of non-parallel independent direct transformations $P_1 \stackrel{p_1, o_1}{\Longleftarrow} K \stackrel{p_2, o_2}{\Longrightarrow} P_2$ such that $(o_1, o_2) \in E'$ for the corresponding matches o_1 and o_2. The critical pair is called strictly confluent, if we have*

1. *Confluence: There are transformations $P_1 \stackrel{*}{\Rightarrow} K'$ and $P_2 \stackrel{*}{\Rightarrow} K'$.*
2. *Strictness: Let N be the common subobject of K, which is preserved by the direct transformations $K \Longrightarrow P_1$ and $K \Longrightarrow P_2$ of the critical pair, then N is also preserved by $P_1 \stackrel{*}{\Rightarrow} K'$ and $P_2 \stackrel{*}{\Rightarrow} K'$, such that the restrictions of the transformatios $K \Rightarrow P_1 \stackrel{*}{\Rightarrow} K'$ and $K \Rightarrow P_2 \stackrel{*}{\Rightarrow} K'$ yield the same morphism $N \to K'$. (See [8] for a more formal version of strict confluence.)*

Theorem 5 (Local Confluence Theorem - Critical Pair Lemma). *Given a typed attributed graph transformation system GTS based on ($\mathbf{AGraphs_{ATG}}$, M), and an E'-M' pair factorization, where M' is the class of all graph part embeddings, then GTS is locally confluent, if all its critical pairs are strictly confluent.*
Local confluence of GTS means that for each pair of direct transformations $H_1 \Leftarrow G \Rightarrow H_2$ there are transformations $H_1 \stackrel{*}{\Rightarrow} X$ and $H_2 \stackrel{*}{\Rightarrow} X$.

Proof idea. This theorem follows from the Local Confluence Theorem in [8] shown for adhesive HLR systems over $(\mathbf{C}, M)$. It requires initial pushouts over M'-morphisms for a class M' "compatible" with M. The proof in [8] is based on completeness of critical pairs shown by using an M-M' pushout pullback decomposition property. In our case M' is the class of all graph part embeddings, which can be shown to satisfy this property, which implies that M' is "compatible" with M. In the proof idea of Thm. 4 we have discussed already how to verify the remaining properties which are required in the general framework of [8]. □

Example 4 (critical pairs). Considering our typed attributed graph transformation system $MethodModelling$ (see Ex. 2) we now analyse its critical pairs. This analysis is supported by AGG. Due to the restriction of attributes in the left-hand side of a production to constants and variables we restrict ourselves to very simple congruence relations on terms of overlapping graphs. Variables may be identified with constants or with other variables. All other congruences are the identities. In the following, the AGG user interface for critical pair analysis is shown. It presents an overview on all critical pairs in form of a table containing all possible pairs of $MethodModelling$ at the bottom of the figure. Applying for example first $addParameter$, $exchangeParameter$ or $checkNewParameter$

and second *checkNewParameter* leads to critical pairs. We consider the two critical pairs of productions *addParameter* and *checkNewParameter* closer. On the top of the figure the left-hand sides of both productions are displayed. The center shows the two overlapping graphs which lead to critical pairs. Both overlapping graphs show the conflict on attribute 'noOfPars' which is increased by production *addParameter* but decreased by *checkNewParameter*. In the left graph both classes are identified, i.e. the new parameter would be of the same type as the the two already existing ones which are equal, while the right graph shows two classes.

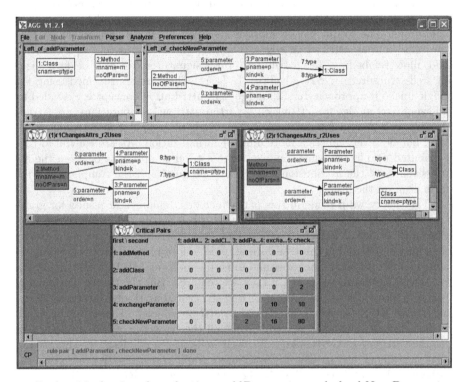

Both critical pairs of productions *addParameter* and *checkNewParameter* are strictly confluent. Applying first *addParameter(mname, "n1", cname, "in")* and then *exchangeParameter(n, n + 1)* and *checkNewParameter()* the resulting graph is isomorphic to applying first *checkNewParameter()* and then *addParameter(mname, "n1", cname, "in")*. The common subgraph N of the conflicting transformations is the result graph applying *checkParameter* to the overlapping graph.

5   Related Work and Conclusions

A variety of approaches to attributed graph transformation [4–6, 11, 12, 16] has already been developed where attributed graphs consist of a graph part and a data part. These approaches are compared with ours in the following. Moreover, general forms of algebraic structures and their transformation have been considered in e.g. [17–19].

A simple form of attributed graphs, where only nodes are attributed, is chosen in [16] and [5]. In [16], nodes are directly mapped to data values which makes relabelling difficult. In [5], nodes are attributed by special edges which are deleted and newly created in order to change attributes. We have defined attributed graphs in the spirit of the later notion of node-attributed graphs, but extended the notion also to edge attribution. In [4,11], attributed graph structures are formulated by algebras of a special signature where the graph part and the data part are separate from each other and only connected by attribution operations. The graph part comprises so-called attribute carriers which play the same role as attribution edges. They assign attributes to graph elements and allow relabelling. I.e. attributed graph signatures can also be formalized by well-structured $AGSIG$-algebras. In [4,11], they are transformed using the single-pushout approach. In [12] and [6] this attribution concept is extended by allowing partial algebras. Using e.g. partial attribution operations, no carriers or edges are needed for attribution. This leads to a slightly more compact notion, but however, causes a more difficult formalization. We are convinced that attributed graphs as defined in Sec. 2 are a good compromise between the expressiveness of the attribution concept on one hand and the complexity of the formalism on the other.

The theory we provide in this paper includes fundamental results for graph transformation which are now available for typed attributed graph transformation in the sense of sections 2 and 4. The general strategy to extend the theory for typed attributed graph transformation is to formulate the corresponding results in adhesive HLR categories and to verify additional HLR properties for the category (**AGraphs_ATG**, M), if they are required. But the theory presented in Sec. 4 is also valid in the context of well-structured attributed graph structure signatures $AGSIG$, which correspond to attributed graph signatures in [4,11]. In fact, the HLR conditions required for all the results in [8] are already valid for the category (**AGSIG-Alg**, M) with well-structured $AGSIG$. This means, that our theory is also true for attributed graph transformation based on the adhesive HLR category (**AGSIG-Alg**, M) with general $AGSIG$ for Thm. 3 and well-structured $AGSIG$ for Thm. 4 and 5.

Future work is needed to obtain corresponding results for extensions of typed attributed graph transformation by further concepts such as application conditions for productions or type graphs with inheritance.

References

1. Ehrig, H.: Introduction to the Algebraic Theory of Graph Grammars (A Survey). In: Graph Grammars and their Application to Computer Science and Biology. Volume 73 of LNCS. Springer (1979) 1–69
2. Ehrig, H., Engels, G., Kreowski, H.J., Rozenberg, G., eds.: Handbook of Graph Grammars and Computing by Graph Transformation, Volume 2: Applications, Languages and Tools. World Scientific (1999)

3. Ehrig, H., Kreowski, H.J., Montanari, U., Rozenberg, G., eds.: Handbook of Graph Grammars and Computing by Graph Transformation. Vol 3: Concurrency, Parallelism and Distribution. World Scientific (1999)
4. Löwe, M., Korff, M., Wagner, A.: An Algebraic Framework for the Transformation of Attributed Graphs. In: Term Graph Rewriting: Theory and Practice. John Wiley and Sons Ltd. (1993) 185–199
5. Heckel, R., Küster, J., Taentzer, G.: Confluence of Typed Attributed Graph Transformation with Constraints. In: Proc. ICGT 2002. Volume 2505 of LNCS., Springer (2002) 161–176
6. Berthold, M., Fischer, I., Koch, M.: Attributed Graph Transformation with Partial Attribution. Technical Report 2000-2 (2000)
7. Ermel, C., Rudolf, M., Taentzer, G.: The AGG-Approach: Language and Tool Environment. In Ehrig, H., Engels, G., Kreowski, H.J., Rozenberg, G., eds.: Handbook of Graph Grammars and Computing by Graph Transformation, Volume 2, World Scientific (1999) 551–603
8. Ehrig, H., Habel, A., Padberg, J., Prange, U.: Adhesive High-Level Replacement Categories and Systems. In: Proc. ICGT 2004. LNCS, Springer (2004) (this volume).
9. Lack, S., Sobociński, P.: Adhesive Categories. In: Proc. FOSSACS 2004. Volume 2987 of LNCS., Springer (2004) 273–288
10. Ehrig, H., Habel, A., Kreowski, H.J., Parisi-Presicce, F.: Parallelism and Concurrency in High-Level Replacement Systems. Math. Struct. in Comp. Science 1 (1991) 361–404
11. Claßen, I., Löwe, M.: Scheme Evolution in Object Oriented Models: A Graph Transformation Approach. In: Proc. Workshop on Formal Methods at the ISCE'95, Seattle (U.S.A.). (1995)
12. Fischer, I., Koch, M., Taentzer, G., Volle, V.: Distributed Graph Transformation with Application to Visual Design of Distributed Systems. In Ehrig, H., Kreowski, H.J., Montanari, U., Rozenberg, G., eds.: Handbook of Graph Grammars and Computing by Graph Transformation, Volume 3, World Scientific (1999) 269–340
13. Bardohl, R.: A Visual Environment for Visual Languages. Science of Computer Programming (SCP) **44** (2002) 181–203
14. Ehrig, H., Prange, U., Taentzer, G.: Fundamental Theory for Typed Attributed Graph Transformation: Long Version. Technical Report TU Berlin. (2004)
15. Ehrig, H., Mahr, B.: Fundamentals of Algebraic Specification 1: Equations and Initial Semantics. Volume 6 of EATCS Monographs on TCS. Springer, Berlin (1985)
16. Schied, G.: Über Graphgrammatiken, eine Spezifikationsmethode für Programmiersprachen und verteilte Regelsysteme. Arbeitsber. des Inst. für math. Maschinen und Datenverarbeitung, PhD Thesis, University of Erlangen (1992)
17. Wagner, A.: A Formal Object Specification Technique Using Rule-Based Transformation of Partial Algebras. PhD thesis, TU Berlin (1997)
18. Llabres, M., Rossello, F.: Pushout Complements for Arbitrary Partial Algebras. In Ehrig, H., Engels, G., Kreowski, H.J., Rozenberg, G., eds.: Theory and Applications of Graph Transformation. Volume 1764., Springer (2000) 131–144
19. Große-Rhode, M.: Semantic Integration of Heterogeneuos Software Specifications. EATCS Monographs on Theoretical Computer Science. Springer, Berlin (2004)

Parallel Independence
in Hierarchical Graph Transformation

Annegret Habel[1] and Berthold Hoffmann[2]

[1] Carl-v.-Ossietzky-Universität Oldenburg, Germany
`habel@informatik.uni-oldenburg.de`
[2] Universität Bremen, Germany
`hof@informatik.uni-bremen.de`

Abstract. Hierarchical graph transformation as defined in [1, 2] extends double-pushout graph transformation in the spirit of term rewriting: Graphs are provided with hierarchical structure, and transformation rules are equipped with graph variables. In this paper we analyze conditions under which diverging transformation steps $H \Leftarrow G \Rightarrow H'$ can be joined by subsequent transformation sequences $H \overset{*}{\Rightarrow} M \overset{*}{\Leftarrow} H'$. Conditions for joinability have been found for graph transformation (called parallel independence) and for term rewriting (known as non-critical overlap). Both conditions carry over to hierarchical graph transformation. Moreover, the more general structure of hierarchical graphs and of transformation rules leads to a refined condition, termed fragmented parallel independence, which subsumes both parallel independence and non-critical overlap as special cases.

1 Introduction

Graph transformation combines two notions that are ubiquitous in computer science (and beyond). Graphs are frequently used as visual models of structured data that consists of entities with relationships between them. Rules allow the modification of data to be specified in an axiomatic way. The book [3] gives a general survey on graph transformation, and [4, 5] describe several application areas.

When graph transformation is used to program or specify systems, it should be possible to group large graphs in a hierarchical fashion so that they stay comprehensible. Many notions of hierarchical graphs have been proposed, and several ways of transforming hierarchical graphs have been studied in the literature. See [6] for a rather general definition. This paper is based on [1], where double-pushout graph transformation [7] has been extended to a strict kind of hierarchical graphs where the hierarchy is a tree, and edges may not connect nodes in different parts of the hierarchy. This is adequate for programming; applications like software modeling may call for a looser notion of hierarchical graphs, e.g., the one in [8]. In [2], transformation rules have been extended to rules with variables [9]. This is done in the spirit of term rewriting [10], a rule-based model for computing with expressions (trees): Rules are equipped

H. Ehrig et al. (Eds.): ICGT 2004, LNCS 3256, pp. 178–193, 2004.
© Springer-Verlag Berlin Heidelberg 2004

with variables that may be instantiated by graphs, so that a single rule application may compare, delete, or copy subgraphs of arbitrary size. Hierarchical graph transformation with variables is the computational model of DIAPLAN, a language for programming with graphs and diagrams that is currently being designed [11].

In general, graph transformation is nondeterministic like other rule-based systems. Several rules may compete for being applied, at different places in a given graph. It is thus important to study under which conditions the result of a transformation sequence is independent of the order in which competing rules are applied. For term rewriting, parallel independence holds if steps have a non-critical overlap [10], and for double pushout graph transformation, the slightly stronger property of direct joinability holds if steps are parallelly independent [12]. These results carry over to hierarchical graph transformation. More precisely, we shall prove that they are special cases of the Fragmented Parallel Independence Theorem.

The paper is organized as follows. Section 2 collects basic notions of graphs and graph morphisms. In Sect. 3, we recall the basic notions of hierarchical graphs and hierarchical graph transformation, and show the relationship to substitution-based graph transformation. In Sect. 4, we discuss how independence results from graph transformation and term rewriting carry over to hierarchical graph transformation, and establish the Fragmented Parallel Independence Theorem. In Sect. 5, we conclude with a brief summary and with some topics for future work.

Acknowledgments

We thank the anonymous referees for their constructive remarks, and for their confidence.

2   Preliminaries

In the following, we recall standard notions of graphs and graph morphisms [7]. As in [9], we distinguish a subset of the label alphabet as variables. Variable edges are placeholders that can be substituted by graphs.

Let $\mathcal{C}$ be a alphabet with a subset $X \subseteq \mathcal{C}$ of *variables* where every symbol l comes with a *rank* $rank(l) \geq 0$.

A *graph* (with variables in X) is a system $G = \langle V_G, E_G, att_G, lab_G \rangle$ with finite sets V_G and E_G of *nodes* (or *vertices*) and *edges*, an *attachment function* $att_G \colon E_G \to V_G^*$ [1], and a *labeling function* $lab_G \colon E_G \to \mathcal{C}$ such that the attachment $att_G(e)$ of every edge e consists of $rank(lab_G(e))$ nodes (that need not be distinct).

---

[1] For a set A, A^* denotes the set of all sequences over A. The empty sequence is denoted by ε. For a mapping $f \colon A \to B$, $f^* \colon A^* \to B^*$ denotes the extension of f with $f^*(\varepsilon) = \varepsilon$ and $f^*(a_1 \ldots a_k) = f(a_1) \ldots f(a_k)$ for $a_1, \ldots, a_k \in V$.

A *graph morphism* $g: G \rightarrow G'$ between two graph G and G' consists of two functions $g_V: V_G \rightarrow V_{G'}$ and $g_E: E_G \rightarrow E_{G'}$ that preserve labels and attachments, that is, $lab_{G'} \circ g_V = lab_G$ and $att_{G'} \circ g_E = g_V^* \circ att_G$. It is *injective* (*surjective*) if g_V and g_E are injective (surjective), and an *isomorphism* if it is both injective and surjective. It is an *inclusion* if g_V and g_E are inclusions.

3   Hierarchical Graph Transformation

In this section, we define hierarchical graphs, hierarchical graph morphisms, and hierarchical graph transformation. For lack of space, we just recall the concepts devised in [1, 2]; a broader discussion of these concepts, and further references to the scientific literature can be found in these papers. At the end of the section, we relate our definitions to their origins, namely double-pushout graph transformation [7], and substitutive graph transformation [9].

A graph becomes hierarchical if its edges contain graphs, the edges of which may contain graphs again, in a nested fashion. Variables may not contain graphs; they are used as placeholders for graphs in transformation rules.

Definition 1 (Hierarchical Graph). The set $\mathcal{H}(X)$ of *hierarchical graphs* (with variables in X) consists of triples $H = \langle \widehat{H}, F_H, cts_H \rangle$ where $\widehat{H}$ is a graph (with variables in X), $F_H \subseteq E_{\widehat{H}}$ is a set of *frame edges* (or just *frames*) that are labeled in $\mathcal{C} \setminus X$, and $cts_H: F_H \rightarrow \mathcal{H}(X)$ is a *contents function* mapping frames to hierarchical graphs.

A hierarchical graph I is a *part* of H if $I = H$, or if I is a part of $cts_H(f)$ for some frame $f \in F_H$. An X-labeled edge in some part I of H is called a *variable edge* of H.

The *skeleton* of a hierarchical graph H is obtained by removing all variable edges from all parts of H; it is denoted by $\underline{H}$. $Var(H)$ denotes the set of *variables* occurring in the parts of H. A hierarchical graph H is *variable-free* if $H = \underline{H}$.

Example 1 (Control flow graphs). In simple control flow diagrams of sequential imperative programs, execution *states* are represented by nodes (depicted as small circles), and execution *steps* are represented by edges: statements (drawn as boxes) are labeled by assignments, and branches (drawn as diamonds) are labeled by conditions. Each step connects one predecessor state to one successor state (for assignments), or to two (for branches, distinguished by "⊕" and "⊖", respectively). Hierarchies are used for representing procedure calls (drawn like assignments, but with doubled vertical lines). They contain control flow graphs of the procedures' bodies. Since procedures may call other procedures, control flow graphs may be nested to arbitrary depth. In Fig. 7 below we show six hierarchical control flow graphs.

Definition 2 (Hierarchical Graph Morphism). A *top hierarchical graph morphism* (*top morphism*, for short) $h: H \rightarrow H'$ between two hierarchical graphs H and H' is a pair $h = \langle \widehat{h}, M \rangle$ where $\widehat{h}: \widehat{H} \rightarrow \widehat{H'}$ is a graph morphism such that

$\widehat{h}(F_H) \subseteq F_{H'}$, and $M = (h_f: cts_H(f) \to cts_{H'}(\widehat{h}(f)))_{f \in F_H}$ is a family of top morphisms between the contents of the frames. A *hierarchical graph morphism* $h: H \to H'$ is a top morphism $h': H \to H''$ between H and some part H'' of H'. A hierarchical graph morphism h is *injective* if the graph morphism $\widehat{h}$ and all morphisms in M are injective; it is an *inclusion* if $\widehat{h}$ and all morphisms in M are inclusions. A top morphism h is *surjective* if $\widehat{h}$ and all morphisms in M are surjective. A top morphism $h: H \to H'$ is an *isomorphism* if it is injective and surjective; then we call H and H' *isomorphic*, and write $H \cong H'$.

Definition 3 (Substitution). A *substitution pair* $x \mapsto \langle H, p \rangle$ consists of a variable $x \in X$ and of a hierarchical graph H with a sequence $p \in V_H^*$ of $rank(x)$ mutually distinct *points*. A finite set

$$\sigma = \{x_1 \mapsto \langle H_1, p_1 \rangle, \ldots, x_n \mapsto \langle H_n, p_n \rangle\}$$

of substitution pairs is a *substitution* if the variables are pairwise distinct. Then $\mathrm{Dom}(\sigma) = \{x_1, \ldots, x_n\}$ is the *domain* of σ.

Let I be a hierarchical graph where the top graph $\widehat{I'}$ of some part I' contains an edge e labeled with x. Then the application of a substitution pair $x \mapsto \langle H, p \rangle$ to e is obtained by replacing I' with a hierarchical graph constructed as follows: Unite $\widehat{I'}$ disjointly with $\widehat{H}$, remove e, identify every point in p with the corresponding attached node in $att_{\widehat{I'}}(e)$, and preserve the contents of the frames. The *instantiation* of a hierarchical graph I according to a substitution σ is obtained by applying all substitution pairs in σ to all edges with a variable label in $\mathrm{Dom}(\sigma)$ simultaneously, and is denoted by $I\sigma$.

Definition 4 (Rule). A *rule* $p = \langle L \leftarrow K \to R \rangle$ consists of two top morphisms with a common domain K. We assume that $K \to L$ and $K \to R$ are inclusions and that $\mathrm{Var}(L) \supseteq \mathrm{Var}(R)$.

The *instance* of a rule p for a substitution σ is defined as $p\sigma = \langle L\sigma \leftarrow \underline{K} \to R\sigma \rangle$, and the *skeleton* of p is given by $\underline{p} = \langle \underline{L} \leftarrow \underline{K} \to \underline{R} \rangle$. A rule p is *variable-free* if $p = \underline{p}$. (We explain in App. A why we take the skeleton $\underline{K}$ of the interface in the instance $p\sigma$, instead of $K\sigma$.)

Definition 5 (Hierarchical Graph Transformation). Consider hierarchical graphs G and H and a rule $p = \langle L \leftarrow K \to R \rangle$. Then G *directly derives* H *through* p if there is a double-pushout

$$
\begin{array}{ccccc}
L\sigma & \longleftarrow & K & \longrightarrow & R\sigma \\
\downarrow{\scriptstyle g} & (1) & \downarrow & (2) & \downarrow \\
G & \longleftarrow & D & \longrightarrow & H
\end{array}
$$

for some substitution σ so that the vertical morphisms are injective. We write $G \Rightarrow_{p,\sigma,g} H$ or $G \Rightarrow_p H$ and call this a *direct derivation* where g is the hierarchical graph morphism $g: L \to G$ defining the *occurrence* of p in G.

A direct derivation $G \Rightarrow_{p,\sigma,g} H$ exists if and only if the occurrence $g: L \to G$ above satisfies the following *hierarchical dangling condition*: *(i)* The graph morphism $\hat{g}$ satisfies the dangling condition for graph morphisms (see [7]), *(ii)* all morphisms in M satisfy the hierarchical dangling condition, and *(iii)* for all deleted frames $f \in F_L \setminus F_I$, the hierarchical morphism $g_f: cts_L(f) \to cts_G(g(f))$ is bijective.

Given a hierarchical graph G, a rule p as above, a substitution σ with $\mathrm{Dom}(\sigma) = \mathrm{Var}(L)$, and an occurrence g satisfying the hierarchical dangling condition, a direct derivation is uniquely determined by the following steps: (1) Remove $g(L\sigma - \underline{K})$ from the part G' of G where $g: L \to G'$ is top, yielding a hierarchical graph D, a top morphism $d: \underline{K} \to D'$ which is the restriction of g, and the inclusion $D' \to G'$. (2) Add $R\sigma$ disjointly to D' and identify the corresponding nodes and edges in $\underline{K}$ and $d(\underline{K})$, yielding a hierarchical graph H' and top morphisms $D \to H$ and $R\sigma \to H$. (3) Obtain H by replacing H' for G' in G.

Remark 1 (Relation to Adhesive High-Level Replacement). Hierarchical graphs without variables and injective hierarchical graph morphisms form a category HiGraphs. We conjecture that the category $\langle$HiGraphs, $\mathcal{M}\rangle$ of hierarchical graphs without variables with the class $\mathcal{M}$ of all injective top morphisms forms an adhesive HLR category. In this case, application of the general results in [13] would yield the Local Church-Rosser Theorems, the Embedding, Extension, and Local Confluence Theorem for $\langle$HiGraphs, $\mathcal{M}\rangle$. The statement no longer holds for the category $\langle$HiGraphs$(X), \mathcal{M}\rangle$ of hierarchical graphs with variables with the class $\mathcal{M}$ of all injective top morphisms.

Example 2 (Transformation of Control Flow Graphs). In Figs. 1 and 2 we show two rules for transforming hierarchical control flow graphs. The rule *loop* removes duplicated code before a loop. The rule *inl* performs "inlining" of a procedure's body for its call. In the figures, the images of the interface's nodes and edges in the left- and right-hand side graphs can be found by horizontal projection.

Figure 7 shows transformations that use the rule *loop* in vertical direction, and the rule *inl* in horizontal direction, respectively. For applying *loop*, the variable D must be instantiated with the assignment "$x := e$" in the right column, and by the procedure call edge containing that assignment in the other columns. For applying *inl*, its variable D must be instantiated with the control flow graphs representing the statements "$x := e$" (in the transformations to the right), and "$y := e'; z := e'''$" (in the transformations to the left), respectively.

A rule is applied by instantiating its variables according to some substitution, and constructing a double-pushout for this instance.

In substitutive graph transformation [9], the application of a rule is determined entirely by instantiation with a substitution.

Definition 6 (Substitutive Graph Transformation). A *substitutive rule* $p^* = \langle L^*, R^* \rangle$ is a pair of hierarchical graphs. Given two hierarchical graphs G, H, G *directly derives* H through p^*, denoted by $G \Rightarrow_{p^*} H$, if there is a

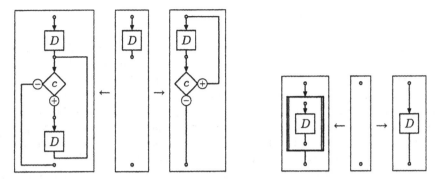

Fig. 1. The rule *loop*   **Fig. 2.** The rule *inl*

substitution σ^* such that $L^*\sigma^* \cong G'$ for some part G' of G, and H equals a copy of G wherein G' is replaced with $R^*\sigma^*$.

Every rule $p = \langle L \leftarrow K \rightarrow R \rangle$ induces a substitutive rule $p^* = \langle L^*, R^* \rangle$ as follows: Extend every part K' of K by a variable edge that is attached to all nodes in $V_{K'}$, and is labeled with a fresh variable label (of rank $|V_{K'}|$), and obtain the hierarchical graphs L^* and R^* by inserting this hierarchical graph for the occurrence of the part K' in L and R, respectively.

Theorem 1 (Substitutive Graph Transformation). *For all hierarchical graphs G, H and all rules p,*

$$G \Rightarrow_p H \text{ if and only if } G \Rightarrow_{p^*} H .$$

Proof Sketch. Without loss of generality, p is a rule with discrete interface $\underline{K}$. (Otherwise, if we consider the modified rule p^- in which all non-variable edges are removed from K, we easily see that $G \Rightarrow_p H \Leftrightarrow G \Rightarrow_{p^-} H$.)

For a start, let us consider the case that the rules are applied on top level. First, let $G \Rightarrow_{p,\sigma,g} H$ be a direct derivation. Define σ^* as the extension of σ by the substitution pair $\{x \mapsto D\}$, where D is the intermediate hierarchical graph of the direct derivation. Then $L^*\sigma^* \cong G$, and $R^*\sigma^* \cong H$ and $G \Rightarrow_{p^*,\sigma^*} H$ is a direct substitutive derivation. Conversely, let $G \Rightarrow_{p^*,\sigma^*} H$ be a direct substitutive derivation. Then $L^*\sigma^* \cong G$ and $R^*\sigma^* \cong H$ for some isomorphisms $g^*: L^*\sigma^* \rightarrow G$ and $h^*: R^*\sigma^* \rightarrow H$. Let σ be the restriction of σ^* to $\text{Dom}(\sigma^*) - \{x\}$ and g be the restriction of g^* to $L\sigma$. Then there is a direct derivation $G \Rightarrow_{p,\sigma,g} H$.

Now, let the direct derivations apply to a part G' of G. Then both kinds of direct derivations construct a graph H wherein G' is replaced by a part H' with a direct top-level derivation.

Theorem 1 shows the close relationship between the double-pushout approach and the substitution-based approach. As a consequence, the main proofs can be done on a substitution-based level.

4   Parallel Independence

The term "parallel independence" has been coined for a criterion of commutativity (or the Local Church-Rosser property) in double-pushout graph transformation (see, e.g., [7]). The related area of term rewriting is about the transformation of terms, or trees, by rules with variables. Commutativity has been studied for term rewriting as well, along with a more general property, called joinability. Commutativity and joinability are important prerequisites for showing that a transformation mechanism has unique normalforms: If all competing direct derivations are commutative (joinable), transformation is strongly confluent (or locally confluent, resp.). Strongly confluent, and terminating locally confluent "abstract reduction systems" do have unique normalforms. (See, e.g., [10].)

We re-phrase commutativity and joinability for hierarchical graph transformation.

Definition 7 (Commutativity and Joinability). A pair of direct derivations $H \Leftarrow G \Rightarrow H'$ of the same hierarchical graph is called *competing* if $H \ncong H'$. Competing direct derivations are

- *commutative* if $H \Rightarrow M \Leftarrow H'$, and
- *joinable* if $H \overset{*}{\Rightarrow} M \overset{*}{\Leftarrow} H'$,

for some hierarchical graph M, respectively. (See Figs. 3 and 4 below.)

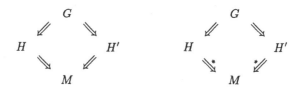

Fig. 3. Commutativity   **Fig. 4.** Joinability

For double-pushout graph transformation it has been shown that commutativity holds if competing direct derivations are parallelly independent of each other (see, e.g., [12, 7, 14]). For term rewriting, the presence of variables in rules has made it necessary to study joinability. Term rewriting steps are joinable if they are non-critically overlapping.

We shall first demonstrate that both criteria, parallel independence as well as non-critical overlaps, carry over to hierarchical graph transformation. However, since hierarchical graphs generalize both graphs and terms, these criteria turn out to be special cases of a more general condition for joinability that will be discussed in the sequel.

General Asumption. In the following, let $H \Leftarrow_{p,\sigma,g} G \Rightarrow_{p',\sigma',g'} H'$ be a pair of competing direct derivations using the rules $p = \langle L \leftarrow K \rightarrow R \rangle$ and $p' = \langle L' \leftarrow K' \rightarrow R' \rangle$.

The morphism g of the rule instance $L\sigma$ in G defines a *skeleton fragment* $g(\underline{L})$ of G that contains an *interface fragment* $g(\underline{K}) \subseteq g(\underline{K})$; also, every variable edge e in L defines a *variable fragment* $g(\sigma(e))$. In the same way, g' defines a skeleton fragment $g'(\underline{L'})$ of G with an interface fragment $g'(\underline{K'})$, and variable fragments $g'(\sigma(e'))$ for the variable edges e' in L'.

Figure 5 below illustrates how "classical" parallel independence carries over to hierarchical graph transformation. Competing direct derivations are parallelly independent if and only if the images of the rules' left-hand side skeletons (the semicircles) overlap only in their skeleton interface fragments (the white areas of the semicircles). The deleted part of the skeleton fragments (drawn as grey semicircles) and their variable fragments (drawn as grey boxes) must be disjoint. In this situation, competing direct derivations leave the occurrence of the respective other rule intact; they commute by a direct derivation using the other rule at the unchanged occurrence.

Figure 6 shows the non-critical overlap of two direct derivations. The left-hand side of one rule must occur completely inside a single variable fragment (of x in the illustration) of the other rule. In this case, the competing direct derivations are not commutative. In general, several steps may be necessary to join them again. Let p be the rule subsuming the occurrence of p' in the variable fragment $g(\sigma(x))$. In this example x occurs twice in p's left hand side. A direct derivation with p leads to a hierarchical graph H wherein $g(\sigma(x))$ will occur as often as x occurs in p's right hand side, say i times. Then H contains $i \geq 0$ occurrences of the left hand side of p'. The occurrences of p' in G and in H are parallelly independent, and can be transformed in 2 and i steps with p', respectively. In the resulting graphs, every variable fragment of x has been transformed in the same way, so that there is a direct derivation with p between the hierarchical graphs, which joins the derivations.

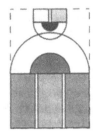   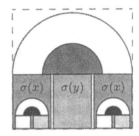

Fig. 5. "Classical" parallel independence   **Fig. 6.** Non-critical overlap

Definition 8 ("Classical" Parallel Independence). A pair of competing direct derivations is *"classically" parallelly independent* if the intersection of $g(L\sigma)$ and $g'(L'\sigma')$ in G is contained in the intersection $g(\underline{K}) \cap g'(\underline{K'})$ of their skeleton interface fragments.

Fact. "Classically" parallelly independent direct derivations are commutative.

Definition 9 (Non-critical Overlap). A pair of competing direct derivations is *non-critically overlapping* if the intersection of $L\sigma$ and $L'\sigma'$ in G consists of items of a single variable fragment, that is, $g(L\sigma) \subseteq g'(\sigma'(e'))$ for some variable edge e' in L' or, vice versa, $g'(L'\sigma') \subseteq g(\sigma(e))$ for some variable edge e in L.

Theorem 2. *Noncritically overlapping derivations are joinable.*

Proof Sketch. Let $H \Leftarrow_{p,\sigma,g} G \Rightarrow_{p',\sigma',g'} H'$ be non-critically overlapping. Without loss of generality, $g'(L'\sigma') \subseteq g(\sigma(e))$ for some variable edge e in L with label, say x. Assume first that g is top. By the Restriction Lemma [7], there is a restricted direct derivation $d'(e)$ of the direct derivation $d': G \Rightarrow_{p'} H'$ to the variable fragment $g(\sigma(e))$ with result, say $H'(e)$. By theorem 1, $G = L^*\sigma^*$ and $H = R^*\sigma^*$. By the Embedding Lemma [7], the direct derivation $d'(e)$ can be embedded into every variable fragment $g(\sigma(e))$ of $G = L^*\sigma^*$ with $lab_L(e') = lab_L(e)$. The embedded derivations are parallelly independent. By parallel independence [12, 7], there is a derivation $L^*\sigma^* \Rightarrow_{p'}^+ L^*\tau^*$ where τ^* is the modification of the substitution σ^* with $\tau^*(x) = H'(e)$ and $\tau^*(y) = \sigma^*(y)$ otherwise. By theorem 1, there is a direct derivation $L^*\tau^* \Rightarrow_p R^*\tau^*$. The direct derivation $d'(e)$ can be embedded into every variable fragment $g(\sigma(e))$ of $R^*\sigma^*$ with $lab_L(e') = lab_L(e)$. Again, the embedded derivations are parallelly independent. Thus, there is a derivation $R^*\sigma^* \Rightarrow_{p'}^* R^*\tau^*$, and, the direct derivations are joinable, see below.

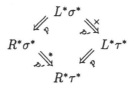

Now, if g is not top, let $\bar{G}$ be the part of G where g is top. Then there are competing derivations $\bar{H} \Leftarrow \bar{G} \Rightarrow \bar{H}'$ that have joining derivation sequences $\bar{H} \overset{*}{\Rightarrow} \bar{M} \overset{*}{\Leftarrow} \bar{H}'$. Since derivations are closed under the part relation, graphs H, H' and M can be constructed by replacing the parts in corresponding to G' in those graphs so that we get the diagram above.

Example 3 (Commuting and Joining Control Flow Derivations). Figure 7 shows several direct derivations of control flow graphs. The graph in the middle of the top row can be transformed in four ways: The rule *loop* applies to the loop in the else part of the top branch, and *inl* can be applied to its procedure call edges. To its left, we see the graph after applying *inl* to the procedure call on the left; beneath it we see the result of applying *loop* to the loop in its else part; the result of applying *inl* twice, to the (isomorphic) procedure calls in that loop is shown on the right.

The *loop* step is "classically" parallelly independent of the left *inl* step, and the result of the commuting steps is shown in the lower left. Both occurrences of the *inl* steps leading to the right are contained in the fragment of the variable D of the *loop* rule; since this variable occurs twice on the left hand side, and once on the right hand side of *loop*, two steps are needed in the top row, and one in

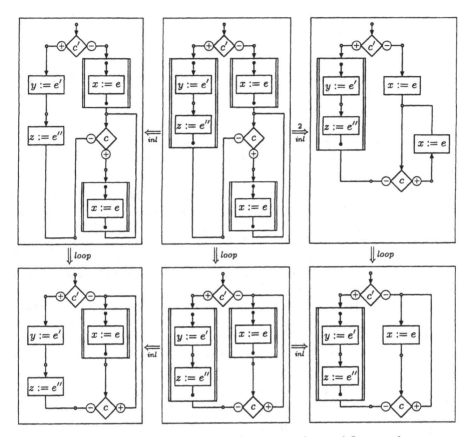

Fig. 7. Parallel independent transformations of control flow graphs

the bottom row, until they lead to the graphs on the right where the *loop* rule can be applied again.

Consider the noncritical overlap illustrated in Fig. 6. For term rewriting, where trees are being transformed, the occurrence of a rule (like p') can only overlap with a single variable fragment (of x), because the occurrence is connected, and the variable fragment is disconnected from other variable fragments. However, graphs need not be connected so that further situations arise in the case of hierarchical graph transformation, which are sketched in Fig. 8.

In the situation on the left, p' overlaps with the skeleton interface fragment, and with two variable fragments of p. The competing derivations would be joinable if the involved fragments of p are preserved in the direct derivation with p. This is the case if they are left intact, i.e., if the involved variables occur once on both sides of p because the skeleton of p is also involved. In the situation on the right, this need not be the case as the skeleton fragment is not involved in the overlap. Here, it suffices when the involved variables have the same number of occurrences on both sides of p.

Thus, whenever the intersection of $L\sigma$ and $L'\sigma'$ in G consists of several variable fragments we have to require that one occurrence induces a decomposition of the other rule into subrules such that the fragments can be transformed separately by the subrules. Furthermore, the transformation must be consistent, i.e, same fragments have to be transformed in the same way, and complete. Furthermore, the transformation must be repetitive, i.e. after the application of the rule a complete parallel transformation must be possible.

A variable edge e in L is *involved* in the direct derivation $G \Rightarrow_{p'} H'$ if the intersection of the skeleton fragment $g'(\underline{L'})$ and the variable fragment $g(\sigma(e))$ is non-trivial, i.e. if the intersection consist not only of points. The label of an involved edge is called an *involved label*.

Definition 10 (Fragmented Direct Derivations). Let $\langle d, d' \rangle$ be a pair of direct derivations. Then d' is *g-decomposable* if there is a decomposition of d' into a non-changing subderivation on the skeleton fragment and subderivations $d'(e)$ on the variable fragments for e in L^*. In this case, we speak of a *g*-decomposition of d'. A *g*-decomposition is *consistent* if $lab_L(e) = lab_L(e')$ implies $\tau(e) = \tau(e')$ for all involved edges. It is *complete* if there is no not-involved edge with involved label. It is *completable* if d' can be extended to a derivation $G \Rightarrow_{p'}^+ I'$ with complete set of involved edges. A *g*-decomposable, consistent, and completable direct derivation is called *g-compatible*. The direct derivation d is *g'-repetitive* if there is a derivation $H \Rightarrow_{p'}^* R^*\tau^*$ of some substitution τ^*. The pair $\langle d, d' \rangle$ is *fragmented* if d' is *g*-compatible and d is *g'*-repetitive or d is *g'*-compatible and d' is *g*-repetitive.

Fact. Every *g*-compatible direct derivation $G \Rightarrow_{p',\sigma',g'} H'$ through a top morphism g' can be extended to a derivation $G \Rightarrow_{p'}^+ L^*\tau^*$ for some substitution τ^*.

Proof Sketch. Let $d': G \Rightarrow_{p'} H'$ be *g*-compatible. Then d' is *g*-decomposable, consistent, and completable. By *g*-decomposability, there is a decomposition of d' into a non-changing subderivation on the skeleton fragment and subderivations $d'(e)$ on the variable fragments for e in L^* such that H' is obtained from L^* by replacing the ordinary variables e in L^* by the result $\tau(e)$ of the subderivation $d'(e)$ and the context variable in L^* by the intermediate hierarchical graph D. By consistency, the replacements define a substitution

$$\tau^* = \{lab_L(e) \mapsto \tau(e) \mid e \in E_L\} \cup \{x \mapsto D\} \ .$$

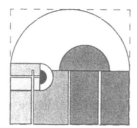 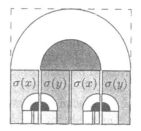

Fig. 8. Fragmented parallel independence

In the case of completeness, $H' = L^*\tau^*$. In the case of completability, d' can be extended to a derivation $G \Rightarrow_{p'}^+ I'$ such that $I' = L^*\tau^*$.

Definition 11 (Fragmented Parallel Independence). A pair of direct derivations $\langle d, d' \rangle$ is *fragmentedly parallel independent* if the skeleton fragments overlap only in the skeleton interface fragments, and if $\langle d, d' \rangle$ is fragmented.

Theorem 3 (Fragmented Parallel Independence). *Every pair of fragmentedly parallel independent direct derivations is joinable.*

Proof Sketch. Let $d: G \Rightarrow_{p,\sigma,g} H$ and $d': G \Rightarrow_{p',\sigma',g'} H'$ be fragmentedly parallel independent. Without loss of generality, assume that d' is g-compatible and d is g'-repetitive. We first consider the case that g is top. By Theorem 1, $G = L^*\sigma^*$ and $H = R^*\sigma^*$. By the g-compatibility of d', there is a derivation $L^*\sigma^* \Rightarrow_{p'}^+ L^*\tau^*$ for some substitution τ^*. By Theorem 1, there is a direct derivation $L^*\tau^* \Rightarrow_p R^*\tau^*$. By g'-repetitiveness of d, there is derivation $R^*\sigma^* \Rightarrow_{p'}^* R^*\tau^*$. Thus, the direct derivations are joinable.

$$
\begin{array}{ccc}
 & L^*\sigma^* & \\
{}^p\swarrow & & \searrow^{d'} \\
R^*\sigma^* & & L^*\tau^* \\
{}_{d'}\searrow & & \swarrow_p \\
 & R^*\tau^* &
\end{array}
$$

The case that g is not top can be reduced to the situation above by the same argument as in the proof of Thm. 2.

Note that fragmented parallel independence subsumes both "classical" parallel independence (illustrated in Fig. 5), and non-critical overlaps (shown in Fig. 6): Only the skeleton fragment of p is involved in the first case, and a single variable fragment is involved in the second case.

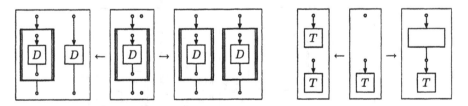

Fig. 9. The rule *fold*   **Fig. 10.** The rule *join*

Example 4 (Fragmented Parallel Independence of Control Flow Graph Transformations). In Figs. 9 and 10, we define two rules that illustrate particular cases of fragmented parallel independence: If some control flow graph D matches the variable fragment of a procedure, *fold* replaces the body D by a call to that procedure, and if a control flow graph ends in two copies of the same subdiagrams

T, *join* redirects one of them to the other. (The empty assignment represents a neutral computation.)

The rule *fold* it a parallel rule of the form $fold = id + inl^{-1}$ (where id is the identical rule) and hence decomposable into two subrules. The rule *join* is not decomposable. Figure 11 shows two fragmentedly parallelly independent direct derivations steps through the rules *fold* and *loop* that overlap in a nontrivial way: The occurrences of the left-hand sides intersect not only in the body of the loop (which is an instantiation of the variable D in *loop*), but also in the "$\ominus$"-successor state of the branch at the bottom. Nevertheless, the direct derivations are joinable, as the *fold* rule divides into two rules, and one of them is just the identity.

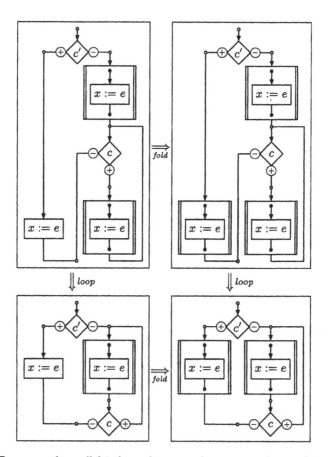

Fig. 11. Fragmented parallel independent transformations of control flow graphs

5 Conclusion

We have studied under which conditions direct transformations of a graph are independent so that they can be joined to a common graph by subsequent transformations. Graphs and rules have been generalized by concepts known from term rewriting: Graphs are equipped with a tree-like hierarchy, as edges may contain graphs which are again hierarchical; rules contain graph variables by which subgraphs of arbitrary size can be compared, deleted, or copied in a single transformation step. Our results combine properties known for plain graph transformation and term rewriting.

To our knowledge, parallel independence of graph transformation has only been studied for the double- and single-pushout approaches. In both cases, neither hierarchies, nor graph variables have been considered. Parallel independence has also been investigated in the more general framework of adhesive high-level replacement systems [15, 13]. It looks as if hierarchical graph transformation without variables is an instance of adhesive high-level replacement; this is not true for hierarchical graph transformation with variables, however.

The study of parallel independence has lead to critical pair lemmata, both for term rewriting and for graph transformation [16]: whenever transformation steps are not parallelly independent, these systems are locally confluent if joinability can be shown for finitely many critical pairs of graphs and terms, respectively. Since parallel independence of hierarchical graph transformation turned out to be a reasonable combination of the results for graph transformation and term rewriting, we shall try to combine these lemmata to obtain a critical pair lemma for hierarchical graph transformation as well.

Furthermore, local confluence implies general confluence if the rules are also terminating. Since termination can be characterized by the finiteness of so-called forward closures of rules, both for term rewriting and for graph transformation [17], we think it may be possible to combine these results to a similar theorem for hierarchical graph transformation. Finally, if we are able to find decidable sufficient criteria for termination, this, together with a critical pair lemma, would allow to decide confluence in restricted cases. This would give immediate benefits for the analysis of DIAPLAN, a language for programming with graphs and diagrams that shall be based on hierarchical graph transformation [11].

References

1. Drewes, F., Hoffmann, B., Plump, D.: Hierarchical graph transformation. Journal of Computer and System Sciences **64** (2002) 249–283
2. Hoffmann, B.: Shapely hierarchical graph transformation. In: Proc. IEEE Symposia on Human-Centric Computing Languages and Environments. IEEE Computer Press (2001) 30–37
3. Rozenberg, G., ed.: Handbook of Graph Grammars and Computing by Graph Transformation. Volume 1: Foundations. World Scientific (1997)

4. Ehrig, H., Engels, G., Kreowski, H.J., Rozenberg, G., eds.: Handbook of Graph Grammars and Computing by Graph Transformation. Volume 2: Applications, Languages and Tools. World Scientific (1999)
5. Ehrig, H., Kreowski, H.J., Montanari, U., Rozenberg, G., eds.: Handbook of Graph Grammars and Computing by Graph Transformation. Volume 3: Concurrency, Parallelism, and Distribution. World Scientific (1999)
6. Busatto, G.: An Abstract Model of Hierarchical Graphs and Hierarchical Graph Transformation. PhD thesis, Universität-Gesamthochschule Paderborn (2002)
7. Ehrig, H.: Introduction to the algebraic theory of graph grammars. In: Graph-Grammars and Their Application to Computer Science and Biology. Volume 73 of Lecture Notes in Computer Science., Springer-Verlag (1979) 1–69
8. Engels, G., Heckel, R.: Graph transformation as a conceptual and formal framework for system modelling and evolution. In: Automata, Languages, and Programming (ICALP 2000). Volume 1853 of Lecture Notes in Computer Science., Springer-Verlag (2000) 127–150
9. Plump, D., Habel, A.: Graph unification and matching. In: Graph Grammars and Their Application to Computer Science. Volume 1073 of Lecture Notes in Computer Science., Springer-Verlag (1996) 75–89
10. Baader, F., Nipkow, T.: Term Rewriting and All That. Cambridge University Press, Cambridge, UK (1998)
11. Hoffmann, B.: Abstraction and control for shapely nested graph transformation. Fundamenta Informaticae 58 (1) (2003) 39–65
12. Ehrig, H., Kreowski, H.J.: Parallelism of manipulations in multidimensional information structures. In: Mathematical Foundations of Computer Science. Volume 45 of Lecture Notes in Computer Science., Springer-Verlag (1976) 284–293
13. Ehrig, H., Habel, A., Padberg, J., Prange, U.: Adhesive high-level replacement categories and systems. In: Graph Transformation (ICGT'04). Lecture Notes in Computer Science, Springer-Verlag (2004)
14. Habel, A., Müller, J., Plump, D.: Double-pushout graph transformation revisited. Mathematical Structures in Computer Science 11 (2001) 637–688
15. Ehrig, H., Habel, A., Kreowski, H.J., Parisi-Presicce, F.: Parallelism and concurrency in high level replacement systems. Mathematical Structures in Computer Science 1 (1991) 361–404
16. Plump, D.: Hypergraph rewriting: Critical pairs and undecidability of confluence. In: Term Graph Rewriting: Theory and Practice. John Wiley, New York (1993) 201–213
17. Plump, D.: On termination of graph rewriting. In: Graph-Theoretic Concepts in Computer Science. Volume 1017 of Lecture Notes in Computer Science., Springer-Verlag (1995) 88–100

A   Use of Skeleton Interfaces in Rule Instances

The "only-if" direction of theorem 1 requires that the interface of a rule instance is its skeleton, and not its instance. Otherwise the rule instance would be applicable to graphs where the substitutive rule does not apply. This shall be illustrated by an example.

The rule $p = \langle L \leftarrow K \rightarrow R \rangle$ shown in Fig. 12 has the interface variable A. If p would b instantiated by the substitution $\sigma = \{A \mapsto \text{•}\boxed{a}\text{•}\}$, the (extended) rule instance $p\sigma = \langle L\sigma \leftarrow K\sigma \rightarrow R\sigma \rangle$ has a direct derivation $d \colon G \Rightarrow_{p\sigma} H$. However, there is no way to extend σ to a substitution σ^* for the substitutive rule $p^* = \langle L^*, R^* \rangle$ so that $L^*\sigma^* \cong G$: The context variable C cannot be instantiated by the substitution pair $D \mapsto$ because that graph is connected to an "inner node" of 'A's substitution. (There is a substitution σ' where $\sigma'(A)$ is a single point, and $\sigma'(C) = \sigma^*(A) \cup \sigma^*(C)$, but the instance $p^*\sigma'$ does not derive H, but a subgraph of H where the right a-edge is missing.)

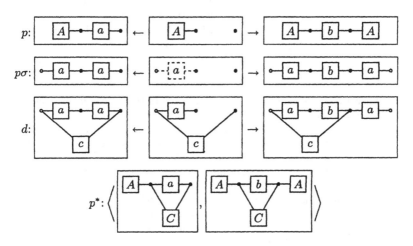

Fig. 12. Skeleton interfaces in instances

Generating Test Cases for Code Generators by Unfolding Graph Transformation Systems[*]

Paolo Baldan[1], Barbara König[2], and Ingo Stürmer[3]

[1] Dipartimento di Informatica, Università Ca' Foscari di Venezia, Italy
baldan@dsi.unive.it
[2] Institut für Formale Methoden der Informatik, Universität Stuttgart, Germany
koenigba@fmi.uni-stuttgart.de
[3] Research and Technology, DaimlerChrysler Berlin, Germany
Ingo.Stuermer@daimlerchrysler.com

Abstract. Code generators are widely used in the development of embedded software to automatically generate executable code from graphical specifications. However, at present, code generators are not as mature as classical compilers and they need to be extensively tested. This paper proposes a technique for systematically deriving suitable test cases for code generators, involving the interaction of chosen sets of rules. This is done by formalising the behaviour of a code generator by means of graph transformation rules and exploiting unfolding-based techniques. Since the representation of code generators in terms of graph grammars typically makes use of rules with negative application conditions, the unfolding approach is extended to deal with this feature.

1 Introduction

The development of embedded software has become increasingly complex and abstraction appears to be the only viable means of dealing with this complexity. For instance, in the automotive sector, the way embedded software is developed has changed in that executable models are used at all stages of development, from the first design phase up to implementation (*model-based development*). Such models are designed with popular and well-established graphical modelling languages such as *Simulink* or *Stateflow* from *The MathWorks*[1]. While in the past the models were implemented manually by the programmers, some recent approaches allow the automatic generation of efficient code directly from the software model via *code generators*. However, at present, they are not as mature as tried and tested C or ADA compilers and their output must be checked with almost the same expensive effort as for manually written code.

---

[*] Research partially supported by EU FET-GC Project IST-2001-32747 AGILE, the EC RTN 2-2001-00346 SEGRAVIS, DFG projects SANDS and the IMMOS project funded by the German Federal Ministry of Education and Research (project ref. 01ISC31D).

[1] See www.mathworks.com

H. Ehrig et al. (Eds.): ICGT 2004, LNCS 3256, pp. 194–209, 2004.
© Springer-Verlag Berlin Heidelberg 2004

One of the main problems in code generator testing is the methodical inability to describe, in a clear and formal way, the mode of operation of the code generator's transformation rules and the interaction between such rules (this is especially true for optimisation rules), a fact which makes it hard to devise effective and meaningful tests. Therefore, an essential prerequisite for testing a code generator is the choice of a formal specification language which describes the code generator's mode of action in a clear way [6].

When dealing with a code generator which translates a graphical source language into a textual target language (e.g. C or Ada), a natural approach consists in representing the generator via a set of graph transformation rules. Besides providing a clear and understandable description of the transformation process, as suggested in [19], this formal specification technique can be used for test case derivation, allowing the specific and thorough testing of individual transformation rules as well as of their interactions. We remark that, while each rule is specified with a single transformation step in mind, it might be quite difficult to gain a clear understanding of how different rules are implemented and how they can interfere over an input graph. Testing all input models triggering any possible application sequence is impractical (if not impossible), because of the large (possibly infinite) number of combinatorial possibilities, and also unnecessary as not all combinations will lead to useful results. It is, however, of crucial importance to select those test cases which are likely to reveal errors in the code generator's implementation.

In this paper we will use unfoldings of graph transformation systems [18, 4] in order to produce a compact description of the behaviour of code generators, which can then be used to systematically derive suitable test cases, involving the interaction of chosen sets of (optimising) rules. Our proposal is based on the definition of two graph grammars: the *generating grammar*, which generates all possible input models for the code generator (Simulink models, in this paper) and the *optimising grammar*, which formalises specific transformation steps within the code generator (here we focus only on optimisations). The structure obtained by unfolding the two grammars describes the behaviour of the code generator on all possible input models. Since the full unfolding is, in general, infinite, the procedure is terminated by unfolding the grammars up to a finite causal depth which can be chosen by the user. Finally, we will show how the unfolded structure can be used to select test cases (i.e., code generator input models), which are likely to uncover an erroneous implementation of the optimisation techniques (as specified within the second graph grammar). The task of identifying sets of rules whose interaction could be problematic and should thus be tested, might require input from the tester. However once such sets are singled out, the proposed technique makes it possible to automatically determine corresponding test cases, namely input models triggering the desired behaviours, straight from the structure produced via the the unfolding procedure.

The behaviour of code generators is naturally represented by graph grammars with negative application conditions [9], while the unfolding approach has been developed only for "basic" double- or single-pushout graph grammars [18, 4].

Hence, a side contribution of the paper is also the generalisation of the unfolding construction to a class of graph grammars with negative application conditions, of which, due to space limitations, we will provide only an informal account.

The rest of the paper is structured as follows. Section 2 gives an overview of the automatic code generation approach and discusses code generator testing techniques. Section 3 presents the class of graph transformation systems used in the paper. Section 4 discusses the idea of specifying a code generator and its possible input models by means of graph transformation rules. Section 5 presents an unfolding-based technique for constructing a compact description of the behaviour of a code generator and Section 6 shows how suitable test cases can be extracted from such a description. Finally, Section 7 draws some conclusions.

2   Automatic Code Generation

In the process of automatic code generation, a graphical model, consisting for instance of dataflow graphs or state charts, is translated into a textual language. First, a working graph free of layout information is created and, in the next step, a gradual conversion of the working graph into a syntax tree takes place. In the individual transformation phases from the working graph to the syntax tree, optimisations are applied in which, for instance, subgraphs are merged, discarded or redrawn. Finally, actual code generation is performed, during which the syntax tree is translated into linear code.

In practice, a complete test in this framework is impossible due to the large or even infinite number of possible input situations. Accordingly, the essential task during testing is the determination of suitable (i.e. error-sensitive) test cases, which ultimately determines the scope and quality of the test.

In the field of compiler testing much research has been done concerning test case design, namely test case generation techniques. We can distinguish two main approaches: *automatic test case generation* and *manual test case generation*. The first approach yields a great number of test cases in a short time and at a relatively low cost. In most cases, as originally proposed by Purdom [17], test programs are derived from a grammar of the source language by systematically exercising all its productions. An overview of this and related approaches is given in [6]. However, the quality of the test cases is questionable because the test case generation process is not guided by the requirements (i.e. the specification).

A different (and more reliable) method is to generate test cases manually with respect to given language standards, like the Ada Conformity Assessment Test Suite (ACATS)[2], or commercial testsuites for ANSI/ISO C language conformance[3]. However, there is no published standard for graphical source languages such as Simulink or Stateflow. Moreover, the manual creation and maintenance of test cases is cost-intensive, time-consuming and also requires knowledge about tool internals.

---

[2] See www.adaic.com

[3] ANSI/ISO FIPS-160 C Validation Suite (ACVS) by Perennial, www.peren.com

A technique for testing a code generator systematically on the basis of graph rewriting rules was proposed in [19]. The graph-rewriting rules *themselves* are used as a blueprint for systematic test case (i.e. model) generation. Additionally, the mentioned paper shows how such models can be used in practice. A two-level hierarchy of testing is proposed: First, suitable input models to be used as test cases for the code generator are determined; then the behaviour of the code generated from such models over specific (suitably chosen) input data is compared with that of the (executable) specification, in order to ensure correctness. A difference with the work in the present paper is that in [19] a test case is selected on the basis of a single rule, while here we will consider the interaction of several rules to derive test cases which can trigger these complex behaviours. Still the methodology proposed in [19] to use the test cases once they are available, can be applied also to the test cases produced via the technique in our paper.

Graph transformation systems have also been used in other ways in connection with code generator specification or verification. For instance, in [11, 16] graph and tree replacement patterns are used for verifying a code generator formally and in [2] graph rewriting rules are used for generating an optimiser. A complete code generator, capable of translating Simulink or Stateflow models into C code, has been specified in [14] with the Graph Rewriting and Transformation language GReAT [12].

3   Graph Transformation Systems

We use hypergraphs, which allow us to conveniently represent functions with n arguments by $(n + 1)$-ary hyperedges (one connection for the result, the rest for the parameters). Moreover we use graph rewriting rules as in the double-pushout approach [8, 10] with added negative application conditions [9]. A rule, apart from specifying a left-hand side graph that is removed and a right-hand side graph that replaces it, specifies also a context graph that is preserved, and forbidden edges that must not occur attached to the left-hand side.

Hereafter Λ is a fixed set of edge *labels* and each label $l \in \Lambda$ is associated with an *arity* $ar(l) \in \mathbb{N}$. Given a set A, we denote by A^* the set of finite sequences of elements of A and for $s \in A^*$, $|s|$ denotes its length.

Definition 1 (Hypergraph). *A (Λ-)hypergraph G is a tuple (V_G, E_G, c_G, l_G), where V_G is a set of nodes, E_G is a set of edges, $c_G: E_G \to V_G^*$ is a connection function and $l_G: E_G \to \Lambda$ is the labelling function for edges satisfying $ar(l_G(e)) = |c_G(e)|$ for every $e \in E_G$. Nodes are not labelled.*

Hypergraph morphisms $\varphi: G \to G'$ and isomorphisms are defined as usual.

Definition 2 (Graph rewriting rules with negative conditions). *A graph rewriting rule r is a tuple $(L \xleftarrow{\varphi_L} I \xrightarrow{\varphi_R} R, N)$ where $\varphi_L: I \to L$ and $\varphi_R: I \to R$ are injective graph morphisms. We call L the* left-hand side*, R the* right-hand side *and I the* context*. We assume that (i) φ_L is bijective on nodes, (ii) L does not contain isolated nodes, (iii) any node isolated in R is in the image of φ_R.*

Furthermore N is a set of injective morphisms $\eta\colon L \to L_\eta$, called negative application conditions, *where (iv) $E_{L_\eta} - \eta(E_L)$ contains a single edge referred to as e_η and (v) L_η does not contain isolated nodes.*

A rule $r = (L \xleftarrow{\varphi_L} I \xrightarrow{\varphi_R} R, N)$ consists of two components. The first component $L \xleftarrow{\varphi_L} I \xrightarrow{\varphi_R} R$ is a graph production, specifying that an occurrence of the left-hand side L can be rewritten into the right-hand side graph R, preserving the context I. Condition (i) stating that φ_L is bijective on nodes ensures that no nodes are deleted. Nodes may become disconnected, having no further influence on rewriting, and one can imagine that they are garbage-collected. Actually, Conditions (ii) and (iii) essentially state that we are interested only in rewriting up to isolated nodes. By (iii) no node is isolated when created and by (ii) nodes that become isolated have no influence on further reductions.

The second component N is the set of negative application conditions. Intuitively, each L_η extends the left-hand side L with an edge e_η which must not be connected to the match of L to allow a rule to be applied. The negative application conditions here are weaker than in [9]. This will allow us to represent negative application conditions by inhibitor arcs in the unfolding (see Section 5).

Definition 3 (Match). *Let $r = (L \xleftarrow{\varphi_L} I \xrightarrow{\varphi_R} R, N)$ be a graph rewriting rule and let G be a graph. Given an injective morphism $\varphi\colon L \to G$, a* falsifying extension *for φ is an injective morphism $\varphi'\colon L_\eta \to G$ such that $\varphi' \circ \eta = \varphi$ for some $\eta \in N$. In this case $\varphi'(e_\eta)$ is called a* falsifying edge *for φ. The morphism φ is called a* match *of r whenever it does not admit any falsifying extension.*

Given a graph G and a match in it, G can be rewritten to H (in symbols: $G \Rightarrow H$), by applying rule r as specified in the double-pushout approach [8].

Definition 4 (Graph grammar). *A* graph grammar *$\mathcal{G} = (\mathcal{R}, G_0)$ consists of a set of rewriting rules $\mathcal{R}$ and a start graph G_0 without isolated nodes. We say that a graph G is* generated *by $\mathcal{G}$ whenever $G_0 \Rightarrow^* G$.*

4   Specifying Code Generation by Graph Transformation

In our setting, code generation starts from an internal graph representation of a Simulink or Stateflow model, free of layout information. Especially the first steps of code generation, involving optimisations that change the graph structure of a model (e.g. for dead code elimination), can be naturally described by graph rewriting rules. In the sequel, the set of optimising rules is called *optimising grammar*, even if we do not fix a start graph. Since our aim is to test the code generator itself, independently of a specific Simulink model, we need some means to describe the set of all possible models that can be given as input to the code generator. In our proposal this is seen as a graph language generated by another grammar, called *generating grammar*.

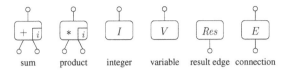

Fig. 1. Edge types for the graph rewriting system (constant folding).

Example: We illustrate the above concepts with an example describing the first steps of code generation, starting from acyclic graphs which represent arithmetic expressions. We will give only excerpts of the two graph rewriting systems: the generating grammar, describing acyclic graphs which represent arithmetic expressions, and the optimising grammar, describing constant folding, i.e., simplification and partial evaluation of arithmetic expressions.

We assume that a maximal depth a is fixed for arithmetic expressions. Then we will use the edge types depicted in Fig. 1 for $1 \leq i \leq a$. Integers and variables are generically represented by I and V edges. Since in our setting we are mainly concerned with structural optimisation steps, we do not consider attributes here: As soon as a test case is generated, it can be equipped with suitable values for all the constants involved. For instance, Fig. 3 shows an acyclic graph representing the arithmetic expression $i_1 + (i_1 * i_2)$ for some arbitrary integers i_1, i_2.

The *generating grammar* $\mathcal{G}_g$, which is depicted in Fig. 2, generates operator, result, integer and variable edges and connects them via E-edges (connecting edges), provided no edge of this kind is present yet. The rules are specified in the form "left-hand side $\Rightarrow$ right-hand side". Edges of the context are drawn with dashed lines and nodes of the context are marked with numbers. Negative application conditions are depicted as crossed-out edges. Note that (CreateConn2) is a rule schema: an E-edge between operator edges is only allowed if the first operator has a smaller (arithmetic) depth than the second one, i.e., if $i < j$, thus ensuring acyclicity. Some of the rules are missing, for example the rule generating a product edge (analogous to the rule (CreateSum)) and several more rules connecting operator edges. The start graph is the empty graph.

The *optimising grammar* $\mathcal{G}_o$ is (partially) presented in Fig. 4: We give rules for reducing the sharing of constants (ConstantSplitting), for removing useless or isolated parts of the graph (KillUselessFunction), (KillLonelyEdge), and for simplifying the graph by evaluating the sum of two integers (ConstantFoldingSum). More specifically, the optimisation corresponding to the last rule computes the integer of the right-hand side as the sum of the two integers of the left-hand side. A requirement for its application is the absence of constant sharing.

5   Unfolding Graph Transformation Systems

The unfolding approach, originally devised for Petri nets [15], is based on the idea of associating to a system a single branching structure, representing all its possible runs, with all the possible events and their mutual dependencies. For

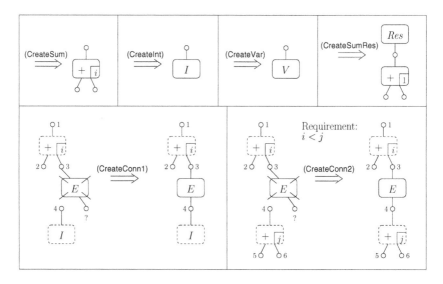

Fig. 2. Rules of the generating grammar (constant folding).

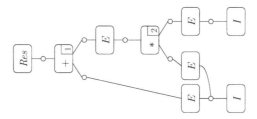

Fig. 3. A graph representing the arithmetic expression $i_1 + (i_1 * i_2)$.

graph rewriting systems, the unfolding is constructed starting from the start graph, considering at each step the rewriting rules which could be applied and then recording in the unfolding possible rule applications and the graph items which they generate (see [18, 4]). Here we sketch how the unfolding construction can be extended to graph grammars with negative application conditions. Space limitations keep us from giving a formal presentation of the theory.

The unfolding of a graph grammar will be represented as a Petri net-like structure. We next introduce the class of Petri nets which plays a basic role in the presentation. Given a set S, we denote by $S^\oplus$ the set of multisets over S, i.e., $S^\oplus = \{m \mid m : S \to \mathbb{N}\}$. A multiset m can be thought of as a subset of S where each $s \in S$ occurs with a multiplicity $m(s)$. When $m(s) \in \{0, 1\}$ for all $s \in S$ the multiset m will often be confused with the set $\{s \in S \mid m(s) = 1\}$.

Definition 5 (Petri net with read and inhibitor arcs). *Let L be a set of transition labels. A Petri net with read and inhibitor arcs is a tuple $N = (S_N, T_N, {}^\bullet(), ()^\bullet, \underline{()}, {}^\smile(), p_N)$ where S_N is a set of places, T_N is a set of transi-*

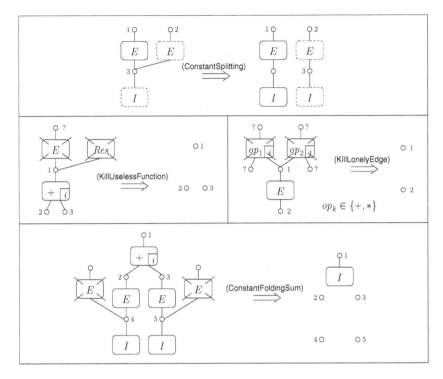

Fig. 4. Rules of the optimising grammar (constant folding).

tions *and* $p_N : T_N \to L$ *is a labelling function. For any transition* $t \in T_N$, $\bullet t$, $t^\bullet$, $\underline{t}$, $\check{t} \in S_N^\oplus$ *denote pre-set, post-set, context and set of inhibitor places of* t.

When $s \in \underline{t}$ we say that t is connected to s via a *read arc*. In this case t may only fire if place s contains a token. This token will not be affected by the firing. On the other hand, when $s \in \check{t}$ we say that t is connected to s via an *inhibitor arc* and t is allowed to fire only if s does *not* contain a token. Read arcs [13] will be used to represent, at the level of Petri nets, the effects arising from the possibility of preserving graph edges in a rewriting step. Inhibitor arcs [1] will be used to model the effects of the negative application conditions that we have at the level of graph grammar rules.

The mutual dependencies between transitions play a crucial role in the definition of the unfolding. Given a net N, the *causality* relation $<_N$ is the least transitive relation such that $t_1 <_N t_2$ if $t_1^\bullet \cap (\bullet t_2 \cup \underline{t_2}) \neq \emptyset$, i.e., if t_1 produces a token consumed or read by t_2.

In ordinary Petri nets, two transitions t_1 and t_2 competing for a resource, i.e., which have a common place in the pre-set, are said to be in conflict. The presence of read arcs leads to an *asymmetric* form of conflict: if a transition t_2 "consumes" a token which is "read" by t_1 then the execution of t_2 prevents t_1 to be executed, while the sequence "t_1 followed by t_2" is legal. The *asymmetric conflict* $\nearrow_N$ can be formally defined by $t_1 \nearrow_N t_2$ if $\underline{t_1} \cap {}^\bullet t_2 \neq \emptyset$ or ${}^\bullet t_1 \cap {}^\bullet t_2 \neq \emptyset$. The last clause includes the ordinary symmetric conflict as asymmetric conflict in both directions, i.e., $t_1 \nearrow_N t_2$ and $t_2 \nearrow_N t_1$. More generally, transitions occurring in a cycle of asymmetric conflicts $t_1 \nearrow_N t_2 \nearrow_N \ldots \nearrow_N t_n$ cannot appear in the same computation since each of them should precede all the others.

The unfolding of a graph grammar with negative application conditions is defined as a Petri graph [5], i.e., a graph with a Petri net "over it", using the edges of the graph as places.

Definition 6 (Petri graph). *Let $\mathcal{G} = (\mathcal{R}, G_0)$ be a graph grammar. A Petri graph (over $\mathcal{G}$) is a tuple $P = (G, N)$ where G is a hypergraph, N is a Petri net with read and inhibitor arcs whose places are the edges of G, i.e., $S_N = E_G$, and the labelling $p_N : T_N \to \mathcal{R}$ of the net maps the transitions to the graph rewriting rules of $\mathcal{G}$. A Petri graph with initial marking is a tuple (P, m_0) where $m_0 \in E_G{}^\oplus$.*

Each transition in the Petri net will be interpreted as an occurrence of a graph production at a given match. Note that Definition 6 does not ask that the pre-set, post-set, context or inhibitor places of a transition t have any relation with the corresponding graph rewriting rule $p_N(t)$, but the unfolding construction presented later will ensure a close relation.

As in Petri net theory, a marking $m \in E_G{}^\oplus$ is called *safe* if any place (edge) contains at most one token. A safe marking m of a Petri graph $P = (G, N)$ can be seen as a graph, i.e., the least subgraph of G including exactly the edges which contain a token in m. Such a graph, denoted by $graph(m)$, is called the *graph generated* by m.

We next describe how a suitable unfolding can be produced from the generating/optimising grammars associated to a code generator. We introduce the criteria and conditions that must be met step by step. At first, the graph grammars are unfolded disregarding the negative application conditions.

Petri graph corresponding to a rewriting rule: Every graph rewriting rule can be represented as a Petri graph without considering negative application conditions: Take both the left-hand side L and the right-hand side R, merge edges and nodes that belong to the context and add a transition, recording which edges are deleted, preserved and created. For instance the Petri graph P corresponding to rule (CreateConn1) in Fig. 2 is depicted in Fig. 5 (see the Petri graph in the middle). Observe that the transition preserves the edges labelled $+$ and I (read arcs are indicated by undirected dotted lines) and produces an edge labelled E. In this case no edges are deleted. In order to distinguish connections of the graph and connections between transitions and places, we draw the latter as dashed lines.

Unfolding step: The initial Petri graph is obviously the start graph G_0 of the generating grammar, with no transitions. At every step, we first search for a match of a left-hand side L, belonging to a graph production r. This match must be potentially coverable (concurrent), i.e., it must not contain items which are causally related and the set of causes of the items in the match must be

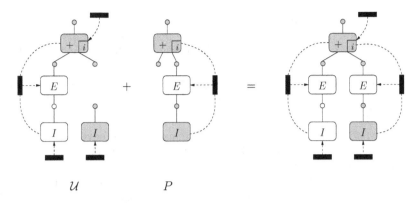

Fig. 5. Performing an unfolding step.

conflict-free (negative application conditions will be taken into account later). We now take the Petri graph P associated to r and merge the edges and nodes of L in P with the corresponding items of the occurrence of L in the partial unfolding. Fig. 5 exemplifies this situation for an incomplete unfolding $\mathcal{U}$ and the Petri graph P representing rule (CreateConn1) of Fig. 2. The left-hand side L which indicates how the merging is to be performed, is marked in grey.

The unfolding of a given graph grammar is usually infinite and thus its construction could continue for an arbitrarily long time. To avoid this we employ two mechanisms, called depth restriction and width restriction.

Depth restriction: The idea consists of truncating the unfolding after a certain depth has been reached. To make this more formal, define the *depth of a transition* t to be the length of the longest sequence $t_0 <_N t_1 <_N \ldots <_N t_n <_N t$, where $<_N$ is the causality relation. The depth of an edge is the depth of the unique transition having such an edge in its post-set. If the edge is in the start graph, then its depth is set to 0. For our purposes it is not necessary to define the depth of a node. Then we fix a parameter k, called *depth restriction*, asking that no items of depth greater than k are ever created by the unfolding construction.

Width restriction: Depth restriction is not sufficient to keep the unfolding finite, since matches of a left-hand side could be unfolded more than once. To stop after a finite number of steps we impose the following conditions:

(1) A rule r which deletes at least one edge, i.e., for which the left-hand side is strictly larger than the context, is applied only once to every match. Note that we would not gain anything from unfolding such a match twice: since at least one token is consumed by firing the corresponding transition and the unfolding is acyclic, the pre-set could never be covered again and thus it would not be possible to fire another copy of the same transition.

(2) A rule r which does not delete any edge, i.e., for which the left-hand side is equal to the context, is unfolded w times for every match, where w is a fixed parameter called *width restriction*. The different copies of this transition can

potentially all be fired since no edge is ever consumed. Actually, since each copy of the transition has an empty pre-set it could possibly be fired more than once, leading to more than one token in a place, a situation which must be avoided in the unfolding where each transition is intended to represent a single occurrence of firing and each place a single occurrence of a token. This problem is solved by introducing a dummy place – initially marked – as the pre-set of such transitions, ensuring that every transition is fired only once.

Generating grammar before optimising grammar: We still have to avoid mixing the two grammars. So far it is still possible to create an unfolding that includes sequences of rewriting steps where rules of the generating grammar are applied to the start graph, followed by the application of rules of the optimising grammar and then again by rules of the generating grammar. Derivations of this kind do not model any interesting situation: in practice, first the model is created, and only then are optimising steps allowed. Hence we impose that whenever there are transitions t_1 and t_2 such that $t_1 <_N t_2$ and $p_N(t_2)$ is a rule of the generating grammar, then also $p_N(t_1)$ must be a rule of the generating grammar. In such a situation we say that $<_N$ is *compatible with the grammar ordering.*

Add inhibitor arcs: In the next step, the final unfolding is obtained by taking every transition t in the Petri graph, labelled by a rule r, considering the corresponding match and adding, for any falsifying edge, an inhibitor arc. Inhibitor arcs are represented by dotted lines with a small circle at one end.

Initial marking: The initial marking contains exactly the edges of the start graph and, in addition, all dummy places that were created during the unfolding.

The structure produced by the above procedure is referred to as *unfolding up to depth k and width w* and denoted by $\mathcal{U}_k^w$.

Example (continued): Fig. 6 shows a part of the unfolding for the grammars of the running example. We assume that the depth restriction k is at least 3, the width restriction is at least 2 and the arithmetic depth a is also at least 2. Table 1 shows the labelling of transitions over rewriting rules and the causal depth of each transition. The depth of every dummy place is 0, while the depth of any other place (edge) is the depth of the transition which has this edge in its post-set. Note that two inhibitor arcs at transitions t_{10} and t_{11} in Fig. 6 are inserted because of the presence of falsifying edges.

The unfolding faithfully represents system behaviour in the following sense.

Proposition 1. *Let G_0 be the start graph of the generating grammar and let $G_0 \Rightarrow^* G$ be a derivation of G such that: (i) the derivation consists of at most k (possibly concurrent) steps, (ii) no rewriting rule is applied more than w times to the same match, (iii) rules of the optimising grammar are applied only after those of the generating grammar. Then there is a reachable marking m in the unfolding truncated at depth k and width w, such that G is isomorphic to graph(m) up to isolated nodes. Furthermore, for every reachable marking m there is a graph G such that $G_0 \Rightarrow^* G$ and G is isomorphic to graph(m), up to isolated nodes.*

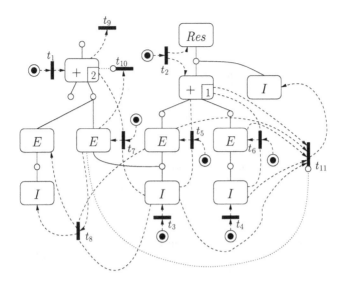

Fig. 6. A part of the unfolding for the example grammars (constant folding).

Table 1. Correspondence between transitions and rules.

transition	rule	depth
t_1	CreateSum	1
t_2	CreateSumRes	1
t_3, t_4	CreateInt	1
t_5, t_6	CreateConn	2

(a) Grammar $\mathcal{G}_{\mathrm{g}}$

transition	rule	depth
t_8	ConstantSplitting	3
t_9	KillUselessFunction	2
t_{10}	KillLonelyEdge	3
t_{11}	ConstantFoldingSum	3

(b) Grammar $\mathcal{G}_{\mathrm{o}}$

6   Generating Test Cases

The application order of optimisation techniques is not fixed a priori, but depends on specific situations within the input graph. Hence, with respect to testing, the situation is quite different from that of imperative programs for which there exists a widely accepted notion of coverage. In order to achieve in our case an adequate coverage for possible optimisation applications (of a single optimisation rule or a combination of different optimisation techniques) we propose to derive (graphical) input models which trigger the application of a single optimisation step or trigger the "combination" of several rules. In the last case the occurrences of the selected rules should be causally dependent on each other or in asymmetric conflict, such that error-prone interactions of rules can be tested.

Test cases triggering such a behaviour can be derived from the unfolding (up to depth k and up to width w) which provides a very compact description of all graphs which can be reached and of all rules which can be applied in a certain number of steps (see Proposition 1).

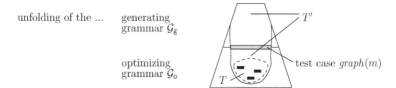

unfolding of the ...   generating
                         grammar $\mathcal{G}_g$

                         optimizing
                         grammar $\mathcal{G}_o$

Fig. 7. Schematical representation of test case generation.

In the following, we denote by $\mathcal{R}_g$ the set of rules of the generating grammar, by G_s its start graph and by $\mathcal{R}_o$ the set of optimising rules.

Definition 7 (Test case). *Given a set of optimising rules $R \subseteq \mathcal{R}_o$, a test case for R is an input model G such that a computation of the optimiser over G can use all the rules in R.*

Take a set R of interesting rules the interaction of which should be tested. The set R can be determined by the tester or by general principles, for instance one could take all sets R up to a certain size. Then proceed as follows:

(1) Take a set T of transitions in $\mathcal{U}_k^w$ (the unfolding up to depth k and up to width w) labelled by rules in R such that (a) for all $r \in R$, there exists a transition $t_r \in T$ such that t_r is labelled by r; (b) for all transitions $t_r \in T$ there exists a transition $t_r' \in T$ which is related to t_r by asymmetric conflict or causal dependency.

(2) Look for a set T' of transitions in $\mathcal{U}_k^w$ such that $T \subseteq T'$, and exactly the transitions of T' can be fired in a derivation of the grammar. Note that not every set T can be extended to such a T', since transitions might be in conflict or block each other by inhibitor arcs.

(3) Take the subset of transitions in T' labelled by rules in $\mathcal{R}_g$ and fire such rules, obtaining a marking m. Then $graph(m)$ is a test case for R.

See Fig. 7 for a schematical representation of the above procedure. Whenever the specification is non-deterministic, we cannot guarantee that the execution over the test case really involves the transformation rules in R, but this is a problem inherent to the testing of non-deterministic systems.

The set T' can be concretely defined by resorting to the notions of configuration and history in the theory of inhibitor Petri nets [7, 3]. Roughly, a *configuration* of $\mathcal{U}_k^w$ is a pair $\langle C, <_C \rangle$ where C is a set of transitions closed under causality and $<_C$ is a partial order including causality $<_\mathcal{U}$, asymmetric conflict $\nearrow_\mathcal{U}$ and a relation $<_p$ which considers the effects of inhibitor arcs: For any place s connected to a transition $t \in C$ by means of an inhibitor arc ($s \in \ {}^{\smile}t$) it chooses if t is executed before the place is filled or after the place is emptied. A configuration $\langle C, <_C \rangle$ can be seen as a concurrent computation, $<_C$ being a computational ordering on transitions, in the sense that the transitions in C can be fired in any total order compatible with $<_C$. A configuration $\langle C, <_C \rangle$ is called *proper* if the partial order $<_C$ is compatible with the grammar ordering, i.e., if $t_1 <_C t_2$ and t_2 belongs to the generating grammar then also t_1 belongs to the generating grammar.

The *history* of a transition t in a configuration $\langle C, <_C \rangle$, denoted by $C[t]$ is the set of transitions which must precede t in any computation represented by C. Formally, $C[t] = \{t' \in C \mid t' \leq_C t\}$. Note that a transition t can have several possible histories in different configurations. This is caused by the presence of read arcs and, even more severely, by inhibitor arcs. With asymmetric conflict only, there is a least history, the set of (proper) causes $\lfloor t \rfloor = \{t' \mid t' \leq_N t\}$, and the history of an event in a given configuration is completely determined by the configuration itself. With inhibitor arcs, in general, there might be several histories of a transition in a given configuration, and even several minimal ones.

Hence, coming back to the problem in step (2) above, the set T' we are looking for can be defined as a proper configuration including T. The choice among the possible configurations T' including T could be influenced by the actual needs of the tester. In many cases the obvious choice will be to privilege configurations with minimal cardinality, since these contain only the events which are strictly necessary to make the rules in T applicable.

Example (continued): We continue with our running example. Assume that we want a test case including the application of rule (ConstantFoldingSum), i.e., $R = \{(\text{ConstantFoldingSum})\}$ (the procedure works in the same way for more than one rule). Transition t_{11} is an instance of this rule and it is contained in several different configurations, for example $H_1 = \{t_2, t_3, t_4, t_5, t_6, t_{11}\}$ which creates only a sum with two integers and corresponding connections, $H_2 = \{t_1, t_2, t_3, t_4, t_5, t_6, t_7, t_8, t_{11}\}$ which creates another E-edge and removes it by (ConstantSplitting) and $H_3 = \{t_1, t_2, t_3, t_4, t_5, t_6, t_7, t_9, t_{10}, t_{11}\}$ which creates another E-edge and another sum and removes them by (KillUselessFunction) and (KillLonelyEdge). All these configurations are histories of t_{11}. Depending on the choice of the history, one obtains the two different test cases (the first for H_1 and the other for H_2 and H_3) in Fig. 8.

To understand how such a test case is derived, consider the history H_3. Causality ($<_\mathcal{U}$), asymmetric conflict ($\nearrow_\mathcal{U}$) and the relation $<_p$ in H_3 are depicted in Fig. 9. Note, for instance, that t_7 is forced to fire before t_{11}, since t_7 corresponds to a rule of grammar $\mathcal{G}_g$, while t_{11} is labelled with a rule of $\mathcal{G}_o$. After firing t_7, transition t_{11} is blocked by an inhibitor arc, and t_{11} can only be enabled by firing t_9 and t_{10}. Thus $t_9 <_p t_{10} <_p t_{11}$ is the only possible choice. Furthermore t_7 and t_9 are in asymmetric conflict ($t_7 \nearrow_N t_9$), since t_9 removes an element of the context of t_7. By taking the subset of rules in H_3 which belong to the generating grammar $\mathcal{G}_g$, namely $t_1, t_2, t_3, t_4, t_5, t_6, t_7$, and firing them, we obtain the test case on the right-hand side of Fig. 8.

7 Conclusion

We have presented a technique for deriving test cases for code generators with a graphical source language. The technique is based on the formalisation of code generators by means of graph transformation rules and on the use of (variants of the) unfolding semantics as a compact description of their behaviour.

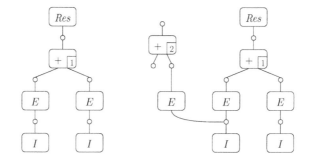

Fig. 8. Generated test cases.

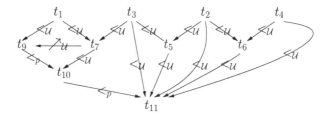

Fig. 9. Causality, asymmetric conflict and relation $<_p$.

The novelty of our approach consists in the fact that we consider graphical models and that we generate a compact description of system behaviour from which we can systematically derive test cases triggering specific behaviours. By using an unfolding technique we can avoid considering all interleavings of concurrent events, thereby preventing combinatorial explosion to a large extent. We believe that this technique can also be very useful for testing programs of visual programming languages.

This paper does not address efficiency issues. Note that the causality relation and asymmetric conflict of an occurrence net can be computed statically without firing the net. Hence, it is important to note that without inhibitor arcs, configurations and histories and hence test cases can be determined in a very efficient way. In the presence of inhibitor arcs, it is necessary to construct suitable relations $<_p$, leading to configurations. Obtaining such relations $<_p$ is quite involved and requires efficient heuristics, which we have already started to develop in view of an upcoming implementation of the test case generation procedure.

Acknowledgements

We would like to thank the anonymous referees for their helpful comments. We are also grateful to Andrea Corradini for comments on an earlier version of this paper.

References

1. T. Agerwala and M. Flynn. Comments on capabilities, limitations and "correctness" of Petri nets. *Computer Architecture News*, 4(2):81–86, 1973.
2. U. Assmann. Graph rewrite systems for program optimization. *TOPLAS*, 22(4):583–637, 2000.
3. P. Baldan, N. Busi, A. Corradini, and G.M. Pinna. Functorial concurrent semantics for Petri nets with read and inhibitor arcs. In *CONCUR'00 Conference Proceedings*, volume 1877 of *LNCS*, pages 442–457. Springer Verlag, 2000.
4. P. Baldan, A. Corradini, and U. Montanari. Unfolding and event structure semantics for graph grammars. In *Proceedings of FoSSaCS '99*, volume 1578 of *LNCS*, pages 73–89. Springer Verlag, 1999.
5. P. Baldan and B. König. Approximating the behaviour of graph transformation systems. In *Proc. of ICGT '02 (International Conference on Graph Transformation)*, pages 14–29. Springer-Verlag, 2002. LNCS 2505.
6. A.S. Boujarwah and K. Saleh. Compiler test case generation methods: a survey and assessment. *Information and Software Technology*, 39(9):617–625, 1997.
7. N. Busi and G.M. Pinna. Non sequential semantics for contextual P/T nets. In *Application and Theory of Petri Nets*, volume 1091 of *LNCS*, pages 113–132. Springer Verlag, 1996.
8. A. Corradini, U. Montanari, F. Rossi, H. Ehrig, R. Heckel, and M. Löwe. Algebraic approaches to graph transformation – part I: Basic concepts and double pushout approach. In G. Rozenberg, editor, *Handbook of Graph Grammars and Computing by Graph Transformation, Vol.1: Foundations*, chapter 3. World Scientific, 1997.
9. H. Ehrig, R. Heckel, M. Korff, M. Löwe, L. Ribeiro, A. Wagner, and A. Corradini. Algebraic approaches to graph transformation – part II: Single pushout approach and comparison with double pushout approach. In G. Rozenberg, editor, *Handbook of Graph Grammars and Computing by Graph Transformation, Vol.1: Foundations*, chapter 4. World Scientific, 1997.
10. H. Ehrig, M. Pfender, and H. Schneider. Graph grammars: An algebraic approach. In *Proc. 14th IEEE Symp. on Switching and Automata Theory*, pages 167–180, 1973.
11. S. Glesner, R. Geiß, and B. Boesler. Verified code generation for embedded systems. In *In Proceedings of the COCV-Workshop (Compiler Optimization meets Compiler Verification)*, volume 65.2 of *ENTCS*, 2002.
12. G. Karsai, A. Agrawal, F. Shi, and J. Sprinkle. On the use of graph transformation in the formal specification of model interpreters. *Journal of Universal Computer Science*, 9(11):1296–1321, 2003.
13. U. Montanari and F. Rossi. Contextual nets. *Acta Informatica*, 32(6), 1995.
14. S. Neema and G. Karsai. Embedded control systems language for distributed processing (ECSL-DP). Technical Report ISIS-04-505, Vanderbilt University, 2004.
15. M. Nielsen, G. Plotkin, and G. Winskel. Petri nets, event structures and domains, Part 1. *Theoretical Computer Science*, 13:85–108, 1981.
16. A. Nymeyer and J.-P. Katoen. Code generation based on formal BURS theory and heuristic search. *Acta Informatica*, 34:597–635, 1997.
17. P. Purdom. A sentence generation for testing parsers. *BIT*, pages 366–375, 1972.
18. L. Ribeiro. *Parallel Composition and Unfolding Semantics of Graph Grammars*. PhD thesis, Technische Universität Berlin, 1996.
19. I. Stürmer and M. Conrad. Test suite design for code generation tools. In *Proc. 18th IEEE Automated Software Engineering (ASE) Conference*, pages 286–290, 2003.

Stochastic Graph Transformation Systems*

Reiko Heckel, Georgios Lajios, and Sebastian Menge

Universität Dortmund, Germany
reiko@upb.de, georgios.lajios@udo.edu
sebastian.menge@uni-dortmund.de

Abstract. To formalize, measure, and predict availability properties, stochastic concepts are required. Reconfiguration and communication in mobile and distributed environments, where due to the high volatility of network connections reasoning on such properties is most important, is best described by graph transformation systems.

Consequently, in this paper we introduce stochastic graph transformation systems, following the outline of stochastic Petri nets. Besides the basic definition and a motivating example, we discuss the analysis of properties expressed in continuous stochastic logic including an experimental tool chain.

1 Introduction

Non-functional requirements concerning the availability of a system, measured in terms of *mean time between failures*, *time to repair*, or the average or maximal *answer time*, play an increasingly important role in mainstream software development. This is largely due to the change of focus from applications running on single machines or reliable local-area networks to Web-based distributed and mobile applications, where connections may be broken or varying in quality, or servers may be temporarily down.

Individual occurrences of failures are generally unpredictable. Therefore, stochastic concepts are required to formalize, measure, and predict availability properties. Specification formalisms providing suitable stochastic extensions include, for example, transition systems (i.e., Markov chains [2, 19]), stochastic Petri nets [18, 5] or process algebras [6, 10]. In order to meet the limitations of available analysis techniques, these formalisms mostly abstract from functional and architectural aspects like application data, changes in the network topology, etc.

However, even simple mobile devices today, like cell phones or PDAs, are equipped with communication and computation power beyond that of stationary computers a few years ago. In order to manage the resulting logic complexity of applications, high-level models of the functionality of the systems are required.

---

* Research funded in part by Deutsche Forschungsgemeinschaft, grant DO 263/8-1 [Algebraische Eigenschaften stochastischer Relationen] and by European Community's Human Potential Programme under contract HPRN-CT-2002-00275, [SegraVis].

H. Ehrig et al. (Eds.): ICGT 2004, LNCS 3256, pp. 210–225, 2004.

This calls for specification techniques which are able to integrate functional and architectural aspects with non-functional (stochastic) requirements.

Therefore, with the approach to stochastic graph transformation presented in this paper, we deliberately do not limit ourselves to models that are easy to analyze. We are convinced that, first of all, a problem-oriented style of modeling is required, because models have to be understood and validated by humans before they can be subject to automated analysis.

The paper is structured as follows. After discussing related work and outlining our approach, in Sect. 3 we present the basic definitions of typed graph transformation systems along with a simple example of a wireless network. Sect. 4 is devoted to definition of Markov Chains and Continuous Stochastic Logic (CSL). In Sect. 5 we make use of stochastic concepts by associating rates of exponential probability distributions with the rules of a graph transformation system and describing the Markov chain generated from it. Sect. 6 concludes the paper with an discussion of tools and relevant theoretical problems.

2   Related Work

The presented approach inherits from two lines of research: *stochastic modeling and analysis* and *graph transformation for mobility*. We discuss both of them in turn.

Stochastic Modeling and Analysis. The underlying model for stochastic analysis is provided by *Markov chains*, i.e., transition systems labeled with probability distributions on transitions [2, 19, 4]. *Stochastic Petri nets* provide a convenient method of describing Markov chains. The reachability graph of the net provides the underlying transition system, and its state transitions are decorated with the probabilities of the net transitions from which they are generated [5]. A similar idea lies behind *stochastic process algebras* where process algebras like CCS or the π-calculus are used to describe the transition system [6].

Generalizing from these examples, the idea of stochastic modeling can be phrased as follows. A state-based formalism is used to specify the desired behavior. From suitable annotations in the specification, probability distributions for the transitions of the generated transition systems are derived which provide the input to stochastic analysis techniques.

Our approach follows the same strategy, replacing Petri nets or process algebra with graph transformation systems. In this way, we obtain a high-level formalism in which both functional and non-functional aspects of mobile systems can be adequately specified.

Graph Transformation for Mobility. Graph transformation systems have been used for describing the semantics of languages for mobility, like the Ambient calculus [13] as well as corresponding extensions of the UML [3]. However, we will be more interested in direct applications to the modeling of mobile and distributed systems.

Generally, we may distinguish approaches which take a strictly local perspective, modeling a distributed system from the point of view of a single node, e.g., [16, 9] from those taking a global point of view, e.g., [24, 14]. In the latter case, each rule specifies the preconditions and effects of a potentially complex protocol with multiple participants, rather than a single operation as in the former case.

Our approach follows the second, global style of specification, with the running example derived from [14]. However, we think that the same combination of graph transformation with stochastic concepts could be applied to the more local style of specification.

3 Typed Graph Transformation

In this section, we provide the basic notions of typed graphs and their transformation according to the so-called algebraic or double-pushout (DPO) approach [12, 8].

Typed Graphs. By a *graph* we mean a directed unlabeled graph $G = \langle G_V, G_E, src^G, tar^G \rangle$ with a set of vertices G_V, a set of edges G_E, and functions $src^G : G_E \to G_V$ and $tar^G : G_E \to G_V$ associating to each edge its source and target vertex. A graph homomorphism $f : G \to H$ is a pair of functions $\langle f_V : G_V \to H_V, f_E : G_E \to H_E \rangle$ preserving source and target.

In this paper, vertices shall represent hardware components, with geographic relations, signals, and communication links as edges. The class of all admissible configurations is defined by means of a *type graph*, i.e., a graph specifying the available types of components and connections. Fixing a type graph TG, an *instance graph* $\langle G, g \rangle$ over TG is a graph G equipped with an attributed graph morphism $g : G \to TG$. A *morphism of typed graphs* $h : \langle G_1, g_1 \rangle \to \langle G_2, g_2 \rangle$ is a graph homomorphism $h : G_1 \to G_2$ that preserves the typing, that is, $g_2 \circ h = g_1$.

Example 1 (mobile system: types and configurations). Figure 1 on the left shows the type graph of our running example, a nomadic wireless network like a mobile phone network or wireless LAN. Stations linked by a geographic neighborhood relation form the static part of the network. To communicate with mobile devices, signals are broadcast and connections may be established. Stations can be broken, indicated by a Boolean attribute $ok : Bool$ with value $ok = false$, otherwise $ok = true$. (For simplicity, we have restricted our presentation to typed graphs, disregarding attributes. However, attributes of a finite data type, like Boolean, can be encoded in the graphical structure.) The idea is that a Device receiving the signal of a Station (because it is in range and the station is not broken) may establish a connection.

On the right, Fig. 1 shows a sample instance graph over the type graph, a state with two stations and two devices with the right device receiving the signal of and maintaining a connection with the right-hand side station.

Graph Transformation. The DPO approach to graph transformation has been developed for vertex- and edge-labeled graphs in [12] and extended to typed graphs in [8].

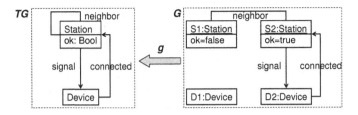

Fig. 1. Type graph TG and sample instance graph $\langle G, g \rangle$.

Given a type graph TG, a *TG-typed graph transformation rule* is a span of injective TG-typed graph morphisms $p = (L \xleftarrow{l} K \xrightarrow{r} R)$, called a *rule span*. The left-hand side L contains the items that must be present for an application of the rule, the right-hand side R those that are present afterwards, and the gluing graph K specifies the "gluing items", i.e., the objects which are read during application, but are not consumed.

A *direct transformation* $G \overset{p(o)}{\Longrightarrow} H$ is given by a *double-pushout (DPO) diagram* $o = \langle o_L, o_K, o_R \rangle$ as shown below, where (1), (2) are pushouts and top and bottom are rule spans.

$$
\begin{array}{ccccc}
L & \xleftarrow{\;l\;} & K & \xrightarrow{\;r\;} & R \\
{\scriptstyle o_L}\big\downarrow & (1) & {\scriptstyle o_K}\big\downarrow\;(2) & & \big\downarrow{\scriptstyle o_R} \\
G & \xleftarrow{\;g\;} & D & \xrightarrow{\;h\;} & H
\end{array}
$$

If we are not interested in the rule and diagram of the transformation we will write $G \overset{t}{\Longrightarrow} H$ or just $G \Longrightarrow H$. We will also identify the transformation step (i.e., the DPO diagram) with the label of the arrow, like in $t = G \overset{t}{\Longrightarrow} H$.

The DPO diagram o is a categorical way of representing the occurrence of a rule in a bigger context. Operationally, it formalizes the replacement of a subgraph in a graph by two gluing diagrams, called pushouts. The left-hand side pushout (1) is responsible for removing the occurrence of $L \setminus l(K)$ in G, resulting in graph D. The right-hand side pushout (2) adds a copy of $R \setminus r(K)$ to D leading to the derived graph H.

In general, we will not be interested in representing the intermediate graph K of a rule separately. We will thus assume that it forms a subgraph of both L and R such that $K = L \cap R$, denoting this rule by $p : L \Rightarrow R$. If they are clear from the context, we will drop the indices L, K, R of the occurrence morphisms.

A *graph transformation system* consists of a type graph and a set of rules which can, in general, be infinite. This can result in an infinite number steps outgoing from a single graph. To avoid this, we will have to make sure that only a finite number of rules is applicable to each finite graph.

A related problem is a consequence of the categorical formalization which defines the derived graph only up to isomorphism. Indeed, for a given rule and occurrence there may exist an infinite (even uncountable) number of results, all isomorphic copies of each other. This is, of course, a disaster for state space analysis.

It has been pointed out [7] that the naive solution of considering arbitrary isomorphism classes is unsatisfactory because the history of vertices and edges becomes confused if there is more than one isomorphism between two graphs. The solution proposed in [7] is based on identifying a canonical isomorphism for each pair of isomorphic graphs in order to establish a notion of identity across different graphs. Instead, we will explicitly remember the elements of interest by extending graphs with variable assignments.

Variable Assignments and Parameterized Rules. The introduction of assignments prepares the ground for interpreting (stochastic) temporal logic over graph transformation systems. Moreover, in order to derive a transition system with meaningful labels, we introduce rules with formal parameters.

Definition 1 (graph transformation with assignments). *A graph transformation system (with parameterized rules)* $\mathcal{G} = \langle TG, P, \pi \rangle$ *consists of*

- *a type graph TG*
- *a set of rule names P*
- *a mapping π associating with every rule name p a TG-typed rule span $s = (L \xleftarrow{l} K \xrightarrow{r} R)$ and a list of formal parameters $e_1, \ldots, e_n \in L \cup R$, i.e.*

$$\pi(p) = \langle e_1 \ldots e_n, s \rangle$$

In this case, we say that $p(e_1 \ldots e_n) : s$ is a rule of $\mathcal{G}$.

Fixing a countable set of variables X, an assignment *in a graph G is a partial mapping $a_G : X \to G_V + G_E$ into the disjoint union of $G's$ vertices and edges.*

The transformation of graphs with assignments *is defined as follows. Given a graph with assignment $\langle G, a_G \rangle$ and a rule $p(e_1 \ldots e_n) : L \xleftarrow{l} K \xrightarrow{r} R$, there exists a transformation $\langle G, a_G \rangle \xRightarrow{p(x_1, \ldots, x_n)} \langle H, a_H \rangle$ with actual parameters $x_1, \ldots, x_n \in X$, whenever*

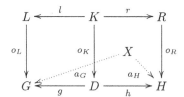

(i) a transformation from G to H via p can be constructed, represented by the double-pushout diagram above;

(ii) assignments a_G, a_H are compatible with the bottom span of the transformation, i.e., there exists an assignment $a_D : X \to D$ such that $g \circ a_D \subseteq a_G$ and $h \circ a_D \subseteq a_H$[1];

(iii) occurrences o_L, o_R are compatible with assignments a_G, a_H, i.e., for all $1 \leq i \leq n$: $e_i \in L$ implies $o_L(e_i) = a_G(x_i)$ and $e_i \in R$ implies $o_R(e_i) = a_H(x_i)$[2].

---

[1] By $f \subseteq f'$ we mean that f' is defined whenever f is, and in these cases they coincide.

[2] Here we mean strong equality, i.e., both sides of the equation need to be defined (and, of course, equal).

A transformation sequence with assignments in $\mathcal{G}$

$$\langle G_0, a_{G_0}\rangle \overset{p_1(x_{11},\ldots,x_{1n_1})}{\Longrightarrow} \langle G_1, a_{G_1}\rangle \overset{p_2(x_{21},\ldots,x_{2n_2})}{\Longrightarrow} \ldots \overset{p_k(x_{k1},\ldots,x_{kn_k})}{\Longrightarrow} \langle G_k, a_{G_k}\rangle$$

is a sequence of consecutive transformation steps with $p_i \in P$, briefly denoted by $\langle G_0, a_{G_0}\rangle \Longrightarrow^*_{\mathcal{G}} \langle G_k, a_{G_k}\rangle$ such that variables are single-assignment, *i.e., if $x_{ij} = x_{i'j'}$ for $i < i'$, then the first is an output and the second is an input parameter and $a_{G_{i+1}}(x_{ij})$ is related to $a_{G_{i'}}(x_{i'j'})$ via the bottom spans of the transformations from G_{i+1} to $G_{i'}$.*

In a transformation rule $p(e_1 \ldots e_n) : s$, the e_i play the role of abstract parameters, i.e., input parameters if $e_i \in L$ and output parameters if $e_i \in R$. Variables in X represent references to graph elements that are used in expressions like $p(x_1 \ldots x_n)$ to denote a step from G to H where rule p is applied at an occurrence which maps each $e_i \in L$ to $a_G(x_i) \in G$ and each $e_i \in R$ to $a_H(x_i) \in H$ (cf. condition (iii) above). Thus, the x_i are *logical variables* used to represent the concrete counterparts of the abstract parameters e_i.

Condition (ii) states that assignments are stable for all elements that are preserved by the transformation. The last condition over transformation sequences ensures that a repeated occurrence of a variable in a label denotes the same graphical object at different points in time.

Now we are ready to define the labeled transition system induced by a graph transformation systems. The labels shall be given by rule names with actual parameters. The state space is not built from concrete graph, but isomorphism classes of graphs with assignments

$$[\langle G, a_G\rangle] = \{\langle H, a_H\rangle \mid \exists \text{ isomorphism } i : G \to H \text{ such that } i \circ a_G = a_H\}.$$

Definition 2 (induced labeled transition system). *Let $\mathcal{G}$ be a graph transformation system, X be a fixed set of variables, and $\langle G_0, a_{G_0}\rangle$ be an initial graph with assignment. The transformations in $\mathcal{G}$ create a labeled transition system $LTS(\mathcal{G}, X, \langle G_0, a_{G_0}\rangle) = \langle L, S, \Rightarrow\rangle$, the* induced labeled transition system, *where*

- *$L = \{p(x_1, \ldots, x_n) | p \in P \wedge x_1, \ldots, x_n \in X\}$ is the set of rule names with actual parameters from X*
- *$S = \{[\langle G_n, a_{G_n}\rangle] | \langle G_0, a_{G_0}\rangle \Longrightarrow^*_{\mathcal{G}} \langle G_n, a_{G_n}\rangle\}$ is the set of isomorphism classes of graphs with assignments reachable from $\langle G_0, a_{G_0}\rangle$,*
- *$\Rightarrow \subseteq S \times L \times S$ is the transformation relation on graphs with assignments, lifted to isomorphism classes, i.e.,*

$$[\langle G, a_G\rangle] \overset{p(x_1\ldots x_n)}{\Longrightarrow} [\langle H, a_H\rangle] \text{ iff } \langle G_0, a_{G_0}\rangle \Longrightarrow^*_{\mathcal{G}} \langle G, a_G\rangle \overset{p(x_1\ldots x_n)}{\Longrightarrow} \langle H, a_H\rangle.$$

Note that the last item implies that all sequences in the transition system from the initial state satisfy the single-assignment condition implicit in the notion of transformation sequences.

Fig. 2. Failure and repair rules.

Example 2 (mobile system: rules and transformations). Graph transformation rules model the failure and recovery of components, their movement, the establishment and loss of connections, etc. Failure and repair of stations are expressed by the rules in Fig. 2.

Devices may move into and out of the sending range of stations, thus loosing and regaining the signal as specified in Fig. 3 in the top. A negative application condition, shown as a crossed out signal edge [15], ensures that there is at most one signal edge between a station and a device. More often than moving out of range entirely, devices should move between cells covered by neighboring stations, as shown in the bottom left of the figure. Finally, when a station is broken, its signal is lost as shown in Fig. 3 on the right.

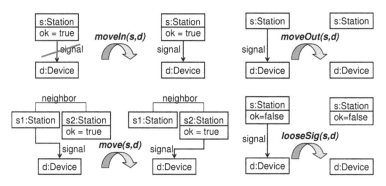

Fig. 3. Movement of devices and loosing the signal due to station failure.

If a device has a signal and no connection, a connection may be established as shown in Fig. 4 on the left. If a device does not have a signal of the station it is connected to, the connection is lost, too, cf. Fig. 4 on the right.

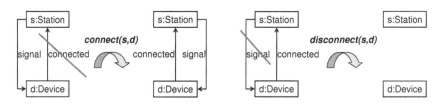

Fig. 4. Establishing a connection and loosing it due to loss of a signal.

In order to ensure continuous connectivity, many networks provide handover protocols. One possible behavior is specified in Fig. 5: If a device does not have a signal of the station it was connected to, but that of a neighboring station, the connection may be handed over to the second station.

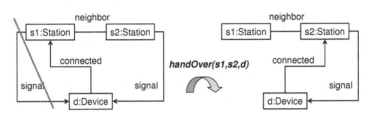

Fig. 5. Handover between two stations.

Figure 6 shows an application of rule *connect(s,d)* with corresponding actual parameters x, y. (As for rules, we skip the intermediate graph D in the sample transformation.) In accordance with the occurrences, the assignments map x to $S2$ and y to $D2$. Graph properties can be expressed by rules with equal left-and right hand sides, whose applicability appears as a loop in the transition system. Figure 6 shows the property rule $con(d)$ with $L = R := P$ indicating whether the device matching the formal parameter d is connected. In graph G this pattern does not find an occurrence. In graph H the property is satisfied if we instantiate d with y (referencing $D2$), but not for x (referencing $D1$).

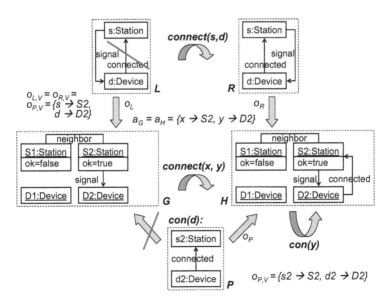

Fig. 6. Transformation step and state property with variable assignment.

4   Markov Chains and Stochastic Logic

After having described how to model mobility with graph transformation systems, and how based on this model non-functional requirements can be expressed, in this section we define Continuous Time Markov Chains (CTMCs) and explain how they can be analyzed. Furthermore, Continuous Stochastic Logic **CSL** is introduced to make assertions on CTMCs.

Markov Chains. First we provide some basic notions adopting the Q-matrix, a kind of "incidence matrix" of the Markov Chain, as elementary notion (cf. [19]).

Definition 3 (Q-matrix). *Let S be a countable set. A Q-matrix on S is a real-valued matrix $Q = Q(s, s')_{s,s' \in S}$ satisfying the following conditions:*

   (i) $0 \le -Q(s, s) < \infty$ for all $s \in S$,
   (ii) $Q(s, s') \ge 0$ for all $s \ne s'$,
   (iii) $\sum_{s' \in S} Q(s, s') = 0$ for all $s \in S$.

The Q-matrix is also called *transition rate matrix*. Some authors (e.g. [2]) use a more general notion of Q-matrix and call the matrices defined above *stable* and *conservative*. The most important notion is captured in the following

Definition 4 (CTMC). *A (homogeneous) Continuous-Time Markov Chain is a pair $\langle S, Q \rangle$ where S is a countable set of states and Q is a Q-matrix on S.*

If $s \ne s'$ and $Q(s, s') > 0$, then there is a transition from s to s'. The transition delay is exponentially distributed with rate $Q(s, s')$. Consequently, the probability that, being in s, the transition $s \to s'$ can be triggered within a time interval of length t is $1 - e^{-Q(s,s')t}$. The *total exit rate* $Q(s) = -Q(s, s)$ specifies the rate of leaving a state s to any other state. If the set $\{s' \mid Q(s, s') > 0\}$ is not a singleton, then there is a competition between the transitions originating in s. The probability that transition $s \to s'$ wins the 'race' is ([2], §1.2, Prop. 2.8)

$$\frac{Q(s, s')}{Q(s)}.$$

$$\begin{pmatrix} -2 & 0 & 2 \\ 4 & -6 & 2 \\ 0 & 2 & -2 \end{pmatrix}$$

Fig. 7. Example CTMC with Q-matrix.

The transition probability matrix $P(t) = (P_{ss'}(t))_{s,s' \in S}$ describes the dynamic behavior of a CTMC. It is the minimal non-negative solution of the equation

$$P'(t) = QP(t), \quad P(0) = I.$$

The (s, s')-indexed entry of $P(t)$ specifies the probability that the system is in state s' after time t if it is in state s at present. Given an initial distribution $\pi(0)$, the *transient solution* $\pi(t) = (\pi_s(t))_{s \in S}$ is then

$$\pi(t) = \pi(0)P(t).$$

In the finite case, $P(t)$ can be computed by the matrix exponential function, $P(t) = e^{Qt}$, but the numerical behavior of the matrix exponential series is rather unsatisfactory [23]. Apart from the transient solution, which specifies the behavior as time evolves, the *steady state* or *invariant distribution* is of great interest.

Definition 5 (invariant distribution). *A map $\pi : S \to [0, 1]$ is an invariant distribution if*

$$\pi Q = 0$$

$$\sum_{s \in S} \pi_s = 1.$$

In the example system shown in figure 7, the vector $\pi = \left(\frac{1}{3}, \frac{1}{6}, \frac{1}{2}\right)$ is an invariant distribution and the transient solution is given by

$$P(t) = \begin{pmatrix} -\frac{1}{3}e^{-6t} + e^{-4t} + \frac{1}{3} & \frac{1}{6} - \frac{1}{2}e^{-4t} + \frac{1}{3}e^{-6t} & \frac{1}{2} - \frac{1}{2}e^{-4t} \\ e^{-4t} - \frac{4}{3}e^{-6t} + \frac{1}{3} & \frac{4}{3}e^{-6t} + \frac{1}{6} - \frac{1}{2}e^{-4t} & \frac{1}{2} - \frac{1}{2}e^{-4t} \\ \frac{1}{3} - e^{-4t} + \frac{2}{3}e^{-6t} & \frac{1}{2}e^{-4t} - \frac{2}{3}e^{-6t} + \frac{1}{6} & \frac{1}{2} + \frac{1}{2}e^{-4t} \end{pmatrix}$$

Note that each row of $P(t)$ converges to π as $t \to \infty$, so the invariant distribution is always approached in the long term and is independent of the initial distribution, a property which can only be assured under certain reachability conditions. We say that a state s' *can be reached* from s, and write $s \rightharpoonup s'$, if there are states $s = s_0, \ldots, s_n = s'$, such that $Q(s_0, s_1) \cdot Q(s_1, s_2) \cdot \ldots \cdot Q(s_{n-1}, s_n) > 0$. If $s \rightharpoonup s'$ and $s' \rightharpoonup s$, then we say that s and s' *communicate*, and write $s \rightleftharpoons s'$.

Definition 6 (irreducible Q-matrix). *A Q-matrix Q is irreducible if $s \rightleftharpoons s'$ for all $s, s' \in S$.*

Non-irreducible matrices can be partitioned into their irreducible components for analysis. We will therefore consider only irreducible matrices. It can be shown that the first condition of Definition 5 is equivalent to $\pi P = \pi$ if Q is finite and irreducible. The entries of the transition probability matrix then converge to the steady state,

$$\lim_{t \to \infty} \pi_{ss'}(t) = \pi_{s'}.$$

In the infinite case, stronger assumptions are necessary (positive recurrence, [19], Th. 3.5.5 and 3.6.2).

A Stochastic Temporal Logic. We use extended Continuous Stochastic Logic **CSL** as presented in [4] to describe properties of CTMCs. Suppose that a labelling function $L : S \to 2^{AP}$ is given, associating to every state s the set of atomic propositions $L(s) \subseteq AP$ that are valid in s. The syntax of **CSL** is:

$$\Phi ::= tt \mid a \mid \neg\Phi \mid \Phi_1 \wedge \Phi_2 \mid \mathcal{S}_{\triangleleft p}(\Phi) \mid \mathcal{P}_{\triangleleft p}(\Phi_1 \mathcal{U}^I \Phi_2)$$

where $\triangleleft \in \{\leq, \geq\}$, $p \in [0,1]$, $a \in AP$ and $I \subset \mathbf{R}$ is an interval. The other boolean connectives are defined as usual, i.e., $ff = \neg tt$, $\Phi \vee \Psi = \neg(\neg\Phi \wedge \neg\Psi)$ and $\Phi \to \Psi = \neg\Phi \wedge \Psi$. The steady-state operator $\mathcal{S}_{\triangleleft p}(\Phi)$ asserts that the steady-state probability of the formula Φ meets the bound $\triangleleft p$. The operator $\mathcal{P}_{\triangleleft p}(\Phi_1 \mathcal{U}^I \Phi_2)$ asserts that the probability measure of the paths satisfying $\Phi_1 \mathcal{U}^I \Phi_2$ meets the bound $\triangleleft p$.[3]

A *path* through a CTMC M is an alternating sequence $\sigma = s_0 t_0 s_1 t_1 \cdots$ with $Q(s_i, s_{i+1}) > 0$. The time stamps t_i denote the sojourn time in state s_i. Let $Path^M$ be the set of all paths through M. Then, $Prob^M$ is defined as $Pr_s\{\sigma \in Path^M \mid \sigma \models \Phi_1 \mathcal{U}^I \Phi_2\}$, where Pr_s denotes the probability measure on sets of paths that start in s, as defined in [4] and $\Phi_1 \mathcal{U}^I \Phi_2$ asserts that Φ_2 will be satisfied at some time instant in the interval I and that at all preceding time instants Φ_1 holds. We define the semantics of CSL as follows:

$$
\begin{aligned}
s &\models tt &\Leftrightarrow& s \in S & s &\models \phi_1 \wedge \phi_2 &\Leftrightarrow& s \models \phi_1 \text{ and } s \models \phi_2 \\
s &\models a &\Leftrightarrow& a \in L(s) & s &\models \mathcal{S}_{\triangleleft p}(\phi) &\Leftrightarrow& \sum_{s \models \phi} \pi_s \triangleleft p \\
s &\models \neg\phi &\Leftrightarrow& s \not\models \phi & s &\models \mathcal{P}_{\triangleleft p}(\phi_1 \mathcal{U}^I \phi_2) &\Leftrightarrow& Prob^M(s, \phi_1 \mathcal{U}^I \phi_2) \triangleleft p
\end{aligned}
$$

CSL Example. Consider the CTMS of Fig. 7 with initial distribution $\pi(0) = (1, 0, 0)$. Define an atomic proposition a to be true in states 1 and 2, i.e. $L(1) = L(2) = \{a\}$, $L(3) = \emptyset$. Then the formula $\mathcal{S}_{\geq 0.5}(a)$ is true, because in the steady-state, a is fulfilled with probability 0.5, whereas $\mathcal{P}_{\geq 0.9}(a\ \mathcal{U}^{[0,1]} \neg a)$ is false, as the probability is only 0.86.

5   Stochastic Graph Transformation Systems

A stochastic graph transformation system associates with each rule name a positive real number representing the rate of the exponentially distributed delay of its application.

Definition 7 (stochastic GTS). *A stochastic graph transformation system* $\mathcal{SG} = \langle TG, P, \pi, \rho \rangle$ *consists of a graph transformation system* $\langle TG, P, \pi \rangle$ *and a function* $\rho : P \to \mathbf{R}^+$ *associating with every rule its application rate* $\rho(p)$.

Example 3 (mobile system: application rates). The application rates for the rules of our mobility example are shown in the following table.

---
[3] The other path and state operators can be derived. Details are given in [4].

rule name p	rate $\rho(p)$	rule name p	rate $\rho(p)$	rule name p	rate $\rho(p)$
repair	500	fail	1	connect	10000
moveIn	1	moveOut	1	disconnect	10000
move	100	looseSig	100000	handOver	100000

For convenience, we only use integer values for rates. We assume that the average time needed to repair a broken host is much smaller than the average time until the next failure (the expected value of the application delay is just the inverse of the rate). If one likes to interpret the proposed values with the unit *per day*, this means that in the mean term, an error occurs once a day and can be repaired in approximately three minutes.

Moving in and out are assumed to happen equally often, but much more scarcely than moving into another cell. When a station fails, the signal is lost almost immediately, so the rate is very high. Connecting, and disconnecting when the signal is lost, are assumed to happen equally fast. Finally, handover has to be possible within a few seconds to guarantee stability of connection.

Exponential distributions are single-parameter distributions that have a wide range of applications in analyzing the reliability and availability of electronic systems. Modeling a component's reliability with an exponential distribution presupposes that the failure rate is constant, which is generally true for electronic components during the main portion of their useful life. This means that the life of a component is independent of its current age (the *memoryless property*).

User mobility and connection duration can also be modeled as exponentially distributed, see [11, 26]. Of course, more detailed and realistic models are not confined to this approach but use other stochastic techniques, too, in order to take into account aspects like speed or direction [27].

From Stochastic GTS to Markov Chains. We now show how a stochastic graph transformation system gives rise to a Markov Chain, so that the analysis techniques described in Sect. 4 can be applied. First, we need an important notion.

Definition 8 (finitely-braching). *Let $LTS = \langle L, S, \Rightarrow \rangle$ a labeled transition system. Let $R(s, s') := \{ p \in P \mid \exists s' : s \overset{p}{\Longrightarrow} s' \}$ be the set of all transitions between s and s'. LTS is called* finitely-branching *iff $R(s, S)$, the set of all rules applicable to s, is finite for all $s \in S$.*

We are now ready for the main result:

Proposition 1 (and Definition: induced Markov chain). *Let $S\mathcal{G} = \langle TG, P, \pi, \rho \rangle$ be a stochastic graph transformation system with start graph $\langle G_0, a_{G_0} \rangle$ and let the induced labeled transition system $LTS(\mathcal{G}, G_0) = \langle L, S, \Rightarrow \rangle$ be finitely-branching. Assume for all $s \in S$ that $\rho(p) = 0$ if $p \in R(s, s)$. We set*[4]

$$Q(s, s') = \begin{cases} \sum\limits_{p \in R(s,s')} \rho(p) & \text{for} \quad s \neq s', \\ -\sum\limits_{t \neq s} Q(s, t) & \text{for} \quad s = s' \end{cases}$$

---

[4] We use the convention $\sum\limits_{\emptyset} = 0$ for the empty sum.

Then $\langle S, Q \rangle$ is a Continuous Time Markov chain, the induced Markov chain of $S\mathcal{G}$.

Proof. We have to show that Q is a Q-matrix on the set of all graphs reachable from the start graph G_0 by applying the transformation rules, and so $\langle S, Q \rangle$ is a CTMC. The finite-branching property is crucial to ensure that condition (i) of Def. 3 is fulfilled, because the sum $\sum_{p \in R(s,s')} \rho(p)$ is finite as is its indexing set.

It also implies that $R(s, s')$ is finite for all $s \neq s'$. So compliance with (ii) is secured by Def. 7 as all $\rho(p)$ have to be finite, and part (iii) of Def. 3 is fulfilled trivially because of the definition of $Q(s, s)$.

The assumption that $\rho(p) = 0$ if $p \in R(s, s)$ for all $s \in S$ is not needed formally, but as we ignore loops, we require them to have infinite delay, so that they will never be applied. $\square$

If the initial graph is finite, the same holds for all graphs in the system since transformations preserve finiteness. In this case, the finite-branching condition is ensured if every finite graph has a finite number of applicable rules only. This is trivial in the case of a finite set of rules in the system. However, there may by occasions where the set of rules is infinite, e.g., when rules with path expressions are regarded as a rule schemata expanding to countably infinite sets of rules. In this case, the condition may be violated if the given graph contains a circle and thus an infinite number of paths.

The initial distribution $\pi(0)$ is given by $\pi_s(0) = 1$ for $s = [\langle G_0, a_{G_0} \rangle]$ and $\pi_s(0) = 0$ else. As discussed above, for assuring existence of a unique steady state solution it is beneficial if the Markov Chain is finite and irreducible, i.e., every state is reachable from every other state in the system. This property, which can be checked on the graph transition system, is typically for non-deterministic models like the one given by our running example. Indeed, this model does not specify an individual application with determined behavior, but rather a whole class of mobile systems with similar structure and behavior, comparable to an architectural style [14]. For the case of infinite systems, analysis is possible under certain conditions (positive recurrence).

Stochastic Logic for Induced Markov Chains. In order to use **CSL** for analyzing stochastic graph transformation systems, we have to define the set of atomic propositions AP as well as the labeling function L.

Definition 9 (stochastic logic over graph transformation systems). *Assume a stochastic graph transformation system $S\mathcal{G} = \langle TG, P, \pi, \rho \rangle$, a set of variables X, and an initial graph with assignment $\langle G_0, a_{G_0} \rangle$. We define*

$$AP = \{p(x_1, \ldots, x_n) | p \in P, x_i \in X\}$$

as the set of all rule names with actual parameters, and the labeling of states

$$L(s) = \{l | s \overset{l}{\Longrightarrow} t\}$$

to be the set of all labels (instantiated rule names) on transitions outgoing from a state.

Thus, we can reason about the applicability of rules to graph elements referenced by variables in X, with the special case of property rules whose left- and right hand sides are copies of a pattern defining a structural graph property (cf. property $con(d)$ in Fig. 6). The transition rates of property rules are set to 0, so that they do not affect the Q-matrix.

In our example, with $X = \{x, y, z\}$ and assignment $a_G = \{x \mapsto D2, y \mapsto D1, z \mapsto S1\}$ into graph G in Fig. 1, this enables us to answer the following questions.

- In the long run, is station z broken at most 1% of the time: $\mathcal{S}_{<0.01}\ z.ok = false$? True.
- Is the overall connectivity of device x: $\mathcal{S}_{>0.5}\ con(y)$? Yes, its 0.555.
- Is the probability of device y being connected before t days from now at least 0.9: $\mathcal{P}_{0.9}\ (true\ \mathcal{U}^{[0,t]}\ con(y))$? True for, e.g., $t \geq 1.25$.

Tool Support. This interpretation of stochastic logic, although semantically satisfactory, has the disadvantage of being "non-standard": The construction of states with assignments is not supported by any graph transformation tool. However, even when analyzing the small example in this paper, we have found that tool support is indispensable in a stochastic setting, much more so than for the analysis of standard transition systems.

A possible way out that worked well in the example is the encoding of assignments into the graphical structure. The idea is to introduce a vertex for each logical variable together with an edge pointing to the referenced graph element. Extending the rules in a similar way and ensuring by means of types that variables are not confused, we obtain a slightly awkward, but semantically equivalent encoding of the transition system.

We have implemented this idea in the GROOVE tool [20] which allows the simulation of graph transformation systems and exports the labeled transition system generated from it. Given this system and providing in a separate file the rule's application rates, we can generate the Markov Chain in the input language of the probabilistic model checker PRISM [17]. Extracting loops via property rules from the transition system we also generate the labeling of states by atomic propositions. Thus we can verify CSL formulas with instantiated graph patterns as atomic propositions against CTMCs generated from stochastic graph transformation systems.

6   Conclusion

We have proposed an approach for analyzing graph transformation systems with stochastic methods. We have shown that under certain conditions, a stochastic graph transformation system induces a Continuous Time Markov Chain. This opens the door to a wide range of applications in modeling, and to tools and numerical methods for analysis.

We have constructed an experimental tool chain consisting of GROOVE and PRISM. Another approach to generate the input to the model checker is followed by [25, 21]. In that work, a transition system specification is generated directly from the given graph grammar. This approach could allow us to benefit from built-in optimizations or on-the-fly techniques, because the construction of the transition system is done inside the model checker. It also allows to use different modeling tools, like AGG or PROGRES [1, 22], which support attributed graph transformation systems. These options are currently under investigation.

Another line of research is the generalization of the basic theory of graph transformation systems concerning independence and critical pairs, concurrency, and synchronization to the stochastic case. In particular, the last issue could be relevant for a compositional translation of graph transformation systems into transition system specifications which are composed of modules combined by synchronous parallel composition.

Acknowledgement

The authors wish to thank Arend Rensink for his help with the GROOVE tool and enlightening discussions on graph transformation and temporal logic.

References

1. The attributed graph grammar system, http://tfs.cs.tu-berlin.de/agg/.
2. William G. Anderson. *Continuous-Time Markov Chains*. Springer, 1991.
3. L. Andrade, P. Baldan, and H. Baumeister. AGILE: Software architecture for mobility. In *Recent Trends in Algebraic Develeopment, 16th Intl. Workshop (WADT 2002)*, volume 2755 of *LNCS*, Frauenchiemsee, 2003. Springer-Verlag.
4. Christel Baier, Boudewijn R. Haverkort, Holger Hermanns, and Joost-Pieter Katoen. Model checking continuous-time markov chains by transient analysis. In *Computer Aided Verification*, pages 358–372, 2000.
5. Falko Bause and Pieter S. Kritzinger. *Stochastic Petri Nets*. Vieweg Verlag, 2nd edition, 2002.
6. Ed Brinksma and Holger Hermanns. Process algebra and markov chains. In J.-P. Katoen E. Brinksma, H. Hermanns, editor, *FMPA 2000*, number 2090 in LNCS, pages 183–231. Springer, 2001.
7. A. Corradini, H. Ehrig, M. Löwe, U. Montanari, and F. Rossi. Note on standard representation of graphs and graph derivations. Technical Report 92-25, FB13, 1992.
8. A. Corradini, U. Montanari, and F. Rossi. Graph processes. *Fundamenta Informaticae*, 26(3,4):241–266, 1996.
9. A. Corradini, L. Ribeiro, and F.L. Dotti. A graph transformation view on the specification of applications using mobile code. In G. Taentzer L. Baresi, M. Pezze and C. Zaroliagis, editors, *Proceedings 2nd Int. Workshop on Graph Transformation and Visual Modeling Technique (GT-VMT 01)*, volume 50.3 of *Elect. Notes in Th. Comput. Sci.* Elsevier Science, 2001.
10. P.R. D'Argenio. *Algebras and Automata for Timed and Stochastic Systems*. IPA Dissertation Series 1999-10, CTIT PhD-Thesis Series 99-25, University of Twente, November 1999.
11. J. Diederich, L. Wolf, and M. Zitterbart. A mobile differentiated services QoS model. In *Proceedings of the 3rd Workshop on Applications and Services in Wireless Networks*, 2003.

12. H. Ehrig, M. Pfender, and H.J. Schneider. Graph grammars: an algebraic approach. In *14th Annual IEEE Symposium on Switching and Automata Theory*, pages 167–180. IEEE, 1973.

13. F. Gadducci and U. Montanari. A concurrent graph semantics for mobile ambients. In St. Brooks and M. Mislove, editors, *Proc. Mathematical Foundations of Programming Semantics, Aarhus*, volume 45 of *Elect. Notes in Th. Comput. Sci.* Elsevier Science, 2001.

14. P. Guo and R. Heckel. Conceptual modelling of styles for mobile systems: A layered approach based on graph transformation. In *Proc. IFIP TC8 Working Conference on Mobile Information Systems, Oslo, Norway*, 2004. To appear.

15. A. Habel, R. Heckel, and G. Taentzer. Graph grammars with negative application conditions. *Fundamenta Informaticae*, 26(3,4):287 – 313, 1996.

16. D. Hirsch and U. Montanari. Consistent transformations for software architecture styles of distributed systems. In G. Stefanescu, editor, *Workshop on Distributed Systems*, volume 28 of *ENTCS*, 1999.

17. M. Kwiatkowska, G. Norman, and D. Parker. PRISM: Probabilistic symbolic model checker. In T. Field, P. Harrison, J. Bradley, and U. Harder, editors, *Proc. 12th Int. Conf. on Modelling Techniques and Tools for Computer Performance Evaluation (TOOLS'02)*, volume 2324 of *LNCS*, pages 200–204. Springer, 2002.

18. G. Conte S. Donatelli M. Ajmone-Marsan, G. Balbo and G. Franceschinis. *Modelling with Generalized Stochastic Petri Nets*. Wiley Series in Parallel Computing. John Wiley and Sons, 1995.

19. James R. Norris. *Markov Chains*. Cambridge University Press, 1997.

20. A. Rensink. The GROOVE simulator: A tool for state space generation. In J.L. Pfaltz, M. Nagl, and B. Böhlen, editors, *Applications of Graph Transformation with Industrial Relevance Proc. 2nd Intl. Workshop AGTIVE'03, Charlottesville, USA, 2003*, volume 3062 of *Lecture Notes in Computer Science*. Springer-Verlag, Berlin, 2004.

21. Ákos Schmidt and Dániel Varró. CheckVML: A tool for model checking visual modeling languages. In Perdita Stevens, Jon Whittle, and Grady Booch, editors, *UML 2003 - The Unified Modeling Language. Model Languages and Applications. 6th International Conference, San Francisco, CA, USA, October 2003, Proceedings*, volume 2863 of *LNCS*, pages 92–95. Springer, 2003.

22. Andy Schürr. PROGRES: a VHL-language based on graph grammars. In *Proc. 4th Int. Workshop on Graph-Grammars and Their Application to Computer Science*, number 532 in LNCS, pages 641–659. Springer-Verlag, 1991.

23. W. Stewart. *Introduction to the Numerical Solution of Markov Chains*. Princeton University Press, 1994.

24. G. Taentzer. Hierarchically distributed graph transformation. In *LNCSWil*, pages 304 – 320, 1996.

25. Dániel Varró. Automated formal verification of visual modeling languages by model checking. *Journal of Software and Systems Modelling*, 2003. Accepted to the Special Issue on Graph Transformation and Visual Modelling Techniques.

26. Oliver T. W. Yu and Victor C. M. Leung. Adaptive resource allocation for prioritized call admission over an ATM-based wireless PCN. *IEEE Journal on Selected Areas in Communications*, 15:1208–1224, 1997.

27. M. Zonoozi and P. Dassanayake. User mobility modeling and characterization of mobility patterns. *IEEE Journal on Selected Areas in Communications*, 15:1239–1252, 1997.

Model Checking Graph Transformations: A Comparison of Two Approaches

Arend Rensink[1], Ákos Schmidt[2], and Dániel Varró[2]

[1] University of Twente
P.O. Box 217, Enschede, 7500 AE, The Netherlands
`rensink@cs.utwente.nl`
[2] Budapest University of Technology and Economics
Department of Measurement and Information Systems
H-1117 Magyar tudósok körútja 2, Budapest, Hungary
`varro@mit.bme.hu`

Abstract. Model checking is increasingly popular for hardware and, more recently, software verification. In this paper we describe two different approaches to extend the benefits of model checking to systems whose behavior is specified by graph transformation systems. One approach is to encode the graphs into the fixed state vectors and the transformation rules into guarded commands that modify these state vectors appropriately to enjoy all the benefits of the years of experience incorporated in existing model checking tools. The other approach is to simulate the graph production rules directly and build the state space directly from the resultant graphs and derivations. This avoids the preprocessing phase, and makes additional abstraction techniques available to handle symmetries and dynamic allocation.

In this paper we compare these approaches on the basis of three case studies elaborated in both of them, and we evaluate the results. Our conclusion is that the first approach outperforms the second if the dynamic and/or symmetric nature of the problem under analysis is limited, while the second shows its superiority for inherently dynamic and symmetric problems.

Keywords: logic properties of graphs and transformations, analysis of transformation systems, semantics of visual techniques, model checking

1   Introduction

Graph transformation [6, 18] represents a rich line of research in computer science. Recently, a wide range of applications have been found especially in the theoretical foundations of diagrammatic specification formalisms such as UML. The main advantage of using graph transformation lies in the fact that not only the (static) program state of these UML-related models can be stored as graphs, but it is quite obvious and natural to define the evolution of these models by transformations on those graphs.

H. Ehrig et al. (Eds.): ICGT 2004, LNCS 3256, pp. 226–241, 2004.

However, software engineers may implant bugs into the system under design even if they use such a high-level and executable specification methodology as graph transformation. In this respect, one has to verify automatically and with mathematical preciseness that the system model fulfills all its requirements.

Model checking is one of the few verification techniques that, in some areas of computer science, have shown their benefits in practice and have been adopted by industry. However, the successes are mainly limited to hardware verification. It has been long recognized that software has features that make the problem inherently harder. Primary among those features is the *dynamic nature* of software, which typically relies heavily upon the dynamic allocation and deallocation of portions of memory to data structures (the heap) and control flow (the stack).

We argue in the paper that the strengths of graph transformation are precisely there where the weaknesses of current model checking approaches lie: namely, in the description of the dynamic nature of software. We have therefore sought to combine the two, by using graph transformations for the specification, and model checking for the verification of systems. This paper describes and compares two, quite different approaches towards this goal, namely, Check-VML [20, 24] and GROOVE [13, 16].

The reason we have chosen these approaches that tackle the model checking problem for graph transformations for a comparison is twofold: a) they represent the two obvious main roads (i.e. to compile graphs into an off-the-shelf tool or to write a state space generator for graphs) b) currently, they have the most extensive tool support.

Related work on model checking graph transformations. The theoretical basics of verifying graph transformation systems by model checking have been studied thoroughly by Heckel et al. in [9] (and subsequent papers). The authors propose that graphs can be interpreted as states and rule applications as transitions in a transition system, which idea is used in both approaches in the paper.

A theoretical framework by Baldan et al. [2] aims at analyzing a special class of hypergraph rewriting systems by a static analysis technique based on approximative foldings and unfoldings of a special class of Petri nets. Recently, this work has been extended in [1] to provide a precise (McMillan-style) unfolding strategy. This is essentially different from both approaches discussed in the current paper in that symmetric situations are only identified on a single path (thus they are might be investigated several times on different paths). But detecting that a certain situation has already been examined on a single path can be much cheaper in general compared to total isomorphism checks (as done in GROOVE).

Dotti et al. [5] use object-based graph grammars for modeling object-oriented systems and define a translation into SPIN to carry out model checking. The main difference (in contrast to CheckVML) is that the authors allow a restricted structure for graph transformation rules that is tailored to model message calls in object-oriented systems. Therefore, CheckVML is more general from a pure graph transformation perspective (i.e. any kind of rules are allowed) However, the framework of [5] relies on higher-level SPIN/Promela constructs (processes and channels), which might result better run-time performance.

Structure of the paper. The rest of the paper is structured as follows. Section 2 introduces the basic concepts of graph transformation systems and model checking on a motivating example. Section 3 and 4 provides an overview of the Check-VML and the GROOVE approach, respectively. We present the results of three case studies in Sec. 5. Finally, Section 6 concludes the paper.

2   Model Checking Graph Transformation Systems

2.1   A Motivating Example: The Concurrent Append Problem

As a motivating running example for the paper, we consider the "Concurrent Append" problem for the Java program listed in Fig. 1, which implements an append method on a list of cells. Given an integer value x as parameter, the program appends a new tail cell to the list if x is not contained in any of the existing cells. An example correctness criterion is that the list of cells must not contain the same value more than once. However, we allow that different threads may access the list concurrently by calling the append method, which might result in undesired race conditions without certain assumptions on atomicity in case of the Java program below.

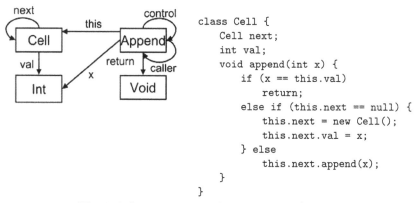

```
class Cell {
    Cell next;
    int val;
    void append(int x) {
        if (x == this.val)
            return;
        else if (this.next == null) {
            this.next = new Cell();
            this.next.val = x;
        } else
            this.next.append(x);
    }
}
```

Fig. 1. A Java program and its metamodel/type graph.

In the paper, we model this problem by using typed graphs [3] (or metamodels in UML terms) for describing the static structure. For instance, the metamodel in Fig. 1 expresses that a node of type Int may be connected to a node of type Cell via an edge of type val (that straightforwardly correspond to the Java attribute val). Furthermore, a next edge is leading from a cell point to the next cell in the list (if there is any). Each invocation of the append method is denoted by an Append node where we register the this pointer (which points to a cell), the caller invocation (which is another Append node), and the return value (of type Void) by edges of corresponding types. Finally, the program counter in each invocation of the append method is denoted by a control loop (self-edge).

All valid instance graphs (or models in UML terms) that represent specific invocations of the append method should comply to this metamodel in a type conforming way for both nodes and edges.

2.2   An Informal Introduction to Graph Transformation

The dynamic behavior of the recursive append method is captured by graph transformation rules. Graph transformation provides a visual, rule and pattern-based manipulation of graph models with solid mathematical foundations [6,18].

A graph transformation rule r consists of a left-hand side (LHS) and a right-hand side (RHS) graph and, potentially, some negative application conditions (NAC) which are traditionally denoted by (red) crosses. Informally, the execution of a rule on a given host graph G (i) finds a matching of the LHS in G, (ii) checks whether the matching can be extended to the matching of NAC (in which case the original matching of the LHS is invalid), (iii) removes all the graph elements from G which has an image in the LHS but not in the RHS, and (iv) creates new graph elements and embeds them into G to provide an image for rule elements that appear only in the RHS but not in the LHS. In other terms, the LHS and NAC graphs denote the precondition while the RHS denotes the postcondition for rule application.

In the paper, we use the rule notation of GROOVE (that is very similar to the notation used in the Fujaba [12]), which abbreviates the different LHS, RHS and NAC rule graphs into a single graph with the following conventions:

- *Reader* nodes and edges (i.e. elements that are part of LHS and RHS) are shown in solid thin (black) lines
- *Eraser* elements (that are part of the LHS but not the RHS) are depicted in dashed (blue) lines.
- *Creator* elements (that are part of the RHS but not the LHS) are depicted in solid thick (green) lines.
- *Embargo* elements (from the NAC) are shown in dotted (red) lines.

A sample graph transformation rule stating how to append a new element to the end of the cell list is depicted in Fig. 2 in both the traditional and the GROOVE notation[1]. The dynamic behavior of this highly recursive append problem is defined by four graph transformation rules (see Figs. 2 and 3).

Append a New Cell. Rule *Append* is responsible for appending a new cell to the list if the control reaches the last cell (see the negative condition inhibiting the existence of a next edge pointing to a Cell) and the value stored at this last cell is not equal to the method parameter. Furthermore, the append method returns one level up in the recursive call hierarchy as simulated by removing the bottom-most Append node and adding a return edge.

---

[1] Note that node identities are not allowed in the GROOVE tool, we only use them for presentation reasons.

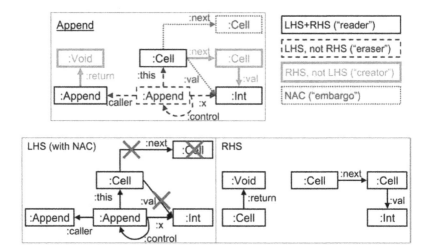

Fig. 2. The *Append* graph transformation rule in different notations.

Go to Next Cell. Rule *Next* checks whether the method parameter is *not* equal to the value stored at the current cell and makes a recursive call then for checking the next cell by generating a new Append node and passing the control to it.

Value Found in List. Rule *Found* checks if the method parameter is equal to the value stored at the current cell and, if so, returns the control to its caller append invocation node p in such a case.

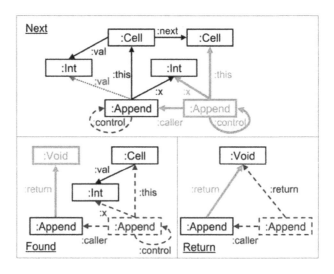

Fig. 3. Additional graph transformation rules for the concurrent append problem.

Return Result. Finally, rule *Return* simply removes an append invocation node (from the stack of recursive calls) if it has already calculated the result.

2.3 The Model Checking Problem of Graph Transformation Systems

The *model checking problem* is to automatically decide whether a certain correctness property holds in a given system by systematically traversing all enabled transitions in all states (thus all possible execution paths) of the system. The correctness properties are frequently formalized as LTL formulae.

In graph transformation systems, a state is a graph, while a transition corresponds to the application of a rule for a certain matching of the left hand side in such a graph. Traversing all enabled transitions then means applying all rules on all possible matchings. During this process, it is important to realize whether a certain state has been investigated before; therefore the model checker has to store all the graphs that it has encountered. Furthermore, ideally a model checker should exploit the symmetric nature of a problem by investigating isomorphic situations only once. The two approaches compared in the paper introduce very different techniques to tackle these problems.

For the current paper, we restrict our investigation to the verification of *safety* and *reachability* properties. A safety property defines a desired property that should always hold on every execution path or (equivalently) an undesired situation which should never hold on any execution paths (which we will call a *danger* property below). A reachability property describes, on the contrary, a desired situation which should be reached along at least one execution path. From a verification point of view, safety and reachability properties are dual: the refutation of a safety property is a counter-example which satisfies the reachability property obtained as the negation of the safety property. On the other hand, if a safety property holds (or a reachability property is refuted) the model checker has to traverse the entire state space.

A safety or reachability property can be interpreted as a special graph pattern (called *property graph* in the sequel) which immediately terminates the verification process if it is matched successfully. We have shown in [14] that the properties expressible in this way are equivalent to the $\exists \neg \exists$ fragment of ($\forall$-free) first order logic with binary predicates. For instance, the property that there exists an element that is shared among two list cells, expressed by the first-order logic property $\exists \, v \colon \mathsf{Int}, c_1, c_2 \colon \mathsf{Cell} \, . \, \mathsf{val}(c_1, v) \wedge \mathsf{val}(c_2, v) \wedge c_1 \neq c_2$ is alternatively encoded in the left graph of Figure 4.

The other property graphs in Fig. 4 are *Isolated* stating that every Int-object is either a method parameter or contained in the list, and *Terminated* expressing that there are no Append-methods left. Since different interleavings of append method calls access the list concurrently, we need model checking to ensure that these properties hold.

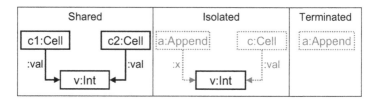

Fig. 4. Danger and reachability property graphs.

3 The CheckVML Approach

Main concepts. The main idea of the CheckVML approach [20, 23, 24] is to exploit off-the-shelf model checker tools like SPIN [11] for the verification of graph transformation systems. More specifically, it translates a graph transformation system parameterized with a type graph and an initial graph (via an abstract transition system representation) into its Promela equivalent to carry out the formal analysis in SPIN. Furthermore, property graphs are also translated into their temporal logic equivalents.

Traditional model checkers are based on so-called *Kripke structures*, which are state-transition models where the structure of a state consists of a subset of a finite universe of propositions. This determines the storage structures used (usually Binary Decision Diagrams or a variant thereof), the logic used to express properties (propositional logic extended with temporal operators, usually LTL or CTL) and the model checking algorithms (automata-based or tableau-based).

Since graph transformation is a meta-level specification paradigm (i.e. it defines how each instance of a type graph should behave) while the Kripke structure (transition system) formalism of Promela is a model-level specification language (i.e. a Promela model describes how a specific model should behave), the main challenge in this approach is *rule instantiation*, i.e. to generate one Promela transition for all the potential application of a graph transformation rule *in a preprocessing phase* at compile time.

The potential benefits of the CheckVML approach are the following:

1. It considers typed and attributed graphs which fits well to the metamodeling philosophy of UML and other modeling languages.
2. The size of the state vector depends only on the dynamic model elements (i.e., elements that can be altered by at least one graph transformation rule) while immutable static parts of a model are not stored in the state vector. This is a typical case for data-flow like systems (dataflow networks, Petri nets, etc).
3. It can be easily adapted to various back-end model checker tools.

The essential disadvantage of the approach is that dynamic model elements (that are not restricted by static constraints) easily blow up both the verification model and state space; moreover, symmetries in graphs can be handled for only very limited cases. Further research is necessitated in these directions.

Graphs and transformation rules. CheckVML uses directed, typed and attributed graphs (or MOF metamodels and models) as model representation (see the example presented in Sec. 2.1-2.2). Inheritance between node types is also supported.

Concerning the rule application strategy, CheckVML prescribes that a matching in the host graph should be an injective occurrence of the LHS (and NAC) graphs. Arbitrary creation and deletion of edges are allowed while there is an *a priori* upper bound for the number of nodes (of a certain type) potentially created during a verification run, which is passed as a parameter to the translator. Moreover, all dangling edges are implicitly removed when deleting a node.

New (unpublished) features. Several new features of CheckVML have been added as an incremental improvement since the previous papers [20, 24]. In order to improve performance, the entire tool has been rewritten, and the translator now uses relational database technology for generating all potential matches.

The main novelty is the automated translation of property graphs into LTL formulae; thus the users do not need SPIN-specific knowledge for stating properties. Since property graphs denote safety or reachability properties, thus this translation should find all potential matchings of this pattern in a similar way as done for instantiating rules.

Furthermore, in order to handle certain isomorphic situations, node identifiers have been ordered and made reusable. When a new node is created, the smallest available identifier is assigned to it, therefore, the same node can be reassigned several times. As a result, certain (but not all) isomorphic host graphs are handled only once.

Input / Output formats. CheckVML uses the GXL format [21] to store all host graphs, rule graphs (LHS, RHS, NAC) and property graphs. An XML configuration file is responsible for declaring the role of a certain graph (rule, host or property), and the user can set several translation parameters as well (e.g. upper bound for nodes of a certain type). In the near future, we plan to port CheckVML to a graph transformation tool with visual graph and rule editing facilities. The AGG tool [8] is a primary candidate due to the similarities between both the graph models and XML formats.

CheckVML generates a Promela model by instantiating rules on the host graph, and the SPIN representation of LTL formulae (which can be copy-pasted into the XSPIN framework). As a result, the users can work with high-level graph models and no (significant) SPIN-specific knowledge is required for modeling. However, counter-examples obtained as results of a verification run are currently available only in SPIN (for instance, in the form of scenarios/sequence diagrams), therefore, SPIN specific knowledge is required for the interpretation of analysis results. In the future, we also plan to investigate the possibilities of back-annotating analysis results so that they could be simulated (played back) in a graph transformation tool. Unfortunately, existing graph transformation tools provide very little support for importing entire execution traces.

Note that the overall ideas behind the CheckVML approach are not restricted to SPIN. In fact, thanks to a recent extension, CheckVML also yields an XML format for the generated transition system. Since the majority of model checker tools use transition systems as the underlying mathematical model (naturally, in their own dialect), this XML output can easily be adapted to various back-end model checkers, e.g. by XSLT scripts.

4   The GROOVE Approach

Main concepts. The idea behind the GROOVE approach (see [15] for further details on the project and downloads) is to use the core concepts of graphs and graph transformations all the way through during model checking. This means that states are explicitly represented and stored as graphs, and transitions as applications of graph transformation rules; moreover, properties to be checked should be specified in a graph-based logic, and graph-specific model checking algorithms should be applied.

This approach implies that very little of the theory and tool development for traditional model checkers can be applied immediately, since the most basic concept, namely the underlying model, has been extended drastically.

Currently only the state space generation part of GROOVE has been fully implemented. However, by the nature of graph transformation, this already implies the ability to express and check safety and reachability of graph properties, since they can be be formulated as rules with an identity morphism. Such a rule is applicable (idempotently) at precisely those states where the property holds. It is then straightforward to use such properties in controlling the state space generation process.

In particular, when treating a safety/danger property as an invariant, the state space generation halts with unexplored states exactly if the property is violated; when treating the inverse of a reachability property as an invariant, it halts precisely if the property is satisfied.

The GROOVE state space generator implements the process described in Sec. 2 to match each newly generated state against existing states up to isomorphism. While an isomorphism check is in principle quite expensive, for the examples we have worked out it stays within practical bounds.

The potential benefits of the GROOVE approach are the following:

1. There is no *a priori* upper bound to the size of the graphs;
2. There is an implicit symmetry check through the identification of isomorphic graphs;
3. No pre- or post-processing is necessary to apply the GROOVE tool to a given graph transformation system, or to translate the results of the model checking back into graphs;
4. Existing graph transformation theory can be directly brought to bear upon the tool, for instance, to discover rule independence or local confluence.

The essential disadvantage of this approach is that the huge body of existing research in traditional model checking is only indirectly applicable. In each of the areas where this applies, we aim to develop alternative techniques that are based directly on graphs.

1. *Storage techniques* (e.g., Binary Decision Diagrams). Rather than storing each graph anew, we store only the differences with the graph that it was derived from, in terms of the nodes and edges added and removed. This does mean that the actual graph has to be reconstructed when it is needed, e.g., for checking isomorphism; to alleviate the resulting time penalty this minimal representation is combined with caching.
2. *State space reduction techniques*, such as partial order reduction and abstraction. For state space reduction, we intend to use confluence properties of graph transformation rules (see advantage 3 above), or graph abstraction in the sense of *shape graphs* (see [19]). A first step towards the latter was reported in [17].
3. *Logics and model checking algorithms.* To replace the propositional logic used in traditional model checking, we have proposed a predicate graph logic in [13] for the purpose of formulating the properties to be checked. Some preliminary ideas on model checking such properties can be found in [4].

Graphs and transformation rules. GROOVE uses untyped, non-attributed, edge-labeled graphs without parallel edges. Node labels are not supported; however, we simulate them using self-edges (which indeed are also depicted by writing the labels inside the nodes). Furthermore, GROOVE implements the single pushout rewrite approach [7] (which means that dangling edges are removed while non-injective matching of the LHSs is allowed). It supports the use of negative application conditions. These can be used to specify, among other things, injectivity constraints; thus we can also simulate transformation systems in which the matchings are intended to be injective.

For the purpose of graph transformation, the lack of typing in GROOVE is not a serious drawback, since type information is not used to control the transformation process (although it may be used to optimize it). The absence of attributes is a potentially greater drawback. The examples presented here have been chosen such that attributes to not play a significant role, and so they can be simulated using ordinary edges. In fact, an extension to "true" attributes is not planned; rather, we plan to interpret data values as a special class of nodes, with ordinary edges pointing to them, as in [10].

Input/output formats GROOVE uses the GXL format [21] to store host graphs and rules. Each rule is saved as a single graph, combining the information in LHS, RHS and NACs by adding structure on the edges (in the form of a prefix) that indicates their role – or, in the case of nodes, by adding special edges for this purpose. A graph transformation system consists of all the rules in a single directory as well as its subdirectories (which are treated as separate namespaces, thus giving rise to a simple hierarchy of rules). In the future we plan to support the special-purpose format GTXL (see [22]).

Table 1. Feature comparison for CheckVML and GROOVE.

	Aspects of comparison	GROOVE	CheckVML
Graph model	Directed graphs	+	+
	Labeled graphs	+	
	Typed and attributed graphs		+
GT rules	NAC	+	+
	Node creation	arbitrary number	a priori upper bound
	Edge creation/removal	+	+
	Dangling edges	removed	removed
	Pattern matching	non-injective	injective
Input / Output	Graphical input (editor)	+	
	XML input	+	+
	Graphical output (trace)	built-in	MSCs in XSPIN
	XML output	+	+
	Property to be proved	graph constraint	graph constraint or LTL in SPIN
		safety / reachability	safety / reachability
Verification	Exploration strategies	extensible library	SPIN
	Symmetry recognition	graph isomorphism	reusable object ids
	Preprocessing	none	translation to SPIN

Alternatively, the GROOVE tool packages a stand-alone graph editor that can be used to construct graphs and rules and save them in the required format, or to read and edit graphs obtained from elsewhere.

State transition systems generated as a result of state space generation are also saved as GXL graphs, in which the nodes correspond to states (hence, graphs) and the edges to rule applications, labeled by the rule names.

State spaces can be generated either using a graphical simulator or using a command-line tool.

- The simulator, described before in [16], supports state space traversal by allowing the user to select and apply rules and matchings, all the while building up the transition system. Alternatively, the user can apply one of the available automatic state space exploration strategies (branching, linear, bounded, invariant). Graphs and transition system can be inspected by showing and hiding edges based on regular expressions over their labels.
- The command-line tool applies a pre-chosen strategy and generates and saves the resulting transition system.

Finally, Table 1 provides a brief summarizing comparison of the two tools.

5 Experimental Comparison

We have carried out three different case studies to compare the two model checking approaches, namely, (1) the Concurrent Append example of the current paper, (2) the dining philosophers problem as discussed in [24], and (3) a mutual exclusive resource allocation example taken from [9]. In the following we briefly describe the salient features of these cases.

Dining philosophers = Symmetries + No dynamic allocation. We have chosen this example because it is a traditional one, which has already been subject of a study for the CheckVML approach. For the purpose of GROOVE, this is an interesting case because with n philosophers, the example obviously has symmetry degree n, and this should then also be the reduction factor in number of states and transitions. On the other hand, the example has no dynamic allocation, and in this sense is not typical of the sort of problem for which we expect a graph transformation-based approach to be superior to traditional model checkers. We checked a safety property stating that no forks are ever held by more than one philosophers.

Concurrent append = Dynamic allocation + No symmetries. This is the running example of the paper. We have chosen it because it combines features that we believe to be typical of the "hard" problems in software verification. On the one hand, it contains dynamic allocation (list cells are created and append method frames are created and deleted), and on the other hand, it specifies concurrent behavior (several append methods are running in parallel). Note that, in the representation chosen here, the example has few non-trivial symmetries. In particular, all Int-objects in the list are distinguished by their value. We checked the property expressing that the list of cells is not allowed to contain the same value more than once.

Mutual exclusion = Dynamic allocation + Symmetries. In this example, processes try to access shared resources by using a token ring. We have chosen this example because it combines dynamic allocation (processes and resources can be created and deleted arbitrarily) and symmetry (processes and resources cannot be distinguished from one another). Moreover, a graph-based description of the protocol is very natural: an argument can be made that the specification of this protocol using graph transformation rules is superior to any other. The verified requirement was that at most one process may be allowed to access each resource at a time.

Of the examples presented here, this is the only one for which the state space is actually infinite (there is no upper bound to the numbers of processes and resources). Therefore, an artificial upper bound has to be imposed for the purpose of state space generation.

Results. In Table 2, we compare (a) the number of states traversed by the model checker during a successful verification run, (b) the number of transitions in the (reachable) state space, (c) the size of memory footprint of the state space, and (d) the execution time for the verification run. Furthermore, we also present the preprocessing time required for CheckVML to translate graph transformation systems into SPIN and the size of state vectors in SPIN.

We have done our best to produce the results of both approaches on an equal basis. We briefly list the characteristics of the experiments:

Memory Usage and Run-Time Performance. Experiments were run on a 3 GHz Pentium IV processor with 1 GB of memory. For the GROOVE experiments, Java Virtual Machine was started with an initial memory size

Table 2. Comparison of verification runs for CheckVML+SPIN and GROOVE.

	entities #	preproc s	vector #bits	states #	transitions #	memory MB	run time s
DinPhil							
CheckVML +	3	3.8	36	57	125	2.6	0.2
SPIN	4	4.5	48	181	554	2.6	0.2
	5	5.0	60	603	2.397	2.6	0.2
	8	6.6	112	25.961	171.058	8.8	0.6
	10	9.1	156	328.503	2.711.200	90.8	7.5
	12			out of memory (for SPIN)			
Groove	3			17	41	0.0	0.1
	4			45	148	0.0	0.2
	5			117	481	0.0	0.5
	8			3.261	21.536	1.7	13.6
	10			32.903	271.634	41.8	199.5
	12			106.329	965.589	74.2	793.3
Append	App : Cell	Append calls and cells initially present in the system					
CheckVML +	2:3 (orig)			out of memory (SPIN)			
SPIN	2:3 (mod)	15.3	200	22	169	2.6	0.5
	2:5 (mod)	117.9	316	86	395	2.6	1.1
	3:5 (mod)	1.021.0	520	3311	5764	37.0	40.0
	rest			out of time (for CheckVML)			
Groove	2:3			57	116	0.0	0.3
	2:5			145	292	0.0	0.6
	3:5			1.125	3.163	0.4	4.4
	3:7			2.716	7.768	1.0	13.0
	4:8			31.104	116.658	12.4	212.1
Mutex	pr:res:new						
CheckVML +	2:2:0	6.1	44	5.772	38.557	2.8	1.3
SPIN	3:2:0	18.5	60	697.004	6.843.310	83.2	14.7
	rest	24,3-180		at least 70 minutes (execution aborted)			
Groove	2:2:0			8.384	15.936	2.3	4.2
	3:2:0			262.054	620.284	79.1	162.6
	3:3:0			out of memory at around 1 million states			
	2:0:2			11.692	22.675	3.1	5.5
	2:0:3			515.134	1.206.935	155.6	361.8

Notation: *pr* is the number of processes initially present in the system
res is the number of resources initially present in the system
new is the upper bound for additional resources and additional processes

of 100 MB and maximum size of 1 GB. Although the space used for the actual storage of the state space is under 200 MB for all the cases reported here, during state space generation the tool heavily relies of caching and limiting the amount of available memory dramatically worsens the run-time performance.

Bounding the State Space. For the mutual exclusion example, we had to put a bound to the state space (as mentioned above). The way this is implemented in both tools is different. In GROOVE, all states which violate the bounding constraint are first generated and added to the transition system, after which the violation is detected and they are ignored for further exploration. In the CheckVML approach, on the other hand, the violation is checked first and hence those states are not generated at all. It turns out that the "spurious" states in the GROOVE results comprise about 85% of the state space and about 25% of the number of transitions.

Activating vs. Creating Nodes. For the original concurrent append example, SPIN failed even on very small examples due to the fact that each node and edge type is dynamic[2]. However, verification times for CheckVML +

---

[2] CheckVML generates the cross-product of all nodes and edges in the preprocessing phase even though the number of edges are only linear in the number of nodes.

SPIN could be reduced by a modeling trick, i.e. altering the models and the rules by adding an *isActive* attribute for each node type and only activating a node by changing this attribute instead of "real" node creation. This way, many graph elements that were originally dynamic are turned into static elements and thus abstracted by CheckVML during preprocessing. This example thus also demonstrated some pros and contras of graph attributes. However, the experimental results for the two approaches are not directly comparable (as denoted by the "(mod)" postfix after the append test cases in Table 2).

Evaluation. Based on Table 2, we come to the following overall conclusions:

- The space needed to store the transition system generated by both tools is comparable. Yet the techniques are very different: for GROOVE it is based on storing the differences between successive states, in terms of nodes and edges added and removed, whereas SPIN (and hence CheckVML) stores states as bit vectors that encode the entire graphs.
- The time needed to generate the states spaces is in a different order of magnitude: on the cases reported here, CheckVML typically takes under a tenth of the time that GROOVE does. For this we offer three possible explanations: (a) SPIN clearly shows the benefits of a more mature technology: over a decade of research has gone into improving its implementation. (b) Over the years, SPIN has been heavily optimized towards its implementation in C, whereas GROOVE has been implemented entirely in Java. (c) The approach taken by GROOVE, involving explicit graph matching and graph isomorphism checks, is inherently more complex.
- For each of the problems studied the GROOVE approach can handle a larger dimension than the CheckVML approach (which dimension is unquestionably significant for the append and mutual exclusion examples). This shows that the potential advantages of the approach, in terms of symmetry checking and dealing with dynamic allocation, also really show up in practice.

6   Conclusions

In the paper, we tackled the problem of model checking graph transformation systems by two different approaches. CheckVML exploits traditional model checking techniques for verification by translating graph transformation systems into SPIN, an off-the-shelf model checker. GROOVE, on the other hand, uses the core concepts of graphs and graph transformations all the way through during model checking.

We compared the two approaches on three case studies having essentially different characteristics concerning the dynamic and symmetric nature of the problem. Our overall conclusion is the following:

- If the problem analyzed lends itself well to be modeled in SPIN; that is, if dynamic allocation and/or symmetries are limited, it is to be expected that the CheckVML approach will always remain superior.

- On the other hand, for problems that are inherently dynamic, the GROOVE approach is a promising alternative.

Our conclusions also imply certain directions for future work. Obviously, CheckVML would yield a much more succinct state vector if further constraints on the metamodels (such as multiplicities) were handled in the preprocessing phase. For GROOVE, it is an interesting issue to make isomorphism checks optional (thus serving as an intelligent compression technique). However, the main line of research should find sophisticated abstraction techniques especially for infinite state graph transformation systems.

References

1. P. Baldan, A. Corradini, and B. König. Verifying finite-state graph grammars: an unfolding-based approach. In *Proc. of CONCUR '04*. Springer-Verlag, 2004. LNCS, to appear.
2. P. Baldan and B. König. Approximating the behaviour of graph transformation systems. In A. Corradini, H. Ehrig, H.-J. Kreowski, and G. Rozenberg (eds.), *Proc. ICGT 2002: First International Conference on Graph Transformation*, vol. 2505 of *LNCS*, pp. 14–29. Springer, Barcelona, Spain, 2002.
3. A. Corradini, U. Montanari, and F. Rossi. Graph processes. *Fundamenta Informaticae*, vol. 26(3/4):pp. 241–265, 1996.
4. D. Distefano, A. Rensink, and J.-P. Katoen. Model checking birth and death. In R. Baeza-Yates, U. Montanari, and N. Santoro (eds.), *Foundations of Information Technology in the Era of Network and Mobile Computing*, vol. 223 of *IFIP Conference Proceedings*, pp. 435–447. Kluwer Academic Publishers, 2002.
5. F. L. Dotti, L. Foss, L. Ribeiro, and O. M. Santos. Verification of object-based distributed systems. In *Proc. 6th International Conference on Formal Methods for Open Object-based Distributed Systems*, vol. 2884 of *LNCS*, pp. 261–275. 2003.
6. H. Ehrig, G. Engels, H.-J. Kreowski, and G. Rozenberg (eds.). *Handbook on Graph Grammars and Computing by Graph Transformation*, vol. 2: Applications, Languages and Tools. World Scientific, 1999.
7. H. Ehrig, R. Heckel, M. Korff, M. Löwe, L. Ribeiro, A. Wagner, and A. Corradini. In *[18]*, chap. Algebraic Approaches to Graph Transformation – Part II: Single pushout approach and comparison with double pushout approach, pp. 247–312. World Scientific, 1997.
8. C. Ermel, M. Rudolf, and G. Taentzer. In *[6]*, chap. The AGG-Approach: Language and Tool Environment, pp. 551–603. World Scientific, 1999.
9. R. Heckel. Compositional verification of reactive systems specified by graph transformation. In *Proc. FASE: Fundamental Approaches to Software Engineering*, vol. 1382 of *LNCS*, pp. 138–153. Springer, 1998.
10. R. Heckel, J. M. Küster, and G. Taentzer. Confluence of typed attributed graph transformation systems. In A. Corradini, H. Ehrig, H.-J. Kreowski, and G. Rozenberg (eds.), *Proc. ICGT 2002: First International Conference on Graph Transformation*, vol. 2505 of *LNCS*, pp. 161–176. Springer, Barcelona, Spain, 2002.
11. G. Holzmann. The model checker SPIN. *IEEE Transactions on Software Engineering*, vol. 23(5):pp. 279–295, 1997.
12. U. Nickel, J. Niere, and A. Zündorf. Tool demonstration: The FUJABA environment. In *The 22nd International Conference on Software Engineering (ICSE)*. ACM Press, Limerick, Ireland, 2000.

13. A. Rensink. Towards model checking graph grammars. In M. Leuschel, S. Gruner, and S. L. Presti (eds.), *Proceedings of the 3rd Workshop on Automated Verification of Critical Systems*, Technical Report DSSE–TR–2003–2, pp. 150–160. University of Southampton, 2003.
14. A. Rensink. Canonical graph shapes. In D. A. Schmidt (ed.), *Programming Languages and Systems – European Symposium on Programming (ESOP)*, vol. 2986 of *LNCS*, pp. 401–415. Springer-Verlag, 2004.
15. A. Rensink. Graphs for object-oriented verification, 2004. See http://www.cs.utwente.nl/~groove.
16. A. Rensink. The GROOVE simulator: A tool for state space generation. In M. Nagl, J. Pfalz, and B. Böhlen (eds.), *Applications of Graph Transformations with Industrial Relevance (AGTIVE)*, vol. 3063 of *LNCS*. Springer-Verlag, 2004.
17. A. Rensink. State space abstraction using shape graphs. In *Automatic Verification of Infinite-State Systems (AVIS)*, ENTCS. Elsevier, 2004. To appear.
18. G. Rozenberg (ed.). *Handbook of Graph Grammars and Computing by Graph Transformations: Foundations*. World Scientific, 1997.
19. M. Sagiv, T. Reps, and R. Wilhelm. Parametric shape analysis via 3-valued logic. *ACM Transactions on Programming Languages and Systems*, vol. 24(3):pp. 217–298, 2002.
20. Á. Schmidt and D. Varró. CheckVML: A tool for model checking visual modeling languages. In P. Stevens, J. Whittle, and G. Booch (eds.), *Proc. UML 2003: 6th International Conference on the Unified Modeling Language*, vol. 2863 of *LNCS*, pp. 92–95. Springer, San Francisco, CA, USA, 2003.
21. A. Schürr, S. E. Sim, R. Holt, and A. Winter. The GXL Graph eXchange Language. http://www.gupro.de/GXL/.
22. G. Taentzer. Towards common exchange formats for graphs and graph transformation systems. In J. Padberg (ed.), *UNIGRA 2001: Uniform Approaches to Graphical Process Specification Techniques*, vol. 44 (4) of *ENTCS*. 2001.
23. D. Varró. Towards symbolic analysis of visual modelling languages. In P. Bottoni and M. Minas (eds.), *Proc. GT-VMT 2002: International Workshop on Graph Transformation and Visual Modelling Techniques*, vol. 72 (3) of *ENTCS*, pp. 57–70. Elsevier, Barcelona, Spain, 2002.
24. D. Varró. Automated formal verification of visual modeling languages by model checking. *Journal of Software and Systems Modeling*, vol. 3(2):pp. 85–113, 2004.

Election, Naming
and Cellular Edge Local Computations
(Extended Abstract)

Jérémie Chalopin[1], Yves Métivier[1], and Wiesław Zielonka[2]

[1] LaBRI, Université Bordeaux I, ENSEIRB,
351 cours de la Libération, 33405 Talence, France
{chalopin,metivier}@labri.fr
[2] LIAFA, Université Paris 7
2, place Jussieu, case 7014, 75251 Paris Cedex 05, France
zielonka@liafa.jussieu.fr

Abstract. We examine the power and limitations of the weakest vertex relabelling system which allows to change a label of a vertex in function of its own label and of the label of one of its neighbours. We characterise the graphs for which two important distributed algorithmic problems are solvable in this model: naming and election.

1 Introduction

The role of the local computation mechanisms is fundamental for delimiting the borderline between positive and negative results in distributed computation. Understanding the power of local computations in different models enhances our understanding of basic distributed algorithms. Yamashita and Kameda [11], Boldi and al. [3], Mazurkiewicz [8] and Chalopin and Métivier [4] characterise families of graphs in which election is possible under different models of distributed computations. Even if these results cover a broad class of models there are still a few natural models which were not yet examined. We consider here one of such models where an elementary computation step modifies the state of one network vertex and this modification depends on its current state and on the state of one of its neighbours. We solve, in this model, two important algorithmic problems: the election problem and the naming problem, which turn out to be not equivalent. We give the characterisation of graphs which admit distributed solutions for both problems in this model.

To this end we find suitable graph morphisms that enable to formulate conveniently the necessary conditions in the spirit of Angluin [1]. It turns out that in our case the relevant morphisms are graph submersions. The presented conditions are also sufficient: algorithms, inspired by Mazurkiewicz [8], are given, that enable to solve the naming and the election problems for corresponding graphs.

H. Ehrig et al. (Eds.): ICGT 2004, LNCS 3256, pp. 242–256, 2004.

1.1   Our Model

A network of processors will be represented as a connected undirected graph $G = (V(G), E(G))$ without self-loop and multiple edges. As usual the vertices represent processors and edges direct communication links. The state of each processor is represented by the label $\lambda(v)$ of the corresponding vertex. An elementary computation step will be represented by relabelling rules of the form given schematically in Figure 1. The computations using uniquely this type of relabelling rules are called in our paper *cellular edge local computations*. Thus an algorithm in our model is simply given by some (possibly infinite but always recursive) set of rules of the type presented in Figure 1. A run of the algorithm consists in applying the relabelling rules specified by the algorithm until no rule is applicable, which terminates the execution. The relabelling rules are applied asynchronously and in any order, which means that given the initial labelling usually many different runs are possible.

Fig. 1. Graphical form of a rule for cellular edge local computations. If in a graph G there is a vertex labelled X with a neighbour labelled Y then applying this rule we replace X by a new label X'. The labels of all other graph vertices are irrelevant for such a computation step and remain unchanged. The vertex of G changing the label will be called *active* and filled with black, the neighbour vertex used to match the rule is called *passive* and marked as unfilled on the figure. All the other vertices of G not participating in such elementary relabelling step are called *idle*.

1.2   Election, Naming and Enumeration

The election problem is one of the paradigms of the theory of distributed computing. It was first posed by LeLann [6]. A distributed algorithm solves the election problem if it always terminates and in the final configuration exactly one processor is marked as *elected* and all the other processors are *non elected*. Moreover, it is supposed that once a processor becomes *elected* or *non elected* then it remains in such a state until the end of the algorithm. Elections constitute a building block of many other distributed algorithms since the elected vertex can be subsequently used to make some centralised decisions, to initialise some other activity, to centralise or to broadcast information etc.

The generic conditions listed above, required for an election algorithm, have a direct translation in our model: we are looking for a relabelling system where each run terminates with exactly one vertex labelled *elected* and all the other vertices labelled as *non elected*. Again we require that no rule allows to change either an *elected* or a *non-elected* label.

The naming problem is another important problem in the theory of distributed computing. The aim of a naming algorithm is to arrive at a final configuration where all processors have unique identities. To be able to give dynamically and in a distributed way unique identities to all processors is very important since many distributed algorithm work correctly only under the assumption that all processors can be unambiguously identified.

The enumeration problem is a variant of the naming problem. The aim of a distributed enumeration algorithm is to attribute to each network vertex a unique integer in such a way that this yields a bijection between the set $V(G)$ of vertices and $\{1, 2, \ldots, |V(G)|\}$.

We also distinguish two kinds of termination: the implicit one that simply means that the algorithm always terminates, and the explicit one that means that at least one node can detect that the algorithm has terminated. Obviously, if we can solve the naming problem with an explicit termination then we can also elect, for example the vertex with the smallest or the greatest identity.

The naming and the election problems are often equivalent for various computational models [8, 4], however this is not the case for our model. It turns out that in our model the class of graphs for which naming is solvable admits a simple and elegant characterisation; unfortunately a similar characterisation for the election problem is quite involved.

1.3   Overview of Our Results

Under the model of cellular edge local computations, we present a complete characterisation of graphs for which naming and election are possible: Theorems 7 and 11. The problems are solved constructively, we present naming and election algorithms that work correctly for all graphs where these problems are solvable. Imposed space limitations do not allow to present the correctness proofs for our algorithms.

1.4   Related Works

The election problem was already studied in a great variety of models [2, 7, 10]. The proposed algorithms depend on the type of the basic computation steps, they work correctly only for a particular type of a network topology (tree, grid, torus, ring with a known prime number of vertices etc.) or it is assumed that some initial extra knowledge is available to processors.

Yamashita and Kameda [11] consider the model where, in each step, one of the vertices, depending on its current label, either changes the label, or sends/receives a message via one of its ports. They proved that there exists an election algorithm for G if and only if the symmetricity of G is equal to 1, where the symmetricity depends on the number of labelled trees isomorphic to a certain tree associated with G ([11], Theorem 1 p. 75).

Mazurkiewicz [8] considers the asynchronous computation model presented in Figure 2. His characterisation of the graphs where enumeration/election are possible is based on the notion of non ambiguous graphs and may be formulated equivalently using coverings [5]. He gives a nice and simple enumeration algorithm for the graphs minimal for the covering relation.

Boldi and al. [3] consider a model where the network is a directed multigraph G and contrary to our model they allow also arc labellings. When a processor is

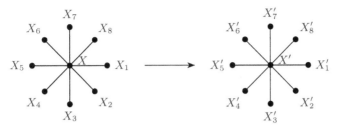

Fig. 2. In the model of Mazurkiewicz [8] a chosen vertex can change its label together with all its neighbours. The relabelling rules have therefore the form presented on this figure. Note that this involves a much greater degree of synchronisation than in the systems that we examine in our paper.

activated, it changes its state depending on its previous state and on the states of its ingoing neighbours; the outgoing neighbours do not participate in such an elementary computation step. They investigate two modes of computation: synchronous and asynchronous while in our paper only asynchronous computations are examined. In their study, they use fibrations which are generalisations of coverings. Boldi and al. [3] prove that there exists an election algorithm in their model for a graph G if and only if G is not properly fibred over another graph H (for the asynchronous case, they only consider discrete fibrations). To obtain this characterisation, they use the same mechanism as Yamashita and Kameda: each node computes its own view and next the node with the weakest view is elected.

In [4], three different asynchronous models are examined. Schematically, the rules of all three models are presented in Figure 3. Note that, contrary to the model we examine in the present paper, all these models allow edge labelling. It turns out that for all models described in Figure 3 naming and election are equivalent. In [4], it is proved that for all models described in Figure 3 the election and naming problems can be solved on a graph G if and only if G is not a covering of any graph H not isomorphic to G, where H can have multiple edges but no self-loop.

We can note that, although the model studied in this paper and model A in Figure 3 seem to be very close, the characterisations of graphs for which the naming problem and the election problem can be solved in these models are very different. The intuitive reason is that if we allow to label the edges then each processor can subsequently consistently identify the neighbours. On the other hand, in the model that we examine here, since edges are no more labelled, a vertex can never know if it synchronises with the same neighbour or another one.

2   Preliminaries

We consider finite, undirected, connected graphs $G = (V(G), E(G))$ with vertices $V(G)$ and edges $E(G)$ without multiple edges or self-loop. Two vertices u and

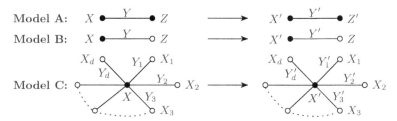

Fig. 3. Elementary relabelling steps for the three models examined in [4].

v are said to be adjacent or neighbours if $\{u, v\}$ is an edge of G (thus u and v are necessarily distinct since no self-loop is admitted) and $N_G(v)$ will stand for the set of neighbours of v. An edge e is incident to a vertex v if $v \in e$ and $I_G(v)$ will stand for the set of all the edges of G incident to v. The degree of a vertex v, denoted $d_G(v)$, is the number of edges incident with v.

A homomorphism between graphs G and H is a mapping $\gamma \colon V(G) \to V(H)$ such that if $\{u, v\} \in E(G)$ then $\{\gamma(u), \gamma(v)\} \in E(H)$. Since our graphs do not have self-loop, this implies that $\gamma(u) \neq \gamma(v)$ whenever u and v are adjacent.

We say that γ is an isomorphism if γ is bijective and γ^{-1} is a homomorphism. A class of graphs will be any set of graphs containing all graphs isomorphic to some of its elements. A graph H is a subgraph of G, noted $H \subseteq G$, if $V(H) \subseteq V(G)$ and $E(H) \subseteq E(G)$. An occurrence of H in G is an isomorphism γ between H and a subgraph H' of G.

For any set S, $|S|$ denotes the cardinality of S while $\mathcal{P}_{\text{fin}}(S)$ is the set of finite subsets of S. For any integer q, we denote by $[1, q]$ the set of integers $\{1, 2, \ldots, q\}$.

Throughout the paper we will consider graphs where vertices are labelled with labels from a recursive label set L. A graph labelled over L is a couple $\mathbf{G} = (G, \lambda)$, where G is an underlying non labelled graph and $\lambda \colon V(G) \to L$ is a (vertex) labelling function. The class of graphs labelled by L will be denoted by $\mathcal{G}_L$.

Let H be a subgraph of G and λ_H the restriction of a labelling $\lambda \colon V(G) \to L$ to $V(H)$. Then the labelled graph $\mathbf{H} = (H, \lambda_H)$ is called a subgraph of $\mathbf{G} = (G, \lambda)$; we note this fact by $\mathbf{H} \subseteq \mathbf{G}$. A homomorphism of labelled graphs is just a labelling-preserving homomorphism of underlying unlabelled graphs.

Submersions are locally surjective graph morphisms:

Definition 1. *A graph G is a submersion of a graph H via a morphism $\gamma \colon G \to H$ if $\forall v \in V(G)$, γ is surjective on the neighbourhood $N_G(v)$, that is $\gamma(N_G(v)) = N_H(\gamma(v))$. The graph G is a proper submersion of H if γ is not an isomorphism; G is submersion-minimal if G is not a proper submersion of any other graph.*

Naturally, submersions of labelled graphs are just submersions of underlying unlabelled graphs preserving the labelling.

For any set $\mathcal{R}$ of edge local relabelling rules of the type described in Figure 1 we shall write $\mathbf{G} \, \mathcal{R} \, \mathbf{G}'$ if $\mathbf{G}'$ can be obtained from $\mathbf{G}$ by applying a rule of $\mathcal{R}$ on some edge of $\mathbf{G}$. Obviously, $\mathbf{G}$ and $\mathbf{G}'$ have the same underlying graph G, only the labelling changes for exactly one (active) vertex. Thus, slightly abusing the

Fig. 4. The labelled graph **G** is a submersion of **H** via the mapping γ which maps each vertex of **G** labelled i to the unique vertex of **H** with the same label i. This submersion is proper and the graph **H** is itself submersion-minimal.

notation, $\mathcal{R}$ will stand both for a set of rules and the induced relabelling relation over labelled graphs. The transitive closure of such a relabelling relation is noted $\mathcal{R}^*$.

The relation $\mathcal{R}$ is called *noetherian* on a graph **G** if there is no infinite relabelling sequence $\mathbf{G}_0 \, \mathcal{R} \, \mathbf{G}_1 \, \mathcal{R} \ldots$, with $\mathbf{G}_0 = \mathbf{G}$. The relation $\mathcal{R}$ is noetherian on a set of graphs if it is noetherian on each graph of the set. Finally, the relation $\mathcal{R}$ is called noetherian if it is noetherian on each graph.

Clearly noetherian relations code always terminating algorithms.

The following simple observation exhibits a strong link between submersions and cellular edge local relabellings. This is a counterpart of the lifting lemma of Angluin [1] adapted to submersions.

Lemma 2 (Lifting Lemma). *Let $\mathcal{R}$ be a cellular edge locally generated relabelling relation and let* **G** *be a submersion of* **H**. *If* **H** $\mathcal{R}^*$ **H**$'$ *then there exists* **G**$'$ *such that* **G** $\mathcal{R}^*$ **G**$'$ *and* **G**$'$ *is a submersion of* **H**$'$.

Proof. It is sufficient to prove the lemma for one step of the relabelling. Let $\varphi : \mathbf{G} \to \mathbf{H}$ be a submersion, $\mathbf{G} = (G, \lambda), \mathbf{H} = (H, \nu)$. Suppose that a cellular edge rule is applied to an active vertex $v \in V(H)$ yielding a new labelling ν' on H. Then, since φ is a submersion, all vertices of $\varphi^{-1}(v)$ are pairwise non adjacent and therefore we can apply the same relabelling rule to all vertices of $\varphi^{-1}(v)$ in G, in any order. This yields a labelling λ' on G such that $\varphi : (G, \lambda') \to (H, \nu')$ remains a submersion. Note that we have simulated here one step relabelling in **H** by several relabellings in **G** that use the same rule. □

3 Enumeration and Naming Problems

We prove that there exists no naming algorithm and no enumeration algorithm on a graph **G** using cellular edge local computations if the graph is not submersion-minimal. The proof is analogous to that of Angluin [1].

Proposition 3. *Let* **G** *be a labelled graph which is not submersion-minimal. There is no naming algorithm for* **G** *and no enumeration algorithm for* **G** *using cellular edge local computations.*

Proof. Let **H** be a labelled graph not isomorphic to **G** such that **G** is a submersion of **H** via φ. For every cellular edge local algorithm $\mathcal{R}$, consider an execution of $\mathcal{R}$ on $\mathbf{H}$ that leads to a final configuration $\mathbf{H}'$. From Lemma 2, there exists an execution of $\mathcal{R}$ on $\mathbf{G}$ such that the final configuration $\mathbf{G}' = (G, \lambda')$ is a submersion of $\mathbf{H}'$. Since $\mathbf{G}'$ is not isomorphic to $\mathbf{H}'$, there exist distinct $v, v' \in V(G)$ such that $\lambda'(v) = \lambda'(v')$. Consequently, $\mathcal{R}$ does not solve either the naming or the enumeration problem on $\mathbf{G}$. $\qquad\square$

3.1   An Enumeration Algorithm

In this section, we describe a Mazurkiewicz-like algorithm $\mathcal{M}$ using cellular edge local computations that solves the enumeration problem on a submersion-minimal graph $\mathbf{G}$.

Each vertex v attempts to get its own number between 1 and $|V(G)|$. A vertex chooses a number and exchanges its number with its neighbours. If a vertex u discovers the existence of another vertex v with the same number, then it compares its *local view* (the numbers of its neighbours) with the local view of v. If the label of u or the local view of u is "weaker", then u chooses another number and broadcasts it again with its local view. At the end of the computation, every vertex will have a unique number if the graph is submersion-minimal.

We consider a graph $\mathbf{G} = (G, \lambda)$ with an initial labelling $\lambda \colon V(G) \to L$. During the computation each vertex $v \in V(G)$ will acquire new labels of the form $(\lambda(v), n(v), N(v), M(v))$, where:

- the first component $\lambda(v)$ is just the initial label (and thus remains fixed during the computation),
- $n(v) \in \mathbb{N}$ is the current *identity number* of v computed by the algorithm,
- $N(v) \in \mathcal{P}_{\mathrm{fin}}(\mathbb{N})$ is the *local view* of v. Intuitively, the algorithm will try to update the current view in such a way that $N(v)$ will consist of current identities of the neighbours of v. Therefore $N(v)$ will be always a finite (possibly empty) set of integers,
- $M(v) \subseteq \mathbb{N} \times L \times \mathcal{P}_{\mathrm{fin}}(\mathbb{N})$ is the current *mailbox* of v. It contains the whole information received by v during the computation.

The fundamental property of the algorithm is based on a total order on the set $\mathcal{P}_{\mathrm{fin}}(\mathbb{N})$ of local views, as defined by Mazurkiewicz [8].

Let $N_1, N_2 \in \mathcal{P}_{\mathrm{fin}}(\mathbb{N})$, $N_1 \neq N_2$. Then $N_1 \prec N_2$ if the maximal element of the symmetric difference $N_1 \bigtriangleup N_2 = (N_1 \setminus N_2) \cup (N_2 \setminus N_1)$ belongs to N_2. Note that in particular the empty set is minimal for $\prec$.

It can be helpful to note that the order $\prec$ is just a reincarnation of the usual lexicographic order. Let $n_1, n_2, \ldots, n_k$ and $m_1, m_2, \ldots, m_l$ be all elements of N_1 and N_2 respectively listed in the decreasing order (decreasing for the usual order over integers): $n_1 > n_2 > \cdots > n_k$ and $m_1 > m_2 > \cdots > m_l$. Then $N_1 \prec N_2$ iff either *(i)* $k \leq l$ and for all i, $1 \leq i \leq k$, $n_i = m_i$ or *(ii)* $n_i < m_i$ where i is the smallest index such that $n_i \neq m_i$.

If $N(u) \prec N(v)$ then we say that the local view $N(v)$ of v is stronger than the one of u (and $N(u)$ is weaker than $N(v)$).

We assume for the rest of this paper that the set of labels L is totally ordered by $<_L$.

Finally, we extend $\prec$ to a total order on $L \times \mathcal{P}_{\text{fin}}(\mathbb{N})$: $(l, N) \prec (l', N')$ if either $l <_L l'$ or ($l = l'$ and $N \prec N'$).

Occasionally we shall use the reflexive closure $\preceq$ of $\prec$.

We describe here the relabelling rules that define the enumeration algorithm.

First of all, to launch the algorithm there is a special initial rule $\mathcal{M}_0$ that just extends the initial label $\lambda(v)$ of each vertex v to $(\lambda(v), 0, \emptyset, \emptyset)$. The rules $\mathcal{M}_1$ and $\mathcal{M}_2$ are close to the rules used by Mazurkiewicz [8]. The first rule $\mathcal{M}_1$ enables a vertex to update its mailbox by looking at the mailbox of one of its neighbours:

(l_1, n_1, N_1, M_1) (l_2, n_2, N_2, M_2) (l_1, n_1, N_1, M_1') (l_2, n_2, N_2, M_2)

$\mathcal{M}_1$:   ●————————○ ⟶ ●————————○

If $M_2 \setminus M_1 \neq \emptyset$ then $M_1' := M_1 \cup M_2$.

The second rule $\mathcal{M}_2$ does not involve any synchronisation with a neighbour vertex. It enables a vertex v to change its identity if the current identity number $n(v)$ is 0 or if the mailbox of v contains a message from a vertex with the same identity but with a stronger label or a stronger local view.

            (l, n, N, M) (l, k, N, M')

$\mathcal{M}_2$:   ● ⟶ ●

If $n = 0$ or there exists $(n, \ell', N') \in M$ such that $(\ell, N) \prec (\ell', N')$ then $k := 1 + \max\{n' \mid \exists (n', \ell', N') \in M\}$ and $M' := M \cup \{(k, \ell, N)\}$.

(In the formula above we assume that max of an empty set is 0.)

The third rule $\mathcal{M}_3$ allows to change the current identity for a vertex v having a neighbour v' with exactly the same current label (all four components should be identical). Moreover, at the same step, the identity $n(v')$ of the neighbour v' of v is inserted into the local view $N(v)$ and at the same time all the elements m of $N(v)$ such that $m < n(v')$ are deleted from the local view. The rationale behind this deletion step is explained in the rule $\mathcal{M}_4$ below.

(l, n, N, M) (l, n, N, M) (l, k, N', M') (l, n, N, M)

$\mathcal{M}_3$:   ●————————○ ⟶ ●————————○

If $n > 0$ and $\forall (n, \ell', N') \in M$, $(\ell', N') \preceq (\ell, N)$ then $k := 1 + \max\{n' \mid \exists (n', \ell', N') \in M\}$, $N' := N \setminus \{m \in N \mid m < n\} \cup \{n\}$ and $M' := M \cup \{(k, \ell, N')\}$.

The fourth rule $\mathcal{M}_4$ enables a vertex v to add the current identity number $n(v')$ of one of its neighbours to its local view $N(v)$. As for the preceding rule, all the elements m belonging to $N(v)$ such that $m < n(v')$ are deleted from the current view.

The intuitive justification for the deletion of all such m is the following. Let us suppose that the vertex v synchronises with a neighbour v' and observes that the current identity number $n(v')$ of v' does not belong to his current view $N(v)$. Then, since the very purpose of the view $N(v)$ is to stock the identity numbers of all the neighbours, we should add $n(v')$ to the view $N(v)$ of v. But now two cases arise. If v synchronises with v' for the first time then adding $n(v')$ to the view of v is sufficient. However, it can also be the case that v synchronised with v' in the past and in the meantime v' has changed its identity number. Then v should not only add the new identity number $n(v')$ to its view but, to remain in a consistent state, we should delete the old identity number of v' from the local view of v. The trouble is that v has no means to know which of the numbers present in its view $N(v)$ should be deleted and it is even unable to decide which of the two cases holds (first synchronisation with v' or not). However, since our algorithm assures the monotonicity of subsequent identity numbers of each vertex, we know that the eventual old identity number of v' is less than the current identity $n(v')$. Therefore, by deleting all $m < n(v')$ from the local view $N(v)$ we are sure to delete all invalid information. Of course, in this way we risk to delete also the legitimate current identities of other neighbours of v from its view $N(v)$. However, this is not a problem since v can recover this information just by (re)synchronising all such neighbours.

$$
\begin{array}{cccc}
(l_1, n_1, N_1, M) & (l_2, n_2, N_2, M) & (l_1, n_1, N_1', M') & (l_2, n_2, N_2, M) \\
\mathcal{M}_4: \quad \bullet\!\!-\!\!-\!\!-\!\!-\!\!-\!\!\circ & & \longrightarrow & \bullet\!\!-\!\!-\!\!-\!\!-\!\!-\!\!\circ
\end{array}
$$

If $n_1 > 0$, $n_2 > 0$, $n_1 \neq n_2$, $\forall (n_1, \ell_1', N_1') \in M$, $(\ell_1', N_1') \preceq (\ell_1, N_1)$
$\forall (n_2, \ell_2', N_2') \in M$, $(\ell_2', N_2') \preceq (\ell_2, N_2)$, and $n_2 \notin N_1$ then
$N_1' := N_1 \setminus \{n' \in N_1 \mid n' < n_2\} \cup \{n_2\}$ and $M' := M \cup \{(n_1, \ell_1, N_1')\}$.

In the following $(\lambda(v), n_i(v), N_i(v), M_i(v))$ will denote the label of a vertex v after the ith computation step of the algorithm $\mathcal{M}$ given above.

The algorithm has some remarkable monotonicity properties:

Lemma 4. *For each step i and each vertex v: (A) $n_i(v) \leq n_{i+1}(v)$, (B) $N_i(v) \preceq N_{i+1}(v)$, and (C) $M_i(v) \subseteq M_{i+1}(v)$. Moreover, there exists at least one vertex v such that at least one of these inequalities/inclusions is strict for v.*

The local knowledge of a vertex v reflects to some extent some real properties of the current configuration:

Lemma 5. *Let $v \in V(G)$. If $(m, \ell, N) \in M_i(v)$ then for some vertex $w \in V(G)$, $n_i(w) = m$. If $n_i(v) \neq 0$ and $(m', \ell', N') \in M_i(v)$ then, for every $1 \leq m \leq m'$, there exist ℓ and N such that $(m, \ell, N) \in M_i(v)$.*

This fact allows to deduce the following properties of the final labelling:

Lemma 6. *Any run ρ of the enumeration algorithm on a connected labelled graph $\mathbf{G} = (G, \lambda)$ terminates and yields a final labelling $(\lambda, n_\rho, N_\rho, M_\rho)$ satisfying the following conditions:*

(1) Let m be the maximal number in the final labelling, $m = \max\{n_\rho(v) \mid v \in V(G)\}$. Then for every $1 \leq p \leq m$ there is some $v \in V(G)$ with $n_\rho(v) = p$,

and for all vertices v, v':

(2) $M_\rho(v) = M_\rho(v')$,

(3) $(\lambda(v), n_\rho(v), N_\rho(v)) \in M_\rho(v')$,

(4) $n_\rho(v) = n_\rho(v')$ implies that $\lambda(v) = \lambda(v')$ and $N(v) = N(v')$,

(5) $n \in N_\rho(v)$ if and only if there exists $w \in N_G(v)$ such that $n_\rho(w) = n$; in this case, $n_\rho(v) \in N_\rho(w)$.

Under the notation of Lemma 6 we can construct the labelled graph $\mathbf{H}_\rho$: the vertices of $\mathbf{H}_\rho$ are integers $n_\rho(v)$, i.e. final identity numbers, each $n_\rho(v)$ labelled by $\lambda(v)$ (this labelling is well defined by Lemma 6 (4)) and with edges naturally inherited from $\mathbf{G}$. In fact, the mapping n_ρ is a submersion from $\mathbf{G}$ to $\mathbf{H}_\rho$. This observation yields:

Theorem 7. *For every graph $\mathbf{G}$, the following statements are equivalent:*

(i) there exists a naming algorithm on $\mathbf{G}$ using cellular edge local computations,

(ii) there exists a naming algorithm with termination detection on $\mathbf{G}$ using cellular edge local computations,

(iii) there exists an enumeration algorithm on $\mathbf{G}$ using cellular edge local computations,

(iv) there exists an enumeration algorithm with termination detection on $\mathbf{G}$ using cellular edge local computations,

(v) the graph $\mathbf{G}$ is a submersion-minimal graph.

4   Election Problem

If we can solve the enumeration problem then we can solve the election problem; once a vertex gets the identity number $|V(G)|$ we declare it *elected*.

Nevertheless, in our model, the enumeration and the election problems are not equivalent. The graph $\mathbf{G}$ in Figure 5 is not submersion-minimal, since the morphism from $\mathbf{G}$ to $\mathbf{H}$ induced by the labelling of $\mathbf{G}$ is locally surjective and therefore neither the enumeration nor the naming problem can be solved on $\mathbf{G}$. But let us execute the preceding algorithm on $\mathbf{G}$. At the end, the vertex labelled 3 in $\mathbf{G}$ will know that it is unique with at least three different neighbours and therefore can declare itself as elected.

Fig. 5. A graph for which we can solve the election problem but not the enumeration problem.

We would like to give here necessary conditions characterising the graphs with solvable election problem. Given a graph $\mathbf{G}$, we denote by $\mathcal{S}_\mathbf{G}$ the set of graphs $\mathbf{H}$ such that there exists a submersion from $\mathbf{G}$ onto $\mathbf{H}$. From Lemma 2, any algorithm $\mathcal{A}$ that solves the election problem on $\mathbf{G}$ using cellular edge local computations will solve the election problem on every graph $\mathbf{H} \in \mathcal{S}_{\mathbf{G}}$.

Remark 8. Consider an algorithm $\mathcal{A}$ that solves the election problem on $\mathbf{G}$. Suppose that there exists a subgraph $\mathbf{G}'$ of $\mathbf{G}$ that is a submersion of a graph $\mathbf{H} \in \mathcal{S}_{\mathbf{G}}$ via a morphism φ. If there exists an execution of $\mathcal{A}$ on $\mathbf{H}$ that elects a vertex $v \in V(H)$ such that $|\varphi^{-1}(v)| > 1$, then there exists an execution of $\mathcal{A}$ on $\mathbf{G}'$ such that the label *elected* appears at least twice. Since each execution of $\mathcal{A}$ on $\mathbf{G}'$ can be extended to an execution of $\mathcal{A}$ on $\mathbf{G}$, there exists an execution of $\mathcal{A}$ over $\mathbf{G}$ that leads to the election of at least two vertices, this is in contradiction with the choice of $\mathcal{A}$. We can therefore define $P_{\mathbf{H}}(\mathbf{G}', \varphi) = \{v \in V(H) \mid |\varphi^{-1}(v)| > 1\}$ and each execution of $\mathcal{A}$ on $\mathbf{H}$ cannot elect a vertex $v \in P_{\mathbf{H}}(\mathbf{G}', \varphi)$.

Consider a graph $\mathbf{H} \in \mathcal{S}_{\mathbf{G}}$. Let $P_{\mathbf{H}}(\mathbf{G})$ be the union of all $P_{\mathbf{H}}(\mathbf{G}', \varphi)$ for φ ranging over all submersions of subgraphs $\mathbf{G}'$ of $\mathbf{G}$ to $\mathbf{H}$ and $C_{\mathbf{H}}(\mathbf{G}) = V(H) \setminus P_{\mathbf{H}}(\mathbf{G})$ (the elements of this set are called the candidates of $\mathbf{H}$ for $\mathbf{G}$). From Remark 8, every election algorithm $\mathcal{A}$ over $\mathbf{G}$ must be such that each execution of $\mathcal{A}$ over $\mathbf{H}$ should elect a vertex in $C_{\mathbf{H}}(\mathbf{G})$. Consequently, if there exists an election algorithm $\mathcal{A}$ on $\mathbf{G}$ then for every graph $\mathbf{H} \in \mathcal{S}_{\mathbf{G}}$, $C_{\mathbf{H}}(\mathbf{G}) \neq \emptyset$.

Suppose that there exist two disjoint subgraphs $\mathbf{G}_1$ and $\mathbf{G}_2$ of $\mathbf{G}$ such that $\mathbf{G}_1$ (resp. $\mathbf{G}_2$) is a submersion of a graph $\mathbf{H}_1 \in \mathcal{S}_{\mathbf{G}}$ (resp. $\mathbf{H}_2 \in \mathcal{S}_{\mathbf{G}}$). Then there does not exist any election algorithm using cellular edge local computations. Indeed, otherwise, there exists an execution of the algorithm on $\mathbf{G}$ such that the label *elected* appears once in $\mathbf{G}_1$ and once in $\mathbf{G}_2$, which is impossible for an election algorithm. Recapitulating:

Proposition 9. *Let $\mathbf{G}$ be a labelled graph such that there exists an election algorithm for $\mathbf{G}$ using cellular edge local computations. Then the following conditions are satisfied:*

1. *for every $\mathbf{H} \in \mathcal{S}_{\mathbf{G}}$, $C_{\mathbf{H}}(\mathbf{G}) \neq \emptyset$,*
2. *there do not exist two disjoint subgraphs $\mathbf{G}_1$ and $\mathbf{G}_2$ of $\mathbf{G}$ such that $\mathbf{G}_1$ (resp. $\mathbf{G}_2$) is a submersion of a graph $\mathbf{H}_1 \in \mathcal{S}_{\mathbf{G}}$ (resp. $\mathbf{H}_2 \in \mathcal{S}_{\mathbf{G}}$).*

4.1 An Election Algorithm

We now consider a graph $\mathbf{G}$ satisfying the conditions of Proposition 9.

Our aim is to present an algorithm such that each execution over $\mathbf{G}$ will detect a graph $\mathbf{H} \in \mathcal{S}_{\mathbf{G}}$ such that there exists a subgraph $\mathbf{G}'$ of $\mathbf{G}$ that is a submersion of $\mathbf{H}$.

To this end we adapt the enumeration algorithm from the preceding section and the termination detection algorithm of Szymansky, Shi and Prywes [9].

The idea is to execute the enumeration algorithm given for a graph and to reconstruct a graph from the mailboxes of the nodes. If the reconstructed graph is an element of $\mathcal{S}_{\mathbf{G}}$, the nodes check if they all agree on this graph.

As in Section 3.1, we start with a labelled graph $\mathbf{G} = (G, \lambda)$. During the computation vertices v will get new labels of the form $(\lambda(v), n(v), N(v), M(v),$

$a(v), H(v))$ representing the following information (again the first component $\lambda(v)$ remains fixed) :

- $n(v) \in \mathbb{N}$ is the *identity number* of the vertex v computed by the algorithm,
- $a(v) \in \mathbb{N}$ is the confidence level of the vertex v,
- $N(v)$ is the *local view* of v. If the vertex v has a neighbour v', relabelling rules will allow v to add the couple $(n(v'), a(v'))$ to $N(v)$. Thus $N(v)$ is always a finite set of couples of integers. For $N \in \mathcal{P}_{\text{fin}}(\mathbb{N}^2)$, we note $\Pi_1(N) = \{n \mid \exists (n, a) \in N\}$ the projection on the first component.
- $M(v) \subseteq \mathbb{N} \times L \times \mathcal{P}_{\text{fin}}(\mathbb{N})$ is the *mailbox* of v and contains the information received by v about the identity numbers existing in the graph and the local views associated with these numbers.
- $H(v)$ is the *history* of the vertex v. If at some computation step $(n, N, M, a) \in H(v)$ then it means that at some previous step the vertex v had a confidence level equal to a for the value M.

The first computation step $\mathcal{S}_0$ replaces just the initial label $\lambda(v)$ by $(\lambda(v), 0, \emptyset, \emptyset, -1, \emptyset)$. The following four rules mimic the rules of the enumeration algorithm:

$(l_1, n_1, N_1, M_1, -1, H_1)$ $(l_2, n_2, N_2, M_2, -1, H_2)$ $(l_1, n_1, N_1, M_1', -1, H_1)$ $(l_2, n_2, N_2, M_2, -1, H_2)$
$\mathcal{S}_1:$ ●────────○ ⟶ ●────────○
If $M_2 \setminus M_1 \neq \emptyset$ then $M_1' := M_1 \cup M_2$.

$(l, n, N, M, -1, H)$ $\qquad (l, k, N, M', -1, H)$
$\mathcal{S}_2:$ ● ⟶ ●
If $n = 0$ or there exists $(n, \ell', K') \in M$ such that $(\ell, \Pi_1(N)) \prec (\ell', K')$ then $k := 1 + \max\{n' \mid \exists(n', \ell', K') \in M\}$ and $M' := M \cup \{(k, \ell, \Pi_1(N))\}$.

$(l, n, N_1, M, -1, H_1)$ $\quad (l, n, N_2, M, -1, H_2)$ $\quad (l, k, N_1', M', -1, H_1)$ $\quad (l, n, N_2, M, -1, H_2)$
$\mathcal{S}_3:$ ●────────○ ⟶ ●────────○
If $n > 0$, $\Pi_1(N_1) = \Pi_1(N_2)$ and $\forall(n, \ell', K') \in M$, $(\ell', K') \preceq (\ell, \Pi_1(N_1))$ then $k := 1 + \max\{n' \mid \exists(n', \ell', K') \in M\}$, $N_1' := N_1 \setminus \{(n_1, a) \in N_1 \mid n_1 < n\} \cup \{(n, -1)\}$ and $M' := M \cup \{(k, \ell, \Pi_1(N_1'))\}$.

$(l_1, n_1, N_1, M, -1, H_1)$ $(l_2, n_2, N_2, M, -1, H_2)$ $(l_1, n_1, N_1', M', -1, H_1)$ $(l_2, n_2, N_2, M, -1, H_2)$
$\mathcal{S}_4:$ ●────────○ ⟶ ●────────○
If $n_1 > 0$, $n_2 > 0$, $n_1 \neq n_2$, $\forall(n_1, \ell_1', K_1') \in M$, $(\ell_1', K_1') \preceq (\ell_1, \Pi_1(N_1))$, $\forall(n_2, \ell_2', K_2') \in M$, $(\ell_2', K_2') \preceq (\ell_2, \Pi_1(N_2))$, and $(n_2, -1) \notin N_1$ then $N_1' := N_1 \setminus \{(n', -1) \in N_1 \mid n' < n_2\} \cup \{(n_2, -1)\}$ and $M' := M \cup \{(n_1, \ell_1, N_1')\}$.

The fifth rule says that if a vertex v detects that all the neighbours it knows have a confidence level $a \geq a(v)$ then it can increment its own confidence level.

To define this rule we need some additional notations. Given a mailbox content M, for each $n > 0$ we define $\pi_n(M)$ as the set of all triples $(n, \ell, N) \in M$ with the first component n. For each non empty set $\pi_n(M)$ we conserve in the mailbox only the triple (n, ℓ, N) with the greatest couple (n, N) for the order $\prec$. This operation gives a new mailbox content that we shall note $u(M)$.

The next step consists in defining a graph G_M. If there exist $(n_1, \ell_1, N_1), (n_2, \ell_2, N_2) \in u(M)$ such that $(n_2, \ell_2) \in N_1$ and $(n_1, \ell_1) \notin N_2$ then we set $G_M = (\emptyset, \emptyset)$. Otherwise, G_M is the graph such that $V(G_M) = \{n \mid (n, \ell, N) \in u(M)\}$ and $E(G_M) = \{\{n_1, n_2\} \mid \exists (n_1, \ell_1, N_1), (n_2, \ell_2, N_2) \in u(M), (n_2, \ell_2) \in N_1$ and $(n_1, \ell_1) \in N_2\}$. The labelling of G_M is inherited from the set M: for $(n, \ell, N) \in u(M)$, $\lambda_M(n) = \ell$. We will denote by $\mathbf{G}_M = (G_M, \lambda_M)$ the corresponding labelled graph.

S_5 :

(l, n, N, M, a, H) $\qquad$ $(l, n, N, M, a+1, H)$

$\bullet \longrightarrow \bullet$

This rule applies whenever $\forall (n, \ell', N') \in M, (\ell', N') \preceq (\ell, \Pi_1(N)), \mathbf{G}_M \in \mathcal{S}_\mathbf{G}$, and $\forall (n', a') \in N, a \leq a'$, and $a \leq |V(\mathbf{G})| + 1$.

The sixth rule enables a node v to update its knowledge of the confidence level of one of its neighbour if the confidence level of this neighbour has increased.

S_6 :

$(l_1, n_1, N_1, M, a_1, H_1)$ $\quad$ $(l_2, n_2, N_2, M, a_2, H_2)$ $\qquad$ $(l_1, n_1, N_1', M, a_1, H_1)$ $\quad$ $(l_2, n_2, N_2, M, a_2, H_2)$

$\bullet \longrightarrow \circ \quad\quad \bullet \longrightarrow \circ$

If $a_1 \geq 0$, $\forall (n_2, \ell_2', N_2') \in M, (\ell_2', N_2') \preceq (\ell_2, \Pi_1(N_2))$, and there exists $(n_2, a) \in N_1$ such that $a_2 > a$ then $N_1' := N_1 \setminus \{(n_2, a)\} \cup \{(n_2, a_2)\}$.

The rule S_7 enables a vertex v to change the value of its mailbox M whenever there exists a neighbour v' that used to have a confidence level a according to M such that $a \geq a(v) - 1$ and such that its mailbox has changed. If a vertex changes its mailbox, then it modifies also its history $H(v)$, so as to remember its former confidence level.

S_7 :

$(l_1, n_1, N_1, M_1, a_1, H_1)$ $\quad$ $(l_2, n_2, N_2, M_2, a_2, H_2)$ $\qquad$ $(l_1, n_1, N_1', M', -1, H_1)$ $\quad$ $(l_2, n_2, N_2, M_2, a_2, H_2)$

$\bullet \longrightarrow \circ \quad\quad \bullet \longrightarrow \circ$

If $\exists (n, l, N) \in M_2 \setminus M_1$ and either $a_1 = 0$ or $(a_1 \geq 0$ and $\exists (n, N, M_1, a) \in H_2, \exists (n, a') \in N_1, a \geq a')$ then $N_1' := \{(n', -1) \mid \exists (n, a) \in N_1\}$, $M' := M_1 \cup M_2$ and $H_1' := H_1 \cup \{(n_1, N_1, M_1, a_1)\}$.

4.2 Correctness of the Election Algorithm

In the following $(\lambda(v), n_i(v), N_i(v), M_i(v), a_i(v), H_i(v))$ will stand for the label of the vertex v after the ith computation step of the election algorithm. The most important property of the algorithm is given in the following proposition. Roughly speaking it states that if the confidence level of vertex v is $|V(G)| + 2$ then $\mathbf{G}$ contains a submersion of $\mathbf{G}_{M(v)}$.

Proposition 10. *If there exists a vertex $v_0 \in V(\mathbf{G})$ and a step i_0 such that $a_{i_0}(v_0) = |V(\mathbf{G})| + 2$, $\mathbf{G}$ contains a submersion $\mathbf{H}$ of $\mathbf{G}_{M_{i_0}(v_0)}$ and for every step $i \geq i_0$ and for every vertex $v \in V(\mathbf{H})$, $M_i(v) = M_{i_0}(v_0)$.*

From Proposition 10 we deduce that if the conditions of Proposition 9 are satisfied then adding the following rule $\mathcal{S}_8$ allows to elect a unique vertex of $\mathbf{G}$: $\mathcal{S}_8$: the label (ℓ, n, N, M, a, H) such that $n = \max\{n \in C_{\mathbf{G}_M}(\mathbf{G})\}$ and $a = |V(\mathbf{G})| + 2$ is replaced by *elected*. The last two rules serve to propagate the information that there is an elected vertex: $\mathcal{S}_9$ allows to transform a label of a vertex with an *elected* neighbour to *non-elected* and $\mathcal{S}_{10}$ propagates the *non-elected* label to all neighbours which are neither *elected* nor *non-elected*.

Summarising we get:

Theorem 11. *There exists an election algorithm over a given graph $\mathbf{G}$ using cellular edge local computations if and only if the following conditions are satisfied:*

1. *for every $\mathbf{H} \in \mathcal{S}_{\mathbf{G}}$, $C_{\mathbf{H}}(\mathbf{G}) \neq \emptyset$,*
2. *there do not exist two disjoint subgraphs $\mathbf{G}_1$ and $\mathbf{G}_2$ of $\mathbf{G}$ such that $\mathbf{G}_1$ (resp. $\mathbf{G}_2$) is a submersion of a graph $\mathbf{H}_1 \in \mathcal{S}_{\mathbf{G}}$ (resp. $\mathbf{H}_2 \in \mathcal{S}_{\mathbf{G}}$).*

5   Examples

If we assume that nodes of a graph $\mathbf{G}$ have unique identifiers then $\mathbf{G}$ is a submersion-minimal graph and the knowledge of its size allows an election.

5.1   Trees, Grids and Complete Graphs

Consider an unlabelled tree T. Since we can colour each tree T with just two colours, if T has at least 2 vertices such colouring yields a submersion of T into the graph K_2 with two vertices and one edge between them. Such a submersion is non trivial if T has at least 3 vertices. Therefore for such trees there does not exist a naming algorithm using cellular edge local computations.

If $\varphi_1 : T \rightarrow K_2$ is a submersion (colouring) of T then exchanging the two colours we get another submersion φ_2 and if T has at least three vertices then for each colour $k \in V(K_2)$ at least one of the sets $\varphi_i(k)$, $i = 1, 2$, has cardinality ≥ 2. Consequently, the election problem cannot be solved for trees with more than 2 vertices.

For the same reasons, square grids, which are also connected and colourable with two colours, do not admit either naming or election algorithms in our model.

Complete graphs are submersion-minimal and therefore admit both naming and election in our model.

5.2   Rings with a Prime Size

First note the following fact:

Proposition 12. *An unlabelled ring of size p is submersion-minimal if and only if p is prime.*

Therefore prime size rings allow both naming and election. This is a quite interesting corollary of our general conditions since our model is the weakest among graph relabelling systems, with the bare minimal synchronisation power. Moreover, contrary to some other algorithms on rings, our enumeration algorithm does not need any sense of direction for computing agents.

References

1. D. Angluin. Local and global properties in networks of processors. In *Proceedings of the 12th Symposium on Theory of Computing*, pages 82–93, 1980.
2. H. Attiya and J. Welch. *Distributed computing: fundamentals, simulations, and advanced topics*. McGraw-Hill, 1998.
3. Paolo Boldi, Bruno Codenotti, Peter Gemmell, Shella Shammah, Janos Simon, and Sebastiano Vigna. Symmetry breaking in anonymous networks: Characterizations. In *Proc. 4th Israeli Symposium on Theory of Computing and Systems*, pages 16–26. IEEE Press, 1996.
4. J. Chalopin and Y. Métivier. Election and local computations on edges (*extended abstract*). In *Proc. of Foundations of Software Science and Computation Structures, FOSSACS'04*, number 2987 in LNCS, pages 90–104, 2004.
5. E. Godard, Y. Métivier, and A. Muscholl. Characterization of Classes of Graphs Recognizable by Local Computations. *Theory of Computing Systems*, (37):249–293, 2004.
6. G. LeLann. Distributed systems: Towards a formal approach. In B. Gilchrist, editor, *Information processing'77*, pages 155–160. North-Holland, 1977.
7. N. A. Lynch. *Distributed algorithms*. Morgan Kaufman, 1996.
8. A. Mazurkiewicz. Distributed enumeration. *Inf. Processing Letters*, 61:233–239, 1997.
9. B. Szymanski, Y. Shy, and N. Prywes. Terminating iterative solutions of simultaneous equations in distributed message passing systems. In *Proc. of the 4th Symposium of Distributed Computing*, pages 287–292, 1985.
10. G. Tel. *Introduction to distributed algorithms*. Cambridge University Press, 2000.
11. M. Yamashita and T. Kameda. Computing on anonymous networks: Part i - characterizing the solvable cases. *IEEE Transactions on parallel and distributed systems*, 7(1):69–89, 1996.

Embedding in Switching Classes
with Skew Gains

Andrzej Ehrenfeucht[2], Jurriaan Hage[1],
Tero Harju[3], and Grzegorz Rozenberg[2,4]

[1] Inst. of Information and Computing Sci., Univ. Utrecht,
P.O.Box 80.089, 3508 TB Utrecht, The Netherlands
jur@cs.uu.nl
[2] Dept. of Computer Science, University of Colorado at Boulder
Boulder, CO 80309, USA
[3] Dept. of Mathematics, University of Turku, FIN-20014 Turku, Finland
[4] Leiden Institute of Advanced Computer Science
P.O. Box 9512, 2300 RA Leiden, The Netherlands

Abstract. In the context of graph transformation we look at the operation of switching, which can be viewed as an elegant method for realizing global transformations of (group-labelled) graphs through local transformations of the vertices.

Various relatively efficient algorithms exist for deciding whether a graph can be switched so that it contains some other graph, the query graph, as an induced subgraph in case vertices are given an identity. However, when considering graphs up to isomorphism, we immediately run into the graph isomorphism problem for which no efficient solution is known. Surprisingly enough however, in some cases the decision process can be simplified by transforming the query graph into a "smaller" graph without changing the answer. The main lesson learned is that the size of the query graph is not the dominating factor, but its cycle rank.

Although a number of our results hold specifically for undirected, unlabelled graphs, we propose a more general framework and give some preliminary results for more general cases, where the graphs are labelled with elements of a group.

1 Introduction

The material in this paper is motivated by a quest for techniques which enable the analysis of certain networks of processors. Our starting point is that the vertices of a directed graph can be interpreted as processors in a network and the edges can be interpreted as the channels/connections between them, labelled with values from some (structured) set, call it Δ, to capture the current state. The dynamics of such a network lies in the ability to change the labellings of the graph which is done by operations performed by the processors. A major aspect of the model here presented is that if a processor performs an input action, it influences the labellings of all incoming edges in the same way; the same holds the output actions which govern the outgoing edges. In other words, we have no

H. Ehrig et al. (Eds.): ICGT 2004, LNCS 3256, pp. 257–270, 2004.

separate control over each edge, only over each processor. On the other hand, actions done by different processors should not interfere with each other, making this model an asynchronous one.

Ehrenfeucht and Rozenberg set forth in [3] a number of axioms they thought should hold for such a network of processors.

A1 Any two input (output) actions can be combined into one single input (output) action.
A2 For any pair of elements $a, b \in \Delta$, there is an input action that changes a into b; the same holds for output actions.
A3 For any channel (i, j), the order of applying an input action to i and an output action to j is irrelevant.

Although each processor i was to have a set of *output actions* Ω_i and a set of *input actions* Σ_i, in [3] (see also [2]) it was derived that under these axioms the input (output) actions of every vertex are the same and form a group. Also, the sets of input and output actions coincide, but an action will act differently on incoming and outgoing edges, as evidenced by the asymmetry in (2) in Section 2. The difference is made explicit by an anti-involution δ, which is an anti-automorphism of order at most two on the group of actions. The notion of anti-involution generalizes that of group inversion. The result of this will be that if a channel between processors i and j is labelled with a, then the channel from j to i will be labelled with $\delta(a)$. The model generalizes the gain graphs of [9] and the voltage graphs of [4].

As we shall see later the graphs labelled with elements from a fixed group Δ (and under some fixed anti-involution of that group), called skew gain graphs in the following, are partitioned into equivalence classes. These equivalence classes capture the possible outcomes of performing actions in the vertices, i.e., the states of the system reachable from a certain "initial" state. The transformation from one skew gain graph to another, is governed by selecting in each vertex an operation, which corresponds to an element of the group. Although the equivalence classes themselves are usually considered static objects, it is not hard to see that there is also a notion of change or dynamics: applying a selector to a skew gain graph yields a new skew gain graph on the same underlying network of processors, but possibly with different labels. For this reason the equivalence classes were called dynamic labelled 2-structures in [3].

Consider now the problem where we have a (target) skew gain graph h which represents our network, and a skew gain graph g, the query graph, which represents a fragment of a network which to us has a special meaning, for instance, it describes a deadlock situation. A question to ask is then: is there a way to transform h by applying a selector, such that in the result we can detect a subgraph similar to g? In terms of the example: is there a possible state in the system, derivable from h which contains a deadlocked subgraph. If the embedding from g into h is known, then this can be (in many cases) efficiently solved by applying the results of Hage [5]. However, the large number of possible embeddings of g into h remains a problem. In fact, we quickly run into the Graph Isomorphism problem which does not have a known efficient solution. In this paper, we seek to alleviate this problem by seeing how we might reduce the skew gain graph g to a different, simpler graph without changing the outcome, i.e. if the reduced graph can be embedded, then so can g.

After introducing our notation for groups, skew gain graphs and switching classes thereof, we continue by formulating a general framework for reasoning about reductions between skew gain graphs, and give a some illustrative and even surprising examples of such reductions. In some cases they work irrespective of the group and involution. In the case of bridging on the other hand, where we can "shorten" the lengths of cycles in our query graph, they generally work only for certain groups. We give examples of these for the group $\mathbf{Z}_2$ and for the group $\mathbf{Z}_3$. The correctness of these reductions follow from rather surprising combinatorial results. We then show how these results can be used to derive an algorithm for the embedding problem, showing that the complexity of the embedding problem depends on the cycle rank of the query graph and not on the number of vertices. Finally, we prove some impossibility results for bridging.

2   Preliminaries

In this paper, we use both elementary group theory and graph theory. In this section we establish notation, and introduce the concept of switching classes of gain graphs with skew gains. For more details on group theory we refer the reader to Rotman [7].

For a group Γ we denote its identity element by 1_Γ. Let Γ be a group. A function $\delta : \Gamma \to \Gamma$ is an *anti-involution*, if it is an anti-automorphism of order at most two, that is, δ is a bijection and for all $x, y \in \Gamma$, $\delta(xy) = \delta(y)\delta(x)$ and $\delta^2(x) = x$. We write (Γ, δ) for a group Γ with a given anti-involution δ.

Define $E_2(V) = \{(u, v) \mid u, v \in V, u \neq v\}$, the set of nonreflexive, directed edges over V. We usually write uv for the edge (u, v), but note $uv \neq vu$. For an edge $e = uv$, the *reverse* of e is $e^{-1} = vu$.

We consider graphs $G = (V, E)$ where the set of edges $E \subseteq E_2(V)$ satisfies the following *symmetry condition*:

$$\text{if } e \in E \text{ then also } e^{-1} \in E.$$

Such graphs can be considered as undirected graphs where the edges have been given a two-way orientation. We use $E(G)$ to denote the edges E of G and similarly $V(G)$ to denote its vertices V.

Two vertices $v, v' \in V(G)$ are *adjacent* in G if $(v, v') \in E(G)$. The *degree* of a vertex is the number of vertices in the graph it is adjacent to. A vertex of degree zero is called *isolated*, a *leaf* has degree one, a *chain vertex* degree two, and all other vertices are called *dense vertices*.

A sequence of vertices $p = (v_1, \ldots, v_k)$, $k > 0$, is a *path* in G if v_i is adjacent to v_{i+1} for $i = 1, \ldots, k-1$ and all vertices are distinct. By $E(p)$ we denote the set of edges $\{(v_1, v_2), \ldots, (v_{k-1}, v_k)\}$. Additionally, p is called a *chain* if all vertices $v_2, \ldots, v_{k-1}$ are chain vertices. The chain p is maximal in G if the endpoints v_1 and v_k are not chain vertices. A *cut edge* in a graph is an edge which is not on any cycle.

Let $G = (V, E)$ be a graph and (Γ, δ) a group with anti-involution. A pair (G, g) where g is a mapping $g : E \to (\Gamma, \delta)$ into the group Γ is called a (Γ, δ)-*gain graph* (on G) (or a *graph with skew gains* or a *skew gain graph*), if g satisfies the following *reversibility condition*

$$g(e^{-1}) = \delta(g(e)) \quad \text{for all } e \in E . \tag{1}$$

In the future we will refer to a skew gain graph (G, g) simply by g unless confusion arises. We adopt in a natural way some of the terminology of graph theory for graphs with skew gains. For instance, every path in G is also a path in g, and we can use $E(g)$ to denote the set of edges of the underlying graph G.

The class of (Γ, δ)-gain graphs on G will be denoted by $\mathbf{L}_G(\Gamma, \delta)$ or simply by $\mathbf{L}_G$. More importantly, $\mathbf{L}(\Gamma, \delta) = \bigcup \{\mathbf{L}_G(\Gamma, \delta) \mid G \text{ is a graph }\}$. A *gain graph* is a $(\Gamma, {}^{-1})$-gain graph; these are also called *inversive* skew gain graphs.

With a path $p = (v_1, \dots, v_k)$ in $g \in \mathbf{L}_G(\Gamma, \delta)$ we can associate the sequence of labels $\lambda(p) = (g(v_1 v_2), \dots, g(v_{k-1} v_k))$. Now, p is an a-path if every value in $\lambda(p)$ is equal to a. Secondly, p is a b-summing path for some $b \in \Gamma$ if $g(v_1 v_2) \cdot g(v_2 v_3) \cdots g(v_{k-1} v_k)$ equals b. (We often denote this fact by writing $g(p) = b$.) In other words, evaluating the product of values found along p using the group operation $\cdot$ of Γ evaluates to the group element b.

Let $g \in \mathbf{L}_G(\Gamma, \delta)$. A set $X \subset V(G)$ is an a-*clique* if for all $x, y \in X$: $x \neq y$ implies $g(x, y) = a$. Also, for $X, Y \subseteq V(G)$, X is said to be a-*connected* to Y, if $X \cap Y = \emptyset$ and $g(x, y) = a$ for all $x \in X, y \in Y$.

A function $\sigma : V \to \Gamma$ is called a *selector*. For each selector σ we associate with g a (Γ, δ)-gain graph g^σ on $G = (V, E)$ by letting, for each $uv \in E$,

$$g^\sigma(uv) = \sigma(u) g(uv) \delta(\sigma(v)) . \tag{2}$$

Example 1.
To illustrate switching, consider g_1 and g_2, the $(\mathbf{Z}_4, id)$-gain graphs of Figure 1(a) and (b) respectively (the group $\mathbf{Z}_4$ is the group of addition modulo 4; the involution is the identity function giving rise to a symmetric graph). The second of these, g_2, can obtained from g_1 by applying the selector σ that maps both 1 and 3 to 3, and both 2 and 4 to 1. For example, the label of the edge $(1, 3)$ is computed

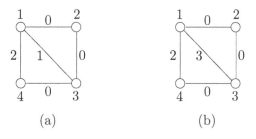

(a)   (b)

Fig. 1. Two elements of $\mathbf{L}_G(\mathbf{Z}_4, id)$.

as follows: $g_2(1,3) = g_1^\sigma(1,3) = \sigma(1)g_1(1,3)\delta(\sigma(3)) = 3+1+\delta(3) = 3+1+3 = 3$, where $+$ is of course addition modulo 4. The path $p = (1,3,4,1)$ is 3-summing in g_1 (here $\lambda(p)$ equals $(1,0,2)$) and 1-summing in g_2 (here $\lambda(p)$ equals $(3,0,2)$). Neither are 0-paths, while $(1,2,3,4)$ is a 0-path in both g_1 and g_2.

The class $[g] \subseteq \mathbf{L}_G(\Gamma,\delta)$ defined by

$$[g] = \{g^\sigma \mid \sigma : V \to \Gamma\}$$

is called the *switching class* generated by g.

It is not difficult to prove that a switching class is an equivalence class of skew gain graphs. The underlying equivalence relation on $\mathbf{L}_G(\Gamma,\delta)$ is that $g \equiv g'$ for $g, g' \in \mathbf{L}_G(\Gamma,\delta)$ if and only if $\exists \sigma : V(G) \to \Gamma$ such that $g' = g^\sigma$. Obviously $g \equiv g$ and if $g_1 \equiv g_2$ then also $g_2 \equiv g_1$, because $g_1^\sigma = g_2$ if and only if $g_1 = g_2^{\sigma^{-1}}$, where the σ^{-1} is such that $\sigma^{-1}(v) = \sigma(v)^{-1}$ for all $v \in V$.

Closure under composition of selectors is something that we would expect in our model: it is a consequence of Axiom A1 of the introduction. If we define the composition of two selectors σ and τ to be $\sigma\tau(v) = \sigma(v)\tau(v)$, then we can prove that for each $g \in \mathbf{L}_G(\Gamma,\delta)$ and selectors σ, τ, $g^{\sigma\tau} = (g^\tau)^\sigma$.

If the group Γ is the cyclic group of order 2, $\mathbf{Z}_2$, then by necessity the involution is the identity function and the skew gain graphs are exactly the undirected simple graphs of, e.g., [1, 6]. Directed graphs are obtained by choosing $\Gamma = \mathbf{Z}_4$ and we take the involution δ to be the group inversion.

3   The General Framework

In the following let Γ be a fixed, but arbitrary group and δ a fixed, but arbitrary involution of Γ.

Let $g \in \mathbf{L}_G(\Gamma,\delta)$ and $h \in \mathbf{L}_H(\Gamma,\delta)$ be skew gain graphs. An injection $\psi : V(G) \to V(H)$ *embeds g into h, denoted by* $g \overset{\psi}{\hookrightarrow} h$, if

$$g(uv) = h(\psi(u)\psi(v)) \text{ for all } uv \in E(G).$$

If we do not care what ψ is, we write $g \hookrightarrow h$ instead. Note that in some definitions of embedding there is also an injection on the labels, but since our application attaches meaning to the labels, we do not allow that here.

The embedding ψ is an *isomorphism* from g to h if $g \overset{\psi}{\hookrightarrow} h$ and $h \overset{\psi^{-1}}{\hookrightarrow} g$. We denote this fact by $g \overset{\psi}{\cong} h$, or, equivalently, $h \overset{\psi^{-1}}{\cong} g$.

The definition of embedding can be extended to switching classes in a natural way:

$$g \hookrightarrow [h] \text{ if and only if there exists } h' \in [h] \text{ such that } g \hookrightarrow h'.$$

In this and the following sections, the central problem is to decide whether the *query* skew gain graph $g \in \mathbf{L}_G(\Gamma,\delta)$ can be embedded in a switch of the *target* skew gain graph $h \in \mathbf{L}_H(\Gamma,\delta)$.

We assume for the remainder of the paper that the target skew gain graph is *total*, meaning that $H = (V, E_2(V))$ for some set of vertices V.

We now come to the definitions central to this paper. We are interested in establishing for a certain query graph g into which other skew gain graph g' it may be transformed so that the ability of embedding g into h is preserved and reflected into g'. More formally, we define $\mathcal{R}_{(\Gamma,\delta)}$ as the set of *embedding equivalent* pairs $(g, g') \in \mathbf{L}(\Gamma, \delta) \times \mathbf{L}(\Gamma, \delta)$ such that

$$\forall h : g \hookrightarrow [h] \iff g' \hookrightarrow [h].$$

Note that in our definition we have left the embedding itself unspecified, meaning that in general we do not care whether g and g' are embedded "in the same place". It also implies that g and g' may have different underlying graphs.

Although we have just defined the largest possible (equivalence) relation relating skew gain graphs from $\mathbf{L}(\Gamma, \delta)$ to each other, it does not give us any concrete information which pairs are actually in the relation for a given group and involution. In the remainder of this paper we shall establish a number of results which either show that some pairs are definitely in this relation, or that some pairs can never be.

Let R be any equivalence relation on $\mathbf{L}(\Gamma, \delta)$. R is an *embedding invariant relation (emir)* if $(g, g') \in R$ implies $(g, g') \in \mathcal{R}_{(\Gamma, \delta)}$.

We now give some examples of emirs that occur in the literaure. The following easy lemma shows that for embedding the identities of the vertices of the query graph are unimportant.

Lemma 1.
For two isomorphic (Γ, δ)-gain graphs g and g' (with isomorphism ϕ from g to g'): if $g \overset{\psi}{\hookrightarrow} h$, then $g' \overset{\psi \cdot \phi^{-1}}{\hookrightarrow} h$.

The second example is that embedding a query graph g is the same as embedding one of its switches:

Lemma 2.
If $g \overset{\psi}{\hookrightarrow} [h]$, then also $g^\sigma \overset{\psi}{\hookrightarrow} [h]$ for any selector $\sigma : V(g) \to \Gamma$.

Note that Lemma 1 implies the existence of an emir R_{IR}: $(g, g') \in R_{\mathrm{IR}}$ if and only if $g \cong g'$. Another example comes from Lemma 2 where it is proved that in fact $\equiv$ is an emir.

We shall now give a slightly more complicated example.

Define R_{DCR} such that $(g, g') \in R_{\mathrm{DCR}}$ if g' can be obtained from g by removing any number of cut edges of g. The symmetric closure of this relation, R_{CR}, is an equivalence relation on (Γ, δ)-gain graphs. So any two g and g' are related if and only if they have exactly the same cycles and the same domain. A basic result from the theory of switching classes proves that this relation is in fact an emir (see for instance [8]):

Theorem 1.
Let H be a graph and let T be a subgraph of H that is a forest. For every (Γ, δ)-gain graph g on T and every h on H: $g \overset{\mathrm{id}}{\hookrightarrow} [h]$.

The proposition states that any acyclic structure can be embedded in a switching class. Note that by removing edges we do not change the size of the domain of the (Γ, δ)-gain graph; this is necessary for establishing embedding invariance.

To combine two emirs into one we can use the join operation: for two emirs R and R' on (Γ, δ), the join of R and R', denoted by $R \vee R'$, is the smallest equivalence relation including both R and R'.

Lemma 3.
If R and R' are emirs, then the join of R and R' is an emir.

The join can be used to combine various emirs into a larger one. For instance, joining an emir such as R_{DCR} with R_{IR} yields an emir that "incorporates" removing of cut edges and isomorphisms. In such a way we can define various emirs and compose these to come as close as possible to the largest of emirs, $\mathcal{R}_{(\Gamma, \delta)}$.

4 Bridging

In this and the coming sections we assume that the group Γ is abelian and that the involution δ is the group inversion $^{-1}$; we will denote the identity of the group simply by 0.

The reason for the restriction is the Cyclic Sum Invariance, which holds for switching classes with abelian groups and involution equal to the group inversion. It means that if one takes any cycle and computes the sum along that cycle for the labels on that path, then this value does not change when the skew gain graph is switched [2]. The gain graphs g_1 and g_2 of Example 1 show that in general this result does not hold, because the cycle $(1, 3, 4, 1)$ sums to different values in each of them, even though they are in the same switching class.

Let $g, g' \in \mathbf{L}(\Gamma, \delta)$ be such that g contains a 0-chain $p = (x_0, \ldots, x_k)$. Then, for integers k and ℓ with $k \geq \ell$, g' is a (k, ℓ)-*bridging* of g, denoted $gB_\ell^k g'$, if g' is a (Γ, δ)-gain graph on $V(g)$ with

$$E(g') = (E(g) - E(p)) \cup E(p') \text{ for } p' = (x_0, \ldots, x_{\ell-1}, x_k)$$

and

$$g'(e) = \begin{cases} g(e), & \text{if } e \in E(g) - E(p) \\ 0, & \text{otherwise} \end{cases}$$

We additionally define B_ℓ^k to be equal to $\left(B_k^\ell\right)^{-1}$ for $k \leq \ell$.

For the following two (Γ, δ)-gain graphs g (left) and g' (right) it holds that $gB_3^5 g'$. In this case $p = (x_0, \ldots, x_5)$:

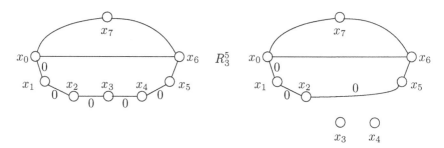

Note that we can assume that the chain is labelled with zeroes, since if it does not we can always switch it so that it does. Also, note that the definition also would have allowed to exclude x_4 and x_5 by choosing $(x_1, \ldots, x_6)$ as our 0-chain.

In what follows we are interested in determining for which groups we can always (i.e., for any (Γ, δ)-gain graph $g \in \mathbf{L}(\Gamma, \delta)$) change bridges of length k to bridges of length ℓ. For this we introduce the following relation $R_\Gamma \subseteq \mathbf{N} \times \mathbf{N}$, where $(k, \ell) \in R_\Gamma$ if and only if B_ℓ^k is an emir on $\mathbf{L}(\Gamma, \delta)$. Obviously, for any group Γ it holds that $(k, k) \in R_\Gamma$ where $k > 0$.

The following lemma couples the concept of bridging to something we can more easily verify. Implicitly we allow the embedding only to be changed on the chain vertices that occur on the bridge.

A (Γ, δ)-gain graph on $\{0, \ldots, n\}$ for some n is an (n, k)-*bridge structure* if it has a 0-path $(0, \ldots, k)$.

The following lemma shows that to decide whether we can bridge paths of length k into ℓ, we can look at total skew gain graphs which have a 0-labelled path $(0, \ldots, k)$ and show that whatever labels are on the other edges, we can always find a 0-summing path from 0 to k of length ℓ.

Lemma 4.
Let k and ℓ be natural numbers, and let $n = \max(k, \ell)$. It holds that $(k, \ell) \in R_\Gamma$ if and only if for every (n, k)-bridge structure g there is a 0-summing path p in g of length ℓ from 0 to k.

Proof. For the if-part we need only note that we can replace a path of length k by one of length ℓ which has the same starting and end-point without changing the sum on any of the cycles of which these vertices are part. Because a bridge is part of a chain, every vertex on the bridge is part of exactly the same cycles.

The only if part follows from the fact if we cannot replace the path of length n by a path of the same sum of length k, then we change the cyclic sum along at least one cycle, which contradicts the Cyclic Sum Invariance.

Theorem 2.
For natural numbers $k_1 > 1$ and $k_2 > 2$: $(k_1, 1) \notin R_\Gamma$ and $(k_2, 2) \notin R_\Gamma$, if Γ is not the trivial group, $\{0\}$.

Proof. Let $a \in \Gamma$ with $a \neq 0$. Let g_2 be a (Γ, δ)-gain graph on $\{0, \ldots, k_2\}$ such that for $1 \leq i \leq k_2 - 1$, $g_2(0, i) = 0$, $g_2(i, k_2) = a$ and all other edges are labelled arbitrarily. Hence for all i, $g_2(0, i, k_2) = a \neq 0$. The same kind of reasoning can be applied to the other case.

Example 2.
If we know that $(5, 3) \in R_\Gamma$, then it is easy to see that $(k, k - (5 - 3)) = (k, k - 2) \in R_\Gamma$ as long as $k - 2 \geq 3$: if g contains a chain of length greater than 5, then we can take any part of this chain of length 5 and reduce it to 3 and thereby reduce the length of the entire chain from k to $k - 2$. We can repeat this process until the chain is not sufficiently long anymore. We conclude that if we prove that $(5, 3) \in R_\Gamma$ then $(k, k - 2) \in R_\Gamma$ for $k \geq 5$ and even $(k, k - 2\ell)$ for $k - 2\ell \geq 3$. Using similar reasoning we conclude that $(3, 5) \in R_\Gamma$ implies that $(k, k + 2\ell)$ for $k \geq 3$.   ◇

In general we have

Lemma 5.
If $(k_1, \ell_1) \in R_\Gamma$ then $(k_2, \ell_2) \in R_\Gamma$ where $\ell_2 = k_2 - (k_1 - \ell_1)m$, $m \geq 1$ and $\ell_2 \geq \ell_1$.

If $\Gamma = \Gamma_1 \times \Gamma_2$ then $(k, \ell) \in R_\Gamma$ implies $(k, \ell) \in R_{\Gamma_i} (i = 1, 2)$, but not vice versa, not even if $\Gamma_1 = \Gamma_2$ (see Theorem 4). The positive result is easy, because the identity of Γ maps to the identities of the factors. Hence the 0-summing paths stay 0-summing in the projection. The following result says that if a bridging is not possible for a given group, it automatically precludes bridging in groups of which it is a factor.

Lemma 6.
If Γ is a group such that $(k, \ell) \notin R_\Gamma$, then this also holds for all groups of which Γ is a factor.

4.1   The Case for $\mathbf{Z}_2$

In view of Theorem 2 it may be surprising that bridgings do exist.

Lemma 7.
$(5, 3) \in R_{\mathbf{Z}_2}$

Proof. Let b be a (5,5)-bridge structure (recall that the path $(0, \ldots, 5)$ is a 0-path, and the other edges are arbitrarily labelled by elements of $\mathbf{Z}_2$). Now, if $b(0, 3) = 0$, then $b(0, 3, 4, 5) = 0$. The same reasoning applies to $(2, 5)$. In the other cases, $b(0, 3) = 1 = b(2, 5)$ and $b(0, 3, 2, 5) = 0$.

Knowing that bridging is possible under $\mathbf{Z}_2$ we can now illustrate that the necessity of the target skew gain graph being total: take a cycle on six vertices – call it c. We can bridge c into c' which consists of two isolated vertices and a cycle on four vertices. Obviously, $c \hookrightarrow [c]$, but $c' \not\hookrightarrow [c]$. The reason is that the target graph, in this case c, does not have all its edges present.

Lemma 8.
$(k, \ell) \notin R_{\mathbf{Z}_2}$ *if* k *and* ℓ *are of opposite parity.*

Proof. Let k and ℓ be of opposite parity. We may assume $\ell, k \geq 3$, because of Theorem 2.

Let $n = \max(k, \ell)$, b be a (n, k)-bridge structure and $V = V(b)$. By Lemma 4, we only need to exhibit one such structure which has no path of length ℓ from 0 to k which sums to 0. For that, choose b such that the sets $K \subseteq V$ and $V - K$ are 0-connected 1-cliques. Here, $K = \{x \mid 0 \leq x \leq k, x \text{ even}\}$. Note that there is a 0-path $(0, 1, \dots, k)$.

We are interested in paths of length ℓ which go from 0 to k and sum to 0. If k is even, then ℓ is odd, and the path is one that starts in K and ends in K. Since we must switch from K to $V - K$ an even number of times, we traverse an odd number of edges within either K and $V - K$. Since these edges each contribute 1 to the sum, and they are the only edges which contribute, the sum along the path equals 1. If k is odd and hence the path starts in K and ends in $V - K$ similar reasoning leads to a sum of 1.

Theorem 2 and Lemmas 5, 7 and 8 lead to the following.

Corollary 1.
If $k \geq \ell > 2$ *then* $(k, \ell) \in R_{\mathbf{Z}_2}$ *if and only if* k *and* ℓ *have the same parity. Also,* $(k, \ell) \in R_{\mathbf{Z}_2}$ *for* $1 \leq \ell \leq 2$ *if and only if* $k = \ell$.

4.2   The Case for $\mathbf{Z}_3$

The following result was quite a surprise.

Lemma 9.
$(6, 4) \in R_{\mathbf{Z}_3}$ *and* $(6, 5) \in R_{\mathbf{Z}_3}$, *but* $(5, 4) \notin R_{\mathbf{Z}_3}$.

Proof. The positive results have been obtained by a computer check of all paths of length 4 and 5, respectively, from 0 to 6 in a $(6, 6)$-bridge structure.

The counterexample for $(5, 4)$ is given in the following figure, where the solid edges are labelled with 0 and the dashed edges (in the direction of the arrow) with 1. The reader may verify that indeed no path from 0 to 6 of length 5 sums to 0.

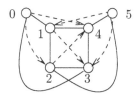

5   An Algorithm for Checking Configuration Containment

In this section we use the fact that $(5, 3) \in R_{\mathbf{Z}_2}$ (Lemma 7) to derive an algorithm for checking that $g \hookrightarrow [h]$ for $g, h \in \mathbf{L}(\mathbf{Z}_2, ^{-1})$ where h is total. The result is mainly based on the following graph theoretical argument which shows that if we consider graphs that do not have any isolated vertex or leaves, and every chain contains a bounded number of chain vertices, then the number of vertices in the graph can be bounded by a constant multiple of the cycle rank of the graph. The *cycle rank* of a graph G is defined as the size of its cycle base, and equals $e - n + k$, where $n = |V(G)|$, $e = |E(G)|$ and k is the number of connected components of G.

Lemma 10. *Let $G = (V, E)$ be a connected graph which has only vertices of degree at least two and at least one dense vertex. If every maximal chain in G has at most $c > 0$ chain vertices, then $|V(G)| \leq 2c\xi$, where ξ is the cycle rank of G.*

Proof. We first make an estimation for graphs which only contain dense vertices. Let $d_G(v)$ denote the degree of the vertex v of G. Then, by the handshaking lemma of graph theory, $2e = \sum_{v \in V} d_G(v) \geq 3n$, since $d_G(v) \geq 3$ for all v. Hence $\xi = e - n + 1 \geq 3n/2 - n + 1 = n/2 + 1$, so that $n \leq 2\xi$ as required. Now, any edge between two dense vertices can be replaced by a chain of a most c chain vertices, which adds to n and e in equal amounts, so that $n \leq 2c\xi$.

Lemma 11.
Let $g \in L(\mathbf{Z}_2, {}^{-1})$ and let ξ be the cycle rank of g. Then, there exists a g' embedding equivalent with g such that $ni(g') \leq 6\xi$, where $ni(g')$ is the number of non-isolated vertices of g'.

Proof. Remove cut-edges, isolated vertices and use the (5,3) bridging to change g into g', which has the property that it consists only of a number of isolated vertices, dense and chain vertices. Neither of these operations change the cycle rank of g. Now apply Lemma 10 to each of the components of the graph (the cycle rank of a disconnected graph equals the sum of the cycle rank of its components) to obtain the given bound for the number of chain and dense vertices, where we use Lemma 7 to limit the number of chain vertices in any chain to 3. We omit in this reasoning components which are simple cycles: connected graphs which have only chain vertices. These, however, can all be reduced to cycles of length at most six, again using Lemma 7.

Finally, we can formulate a bound on the time complexity of the embedding problem for $\mathbf{Z}_2$ as follows:

Theorem 3.
Let $g, h \in L(\mathbf{Z}_2, {}^{-1})$ where h is total, $n = |V(h)|$ and ξ is the cycle rank of g. It can be decided in $O(n^{6\xi+2})$ time whether $g \hookrightarrow [h]$.

Proof. After checking that $|V(g)| \leq h$, we can find an embedding equivalent g' such that $ni(g') \leq 6\xi$ through Lemma 11. Now, we actually remove the isolated vertices from g' (we have already checked that g does not have more vertices than h does). The number of possible injections from g' into h is bounded by $n^{6\xi}$, for each of which we have to do at most $O(n^2)$ work to see if under the injection, we can switch h so that it contains g' (using the results of [5]). The preprocessing of g, which consists of removing leaves, isolated vertices and shortening chains, can easily be done in time $O(n^2)$.

6   Some Impossibility Results

In this section we are interested in determining, given a natural number ℓ, for which finitely generated abelian group Γ it holds for every $k > \ell$, that $(k, \ell) \notin R_\Gamma$. In Lemma 2 we found two such examples, $\ell = 1$ and $\ell = 2$, in which case impossibility was obtained for all groups. Since we have already treated the cases for $\ell \leq 2$, we assume $\ell \geq 3$, and hence $k > 3$. From Lemma 6 and the fundamental result on finitely generated abelian groups, it follows that we can restrict ourselves to solving this question for the cyclic groups (of order a prime power) and $\mathbf{Z}$.

Since we are interested in proving the impossibility of bridging, we have to show that we can always find (k, k)-bridging structures in which there is no 0-summing path of length ℓ from 0 to k.

First we investigate which edges in the bridge structures *must* be labelled with a non-identity element. These are exactly the edges that are on a path of length ℓ from 0 to k which traverse only edges on the path $(0, \ldots, k)$, except for one edge which has an undetermined label. We observe that these edges are those of the form

$$(i, i + (k - \ell + 1)), \quad i = 0, \ldots, \ell - 1. \tag{3}$$

We shall next prove that the only bridging $(k, 3)$ for $k > 3$ occurs if the group is trivial or the group is $\mathbf{Z}_2$. The main technique used here is to generate a family of skew gain graphs, depending on k, which contains a large 0-clique, and only relatively few other edges. Parts of the paths in the 0-clique contribute nothing to the sum along a path, so only the values on the other edges really matter. To simplify the proof any vertex outside X is connected in a uniform way to all vertices in X and (by reversibility) the other way around. In 2-structures jargon, such a set X is called a *clan* (see [2]). The next theorem is a typical example of this kind and can be viewed as an illustration of the proof technique.

Theorem 4.
If for $k > 3, (k, 3) \in R_\Gamma$ for a finitely generated abelian group, then Γ is either $\mathbf{Z}_2$ or the trivial group.

Proof. Like in Lemma 2 the idea is to find a (k, k)-bridge structure which does not exhibit a 0-summing path of length $\ell = 3$ from 0 to k. Because of Lemma 6 and the fundamental theorem on finitely generated abelian groups, we start by considering the cyclic groups of order larger than two and the group $\mathbf{Z}$ of integer addition.

Consider the following graph in which all edges whose value is as yet unknown are labelled with a variable label a_i for some i, and the vertex X represents a 0-clique on $k - \ell$ vertices.

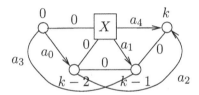

By (3), a_0, a_1 and a_2 should be labelled by values different from 0. It is easily seen that also $a_4 \neq 0$ (for paths through X). We also find that $a_0 \neq a_4{}^{-1}$, because of the path $(0, k-2, x, k)$ where $x \in X$. In fact if we set $a_3 = 0$, a_0, a_2 and a_4 to the generator of the group, 1, and a_3 to 1^{-1} there is no path of length $\ell = 3$ which sums to 0. It is important to note that since the group has order at least three, $1 \neq 1^{-1}$.

Since a $(k, 3)$ bridging existed for $\mathbf{Z}_2$, we should also show that such a bridging is not possible for $\mathbf{Z}_2 \times \mathbf{Z}_2$. Taking the same graph as our starting point, we choose a_3 the identity $(0, 0)$ and set $a_0 = a_1 = a_2 = (0, 1)$ and $a_4 = (1, 0)$. Again, the reader can verify (there are only a finite number of cases), that no path of length 3 from 0 to k sums to $(0, 0)$.

7   Conclusions and Future Work

Taking the model of Ehrenfeucht and Rozenberg as our starting point, we have considered the embedding problem in detail. We have set up a framework to establish results about reducing query skew gain graphs to smaller ones and proved some general results in this matter. Then we concentrated on bridging, which, for $\mathbf{Z}_2$ at least, results in an algorithm for the embedding problem which is dominated not by the size of the query graph, but by its cycle rank, corresponding to the general intuition in switching classes that cycles make life harder.

We have not completed a full investigation of all possible bridgings for all possible finitely generated abelian groups, although we have the full picture for $\mathbf{Z}_2$ and $\mathbf{Z}_3$. We do conjecture that for every such group Γ there is for every ℓ a k such that $(k, \ell) \notin R_\Gamma$. Note by the way, that bridging is just one possible reduction strategy and others might exist. In that sense, the research in this area is still very much open, especially for non-abelian groups where bridging is not an option.

Acknowledgements

We thank Hans Bodlaender for some helpful discussions.

References

1. A. Ehrenfeucht, J. Hage, T. Harju, and G. Rozenberg. Complexity issues in switching of graphs. In H. Ehrig, G. Engels, H.-J. Kreowski, and G. Rozenberg, editors, *Theory And Application Of Graph Transformations - TAGT '98*, volume 1764 of *Lecture Notes in Computer Science*, pages 59–70, Berlin, 2000. Springer Verlag.

2. A. Ehrenfeucht, T. Harju, and G. Rozenberg. *The Theory of 2-Structures*. World Scientific, Singapore, 1999.
3. A. Ehrenfeucht and G. Rozenberg. Dynamic labeled 2-structures. *Mathematical Structures in Computer Science*, 4:433–455, 1994.
4. J. L. Gross and T. W. Tucker. *Topological Graph Theory*. Wiley, New York, 1987.
5. J. Hage. The membership problem for switching classes with skew gains. *Fundamenta Informaticae*, 39(4):375–387, 1999.
6. J. Hage, T. Harju, and E. Welzl. Euler graphs, triangle-free graphs and bipartite graphs in swithing classes. *Fundamenta Informaticae*, 58(1):23–37, November 2003.
7. J. J. Rotman. *The Theory of Groups*. Allyn and Bacon, Boston, 2nd edition, 1973.
8. T. Zaslavsky. Signed graphs. *Discrete Applied Math.*, 4:47–74, 1982. Erratum on p. 248 of volume 5.
9. T. Zaslavsky. Biased graphs. I. Bias, balance, and gains. *J. Combin. Theory, Ser. B*, 47:32–52, 1989.

Synchronizers for Local Computations

Yves Métivier, Mohamed Mosbah, Rodrigue Ossamy, and Afif Sellami

LaBRI - University of Bordeaux1
351 Cours de la Libération
33405 - Talence
{metivier,mosbah,ossamy,sellami}@labri.fr

Abstract. A synchronizer is intended to allow synchronous algorithms to be executed on asynchronous networks. It is useful because designing synchronous algorithms is generally much easier than designing asynchronous ones. In this paper, we provide synchronization protocols described as local computations. We obtain a general and an unified approach for handling synchrony in the framework of local computations.

1 Introduction

Distributed algorithms are studied under various models. One fundamental criterium is synchronous or asynchronous [1, 3, 7, 11]. In a synchronous model, we assume that there is a global clock and that the operations of components take place simultaneously: at each clock tick an action takes place on each process. This is an ideal timing model that does not represent what really happens in most distributed systems. In fact, synchronization in networks may be viewed as a control structure which enables to control relative steps of different processes; it may be illustrated by the following examples:

1. the processes execute actions in lock-steps called pulses or rounds. In a pulse, a process p executes the following sequence of discrete steps:
   (a) p sends a message,
   (b) p receives some messages
   (c) p performs local computations;
2. another assumption is that computation events of pulse p appear after computation events of pulse $p - 1$ and all messages sent in pulse p are delivered before computations events of pulse $p + 1$;
3. if each process is equipped with a counter for local computations, we may assume that the difference between two counters is at most 1; and more generally, for a given non negative integer k we may assume that the difference between any two counters is at most k;
4. some algorithms need some synchronization barriers applied to a group of processes, it means that all group members are blocked until all processes of the group have reached this barrier.

In the asynchronous model there is no global clock, separate components take steps at arbitrary relative speeds. It is assumed that messages are delivered,

H. Ehrig et al. (Eds.): ICGT 2004, LNCS 3256, pp. 271–286, 2004.

processes perform local computations and send messages, but no assumption is made about how long it may take. There exist also intermediate models like models assuming the knowledge of bounds on the relative speeds of processes or links. This paper presents several methods for the simulation of synchrony on asynchronous distributed systems by means of local computations.

1.1   The Model

We consider networks of processes with arbitrary topology. A network is represented as a connected, undirected graph where vertices denote processes and edges denote direct communication links. Labels are attached to vertices and edges. The identities of the vertices, a distinguished vertex, the number of processes, the diameter of the graph or the topology are examples of labels attached to vertices; weights, marks for encoding a spanning tree or the sense of direction are examples of labels attached to edges.

Labels are modified locally, that is in general, on star graphs or on edges of the given graph, according to certain rules depending on the subgraph only (*local computations*). The relabelling is performed until no more transformation is possible, i.e., until a normal form is obtained.

The model of local computations has several interests:

- it gives an abstract model to think about some problems in the field of distributed computing independently of the wide variety of models used to represent distributed systems [5],
- it is easier to understand and to explain problems, to compute their solutions or to obtain results of impossibility,
- any positive solution in this model may guide the research of a solution in a weaker model or be implemented in a weaker model using randomized algorithms,
- this model gives nice properties and examples using classical combinatorial material.

1.2   Synchronizers

A synchronous distributed system is organized as a sequence of pulses: in a pulse each process performs a local computation. In an asynchronous system the speed of processes can vary, there is no bounded delay between consecutive steps of a process. A synchronizer is a mechanism that transforms an algorithm for synchronous systems into an algorithm for asynchronous systems.

As the non-determinism in synchronous systems is weaker, in general, algorithms for synchronous systems are easier to design and to analyze than those for asynchronous networks. In asynchronous systems, it is difficult to deal with the absence of a global synchronization of processes. Consequently, it is useful to have a general method to transform an algorithm for synchronous networks into an algorithm for asynchronous networks. Therefore, it becomes possible to design a synchronous algorithm, test it and analyze it and then use the standard method to implement it on an asynchronous network. A synchronizer operates by generating a sequence of local clock *pulses* at each process. An introduction and the main results about synchronizers may be found in [1, 3, 7, 9, 11].

1.3   The Main Results

In this paper we focus on efficient synchronizers where the aim is to reduce the variation of pulses between the processes of the network. We discuss several synchronizers according to the assumptions made on the network. We consider mainly three classes of networks. The first class consists of networks whose sizes are known. That is, each process has a local knowledge containing the size of the network. For this class of networks, we present two types of synchronizers based on two different techniques. The first one uses the SSP algorithm [12] that is commonly used to detect stable properties. The second one is a randomized procedure based on the use of random walks on graph. The final class of networks we deal with is the class of tree-shaped networks. In fact, if the network is a tree, the synchronization protocol begins by electing a vertex which will afterwards ensure the synchronization of the whole network.

1.4   Summary

The paper is organized as follows. Section 2 recalls briefly several definitions of local computations and introduces their use to describe distributed algorithms. Section 3 presents important properties of synchronizers. A simple synchronizer is given in Section 4. Section 5 deals with synchronizers for networks with known size and   Section 6 presents a synchronizer for trees. A general method for building and using synchronizers is discussed in Section 7.

2   Definitions and Notations

2.1   Undirected Graphs

We only consider finite, undirected and connected graphs without multiple edges and self-loops. If G is a graph, then $V(G)$ denotes the set of vertices and $E(G)$ denotes the set of edges; two vertices u and v are said to be adjacent if $\{u, v\}$ belongs to $E(G)$. The distance between two vertices u, v is denoted $d(u, v)$.

Let v be a vertex and k a non negative integer, we denote by $B_G(v, k)$, or briefly $B(v, k)$, the centered ball of radius k with center v, i.e., the subgraph of G defined by the vertex set $V' = \{v' \in V(G) \mid d(v, v') \leq k\}$ and the edge set $E' = \{\{v_1, v_2\} \in E(G) \mid d(v, v_1) < k \text{ and } d(v, v_2) \leq k\}$.

2.2   Graph Relabelling Systems
       for Encoding Distributed Computation

In this subsection, we illustrate, in an intuitive way, the notion of graph relabelling systems by showing how some algorithms on networks of processes may be encoded within this framework [6]. As usual, such a network is represented by a graph whose vertices stand for processes and edges for (bidirectional) links between processes. At every time, each vertex and each edge is in some particular state and this state will be encoded by a vertex or edge label. According to its own state and to the states of its neighbours (or a neighbour), each vertex may decide to realize an elementary *computation step*. After this step, the states of this vertex, of its neighbours and of the corresponding edges may have changed according to some specific *computation rules*. Let us recall that graph relabelling systems satisfy the following requirements:

(C1) they do not change the underlying graph but only the labelling of its components (edges and/or vertices), the final labelling being the result,

(C2) they are local, that is, each relabelling changes only a connected subgraph of a fixed size in the underlying graph,

(C3) they are locally generated, that is, the applicability condition of the relabelling only depends on the local context of the relabelled subgraph.

A precise description and definition of local computations can be found in [4]. We recall here only the description of local computations, we explain the convention under which we will describe graph relabelling systems later. If the number of rules is finite then we will describe all rules by their preconditions and relabellings. We will also describe a family of rules by a generic rule scheme ("meta-rule"). In this case, we will consider a generic star-graph of generic center v_0 and of generic set of vertices $B(v_0, 1)$. If $\lambda(v)$ is the label of v in the precondition, then $\lambda'(v)$ will be its label in the relabelling. We will omit in the description labels that are not modified by the rule. This means that if $\lambda(v)$ is a label such that $\lambda'(v)$ is not explicitly described in the rule for a given v, then $\lambda'(v) = \lambda(v)$. In all the examples of graph relabelling systems that we consider in this paper the edge labels are never changed.

An Example

The following relabelling system performs the election algorithm in the family of tree-shaped network. The set of labels is $L = \{N, elected, non\text{-}elected\}$. The initial label on all vertices is $l_0 = N$.

R1 : **Pruning rule**
    Precondition :
- $\lambda(v_0) = N$,
- $\exists! \, v \in B(v_0, 1), v \neq v_0, \lambda(v) = N$.

    Relabelling :
- $\lambda'(v_0) := non\text{-}elected$.

R2 : **Election rule**
    Precondition :
- $\lambda(v_0) = N$,
- $\forall v \in B(v_0, 1), v \neq v_0, \lambda(v) \neq N$.

    Relabelling :
- $\lambda'(v_0) := elected$.

Let us call a pendant vertex any vertex labelled N having exactly one neighbour with the label N. There are two meta-rules $R1$ and $R2$. The meta-rule $R1$ consists in cutting a pendant vertex by giving it the label *non-elected*. The label N of a vertex v becomes *elected* by the meta-rule $R2$ if the vertex v has no neighbour labelled N. A complete proof of this system may be found in [6].

3   Synchronizer Properties

Throughout the rest of this work, several distributed synchronization protocols will be described.

The operations of processes take place in a sequence of discrete steps called pulses: we represent a pulse by a counter, then we associate to each process a pulse number which is initialized to 0 or 1. At each step, a process goes from pulse i to pulse $i + 1$.

All of these protocols involve synchronizing the system at *every synchronous round*. This is necessary because the protocols are designed to work for arbitrary synchronous algorithms. All the synchronizers we will build are "global", in the sense that they involve synchronization among arbitrary nodes in the whole network. To preserve this "global" synchronization, each synchronizer has to satisfy some properties. The essential property we seek to preserve in translating a generic synchronous algorithm $\mathcal{A}_s$ into an asynchronous algorithm $\mathcal{A}_{as}$ is that the pulse difference between two arbitrary nodes is at most 1 (*Main Theorem*). In order to ensure that this property holds for all nodes and at any time or pulse, we begin by requiring *pulse compatibility* in the network. This means that a node can only increase its pulse, when it is sure that there is no node in the network that is still in a lower pulse. This property is guaranteed by the validity of Theorem 2. Furthermore, we will strengthen our synchronization assumption by forcing *pulse convergence* at any time. By *pulse convergence* we mean the fact that all the vertices of a network have simultaneously to be in pulse π before any node starts the pulse $\pi + 1$. Pulse convergence is stated by Theorem 1. Thus, the correctness of Theorem 3 (*Main Theorem*) can be deduced from the pulse compatibility and the pulse convergence.

The correctness of our synchronizers will then depend on the validity of the following three theorems:

Theorem 1 (Convergence Theorem). *Let π ($\pi > 0$) be the maximum pulse that has been reached so far. After a finite number of steps $T_\pi \geq 0$, all the vertices of G are in the same pulse π.*

Theorem 2 (Pulse compatibility Theorem). *A vertex u in G changes its pulse $p(u)$ only when there is no node v in G such that $v \neq u$ and $p(v) < p(u)$.*

Theorem 3 (Main Theorem). *At any time t, the pulse difference between two vertices v and u of a network G is at most 1.*

4   A Simple Synchronizer

We recall here the synchronization as presented and used in [10]. On each vertex v of a graph there is a counter $p(v)$, the initial value of $p(v)$ is 0. At each step the value of the counter $p(v_0)$ depends on the value of the counters of the neighbours of v_0 more precisely if $p(v_0) = i$ and if for each neighbour v of v_0 $p(v) = i$ or $p(v) = i + 1$ then v_0 is considered as *safe* and the new value of $p(v_0)$ is $i + 1$.

R1 : **The synchronization rule**
  Precondition :
  - $p(v_0) = i$,
  - $\forall v \in B(v_0, 1)$ $p(v) = i$ or $p(v) = i + 1$.

  Relabelling :
  - $p(v_0) := i + 1$.

A simple induction on the distance between vertices proves that:

Proposition 1. *For all vertices v_1 and v_2, $|p(v_1) - p(v_2)| \leq d(v_1, v_2)$.*

Remark 1. This synchronization does not need any knowledge on the graph, in particular on its size.

To implement this synchronization, a counter modulo 3 is sufficient: each process needs to compare the value of its counter to the value of each neighbour. More precisely, for each process v_0 and for each neighbour v of v_0 we determine if : $p(v_0) = p(v) - 1$, or $p(v_0) = p(v)$, or $p(v_0) = p(v) + 1$. Finally, the synchronization may encoded by:

R1 : **The synchronization rule**
  Precondition :
  - $p(v_0) = i$,
  - $\forall v \in B(v_0, 1)$ $p(v) = i \bmod 3$ or $p(v) = (i + 1) \bmod 3$.

  Relabelling :
  - $p(v_0) := (i + 1) \bmod 3$.

The α synchronizer [11] is similar to the synchronizer presented in this section. In fact, the precondition of rule R_1 expresses that the vertex v_0 is safe. In contrast to the α synchronizer, a vertex v_0 generates the next pulse as soon as it is safe. It does not wait until all its neighbours become safe. This synchronizer does not guarantee the *pulse compatibility theorem* and the *Convergence theorem*. Thus, it does not satisfy Theorem 3. It is therefore not appropriated to perform a "global" synchronization in a network. The next section will be devoted to the presentation of synchronizers that are able to preserve a "global" synchronization at every pulse.

5   Graphs with Known Size

5.1   The SSP Algorithm

We describe in our framework the algorithm by Szymanski, Shi and Prywes (the SSP algorithm for short) and then the synchronizer.

We consider a distributed algorithm which terminates when all processes reach their local termination conditions. Each process is able to determine only its own termination condition. The SSP algorithm detects an instant in which the entire computation is achieved.

Let G be a graph such that a boolean predicate $P(v)$ and an integer $a(v)$ is associated with each node v in G. Initially $P(v)$ is *false*, the local termination condition is not reached, and $a(v)$ is equal to -1. Transformations of the value of $a(v)$ are defined by the following rules.

Each local computation acts on the integer $a(v_0)$ associated to the vertex v_0; the new value of $a(v_0)$ depends on values associated to vertices of $B(v_0, 1)$. More precisely, let v_0 be a vertex and let $\{v_1, ..., v_d\}$ the set of vertices adjacent to v_0.

- If $P(v_0) = false$ then $a(v_0) = -1$;
- if $P(v_0) = true$ then $a(v_0) = 1 + Min\{a(v_k) \mid 0 \leq k \leq d\}$.

We consider the following assumption:
for each node v the value of $P(v)$ eventually becomes true and remains true for ever.

We will use the following notation. Let $(\mathbf{G}_i)_{0 \leq i}$ be a relabelling chain associated to SSP's algorithm. We denote by $a_i(v)$ (resp. $P_i(v)$) the integer (resp. the boolean) associated to the vertex v of $\mathbf{G}_i$. We have [12]:

Proposition 2. *Let $(\mathbf{G}_i)_{0 \leq i \leq n}$ be a relabelling chain associated to SSP's algorithm; let v be a vertex of G, we suppose that $h = a_i(v) \geq 0$. Then :*

$$\forall w \in V(G) \quad d(v, w) \leq h \Rightarrow a_i(w) \geq 0.$$

From this property we deduce that if a node knows the size of the graph (or a bound of the size, or the diameter or a bound of the diameter) then it can detect when $P(v)$ is true for all vertices of the graph.

5.2   A Synchronizer by Using the SSP Computations

In this section we introduce a new synchronization protocol. This protocol is based on the SSP algorithm. Let u be a vertex, the integer $p(u)$ denotes the value of the pulse associated to the vertex u. In our assumption, a vertex v satisfies the stable properties, if there is no node $u \in B_G(v, 1)$ such that $p(u) \neq p(v)$. Before giving a formal description of this algorithm, we have to explain the way it works.

Let G be a graph with diameter D. To each node v of G, we associate two integers $p(v)$ and $a(v)$, where $p(v)$ denotes the pulse and $a(v)$ denotes the SSP value for the vertex v. Initially, $p(v)$ and $a(v)$ are respectively set to 1 and 0.

A vertex v can start the next pulse ($p(v) = p(v) + 1$) and its counter $a(v)$ is reset to 0 when it detects that the value of the pulse of all the vertices is equal to $p(v)$. Now, we give a formal description of the above arguments.

Consider a labelling function λ, where $\lambda : V \rightarrow [1..\infty] \times [0..D]$. Initially all vertices are labelled $(1,0)$. The graph rewriting system is $\mathcal{R}_1 = (L_1, I_1, P_1)$ defined by $L_1 = \{[1..\infty] \times [0..D]\}$, $I_1 = \{(1,0)\}$, $P_1 = \{R_1, R_2\}$ where R_1 and R_2 are the relabelling rules described below. Consider a vertex v_0, the SSP synchronization is defined by:

R1 : **The observation rule**
  Precondition :
   • $\lambda(v_0) = (p(v_0), a(v_0))$,
   • $a(v_0) < D$,
   • $\forall v \in B(v_0, 1)\ p(v) \geq p(v_0)$,
   • $a(v_0) = Min\{a(v) \mid v \in B(v_0, 1)\ \text{and}\ p(v) = p(v_0)\}$.
  Relabelling :
   • $\lambda'(v_0) := (p(v_0), a(v_0) + 1)$.

R2 : **The changing phase rule**
  Precondition :
   • $\lambda(v_0) = (p(v_0), D)$.
  Relabelling :
   • $\lambda'(v_0) := (p(v_0) + 1, 0)$.

We will use the following notation. Let $(\mathbf{G}_i)_{0 \leq i}$ be a relabelling chain associated to the SSP synchronization. We denote by $p_i(v)$ (resp. $a_i(v)$) the pulse (resp. the integer) associated to the vertex v of $\mathbf{G}_i$.

For the correctness of this algorithm we state some invariants.

Fact 1 $p_{i+1}(v) \geq p_i(v)$.

Fact 2 *If $p_{i+1}(v) = p_i(v) + 1$ then $a_i(v) = D$.*

Lemma 1. *If $p_i(v) = \pi$ and $a_i(v) = h$ then:*

$$\forall w \in V(G) \quad d(v, w) \leq h \Rightarrow p_{i-h}(w) \geq \pi.$$

Proof. We show the lemma by induction on i. If $i = 0$ the property is obvious.

First we assume that $p_{i+1}(v) = p_i(v) = \pi$ and $a_{i+1}(v) = a_i(v) = h$. By the inductive hypothesis: $d(v, w) \leq h \Rightarrow p_{i-h}(w) \geq \pi$. From Fact 1, $p_{i-h+1}(w) \geq p_{i-h}(w) \geq \pi$. Thus $p_{(i+1)-h}(w) \geq \pi$.

Now we assume that $p_{i+1}(v) = p_i(v) = \pi$ and $a_{i+1}(v) = a_i(v) + 1 = h$. If $d(v, w) \leq h = a_i(v) + 1$ then let u such that $d(v, u) = 1$ and $d(u, w) = a_i(v) = h - 1$.

We have $a_{i+1}(v) = a_i(v) + 1 \Rightarrow p_i(u) \geq p_i(v) \geq \pi$ and $a_i(u) \geq h - 1$ (by the precondition of the observation rule). By the inductive hypothesis applied to the vertex u, $p_{i-(h-1)}(w) \geq \pi$ and finally $p_{(i+1)-h}(w) \geq \pi$.

The last case is $p_{i+1}(v) = p_i(v) + 1$, necessary $a_{i+1}(v) = 0$ and this achieves the proof.

From this lemma and Fact 1, it follows:

Corollary 1. *If* $p_i(v) = \pi$ *and* $a_i(v) = h$ *then* $d(v, w) \leq h \Rightarrow p_i(w) \geq \pi$.

Lemma 2. *If* $p_i(v) = \pi$ *and* $p_i(w) = \pi + 1$ *then* $\forall u \in V(G)$ $(p_i(u) = \pi$ *or* $p_i(u) = \pi + 1)$.

Proof. Let j be such that $p_j(w) = \pi$ and $p_{j+1}(w) = \pi + 1$, by the precondition of the phase rule and the previous corollary:

$$\forall u \in V(G) \quad p_j(u) \geq \pi.$$

For the same reasons, as $p_i(v) = \pi$, there does not exist u such that $p_i(u) > \pi+1$.

5.3   Randomized Synchronization Algorithm

The main idea of this algorithm is based on the use of *random walks* in a network. Initially, each node (process) gets a *token*. At each step, a node that has a token passes it randomly to one of its neighbors. When more than one token meet at one node, they merge to one token. In a connected undirected graph, with high probability, all tokens will merge to one token and the node, that gets it, starts the next synchronization pulse.

For our purpose, we have slightly modified the above described algorithm. The motivation of this departure from the main idea is well grounded since it can be quite laborious for a node, that has a token, to know that there is no other token in the whole network. In order to avoid this kind of problem, we represent our tokens as natural numbers and the action of merging tokens is now done by adding the numbers corresponding to these tokens. Moreover, a node does not only pass one token to one of its neighbors. Rather, it first merges all the tokens of its neighborhood to one and passes the resulting number to one of its neighbors. At the beginning of each pulse, each vertex produces a new token with value 1. As soon as we have one token left, we try (if possible) to broadcast this information through the whole graph (rule R_3). Thus, more than one node could start the next pulse. If a node u has a token $c(u)$ such that $c(u) = |V|$, then u is allowed to start the next synchronization pulse. We assume that each node knows the size of the network. A formal description of the algorithm is done below.

Let G be a graph with n vertices. Consider a labelling function λ, where $\lambda : V \rightarrow [1..\infty] \times [0..n]$. The first item of the label of a node u represents the pulse of u and the second represents the minimum number of vertices that are in the same pulse as u. Initially, at least one vertex is labeled $(1, 1)$ and the others are $(0, 0)$-labeled. The graph rewriting system is $\mathcal{R}_2 = (L_2, I_2, P_2)$ defined by $L_2 = \{[1..\infty] \times [0..n]\}$, $I_2 = \{\{(1, 1)\}\}$, $P_2 = \{R_1, R_2, R_3, R_4\}$ where R_1, R_2, R_3 and R_4 are the relabelling rules listed below. Let v_0 be a vertex, the randomized synchronization algorithm (RS algorithm in the sequel) is described as follow.

R1 : **The convergence rule**
    Precondition :
- $\lambda(v_0) = (p(v_0), c(v_0))$,
- $c(v_0) < |V|$,
- $\exists v \in B(v_0, 1)\ p(v) = p(v_0) + 1$.

    Relabelling :
- $\lambda'(v_0) := (p(v_0), |V|)$.

In the *convergence rule*, each vertex v_0 sets the value of its token $c(v_0)$ to $|V|$ as soon as there exists one vertex $v \in B_G(v_0, 1)$ that is in a greater pulse than $p(v_0)$.

R2 : **The phase rule**
    Precondition :
- $\lambda(v_0) = (p(v_0), c(v_0))$,
- $c(v_0) = |V|$.

    Relabelling :
- $\lambda'(v_0) := (p(v_0) + 1, 1)$.

In the *phase rule*, each vertex v_0 that has a token $c(v_0)$ such that $c(v_0) = |V|$ increases its pulse number and sets the value of its token to 1.

R3 : **The propagation rule**
    Precondition :
- $\lambda(v_0) = (p(v_0), c(v_0))$,
- $c(v_0) < |V|$,
- $\exists v \in B(v_0, 1)\ c(v) = |V|$ and $p(v) = p(v_0)$.

    Relabelling :
- $\lambda'(v_0) := (p(v_0), |V|)$.

In the *propagation rule*, a vertex v in the neighborhood of v_0 is in the same pulse $p(v_0)$ and $c(v) = |V|$ then the token of v_0 is set to $|V|$. This rule has only the final aim to broadcast the token information $c(v) = |V|$.

R4 : **The collecting rule**
    Precondition :
- $\lambda(v_0) = (p(v_0), c(v_0))$,
- $c(v_0) < |V| \wedge c(v_0) > 0$,
- $\forall v \in B(v_0, 1)\ p(v) = p(v_0)$.

    Relabelling :
- $w_0 := choose\ at\ random\ v \in B(v_0, 1)(v \neq v_0)$,
- $\forall v \in B(v_0, 1)(v \neq v_0)\lambda'(v) := (p(v), 0)$,
- $S := \sum_{v \in B(v_0, 1)} c(v)$,
- $\lambda'(w_0) := (p(w_0), S)$.

In the *collecting rule*, each vertex in the neighborhood of v_0 is in the same pulse $p(v_0)$. If v_0 has a token of value $c(v_0) < |V|$, then the sum of all the tokens in $B_G(v_0, 1)$ is computed and the resulting token is send to one neighbor w_0 of v_0. Note that the choice of w_0 is done randomly. The tokens of all the vertices $v \in B_G(v_0, 1)$ such that $v \neq w_0$ are set to 0.

We now turn our attention to the correctness of the described algorithm. Therefore, we first state some useful properties that will help us to show the validity of our synchronization assumption. One has also to notice that a node that does not have a token gets a *pseudo-token* with value 0. Let $(\mathbf{G}_i)_{0 \leq i}$ be a relabelling chain associated to a run of the RS algorithm. We denote by $p_i(v)$ (resp. $c_i(v)$) the pulse (resp. the integer) associated to the vertex v of $\mathbf{G}_i$. We assume that initially exactly one vertex of $\mathbf{G}_0$ is labeled $(1,1)$ and all the other vertices are labeled $(0,0)$. As from now, we are going to pay attention to the validity of the properties of our algorithm.

First we have the two following facts.

Fact 3 $\forall v \in V(G) \quad \forall i \quad p_{i+1}(v) \geq p_i(v)$.
$\forall v \in V(G) \quad \forall i \quad p_{i+1}(v) = p_i(v) + 1 \Rightarrow c_i(v) = |V|$.

Lemma 3. *Let π be a pulse, we have:*

1. *If there exists v_0 such that $0 < c_i(v_0) < |V|$ and $p_i(v_0) = \pi$ then:*

$$\sum_{\{v \mid p_i(v) = \pi\}} c_i(v) = Card\{v \mid p_i(v) = \pi\}.$$

2. *If there does not exist v_0 such that $0 < c_i(v_0) < |V|$ and $p_i(v_0) = \pi$ then:*

$$\forall v \quad such \ that \quad p_i(v) = \pi \quad either \quad c_i(v) = 0 \quad or \quad c_i(v) = |V|.$$

   Furthermore if for all v such that $p_i(v) = \pi$ we have $c_i(v) = 0$ then there exists w such that $p_i(w) = \pi + 1$.
3. $(\exists v, w \ such \ that \ p_i(v) = p_i(w) + 1) \Rightarrow (c_i(w) = 0 \ or \ c_i(w) = |V|)$.
4. *if $\exists v, w$ such that $p_i(v) = p_i(w) + 1$ then*

$$\sum_{\{u \mid p_i(u) = p_i(v)\}} c_i(u) = Card\{u \mid p_i(u) = p_i(v)\}.$$

5. *If there exists a vertex v such that $c_i(v) = |V|$ then for all vertex u we have: $p_i(u) \geq p_i(v)$.*
6.

$$\forall u, v \quad |p_i(u) - p_i(v)| \leq 1.$$

Proof. The proof is by induction on i. We assume that all the properties are true at step i and then we examine what happens according to the rewriting rule which is applied.

For the validity of our synchronization assumption we have to show that each node of G has enough knowledge of the whole graph to decide, if necessary, to change its pulse. This knowledge can only be achieved if we are able to ensure that after a finite time, and with high probability, there is only one token left in the whole network. Due to the lack of space, the main results of random walks and the proofs of Theorem 2, Theorem 1 and Theorem 3 can be found in [8].

6   Trees

All the synchronization protocols developed so far were dedicated to any type of graphs. These protocols need adequate knowledge of some graph characteristics to be exact and faultless. Now we introduce a new methodology devoted to trees. Although this methodology is restricted to trees, it has the advantage that we do not need to have more knowledge (size, diameter or existence of a distinguished node).

The main idea of this protocol resides in the use of an election algorithm [2] to decide which node should start the next pulse. This algorithm solves the problem of the election in anonymous trees using graph relabeling systems. At the beginning of the synchronization protocol, all vertices are in the same pulse. Their labels have two items. The first one is needed for the election algorithm and the second one represents the pulse number. We take advantage of the election algorithm to choose a vertex u that starts the next pulse. Furthermore, all the nodes v, that have a node w in their neighborhood such that $p(w) > p(v)$, increase their pulse. When a node increases its pulse, its *election*-label is set back to the initial value. A formalized description of this synchronization protocol is given below.

Let T be a tree. Initially all vertices are labeled $(N, 1)$. The graph rewriting system is $\mathcal{R}_4 = (L_4, I_4, P_4)$ defined by $L_4 = \{\{L, N\} \times [1..\infty]\}$, $I_4 = \{(N, 1)\}$, $P_4 = \{R_1, R_2, R_3\}$ where R_1, R_2 and R_3 are the relabeling rules given below.

R1 : **Leaf elimination rule**
   Precondition :
   - $\lambda(v_0) = (label(v_0), p(v_0))$,
   - $label(v_0) = N$,
   - $\exists v \in N(v_0) \, label(v) = N$, $p(v) = p(v_0)$, and
     $\forall w \in N(v_0)/v \, p(w) = p(v_0)$, $label(w) = L$
   Relabelling :
   - $\lambda'(v_0) := (L, p(v_0))$.

R2 : **Tree election rule**
   Precondition :
   - $\lambda(v_0) = (label(v_0), p(v_0))$,
   - $label(v_0) = N$,
   - $\forall v \in N(v_0) \, p(v) = p(v_0)$, $label(v) = L$
   Relabelling :
   - $\lambda'(v_0) := (N, p(v_0) + 1)$.

R3 : **The propagation rule**
   Precondition :
   - $\lambda(v_0) = (label(v_0), p(v_0))$,
   - $label(v_0) = L$,
   - $\exists v \in N(v_0) \, p(v) > p(v_0), label(v) = N$
   Relabelling :
   - $\lambda'(v_0) := (N, p(v_0) + 1)$.

The proofs of convergence theorem and of pulse compatibility can be found in [8]. Note that the synchronization protocol presented in this section can be extented to graphs with a distringuished vertex. The idea is to compute a spanning tree, then to run the protocol on the computed tree (all the details can be found in [8]).

7   Building Synchronizers

The main goal of this section is to give a methodology, that should transform the protocols introduced in previous sections in operative synchronizers. All the developed protocols assume the existence of a *pulse generator* at each node of the network. This means, that a node v has a pulse variable $p(v)$, and it is supposed to generate a sequence of local *clock pulses* by increasing the value of $p(v)$ from time to time (i.e. $p(v) = 0, 1, 2, ...$). These pulses are supposed to simulate the ticks of the global clock in the synchronous setting. Obviously, the use of these protocols as stand-alone applications will give nothing in an asynchronous environment. For this reason, we introduce some definitions that should help us to ensure some guarantee about the relationship between the pulse values at neighboring nodes at various moments during the execution.

Definition 1. *We denote by $t(v, p)$ the physical time in which v has increased its pulse to p. We say that v is at pulse $p(v) = p$ (or at pulse p) during the time interval $\tau(v, p) =]t(v, p), t(v, p + 1)[$.*

Since our network is fully asynchronous, we are not able to force all the vertices to maintain the same pulse at all times. However, we know that it is possible to guarantee a weaker form of compatibility between the pulses of neighboring nodes in the network. This form of compatibility is stated in Definition 2.

Definition 2 (Pulse Compatibility). *If node v sends an original message m to a neighbor w during its pulse $p(v) = p$, then m is received at w during its pulse $p(w) = p$ as well.*

7.1   Methodology to Construct a Synchronizer

To build a functioning synchronizer from each of our protocols, we have to change their specifications such that, given an algorithm Π_S written for a synchronous network and a protocol ν, it should be possible to *combine* Π_S on top of ν to yield a protocol $\Pi_A = \nu(\Pi_S)$ that can be executed on an asynchronous network. Π_A has two components: The *original component* and the *synchronization component*. Each of these components has its own local variables and messages at every vertex. The *original component* consists of the local variables and the messages of the original protocol Π_S, whereas the *synchronization component* consists of local synchronization variables and synchronization messages.

As from now, we are going to show that the changes needed to build a synchronizer from any of our protocols can be done effortless. Further, we will show that it is possible to take advantage of these changes to prove the correctness of the so constructed synchronizers. This proof will be done with respect to the definition given by David Peleg [9].

Conceptually, the modifications of our protocols are done in two steps. The first step affects the label attached to each vertex. In fact, we add three new items to each label. The first item consists of a *boolean* variable S. This variable decides which component(*original component* or *synchronization component*) is *active* on each vertex. Only the rules of the *active* component can be applied on a vertex v. The other items are two buffers B_p and B_{p-1} that represent the messages that a node v, at pulse $p(v)$, has to send respectively in pulses $p(v)$ and $p(v) - 1$. We claim that a protocol μ that is generated from the above modifications satisfies all the specifications of a synchronizer. In a general way, a protocol $\Pi_A = \mu(\Pi_S)$ will work in two phases.

Initially, the item S_v is set to *true* for all nodes v and therefore the *synchronization component* is active(Phase one). As soon as a node v increases its pulse $p(v)$, its variable S_v is set to *false*. Hence, only rules from Π_S can be applied on v. The *original component* is active(phase two). Each application of a rule of Π_S sets back the state of S_v to *true*. Thus the synchronization component can begin to synchronize v with all the other nodes. This cycle is done until Π_S stops Π_A with any specific request.

7.2   A General Overview

Both phases depicted in the previous section are summarized in Fig. 1. The rule R_1 represents phase one and R_2 represents phase two. Δ describes the set of rules used in the synchronization protocol(ν) and Ω is the set of relabeling rules used in the algorithm Π_S. The first rule means that as long as $S_v = True$ holds, it is only allowed to synchronize v with all the nodes in the network. As soon as v changes its pulse, the second phase is started for v. In this phase, it is no more allowed to synchronize but it is permitted to execute rules of the algorithm Π_S. After the execution of Π_S, rule one is active once again, and the cycle goes on. All our synchronizers will have to use a rule that plays the role of R_2. This rule should ensure a good interaction between the two protocols ν and Π_S. The rule R_2 is executed only if the subgraph pictured between brackets in Fig. 1 does not exist in the neighborhood of v.

Correctness of Our Synchronizers. We now introduce some meaningful properties that can all be guaranteed from the synchronizers we can build.

Definition 3 (Readiness property). *A node v is ready for pulse p, denoted $Ready(v,p)$, once it has already received all messages of the algorithm sent to it by its neighbors during their pulse number $p - 1$.*

Definition 4 (Readiness rule). *A node v is allowed to generate its pulse p once it is finished with its required original actions for pulse $p-1$ and $Ready(v,p)$ holds.*

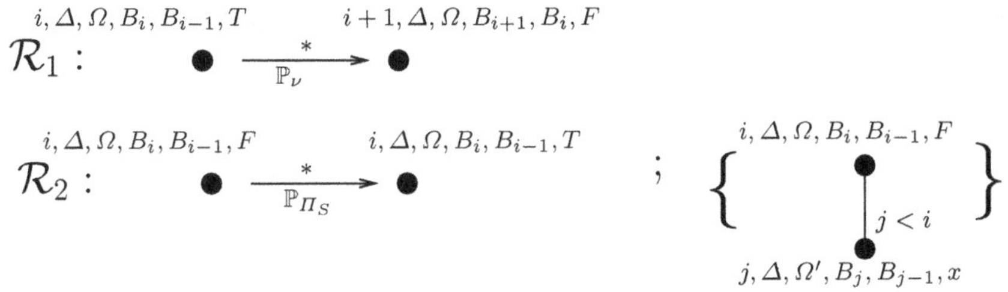

$$\mathcal{R}_1: \quad \overset{i,\Delta,\Omega,B_i,B_{i-1},T}{\bullet} \xrightarrow[\mathbb{P}_\nu]{*} \overset{i+1,\Delta,\Omega,B_{i+1},B_i,F}{\bullet}$$

$$\mathcal{R}_2: \quad \overset{i,\Delta,\Omega,B_i,B_{i-1},F}{\bullet} \xrightarrow[\mathbb{P}_{\Pi_S}]{*} \overset{i,\Delta,\Omega,B_i,B_{i-1},T}{\bullet} \quad ; \quad \left\{ \begin{array}{c} \overset{i,\Delta,\Omega,B_i,B_{i-1},F}{\bullet} \\ \Big\downarrow {\scriptstyle j<i} \\ \underset{j,\Delta,\Omega',B_j,B_{j-1},x}{\bullet} \end{array} \right\}$$

Fig. 1. A general method to construct a synchronizer.

Lemma 4. *Let μ be a synchronizer built as described in section 7.1. μ always satisfies the Readiness rule.*

Proof. The proof is an immediate consequence of the use of rule R_2 (*see Figure 1*). A node v is allowed to generate its pulse p, if and only if $S_v = T$ and v satisfies the conditions required from the *synchronization protocol*. The only possibility for S_v to become *True* is the execution of rule R_2. Thus, v has executed its required Π_S-actions for pulse $p-1$ and $Ready(v,p)$ holds.

Definition 5 (Delay rule). *If a node v receives in pulse p a message sent to it from a neighbor w during some later pulse $p' > p$ of w, then v declines consuming it and temporarily stores it in a buffer. It is allowed to process it only once it has already generated its pulse p'.*

Lemma 5. *Let μ be a synchronizer built as described in section 7.1. μ always satisfies the Delay rule.*

Proof. Let v be a node that has received in pulse p a message m' sent to it from a neighbor w during pulse $p' > p$. All our synchronization protocols guaranteed that after a finite number of steps, v and w will be in the same pulse p'. On the other hand, v can only consume the messages contained in the buffer B_p^w. Such a buffer always exists. Indeed, the pulse difference between two nodes in the whole network is maximal 1. This means that v will be able to consume m' as soon as $p(v) = p'$ holds.

Lemma 6. *[9] A synchronizer imposing the readiness and delay rules guarantees pulse compatibility.*

The above lemma states easily the reasons why all our synchronizers guarantee the principle of pulse compatibility introduced in definition 2. Peleg showed in [9] an essential relationship between the concept of pulse compatibility and the correctness of a synchronizer. One of the interesting parts of his work was announced as the lemma below.

Lemma 7. *[9] If synchronizer μ guarantees pulse compatibility, then it is correct.*

Lemma 8. *Let μ be a synchronizer built as described in section 7.1. μ is correct.*

Proof. The proof is deduced from the lemmas 4, 5 and according to the lemmas 6 and 7.

References

1. H. Attiya and J. Welch. *Distributed computing*. McGraw-Hill, 1998.
2. M. Bauderon, Y. Métivier, M. Mosbah, and A. Sellami. From local computations to asynchronous message passing systems. Technical Report RR-1271-02, University of Bordeaux 1, 2002.
3. K. M. Chandy and J. Misra. *Parallel program design - A foundation*. Addison-Wesley, 1988.
4. E. Godard, Y. Métivier, and A. Muscholl. Characterizations of Classes of Graphs Recognizable by Local Computations. *Theory of Computing Systems*, pages 249–293, 2004.
5. L. Lamport and N. Lynch. Distributed computing: models and methods. *Handbook of theoretical computer science*, B:1157–1199, 1990.
6. I. Litovsky, Y. Métivier, and E. Sopena. Graph relabelling systems and distributed algorithms. In H. Ehrig, H.J. Kreowski, U. Montanari, and G. Rozenberg, editors, *Handbook of graph grammars and computing by graph transformation*, volume 3, pages 1–56. World Scientific, 1999.
7. N. A. Lynch. *Distributed algorithms*. Morgan Kaufman, 1996.
8. Y. Métivier, M. Mosbah, R. Ossamy, and A. Sellami. Synchronizers for local computations. Technical Report 1322-04, LaBRI - University of Bordeaux 1, 2004.
9. D. Peleg. *Distributed computing - A locality-sensitive approach*. SIAM Monographs on discrete mathematics and applications, 2000.
10. P. Rosenstiehl, J.-R. Fiksel, and A. Holliger. Intelligent graphs. In R. Read, editor, *Graph theory and computing*, pages 219–265. Academic Press (New York), 1972.
11. G. Tel. *Introduction to distributed algorithms*. Cambridge University Press, 2000.
12. B. Szymanski Y. Shi and N. Prywes. Terminating iterative solutions of simultaneous equations in distributed message passing systems. In *4th International Conference on Distributed Computing Systems*, pages 287–292, 1985.

Constraints and Application Conditions: From Graphs to High-Level Structures

Hartmut Ehrig[1], Karsten Ehrig[1],
Annegret Habel[2], and Karl-Heinz Pennemann[2]

[1] Technische Universität Berlin, Germany
{ehrig,karstene}@cs.tu-berlin.de
[2] Carl v. Ossietzky Universität Oldenburg, Germany
{habel,k.h.pennemann}@informatik.uni-oldenburg.de

Abstract. Graph constraints and application conditions are most important for graph grammars and transformation systems in a large variety of application areas. Although different approaches have been presented in the literature already there is no adequate theory up to now which can be applied to different kinds of graphs and high-level structures. In this paper, we introduce an improved notion of graph constraints and application conditions and show under what conditions the basic results can be extended from graph transformation to high-level replacement systems. In fact, we use the new framework of adhesive HLR categories recently introduced as combination of HLR systems and adhesive categories. Our main results are the transformation of graph constraints into right application conditions and the transformation from right to left application conditions in this new framework.

1 Introduction

According to the requirements of several application areas the rules of a graph grammar have been equipped in [4] by a very general notion of application conditions. In a subsequent paper [8], the notion of application conditions is restricted to contextual conditions like the existence or non-existence of certain nodes and edges or certain subgraphs in the given graph. In [9], the authors introduce graphical consistency constraints, also called graph constraints, that express very basic conditions on graphs as e.g. the existence or uniqueness of certain nodes and edges in a graphical way.

Basic results for graph constraints and application conditions have been shown in [9,10] first for the single and later in the double pushout approach for different kinds of graphs. Unfortunately there is no adequate theory up to now which can be applied not only to graphs but also to high-level structures in the sense of [5].

A new version of high-level replacement systems, called adhesive HLR systems, has been introduced in [6] combining HLR systems in the sense of [5] and adhesive categories (see [11]). This new framework has been used not only to reformulate the basic results like local Church Rosser, Parallelism and Concurrency

H. Ehrig et al. (Eds.): ICGT 2004, LNCS 3256, pp. 287–303, 2004.

Theorem from [5], but also to present an improved version of the Embedding Theorem [3] and the local Confluence Theorem, known as Critical Pair Lemma [12]. Moreover it can be applied to all kinds of graphs and Petri nets satisfying the HLR1 and HLR2 conditions in [5] and also to typed attributed graphs in [7].

In this paper we use adhesive HLR categories and systems to improve and generalize the basic notions and results for constraints and application conditions from graphs to high-level structures. For this purpose we present an improved notion of graph constraints, based on positive and negative atomic constraints, and of application conditions, based on atomic conditional conditions. In our main theorems we show how to transform constraints into right application conditions, and right into left application conditions in the framework of adhesive HLR systems. As additional condition we only need finite coproducts and a suitable E-M-factorization which is valid in all our example categories.

The paper is organized as follows. In section 2 we present our improved notions of graph constraints and application conditions. In section 3 we give a short introduction of adhesive HLR categories together with some basic properties. Then we generalize graph constraints and application conditions to the framework of adhesive HLR categories. In section 4, we present the main results for graphs and high-level structures and give several illustrating examples for graphs and place transition nets. A conclusion including further work is given in section 5.

2   Constraints and Application Conditions for Graphs

In the following, we assume that the reader is familiar with the notions of graphs and graph morphisms, see e.g. [3, 2]. Graph constraints, first investigated by [9], allow to express basic conditions on graphs as e.g. the existence or uniqueness of certain nodes and edges in a graphical way.

Definition 1 (graph constraint). *An* atomic graph constraint *is of the form* $PC(a)$ *or* $NC(a)$ *where* $a: P \to C$ *is an arbitrary graph morphism. It is said to be a* positive *or* negative *atomic graph constraint, respectively. A* graph constraint *is a Boolean formula over atomic graph constraints, i.e. every atomic graph constraint is a graph constraint and, for every graph constraint* c, $\neg c$ *is a graph constraint and, for every index set* I *and every family* $(c_i)_{i \in I}$ *of graph constraints,* $\wedge_{i \in I} c_i$ *and* $\vee_{i \in I} c_i$ *are graph constraints. A graph* G *satisfies* $PC(a)$ $(NC(a))$, *written* $G \models PC(a)$ $(NC(a))$, *if for every injective morphism* $p: P \to G$ *there exists (does not exist) an injective morphism* $q: C \to G$ *such that* $q \circ a = p$.

$$P \xrightarrow{a} C \qquad\qquad P \xrightarrow{a} C$$

Fig. 1. Satisfiability of atomic constraints.

A graph G satisfies a graph constraint $\neg c$, written $G \models \neg c$, if and only if G does not satisfy the graph constraint c. G satisfies $\wedge_{i \in I} c_i$ ($\vee_{i \in I} c_i$), written $G \models \wedge_{i \in I} c_i$ ($\vee_{i \in I} c_i$), if and only if G satisfies all (some) graph constraints c_i with $i \in I$.

Example 1. Examples of graph constraints:

$\mathrm{PC}(\bigcirc\,\bigcirc \rightarrow \bigcirc)$	There exists at most one node.
$\mathrm{PC}(\bigcirc \rightarrow \text{\reflectbox{8}})$	Every node has a loop.
$\mathrm{NC}(\bigcirc \rightarrow \text{\reflectbox{8}})$	The graph is loop-free.
$\neg\mathrm{PC}(\bigcirc \rightarrow \text{\reflectbox{8}})$	There exists a node without loop.
$\mathrm{PC}(\emptyset \rightarrow \text{\reflectbox{8}})$	There exists a node with a loop.
$\mathrm{NC}(\emptyset \rightarrow \text{\reflectbox{8}})$	There exists no node with a loop.
$\bigwedge_{k=1}^{\infty} \mathrm{NC}(\emptyset \rightarrow C_k)$	The graph is acyclic (C_k denotes a cycle of length k).

Remark. The definition of graph constraints generalizes the one in [9], because we allow negative atomic constraints and non-injective a.

Fact. If a is non-injective and $G \models \mathrm{PC}(a)$, then there is no injective $p: P \rightarrow G$.

Proof. Assume there is an injective $p: P \rightarrow G$. Then $G \models \mathrm{PC}(a)$ implies the existence of an injective $q: C \rightarrow G$ with $q \circ a = p$. This implies a injective (contradiction).

Fact. If a is non-injective, then $G \models \mathrm{NC}(a)$.

Proof. Assume $G \not\models \mathrm{NC}(a)$. Then there exist injective $p: P \rightarrow G$ and $q: C \rightarrow G$ with $q \circ a = p$. The injectivity of p implies the injectivity of a (contradiction).

Application conditions for graph replacement rules were first introduced in [4]. In a subsequent paper [8], a special kind of application conditions were considered which can be represented in a graphical way. In particular, contextual conditions like the existence or non-existence of certain nodes and edges or certain subgraphs in the given graph can be expressed. In [9] so-called conditional application conditions were introduced.

Definition 2 (application condition over a graph). *An* (conditional) atomic application condition *over a graph L is of the form* $\mathrm{P}(x, \vee_{i \in I} x_i)$ *or* $\mathrm{N}(x, \wedge_{i \in I} x_i)$ *where $x: L \rightarrow X$ is an arbitrary graph morphism and $x_i: X \rightarrow C_i$ with $i \in I$ injective graph morphisms. It is said to be a* positive *or* negative atomic application condition, *respectively. An* application condition *over L is a Boolean formula over atomic application conditions over L, i.e. every atomic application condition is an application condition and, for every application condition* acc, $\neg$ acc *is an application condition and, for every index set I and every family ($\mathrm{acc}_i)_{i \in I}$ of application conditions, $\wedge_{i \in I}$ acc_i and $\vee_{i \in I}$ acc_i are application conditions.*

Fig. 2. Satisfiability of atomic application conditions.

A match $m: L \rightarrow G$ satisfies $\mathrm{acc}_L = \mathrm{P}(x, \vee_{i \in I} x_i)$ $(\mathrm{N}(x, \wedge_{i \in I} x_i))$, written $m \models \mathrm{acc}_L$, if for all injective morphisms $p: X \rightarrow G$ with $p \circ x = m$ there exists (does not exist) $i \in I$ and an injective morphism $q_i: C_i \rightarrow G$ with $q_i \circ x_i = p$.

A match $m: L \rightarrow G$ satisfies an application condition of the form $\neg$ acc, written $m \models \neg$ acc, if and only if m does not satisfy the application condition acc. A match m satisfies $\wedge_{i \in I} \mathrm{acc}_i$ $(\vee_{i \in I} \mathrm{acc}_i)$, written $m \models \wedge_{i \in I} \mathrm{acc}_i$ $(\vee_{i \in I} \mathrm{acc}_i)$, if and only if m satisfies all (some) acc_i with $i \in I$.

Remark. The definition of an application condition slightly generalizes the ones in [8, 9]. Let us consider the well-known negative application condition $\mathrm{NAC}(x)$, where $x: L \rightarrow X$ is a graph morphism. A match $m: L \rightarrow G$ satisfies $\mathrm{NAC}(x)$, written $m \models \mathrm{NAC}(x)$, if there does not exist an injective morphism $p: X \rightarrow G$ with $p \circ x = m$. $\mathrm{NAC}(x)$ is equivalent to $\mathrm{P}(x, \vee_{i \in I} x_i)$ for $I = \emptyset$ and hence a special case of positive atomic application conditions.

Example 2. Examples of application conditions and their meaning for an injective match $m: L \rightarrow G$:

$\mathrm{NAC}(\underset{1}{\bigcirc}\,\underset{2}{\bigcirc} \rightarrow \underset{1}{\bigcirc}\text{-}\underset{2}{\bigcirc})$ There is no edge from node $m(1)$ to node $m(2)$.

$\bigwedge_{k=1}^{\infty} \mathrm{NAC}(\underset{1}{\bigcirc}\,\underset{2}{\bigcirc} \rightarrow \mathrm{P}_k)$ There is no path connecting node $m(1)$ and node $m(2)$. (P_k denotes a path of length k)

$\mathrm{P}(\underset{1}{\bigcirc}\,\underset{2}{\bigcirc} \rightarrow \underset{1}{\bigcirc}\text{-}\underset{2}{\bigcirc} \rightarrow \underset{1}{\bigcirc}\text{⇄}\underset{2}{\bigcirc})$ If there is an edge from $m(1)$ to $m(2)$, then there also is an edge from $m(2)$ to $m(1)$.

A rule $p = \langle L \leftarrow K \rightarrow R \rangle$ consists of two injective graph morphisms with a common domain K. Given a rule p and a graph morphism $K \rightarrow D$, a direct derivation consists of two pushouts (1) and (2). We write $G \Rightarrow_{p,m,m^*} H$ and say that $m: L \rightarrow G$ is the match and $m^*: R \rightarrow H$ is the comatch of p in H.

$$
\begin{array}{ccccc}
L & \longleftarrow & K & \longrightarrow & R \\
\downarrow m & (1) & \downarrow & (2) & \downarrow m^* \\
G & \longleftarrow & D & \longrightarrow & H
\end{array}
$$

Definition 3 (application condition for a rule). An application condition $A(p) = (A_L, A_R)$ for a rule $p = \langle L \leftarrow K \rightarrow R \rangle$ consists of a left application condition A_L over L and a right application condition A_R over R. A direct derivation $G \Rightarrow_{p,m,m^*} H$ satisfies an application condition $A(p) = (A_L, A_R)$, if

$$m \models A_L \quad and \quad m^* \models A_R.$$

3   Constraints and Application Conditions for High-Level Structures

The main idea of high-level replacement systems is to generalize the concepts of graph replacement from graphs to all kinds of structures which are of interest in Computer Science and Mathematics. In the following, we will consider constraints and application conditions in adhesive HLR-categories (see [6]) and prove our transformation results on this general level.

Definition 4 (adhesive HLR-category). *A category **C** with a morphism class M is called adhesive HLR category, if 1) M is a class of monomorphisms closed under compositions and decompositions ($g \circ f \in M$, $g \in M$ implies $f \in M$), 2) **C** has pushouts and pullbacks along M-morphisms, i.e. pushouts and pullbacks, where at least one of the given morphisms is in M, and M-morphisms are closed under pushouts and pullbacks, and 3) pushouts in **C** along M-morphisms are VK-squares, i.e. for any commutative cube in **C** where we have the pushout with $m \in M$ in the bottom and the back faces are pullbacks, it holds: the top is pushout $\Leftrightarrow$ the front faces are pullbacks.*

$$
\begin{array}{ccc}
 & C' \xrightarrow{\ f'\ } B' & \\
\overset{m'}{\nearrow} \ \downarrow g' \quad \overset{n'}{\nearrow} \ | \ b & & \\
A' \xrightarrow{\ c\ } D' & & \\
a \ |_{\overset{m}{\nearrow}} \ C \xrightarrow{\ f\ } B & & \\
A \xrightarrow{\ g\ } D & &
\end{array}
$$

Example 3. All examples of adhesive categories defined in [11] are adhesive HLR categories for the class M of all monomorphisms. As shown in [11] this includes the categories **Sets** of sets, **Graphs** of graphs and several variants of graphs like typed, labelled and hypergraphs. Moreover this includes the category **PT-Net** of place transition nets considered in [5] already. The following categories are important examples of adhesive HLR categories where M is not the class of all monomorphisms: the category $\langle$**AGraphs**$_{\mathbf{ATG}}, M\rangle$ of typed attributed graphs with type graph ATG and class M of all injective morphisms with isomorphism on the data part is (see [7]), the category $\langle$**AHL-Net**$, M\rangle$ of algebraic high level nets with class M of all strict injective net morphisms, and the category $\langle$**Spec**$, M\rangle$ of algebraic specifications with class M of all strict injective specification morphisms [5].

Fact (HLR properties of adhesive HLR categories). Given an adhesive HLR-category $\langle \mathbf{C}, M \rangle$, the following HLR conditions are satisfied.

1. Pushouts along M-morphisms are pullbacks.
2. Pushout-pullback decomposition: If the diagram (1)+(2) is a pushout, (2) a pullback, and $l, w \in M$, then (1) and (2) are pushouts and also pullbacks.

$$A \xrightarrow{b} B \xrightarrow{r} E$$

$$l \downarrow \quad (1)\ s \downarrow \quad (2) \quad \downarrow v$$

$$C \xrightarrow{u} D \xrightarrow{w} F$$

3. Uniqueness of pushout complements for M-morphisms: Given $b: A \to B$ in M and $s: B \to D$ then there is up to isomorphism at most one C with $l: A \to C$ and $u: C \to D$ such that diagram (1) is a pushout.

Proof. See [6, 11].

General assumption. In the following, we assume that $\langle \mathbf{C}, M \rangle$ is an adhesive HLR category with binary coproducts and epi-M-factorizations, that is, for every morphism there is an epi-mono-factorization with monomorphism in M.

We will consider structural constraints and application conditions in our general framework. Structural constraints, short constraints, correspond to graph constraints in section 2, but not necessarily to logical constraints defined by predicate logic.

Definition 5 (constraints). *An* atomic constraint *is of the form* $\mathrm{PC}(a)$ *or* $\mathrm{NC}(a)$ *where* $a: P \to C$ *is an arbitrary morphism.* $\mathrm{PC}(a)$ *is said to be* positive *and* $\mathrm{NC}(a)$ negative. *A* constraint *is a Boolean formula over atomic constraints. An object G satisfies* $\mathrm{PC}(a)$ ($\mathrm{NC}(a)$), *written* $G \models \mathrm{PC}(a)$ ($\mathrm{NC}(a)$), *if for every morphism* $p: P \to G$ *in M there exists (does not exist) a morphism* $q: C \to G$ *in M such that* $q \circ a = p$ *(see figure 1). Satisfiability of arbitrary constraints is defined in the usual way (see definition 1).*

Definition 6 (application condition over an object). *An* atomic applica-tion condition *over an object L is of the form* $\mathrm{P}(x, \vee_{i \in I} x_i)$ *or* $\mathrm{N}(x, \wedge_{i \in I} x_i)$ *where* $x: L \to X$ *is an arbitrary morphism and* $x_i: X \to C_i$ *with* $i \in I$ *are morphisms in M. It is said to be a* positive *or* negative *atomic application condition, re-spectively. An* application condition *over L is a Boolean formula over atomic application conditions over L. A match $m: L \to G$ satisfies* $\mathrm{acc}_L = \mathrm{P}(x, \vee_{i \in I} x_i)$ ($\mathrm{N}(x, \wedge_{i \in I} x_i)$), *written* $m \models \mathrm{acc}_L$, *if for all morphisms* $p: X \to G$ *in M with* $p \circ x = m$ *there exists (does not exist) $i \in I$ and a morphism* $q_i: C_i \to G$ *in M with* $q_i \circ x_i = p$ *(see figure 2). Satisfiability of arbitrary application conditions is defined in the usual way (see definition 2).*

The special case of negative atomic application conditions $\mathrm{NAC}(x)$ and gen-eral application conditions for rules (see definition 3) are defined as in the graph case.

General Remark. In the case $I = \emptyset$ we have $\mathrm{P}(x, \vee_{i \in I} x_i) \equiv \mathrm{NAC}(x)$, where $m \models \mathrm{NAC}(x)$ means that there is no $p \in M$ with $p \circ x = m$. Moreover we have $\mathrm{N}(x, \wedge_{i \in I} x_i) \equiv \mathrm{true}$ for $I = \emptyset$, because $m^* \models \mathrm{N}(x, \wedge_{i \in \emptyset} x_i) \Leftrightarrow \forall p(p \in M \wedge p \circ x = m^* \Rightarrow \neg(\exists i \in I = \emptyset...)) \Leftrightarrow \forall p\ \mathrm{true} \Leftrightarrow \mathrm{true}$.

4   Main Results for Graphs and High-Level Structures

In the following, we will show that arbitrary constraints can be transformed into right application conditions and that right application conditions can be transformed in left application conditions. We first show that positive and negative atomic constraints can be transformed into right application conditions.

Lemma 1 (transformation of positive atomic constraints into right application conditions). *Given a positive atomic constraint* $\mathrm{PC}(a)$ *with* $a\colon P \to C$ *and comatch* $m^*\colon R \to H$*. Then there is a right application condition* $T(\mathrm{PC}(a))$ *such that* $m^* \models T(\mathrm{PC}(a)) \Leftrightarrow H \models \mathrm{PC}(a)$.

Construction.   Let $T(\mathrm{PC}(a))$ be the right application condition

$$T(\mathrm{PC}(a)) = \wedge_S \mathrm{P}(R \xrightarrow{s} S, \vee_{i \in I}(S \xrightarrow{t_i \circ t} T_i))$$

1. The conjunction $\wedge_S$ ranges over all "gluings" S of R and P in figure 3(a). More precisely over all triples $\langle S, s, p \rangle$ with arbitrary $s\colon R \to S$ and $p\colon P \to S$ in M such that the pair $\langle s, p \rangle$ is jointly epimorphic. For each such triple $\langle S, s, p \rangle$ we construct the pushout (1) of p and a leading to $t\colon S \to T$ and $q\colon C \to T$.
2. The disjunction $\vee_{i \in I}$ ranges over all $S \xrightarrow{t_i \circ t} T_i$ with epimorphism t_i such that $t_i \circ t$ and $t_i \circ q$ are in M. For $I = \emptyset$ we have $T(\mathrm{PC}(a)) = \wedge_S \mathrm{NAC}(R \xrightarrow{s} S)$.

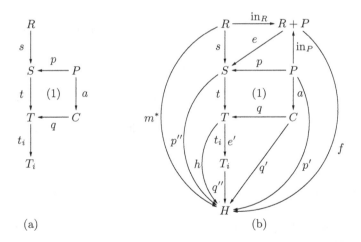

(a)   (b)

Fig. 3. Construction of $T(\mathrm{PC}(a))$/Correspondence of $T(\mathrm{PC}(a))$ and $\mathrm{PC}(a)$.

Proof. See appendix A.

Example 4. Consider the positive atomic graph constraint $\mathrm{PC}(\bigcirc \to \text{\Large 8})$ (see example 1) and the rule $p = \langle \bigcirc\bigcirc \leftarrow \bigcirc\bigcirc \to \bigcirc\!\!-\!\!\bigcirc \rangle$. According to the construction in lemma 1 the graph constraint can be transformed into the following

conjunction of right positiv atomic application conditions $\wedge_{i=1}^{4} P(O\!-\!O \rightarrow S_i,$
$S_i \rightarrow T_i) \wedge P(O\!-\!O \rightarrow S_5, \vee_{j=1}^{2} S_5 \rightarrow T_{5j})$ with S_i, T_i, T_{5j} as shown below. The
condition expresses the positiv atomic application condition "Every node out-
side (see T_1, T_4) and inside (see T_2, T_3, T_{51}, T_{52}) the comatch must have a loop.",
where S_1, S_2, S_3 correspond to injective and S_4, S_5 to non-injective comatches.
Altogether this condition means that for each comatch $m^* : R \rightarrow H$ each node
of H must have a loop, which is equivalent to $H \models \mathrm{PC}(O \rightarrow \mathcal{8})$.

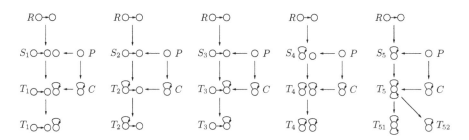

**Lemma 2 (transformation of negative atomic constraints into appli-
cation conditions).** *Given a negative constraint* $\mathrm{NC}(a)$ *with* $a: P \rightarrow C$ *and
comatch* $m^*: R \rightarrow H$. *Then there is an application condition* $T(\mathrm{NC}(a))$ *such
that* $m^* \models T(\mathrm{NC}(a)) \Leftrightarrow H \models \mathrm{NC}(a)$.

Construction. Let $T(\mathrm{NC}(a))$ be the following right application condition

$$T(\mathrm{NC}(a)) = \wedge_S \mathrm{N}(R \xrightarrow{s} S, \wedge_{i \in I}(S \xrightarrow{t_i \circ t} T_i)),$$

where the morphisms $s: R \rightarrow S$ and $t_i \circ t: S \rightarrow T_i$ are the same as in the construction of lemma 1 and $m^* \models T(\mathrm{NC}(a))$ is now defined by: For all $\langle S, s, p \rangle$ as
given in the construction, all $p'': S \rightarrow H$ in M with $p'' \circ s = m^*$ there is *no* $i \in I$
and $q'': T_i \rightarrow H$ in M with $q'' \circ t_i \circ t = p''$. For $I = \emptyset$ we have, according to the
general remark after definition 6, $m^* \models T(\mathrm{NC}(a)) \Leftrightarrow$ true.

Proof. See appendix A.

Example 5. According to example 3 we now consider place transition nets. Con-
sider the negative atomic net constraint $\mathrm{NC}(\emptyset \rightarrow O\!\!-\!\!\square)$. H satisfies this constraint if H contains no subnet of the form $O\!\!-\!\!\square$, where we call such a place a
"sink place". Consider the rule $p = \langle O\!\!\Leftarrow\!\!\square\!\!\Rightarrow\!O \leftarrow O\,O \rightarrow O\!\!-\!\!\square\!-\!O \rangle$. According to the construction in lemma 2 the constraint can be transformed into the
application condition $\mathrm{N}(R \rightarrow S_1, \wedge_{i=1}^{3} S_1 \rightarrow T_{1i}) \wedge \mathrm{N}(R \rightarrow S_2, \wedge_{i=1}^{2} S_2 \rightarrow T_{2i})$
where R, S_1, S_2, T_{ij} are given below. The condition means "No sink place is al-
lowed to be outside or inside the comatch, e.g. no sink place is allowed in H.",
where S_1 takes care of injective and S_2 of non-injective comatches $m^*: R \rightarrow H$.

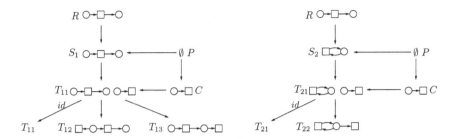

The transformation in lemma 1 and 2 can be extended to arbitrary constraints.

Theorem 1 (transformation of constraints into application conditions).
Given a constraint c and a comatch $m^*\colon R \to H$. Then there is an application condition $T(c)$ such that $m^* \models T(c) \Leftrightarrow H \models c$.

Proof. For atomic constraints, the transformation is given in the proof of lemma 1 and 2, respectively. For arbitrary constraints, the transformation is inductively defined as follows: $T(\neg c) = \neg T(c)$, $T(\wedge_{i \in I} c_i) = \wedge_{i \in I} T(c_i)$ and $T(\vee_{i \in I} c_i) = \vee_{i \in I} T(c_i)$. Now the proof of the statement is straightforward.

In the following, we will show that arbitrary right application conditions can be transformed into left application conditions. For this purpose, we first show that right positive and then right negative atomic application conditions can be transformed into corresponding left atomic application conditions.

Lemma 3 (transformation from right positive atomic to left positive application conditions). *Given a rule $p = \langle L \leftarrow K \to R \rangle$ and a right positive atomic application condition acc_R then there is a left positive atomic application condition acc_L such that for all direct derivations $G \Rightarrow_{p,m,m^*} H$ we have:*

$$m \models \mathrm{acc}_L \Leftrightarrow m^* \models \mathrm{acc}_R.$$

Construction. Let $\mathrm{acc}_R = \mathrm{P}(R \xrightarrow{x} X, \vee_{i \in I}(X \xrightarrow{x_i} C_i))$ be a right positive atomic application condition in figure 4. Then we construct a left positive atomic application condition $\mathrm{acc}_L = p^{-1}(\mathrm{acc}_R) = \mathrm{P}(L \xrightarrow{y} Y, \vee_{i \in I'}(Y \xrightarrow{y_i} D_i))$ with $I' \subseteq I$ or $p^{-1}(\mathrm{acc}_R) = \mathrm{true}$ as follows:

1. If the pair $\langle r\colon K \to R, x\colon R \to X \rangle$ has a pushout complement, define $y\colon L \to Y$ by two pushouts (1) and (2), otherwise $p^{-1}(\mathrm{acc}_R) = \mathrm{true}$.
2. For each $i \in I$, if the pair $\langle r^*\colon Z \to X, x_i\colon X \to C_i \rangle$ has a pushout complement, then $i \in I'$ and $y_i\colon Y \to D_i$ is defined by two pushouts (3) and (4), otherwise $i \notin I'$. Since pushout complements of M-morphisms (if they exist) are unique, the construction yields a unique result up to isomorphism.

Proof. See appendix A.

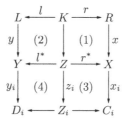

Fig. 4. Transformation of application conditions.

Example 6. Consider the rule $p = \langle \bigcirc\!\!-\!\!\bigcirc \leftarrow \bigcirc\bigcirc \rightarrow \bigcirc\bigcirc \rangle$ and – according to the remark after definition 2 – the right application condition $\mathrm{acc}_R = \mathrm{NAC}(\bigcirc\bigcirc \rightarrow \bigcirc\!\!-\!\!\bigcirc)$ (see example 2), meaning that an edge between the nodes in the comatch must not exist. According to the construction of lemma 3 with $I = \emptyset$ the right atomic application condition is transformed into the left atomic application condition $\mathrm{acc}_L = \mathrm{NAC}(\bigcirc\!\!-\!\!\bigcirc \rightarrow \bigcirc\!\!\supset\!\!\bigcirc)$ with $I = \emptyset$ meaning that two parallel edges between the nodes in the match must not exist.

$$
\begin{array}{ccc}
L & K & R \\
\bigcirc\!\!-\!\!\bigcirc \leftarrow \bigcirc\,\bigcirc \rightarrow \bigcirc\,\bigcirc \\
\downarrow \quad\quad \downarrow \quad\quad \downarrow \\
\bigcirc\!\!\supset\!\!\bigcirc \leftarrow \bigcirc\!\!-\!\!\bigcirc \rightarrow \bigcirc\!\!-\!\!\bigcirc \\
Y \quad\quad Z \quad\quad X
\end{array}
$$

Remark. 1. Dually we can construct from acc_L a right atomic application condition $\mathrm{acc}_R = p(\mathrm{acc}_L)$ such that $m \models \mathrm{acc}_L \Leftrightarrow m^* \models p(\mathrm{acc}_L)$.
2. For $I = \emptyset$, acc_R is a negative atomic application condition, i.e. $\mathrm{acc}_R \Leftrightarrow \mathrm{NAC}(x)$. In this case $p^{-1}(\mathrm{acc}_R)$ is either true or also a negative atomic application condition, i.e. $p^{-1}(\mathrm{acc}_R) \Leftrightarrow \mathrm{NAC}(y)$. For $I \neq \emptyset$, acc_R is a "real" atomic application condition and $p^{-1}(\mathrm{acc}_R)$ may be either true, a negative atomic application condition (if $I' = \emptyset$) or also a "real" atomic application condition.
3. Since x_i ($i \in I$) and also r and r^* are in M, also z_i and y_i are in M in (3) and (4) respectively (M-morphisms are closed under pushouts and pullbacks).

Lemma 4 (transformation from right negative atomic to left negative application conditions). *Given a rule $p = \langle L \leftarrow K \rightarrow R \rangle$ and a right negative atomic application condition acc_R then there is a left negative atomic application condition acc_L such that for all direct derivations $G \Rightarrow_{p,m,m^*} H$ we have: $m \models \mathrm{acc}_L \Leftrightarrow m^* \models \mathrm{acc}_R$.*

Construction. Let $\mathrm{acc}_R = \mathrm{N}(R \xrightarrow{x} X, \wedge_{i \in I}(X \xrightarrow{x_i} C_i))$ be a right negative atomic application condition. Then we construct a left negative atomic application condition $\mathrm{acc}_L = p^{-1}(\mathrm{acc}_R)$ as follows:

1. If $I = \emptyset$ then $\mathrm{acc}_R = \mathrm{true}$ and we define $\mathrm{acc}_L = \mathrm{true}$.
2. If $I \neq \emptyset$ and $\langle r, x \rangle$ has no pushout complement then $\mathrm{acc}_L = \mathrm{true}$.

3. If $I \neq \emptyset$ and $\langle r, x \rangle$ has a pushout complement then define $y \colon L \to Y$ by two pushouts (1) and (2) in figure 4. Moreover for each $i \in I$ if $\langle r^*, x_i \rangle$ has a pushout complement then $i \in I'$ and $y_i \colon Y \to D_i$ is defined by pushouts (3) and (4) in figure 4, otherwise $i \notin I'$. Now define

$$\mathrm{acc}_L = p^{-1}(\mathrm{acc}_R) = \mathrm{N}(L \xrightarrow{y} Y, \wedge_{i \in I'} (Y \xrightarrow{y_i} D_i))$$

where $\mathrm{acc}_L = \mathrm{true}$ in the case $I' = \emptyset$.

Proof. See appendix A.

Example 7. Consider the same rule p as in example 5 and the following right negative atomic application condition $\mathrm{acc}_R = \mathrm{N}(R \to X, X \to C)$ which corresponds to $\mathrm{N}(R \to S_2, S_2 \to T_{22})$ in example 5. $H \models \mathrm{acc}_R$ means that for each non-injective comatch $m^* \colon R \to H$ the place in $m^*(R)$ must not be a sink place. According to the general construction in figure 4 we obtain the following left negative atomic application condition $\mathrm{acc}_L = \mathrm{N}(L \to Y, Y \to D)$. $G \models \mathrm{acc}_L$ means that for each non-injective match $m \colon L \to G$ the place in $m(L)$ must not be a sink place. Note that a non-injective match $m \colon L \to G$ can only identify the places, because otherwise the gluing condition would be violated.

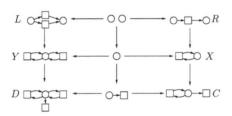

The transformation in lemma 3 and 4 can be extended to arbitrary right application conditions.

Theorem 2 (transformation from right to left application conditions).
Given a rule $p = \langle L \leftarrow K \to R \rangle$ and right application condition acc_R then there is a left application condition acc_L such that for all direct derivations $G \Rightarrow_{p,m,m^*} H$ we have: $m \models \mathrm{acc}_L \Leftrightarrow m^* \models \mathrm{acc}_R$.

Proof. For right atomic application conditions, the transformation is defined as in the proof of lemma 3 and 4, respectively. For arbitrary right application conditions, the transformation is defined as follows: $p^{-1}(\neg \mathrm{acc}_R) = \neg p^{-1}(\mathrm{acc}_R)$, $p^{-1}(\wedge_{i \in I} \mathrm{acc}_{iR}) = \wedge_{i \in I} p^{-1}(\mathrm{acc}_{iR})$, and $p^{-1}(\vee_{i \in I} \mathrm{acc}_{iR}) = \vee_{i \in I} p^{-1}(\mathrm{acc}_{iR})$. Now the proof of the statement is straightforward.

5   Conclusion

In the present paper we have introduced a general notion of constraints and application conditions that is more expressive than previous ones in the graph case and is formulated now for high-level structures in the new framework of adhesive HLR categories (see [6]). It is shown that constraints can be transformed into right application conditions for rules and that right application conditions can be transformed into left ones. As a consequence, we have a mechanism to integrate constraints into rules and to ensure that the constraints remain satisfied.

Further topics could be the followings.

(1) Extension of the notions of constraints and application conditions: Although the constraints are more general than the ones in [9], there are constraints which cannot be expressed up to now. E.g. the constraints like "Every node has an outgoing or incoming edge" and "There exists a node such that all outgoing edges are labelled by a" could be expressed if one would extend the concepts by alternative or conditional atomic graph constraints and existential satisfaction of graph constraints. Also the extension of constraints from statical (propositional logic) to dynamical (temporal logic) constraints [10] and a transformation between logical and graphical constraints (e.g. OCL-Constraints [1]) is interesting.

(2) Extensions of the theory: In [8], the local Church Rosser theorems I and II are proved for single-pushout rules with negative atomic application conditions. These results are also valid for the double-pushout rules with arbitrary application conditions in high-level structures provided that the notion of independence is extended such that the induced matches satisfy the corresponding application conditions. Moreover, it would be important to generalize the results in [6] to rules with application conditions.

(3) Applications of the theory: The theory can be applied already not only to graph transformations over labelled graphs (see [3, 2]) but also to several variants of graphs like typed attributed graphs and hypergraphs and also to Petri nets (see [5, 6]) and examples 4-7. For building up Petri nets satisfying some net constraints, one could integrate the constraints as application conditions into the rules. Another important application of adhesive HLR categories and corresponding systems is typed attributed graph transformation as presented in [7], where also a slight extension of the general assumption in section 3 seems to be useful.

References

1. Paolo Bottoni, Manuel Koch, Manuel Parisi-Presicce, and Gabriele Taentzer. Consistency checking and visualization of ocl constraints. In *UML 2000*, volume 1939 of *Lecture Notes in Computer Science*, pages 294–308. Springer-Verlag, 2000.
2. Andrea Corradini, Ugo Montanari, Francesca Rossi, Hartmut Ehrig, Reiko Heckel, and Michael Löwe. Algebraic approaches to graph transformation. Part I: Basic concepts and double pushout approach. In *Handbook of Graph Grammars and Computing by Graph Transformation*, volume 1, chapter 3, pages 163–245. World Scientific, 1997.
3. Hartmut Ehrig. Introduction to the algebraic theory of graph grammars. In *Graph-Grammars and Their Application to Computer Science and Biology*, volume 73 of *Lecture Notes in Computer Science*, pages 1–69. Springer-Verlag, 1979.

4. Hartmut Ehrig and Annegret Habel. Graph grammars with application conditions. In G. Rozenberg and A. Salomaa, editors, *The Book of L*, pages 87–100. Springer-Verlag, Berlin, 1986.

5. Hartmut Ehrig, Annegret Habel, Hans-Jörg Kreowski, and Francesco Parisi-Presicce. Parallelism and concurrency in high level replacement systems. *Mathematical Structures in Computer Science*, 1:361–404, 1991.

6. Hartmut Ehrig, Annegret Habel, Julia Padberg, and Ulrike Prange. Adhesive high-level replacement categories and systems. In *Graph Transformation (ICGT'04)*, Lecture Notes in Computer Science. Springer-Verlag, 2004.

7. Hartmut Ehrig, Ulrike Prange, and Gabriele Taentzer. Fundamental theory of typed attributed graph transformation. In *Graph Transformation (ICGT'04)*, Lecture Notes in Computer Science. Springer-Verlag, 2004.

8. Annegret Habel, Reiko Heckel, and Gabriele Taentzer. Graph grammars with negative application conditions. *Fundamenta Informaticae*, 26:287–313, 1996.

9. Reiko Heckel and Annika Wagner. Ensuring consistency of conditional graph grammars — a constructive approach. In *SEGRAGRA 95*, volume 2 of *Electronic Notes in Theoretical Computer Science*, pages 95–104, 1995.

10. Manuel Koch and Francesco Parisi-Presicce. Describing policies with graph constraints and rules. In *Graph Transformation (ICGT 2002)*, volume 2505 of *Lecture Notes in Computer Science*, pages 223–238. Springer-Verlag, 2002.

11. Stephen Lack and Paweł Sobociński. Adhesive categories. In *Proc. of Foundations of Software Science and Computation Structures (FOSSACS'04)*, volume 2987 of *Lecture Notes in Computer Science*, pages 273–288. Springer-Verlag, 2004.

12. Detlef Plump. Hypergraph rewriting: Critical pairs and undecidability of confluence. In *Term Graph Rewriting: Theory and Practice*, pages 201–213. John Wiley, New York, 1993.

A   Appendix

In this appendix we present the proofs of all lemmata given in section 4.

Proof of Lemma 1. 1. Let $m^* \models T(\mathrm{PC}(a))$. We have to show $H \models \mathrm{PC}(a)$, i.e. for all morphisms $p': P \to H$ in M there is a morphism $q': C \to H$ in M with $q' \circ a = p'$. Given a morphism $p': P \to H$ in M and a comatch $m^*: R \to H$, we construct the coproduct $R + P$ with injections in_R and in_P in figure 3(b). By the universal property of coproducts, there is a unique morphism $f: R + P \to H$ with $f \circ \mathrm{in}_R = m^*$ and $f \circ \mathrm{in}_P = p'$. Now let $f = p'' \circ e$ be an epi-mono factorization of f with epimorphism e and monomorphism p'' in M, and define $s = e \circ \mathrm{in}_R$ and $p = e \circ \mathrm{in}_P$. Then the pair $\langle s, p \rangle$ is jointly epimorphic, because e is an epimorphism, and p is in M, because $p'' \circ p = p'' \circ e \circ \mathrm{in}_P = f \circ \mathrm{in}_P = p'$ is in M (M-morphisms are closed under decomposition). Hence $\langle S, s, p \rangle$ belongs to the conjunction $\wedge_S$ of $T(\mathrm{PC}(a))$. Moreover we have $p'' \circ s = p'' \circ e \circ \mathrm{in}_R = f \circ \mathrm{in}_R = m^*$ with monomorphism p'' in M.

In the case $I \neq \emptyset$, $m^* \models T(\mathrm{PC}(a))$ implies existence of $i \in I$ and $q'': T_i \to H$ in M with $q'' \circ t_i \circ t = p''$. Now let $q' = q'' \circ t_i \circ q$ then q' is in M, because q'' is in M by construction and $t_i \circ q$ is in M by step 2 in the construction (M-morphisms are closed under decompositions). Finally we have $H \models \mathrm{PC}(a)$, because $q' \circ a = q'' \circ t_i \circ q \circ a = q'' \circ t_i \circ t \circ p = p'' \circ p = p'$.

In the case $I = \emptyset$, the existence of $p'' \in M$ with $p'' \circ s = m^*$ contradicts $m^* \models T(\mathrm{PC}(a)) = \wedge_S \mathrm{NAC}(s)$. Hence our assumption to have a $p': P \to H$ in M is false, which implies $H \models \mathrm{PC}(a)$.

2. Let $H \models \mathrm{PC}(a)$. We have to show $m^* \models T(\mathrm{PC}(a))$, i.e. for all triples $\langle S, s, p \rangle$ constructed in step 1 and all monomorphisms $p'': S \to H$ in M with $p'' \circ s = m^*$ we have to find $i \in I$ and a morphism $q'': T_i \to H$ in M with $q'' \circ t_i \circ t = p''$. Given $\langle S, s, p \rangle$ and p'' in M as above we define $p' = p'' \circ p: P \to H$. Then p' is in M, because p and p'' are in M, and $H \models \mathrm{PC}(a)$ implies $q': C \to H$ in M with $q' \circ a = p'$. Hence $p'' \circ p = p' = q' \circ a$. The universal property of pushouts implies the existence of a unique morphism $h: T \to H$ with $h \circ t = p''$ and $h \circ q = q'$. Now let $h = q'' \circ e'$ be a epi-mono factorization of h with epimorphism e' and monomorphism q'' in M. Then $q'' \circ e' \circ t = h \circ t = p''$ in M implies $e' \circ t$ is in M and $q'' \circ e' \circ q = h \circ q = q'$ in M implies $e' \circ q$ in M (M closed under decompositions). Hence according to construction step 2 $e' \circ t$ belongs to the family $(S \overset{t_i \circ t}{\to} T_i)_{i \in I}$ of $T(\mathrm{PC}(a))$ such that $e' = t_i: T \to T_i$ for some $i \in I$. In the case $I \neq \emptyset$ we have q'' in M and $q'' \circ t_i \circ t = q'' \circ e' \circ t = h \circ t = p''$ implies $m^* \models T(\mathrm{PC}(a))$. In the case $I = \emptyset$ we have a contradiction which means that our assumption to have $p'' \in M$ with $p'' \circ s = m^*$ is false. This implies $m^* \models \wedge_S \mathrm{NAC}(s) = T(\mathrm{PC}(a))$.

Remark. The proof of lemma 1 in both directions does not require that $a: P \to C$ is in M. If a is not in M, however, then t is not in M. (In fact, p in M in pushout (1) implies (1) pushout and pullback such that t in M would imply a in M). Hence there is no t_i in M s.t. $t_i \circ t$ is in M. This implies $I = \emptyset$, s.t. $T(\mathrm{PC}(a)) = \wedge_S \mathrm{NAC}(s)$. Hence we have for a not in M or $T = \emptyset$ the equivalence

$$m^* \models \wedge_S \mathrm{NAC}(s) \Leftrightarrow H \models \mathrm{PC}(a).$$

Proof of Lemma 2. 1. Let $m^* \models T(\mathrm{NC}(a))$ and $I \neq \emptyset$. The claim $H \not\models \mathrm{NC}(a)$ implies the existence of morphisms $p': P \to H$ and $q': C \to H$ in M with $q' \circ a = p'$ and will lead to a contradiction. Given $p': P \to H$ in M as above and m^* we construct the coproduct $R + P$. This leads to a unique $f: R + P \to H$ with $f \circ \mathrm{in}_R = m^*$ and $f \circ \mathrm{in}_p = p'$. Now let $f = p'' \circ e$ an epi-mono-factorization of f with epimorphism e and monomorphism p'' in M and define $s = e \circ \mathrm{in}_R$ and $p = e \circ \mathrm{in}_P$. Similar to part 1 of the proof of lemma 1 we have that $\langle S, s, p \rangle$ belongs to the family $\wedge_S$ of $T(\mathrm{NC}(a))$ and we have $p'' \circ p = p'' \circ e \circ \mathrm{in}_P = f \circ \mathrm{in}_P = p'$. Moreover we have $p'' \circ s = p'' \circ e \circ \mathrm{in}_R = f \circ \mathrm{in}_R = m^*$ with monomorphism p'' in M and $q' \circ a = p'$. Now pushout (1) in the figure 3, implies existence of $h: T \to H$ with $h \circ t = p''$ and $h \circ q = q'$. Now let $h = q'' \circ e'$ epi-mono-factorization of h with epimorphism e' and monomorphism q'' in M. Then, by the decomposition property of M, $p'' = h \circ t = q'' \circ e' \circ t$ in M implies $e' \circ t$ in M and $q' = h \circ q = q'' \circ e' \circ q$ implies $e' \circ q$ in M. Hence $e' \circ t$ belongs to the family $t_i \circ t$ in the construction of $T(\mathrm{NC}(a))$. Hence there is $i \in I$ with $e' = t_i: T \to T_i$ and q'' in M with $q'' \circ t_i \circ t = q'' \circ e' \circ t = h \circ t = p''$.

Our assumption $m^* \models T(\mathrm{NC}(a))$ implies that for all $\langle S, s, p \rangle$ as in the construction and all $p'': S \to H$ in M with $p'' \circ s = m^*$ as given above there is no

$i \in I$ and $q'': T_i \rightarrow H$ in M with $q'' \circ t_i \circ t = p''$. This is a contradiction to existence of $i \in I$ and q'' constructed above.

2. Let $H \models \text{NC}(a)$ and $I \neq \emptyset$. The claim $m^* \not\models T(\text{NC}(a))$ will lead to a contradiction. For $I \neq \emptyset$, $m^* \not\models T(\text{NC}(a))$ implies the existence of $\langle S, s, p \rangle$ as in the construction of $T(\text{NC}(a))$, and existence of $p'': S \rightarrow H$ in M with $p'' \circ s = m^*$, existence of $i \in I$ with $t_i \circ t, t_i \circ q$ in M, and existence of $q'': T_i \rightarrow H$ in M with $q'' \circ t_i \circ t = p''$. Now let $p' = p'' \circ p$, then p, p'' in M implies p' in M. Further let $q' = q'' \circ t_i \circ q$, then q'', $t_i \circ q$ in M implies q' in M. This implies $q' \circ a = q'' \circ t_i \circ q \circ a = q'' \circ t_i \circ t \circ p = p'' \circ p = p'$. Hence we have in contradiction to $H \models \text{NC}(a)$ p' and q' in M with $q' \circ a = p'$. For $I = \emptyset$, $m^* \not\models T(\text{NC}(a))$ implies existence of $p'' \in M$ with $p'' \circ s \neq m^*$. But this contradicts $H \models \text{NC}(a)$.

3. Let $I = \emptyset$. Then $m^* \models T(\text{NC}(a)) \Leftrightarrow$ true because $T(\text{NC}(a)) =$ true in this case. The claim $H \not\models \text{NC}(a)$ leads according to part 1 of the proof to $I \neq \emptyset$ which contradicts $I = \emptyset$. Hence we have $H \models \text{NC}(a)$. This means we have for $I = \emptyset$ $m^* \models T(\text{NC}(a)) \Leftrightarrow H \models \text{NC}(a) \Leftrightarrow$ true.

Remark. The proof of lemma 2 does not require that a is in M. If a is not in M we have again $I = \emptyset$ (as in remark after lemma 1).

Proof of Lemma 3. Let $G \underset{p,m,m^*}{\Longrightarrow} H$ be any direct derivation.

Case 1. The pair $\langle r: K \rightarrow R, x: R \rightarrow X \rangle$ has no pushout complement. Then $p^{-1}(\text{acc}_R) =$ true and $m \models p^{-1}(\text{acc}_R)$. We have to show $m^* \models \text{acc}_R$. This is true, because there is no $p: X \rightarrow H$ with $p \in M$ and $p \circ x = m^*$. Otherwise, since the pair $\langle r, m^* \rangle$ has a pushout complement, the pair $\langle r, x \rangle$ would have a pushout complement in contradiction to case 1 (pushout-pullback decomposition, $r, p \in M$).

Case 2. The pair $\langle r: K \rightarrow R, x: R \rightarrow X \rangle$ has a pushout complement and $I \neq \emptyset$.

Case 2.1. $m \models p^{-1}(\text{acc}_R)$. We have to show $m^* \models \text{acc}_R$, i.e. given a morphism $p: X \rightarrow H$ in M with $p \circ x = m^*$ we have to find an $i \in I$ and a morphism $q: C_i \rightarrow H$ in M with $q \circ x_i = p$. From the double pushout for $G \Rightarrow_{p,m,m^*} H$ and $p \circ x = m^*$ we obtain the following decomposition in pushouts (1), (2), (5), (6): First (5) is constructed as pullback of p and d_1 leading to pushouts (1) and (5) (pushout-pullback decomposition lemma, r, p in M), with same square (1) as in the construction because of uniqueness of pushout complements for M-morphisms. Then (2) is constructed as pushout and we have $p': Y \rightarrow G$ with $p' \circ y = m$ and pushout (6) induced by the pushouts (2) and (2) + (6). Since p is in M, z and p' are in M (M-morphisms are closed under pullbacks and pushouts).

In the case $I' = \emptyset$ we have no $p: X \rightarrow H$ with $p \in M$ and $p \circ x = m^*$, because this would imply $p': Y \rightarrow G$ with $p' \in M$ and $p' \circ y = m$ violating $m \models p^{-1}(\text{acc}_R)$. Having no p with $p \circ x = m^*$, however, implies $m^* \models \text{acc}_R$. In the case $I' \neq \emptyset$ we have by $m \models p^{-1}(\text{acc}_R)$ an $i \in I' \subseteq I$ with $y_i: Y \rightarrow D_i$ and $q': D_i \rightarrow G$ in M with $q' \circ y_i = p'$. Now we are able to decompose pushouts (6) and (5) into pushouts (4)+(8) and (3)+(7) respectively using the same technique as above

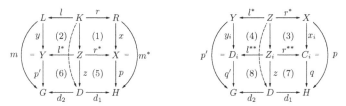

Fig. 5. Decomposition of pushouts.

now from left to right (pushout-pullback decomposition, $l^*, q' \in M$) leading to a morphism $q: C_i \to H$ in M with $q \circ x_i = p$. This implies $m^* \models \mathrm{acc}_R$.

Case 2.2. $m^* \models \mathrm{acc}_R$. We have to show $m \models p^{-1}(\mathrm{acc}_R)$. Due to case 2 we have $p^{-1}(\mathrm{acc}_R) \neq \mathrm{true}$. Hence for each morphism $p': Y \to G$ in M with $p' \circ y = m$ we have to find an $i \in I'$ and a morphism $q': D_i \to G$ in M with $q' \circ y_i = p'$. Given a morphism p' in M with $p' \circ y = m$ we can construct pushouts (1), (2), (5), (6) as above, where this time we first construct (6) as pullback leading in the right-hand side to a morphism $p: X \to H$ in M with $p \circ x = m^*$. Now $m^* \models \mathrm{acc}_R$ implies the existence of an $i \in I$ and a morphism $q: C_i \to H$ in M with $q \circ x_i = p$. Due to pushout (5) the pair $\langle r^*, p \rangle$ has a pushout complement, so that this is also true for $x_i: X \to C_i$ with $q \circ x_i = p$. Hence we have an $i \in I'$ and can decompose pushouts (5) and (6) into pushouts (3)+(7) and (4)+(8) from right to left leading to a morphism $q': D_i \to G$ in M with $q' \circ y_i = p'$. This implies $m \models p^{-1}(\mathrm{acc}_R)$.

Case 3. The pair $\langle r, x \rangle$ has a pushout complement, but $I = \emptyset$.

Case 3.1. $m^* \not\models \mathrm{acc}_R = \mathrm{NAC}(x)$ implies $p \in M$ with $p \circ x = m^*$. As shown in case 2.1 we obtain $p' \in M$ with $p' \circ y = m$ which implies $m \not\models \mathrm{NAC}(y)$.

Case 3.2. $m \not\models p^{-1}(\mathrm{acc}_R) = \mathrm{NAC}(y)$ implies in a similar way $m^* \not\models \mathrm{NAC}(x)$ using the construction in case 2.2.

Proof of Lemma 4. Let $G \underset{p,m,m^*}{\Longrightarrow} H$ be any direct derivation. We have to show

$$(\star) \quad m \models \mathrm{acc}_L \Leftrightarrow m^* \models \mathrm{acc}_R.$$

Case 1. $I = \emptyset$. Then $\mathrm{acc}_R = \mathrm{acc}_L = \mathrm{true}$ which implies $(\star)$.

Case 2. $I \neq \emptyset$ and $\langle r, x \rangle$ has no pushout complement then $\mathrm{acc}_L = \mathrm{true}$. We have to show $m^* \models \mathrm{acc}_R$. Assume $m^* \not\models \mathrm{acc}_R$. Then there is $p \in M$ with $p \circ x = m^*$. Since $\langle r, m^* \rangle$ has a pushout complement the pushout-pullback decomposition lemma with $r, p \in M$ implies that also $\langle r, x \rangle$ has a pushout complement. Contradiction. Hence $m^* \models \mathrm{acc}_R$.

Case 3. Let $I = \emptyset$ and $I' = \emptyset$ and $\langle r, x \rangle$ has a pushout complement. In this case we have $\mathrm{acc}_L = \mathrm{true}$ and we have to show $m^* \models \mathrm{acc}_R$. Assume $m^* \not\models \mathrm{acc}_R$.

Then there is $p \in M$ with $p \circ x = m^*$ and $i \in I$, $q_i \in M$ with $q_i \circ x_i = p$. This implies $m^* = q_i \circ x_i \circ x$. Since $\langle r, m^* \rangle$ has a pushout complement and $r, q_i, x_i \in M$ the pushout-pullback decomposition lemma implies that also $\langle r^*, q_i \circ x_i \rangle$ and $\langle r^*, x_i \rangle$ has a pushout complement. Contradiction to $I' = \emptyset$. Hence $m^* \models \mathrm{acc}_R$.

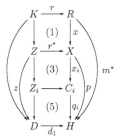

Case 4. Let $I \neq \emptyset$ and $I' = \emptyset$ and $\langle r, x \rangle$ has a pushout complement. In this case we use the following negations:

a) $m \not\models \mathrm{acc}_R = N(L \xrightarrow{y} Y, \wedge_{i \in I'}(Y \xrightarrow{y_i} D_i)) \Leftrightarrow$
$\exists p' : Y \to G$, $p' \in M$, $p' \circ y = m$ and $\exists i \in I'$ $\exists q_i' : D_i \to G$, $q_i' \in M, q_i' \circ y_i = p'$.

b) $m^* \not\models \mathrm{acc}_R = N(R \xrightarrow{x} X, \wedge_{i \in I}(X \xrightarrow{x_i} C_i)) \Leftrightarrow$
$\exists p : X \to H$, $p \in M$, $p \circ x = m^*$ and $\exists i \in I$ $\exists q_i : C_i \to H$, $q_i \in M$, $q_i \circ x_i = p$.

Case 4.1. $m \not\models p^{-1}(\mathrm{acc}_R) = \mathrm{acc}_L$ and we have to show $m^* \not\models \mathrm{acc}_R$. By $m \not\models \mathrm{acc}_L$ we have $p' \in M$, $i \in I'$ and $q_i' \in M$ as given in a). From the double pushout of $G \Rightarrow_{p,m,m^*} H$ and $m = p' \circ y$ with $p' \in M$ we can construct pushouts (2), (6), (1), (5) in figure 5 by the pushout-pullback decomposition lemma with $l, p' \in M$ leading to the commutative diagram in figure 5 with $p \in M$, $p \circ x = m^*$. Using $q_i' \circ y_i = p'$ we are able to decompose the pushouts (6) and (5) to pushouts (4), (8) and (3), (7) (by the pushout-pullback decomposition lemma with $l^*, q_i' \in M$) leading to morphisms $q_i : C_i \to H$, $q_i \in M$ with $q_i \circ x_i = p$ (see figure 5). This implies $m^* \not\models \mathrm{acc}_R$ as given in b).

Case 4.2. Let $m^* \not\models \mathrm{acc}_R$ and we have to show $m \not\models \mathrm{acc}_L$. By $m^* \not\models \mathrm{acc}_R$ we have $p \in M$, $i \in I$, and $q_i \in M$ as given in b). From the double pushout of $G \Rightarrow_{p,m,m^*} H$ and $m^* = p \circ x$ with $p \in M$ we can construct pushouts (1), (5), (2), (6) in figure 5 leading to $p \in M$ with and $p \circ x = m^*$. Using $q_i \circ x_i = p$ we are able to decompose the pushouts (5) and (6) to pushouts (3), (7) and (4), (8) in figure 5 with $q' \in M$ and $q' \circ y_i = p'$. This implies $m \not\models \mathrm{acc}_R$ as given in a).

Specification Matching of Web Services Using Conditional Graph Transformation Rules*

Alexey Cherchago and Reiko Heckel

University of Paderborn, Germany
{cherchago,reiko}@upb.de

Abstract. The ability of applications to dynamically discover required services is a key problem for Web Services. However, this aspect is not adequately supported by current Web Services standards. It is our objective to develop a formal approach allowing the automation of the discovery process. The approach is based on the matching of requestor's requirements for a useful service against service descriptions.

In the present paper, we concentrate on behavioral compatibility. This amounts to check a relation between provided and required operations described via operation contracts. Graph transformation rules with positive and negative application conditions are proposed as a visual formal notation for contract specification. We establish the desired semantic relation between requestor and provider and prove the soundness and completeness of a syntactic notion of matching w.r.t. this relation.

1 Introduction

The Web Services platform provides the means to adopt the World Wide Web for application integration based on standards for communication, interface description, and discovery of services. The prosperity of this technology strongly depends on the ability of applications to discover useful services and select those that can be safely integrated with existing components. Much work has been done to achieve this aim. The interface of an offered service can be specified in the Web Service Description Language (WSDL). This specification along with some keywords characterizing the service can be published at a UDDI-registry which serves as a central information broker and supplies this information to potential clients. However, current techniques do not support the automation of checking behavioral compatibility of the requestor's requirements with service descriptions.

In our work the compatibility of provided and required services is defined via the compatibility of operations constituting the service interfaces: For all *required operations* it is necessary to find structurally and behaviorally compatible *provided operations*. Structural compatibility requires a correspondence

---

* Research funded in part by European Community's Human Potential Programme under contract HPRN-CT-2002-00275, [SegraVis].

H. Ehrig et al. (Eds.): ICGT 2004, LNCS 3256, pp. 304–318, 2004.

between provided and required operation signatures. This can be checked using techniques developed for retrieving functions and components from software component libraries [19].

We shall concentrate on the second problem – *behavioral compatibility*. Service requestor and provider specify the behavior of operations by means of *contracts* [7], classically expressed in some form of logic. Instead, we propose to use graph transformation rules with positive and negative application conditions for this purpose. They have the advantage of blending intuitively with standard modeling techniques like the UML, providing contracts with an operational (rather than purely descriptive) flavor.

Since our approach is presented at the level of models, it can be used for different target technologies and languages, provided that they implement a Service-oriented Architecture (SOA). Web services represent the most prominent target platform, but there is no direct link of our approach to, say, XML-based languages. However each practical realization of the theoretical concepts requires a mapping from model-level concepts to a concrete technology. While developing such a mapping is outside the scope of this paper, we will briefly discuss the steps to be taken in the conclusion.

The classical interpretation of rules (based, e.g., on the double-pushout (DPO) approach to graph transformation [6]), however, is not adequate for contracts. It assumes that nothing is changed in the transformation beyond what is explicitly specified in the rule. A contract, however, represents a potentially incomplete specification of an operation. *Graph transitions* have been proposed to provide a looser interpretation of graph transformation rules. The double-pullback (DPB) approach [11] defines graph transitions as a generalization of DPO transformations, allowing additional changes that are not encoded in the rules.

Moreover, in order to increase the expressiveness of our graphical contract language, we consider rules with positive and negative application conditions. Negative conditions are well-known to increase the expressive power of rules [9]. In the classic approaches, positive application conditions can be encoded by extending both the left- and the right-hand side of a rule by the required elements: they become part of the context. This is no longer possible, however, in the presence of unspecified effects. In fact, the implicit frame condition, that all elements present before the application that are not explicitly deleted are still present afterwards, is no longer true. Thus, an element matched by a positive application condition may disappear, while an element which is shared between the left- and the right-hand side must be preserved.

Based on the notion of graph transition we will define an operational understanding of what it means for a provider rule to match the requestor's requirements. This shall be captured in a *semantic compatibility relation*. Since such a relation, being based on an infinite set of transitions, can not be computed directly, we introduce a *syntactic matching relation* which provides necessary and sufficient conditions at the level of rules.

After presenting in the next section the basic ideas of service specification, in Section 3 we will discuss the service specification matching, concentrating on the compatibility between provided and required operation contracts. The formalization of these notions, together with a theorem ensuring the soundness and completeness of the syntactic relation w.r.t. the semantic one, will be given in Section 4. In Sections 5 and 6 we discuss related approaches, conclude, and list issues for future work.

2   Service Specification

In this section we consider the basic ingredients of service specifications and introduce a simple scenario of a Web service for booking a hotel room, which may be required, e.g., for a travel booking system. The scenario is not intended to be complete, but it keeps step with standardization efforts in the travel industry (see, e.g., [1]) and allows to exemplify main ideas of our approach.

We start with the data model of the sample scenario expressed by the UML class diagram in Fig. 1: A customer (class Customer) intends to book a room (class Room) in a hotel (class Hotel). A booking information (class BookingInfo) and possibly a business license code (class LicenseInfo) of a customer (e.g. travel bureau) are required to make a reservation. The result of a booking process is represented by an acknowledgment in the form of a reservation tag (class ReservTag) and/or a reservation document (class Reservation) containing all details of the reservation. To avoid additional complications, we assume that service requestor and provider are working with the same data model, agreed upon in advance.

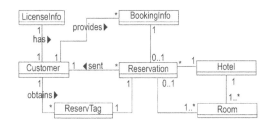

Fig. 1. Data model of the hotel reservation scenario.

According to [14], a Web service is represented by an interface describing a collection of operations that are network-accessible through standardized XML messaging. An example of provided and required interfaces in UML notation is given in Fig. 2.

An interface contains only structural information about operations. The behavior of these operations shall be specified by *contracts*. A contract consists of a pre-condition specifying the system state before some behavior is executed and a post-condition describing the system state after the execution of the behavior.

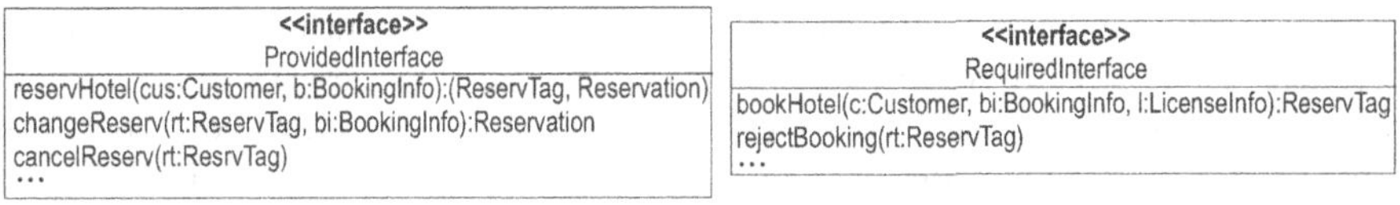

<<interface>> ProvidedInterface
reservHotel(cus:Customer, b:BookingInfo):(ReservTag, Reservation)
changeReserv(rt:ReservTag, bi:BookingInfo):Reservation
cancelReserv(rt:ResrvTag)
...

<<interface>> RequiredInterface
bookHotel(c:Customer, bi:BookingInfo, l:LicenseInfo):ReservTag
rejectBooking(rt:ReservTag)
...

Fig. 2. Provided and required interfaces.

There are different approaches employing formal techniques (e.g., description logic [15], algebraic specification languages [20], etc.) to contract specification (see Section 5). The main obstacle of these approaches is their lack of usability in the software industry, where knowledge and skills in the application of logic formalisms are scarce. Instead, we seek a notation that is close to the standard software modeling languages (e.g., UML) and has, at the same time, a formal background allowing to provide automation. This visual formal notation for contracts is introduced by *typed graph transformation* [3].

In this context, a class diagram is considered as a directed graph, whose vertices contain type declarations. Their relation with object diagrams representing run-time states is expressed by the notion of a *type graph* (TG) and corresponding *instance graphs* [3]. A graph transformation rule $p : L \Rightarrow R$ consists of a pair of TG-typed instance graphs L,R with compatible structure, i.e., such that edges that appear in both L and R are connected to the same vertices in both graphs, vertices with the same name have the same type, etc. The left-hand side L denotes the pre-condition of the rule and the right-hand side R represents the post-condition (cf. the part of Fig. 3 marked by the dashed rectangle). The effect encoded in the rule is defined by the items which have to be deleted (exist only in L), created (exist only in R), and preserved (exist in both L and R).

In addition, rules are equipped with positive and negative application conditions specifying required and forbidden contexts. A positive application condition contains elements whose presence is required by the rule, but without specifying wether these elements shall be preserved or deleted. This is possible because, seeing contracts as incomplete specifications of operations, we adopt a loose semantics for rules which permits unspecified changes. Negative application conditions represent structures which must not be present when the rule is applied. Formally, a *graph transformation rule with application condition* $\hat{p} : \{\hat{L}_P, \hat{L}_N\} \supseteq L \Rightarrow R$ contains, in addition to TG-typed instance graphs L and R, graphs $\hat{L}_{P/N}$ specifying extensions of L by the required or forbidden elements.

Fig. 3 shows the graph transformation rules for the required operation bookHotel() and the provided operation reservHotel(). $\hat{L}_{rP}$ and $\hat{L}_{pP}$ represent the patterns needed for the rule application along with the input parameters of the operations: information about a customer (verticies c and cus) and booking details (verticies bi). While the graphs $\hat{L}_{pP}$ and L_p in the provider rule are identical, the graph $\hat{L}_{rP}$ in the requestor rule additionally contains the vertex l:LicenseInfo denoting a business license code of the customer. This vertex being the input parameter of the operation bookHotel() is required to be present, but does not participate in the following computation.

The negative patterns of both rules $\hat{L}_{rN}$ and $\hat{L}_{pN}$ prevent us from booking a room which is already reserved by another customer. The reservation of a room is shown by newly created vertices res, rs, rt and the corresponding edges between them in the right-hand sides of the rules. The extra association sent in the lower rule reflects the fact that the reservation document is sent to the customer.

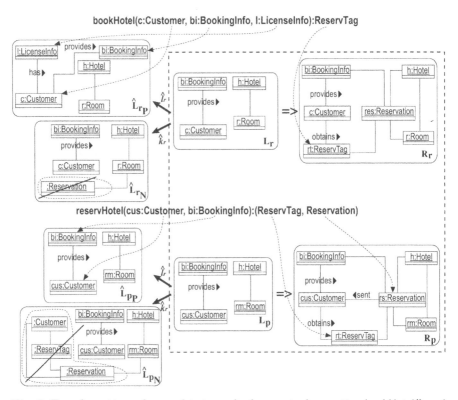

Fig. 3. Transformation rules on object graphs for required operation bookHotel() and provided operation reservHotel().

To summarize, a service specification consists of a data model, structural (operation signatures) and behavioral (operation contracts) specifications of operations constituting a service. In the next section we discuss specification matching and consider an example of matching required and provided operation contracts.

3   Specification Matching

In general, specification matching has to deal with all three aspects of specifications, i.e., data types, signatures, and contracts. For simplicity, we ignore the matching of data models and discuss the matching of signatures only briefly (see [19] for a general discussion).

As an example, consider the relation between the required operation bookHotel() and the provided operation reservHotel() whose signatures and contracts are depicted in Fig. 3. The signatures of the two operations differ in several ways. The required operation has an extra input parameter l:LicenseInfo, while the provided operation contains an extra output parameter of type Reservation. This does not contradict their compatibility, because any input of the requestor which is not required may simply be ignored by the provider. Similarly, any unexpected output of the provider may be skipped by the requestor.

To determine the relation between signatures and contracts, we require that input and output parameters of each operation are represented by vertices with corresponding types in the rules. These dependencies are indicated by the dashed arrows in Fig. 3.

Now we consider behavioral compatibility, i.e., the compatibility of preconditions and effects. Pre-conditions are captured by the patterns $\hat{L}_P$ and $\hat{L}_N$ of positive and negative constraints. In order to perform an operation, the provider requires input data from the requestor satisfying certain conditions. In the provider rule of Fig. 3, the input data consists of customer and booking information. The requestor has to be prepared to deliver this data and to guarantee the required properties. This can be expressed by a graph homomorphism from $\hat{L}_{pP}$ to $\hat{L}_{rP}$.

The restrictions towards the applicability of the provided operation are also described via undesired patterns of negative constraints. If the provided operation imposes more restrictive constraints than this is anticipated by the required one, this represents an unexpected limitation. To avoid this situation, one should check the existence of a graph homomorphism from $\hat{L}_{rN}$ to $\hat{L}_{pN}$.

A requestor expects some clearly specified benefit form the invocation of a service. The effect of the provided operation must not be smaller than specified by the requestor. That means, the requestor rule $p_r : L_r \Rightarrow R_r$ must be embedded in the provider $p_p : L_p \Rightarrow R_p$ as it is the case with the rules in Fig. 3. For example, the operation reservHotel() additionally creates the edge sent, denoting the delivery of a reservation document to the customer. This vertex is not presented in the requestor contract, because it is regarded as sufficient to obtain a reservation tag. Nevertheless, the effect of the provided operation fits the clients requirements.

Next, we will present a formalization for the intuitive ideas obtained from the example.

4   Formalization

Contract matching can be formalized as a relation between graph transformation rules. In this section, we define two such relations, a semantic one based on the operational interpretation of rules, and a syntactic one which provides necessary and sufficient conditions for the semantic relation. First, however, we review some of the basic notions of the *double-pushout* (DPO) [6] and *double-pullback* (DPB) [11] approach to graph transformation.

4.1   Graph Transformation

Given a graph TG, called *type graph*, a TG-*typed (instance) graph* consists of a graph G together with a typing homomorphism $g : G \to TG$ (cf. Fig. 4 on the left) associating with each vertex and edge x of G its type $g(x) = t$ in TG. In this case, we also write $x : t \in G$. A TG-typed graph morphism between two TG-typed instance graphs $\langle G, g \rangle$ and $\langle H, h \rangle$ is a graph morphism $f : G \to H$ which preserves types, that is, $h \circ f = g$.

The DPO approach to graph transformation has originally been developed for vertex- and edge-labelled graphs [6]. Here, we discuss the typed version [3].

Graph transformation rules, also called graph productions, are specified by pairs of injective graph morphisms ($L \xleftarrow{l} K \xrightarrow{r} R$), called *rule spans*. The *left-hand side* L contains the items that must be present for an application of the rule, the *right-hand side* R those that are present afterwards, and the *interface graph* K specifies the "gluing items", i.e., the objects which are read during application, but are not consumed.

Fig. 4. Typed graph and graph morphism (left) and double-pushout (or -pullback) diagram (right).

Definition 1. (rule, graph transformation system) *A rule span typed over TG, in short TG-typed rule span, $s = (L \xleftarrow{l} K \xrightarrow{r} R)$ is a span of injective TG-typed graph morphisms.*

A graph transformation system $GTS = \langle TG, P, \pi \rangle$ consists of a type graph TG, a set of rule names P, and a mapping π associating with each rule name p a TG-typed rule span $\pi(p)$. If $p \in P$ is a rule name and $\pi(p) = s$, we say that $p : s$ is a rule of GTS.

In DPO, transformation of graphs is defined by a pair of pushout diagrams, a so-called double-pushout construction (cf. Fig. 4 on the right). Operationally speaking that means: the elements of G matched by $L \setminus l(K)$ are removed, and a copy of $R \setminus r(K)$ is added to D.

4.2   Graph Transitions

The DPO approach ensures that the changes to the given graph H are exactly those specified by the rule. However, operation contracts represent specifications that are, in general, incomplete, that is, additional effects should be allowed in the transformation. Therefore, a more liberal notion of rule application is required which ensures that *at least* the elements of G matched by $L \setminus l(K)$ are removed, and *at least* the elements matched by $R \setminus r(K)$ are added. This interpretation is supported by the double-pullback (DPB) approach to graph transformation [11], which generalizes DPO by allowing additional, unspecified changes. Formally, graph transitions are defined by replacing the double-pushout diagram of a transformation step with a double-*pullback*.

Definition 2. (graph transition) *Let $p : s$ be a rule span with $s = (L \xleftarrow{l} K \xrightarrow{r} R)$. Then, a graph transition from G to H via p, denoted by $G \overset{p/d}{\rightsquigarrow} H$, is a diagram like the right part of Fig. 4 where both (1) and (2) are pullback squares. A graph transition (or briefly transition) is called* injective *if both g and h are injective graph morphisms. It is called* faithful *if it is injective, and the morphisms d_L and d_R satisfy the following condition: for all $x, y \in L$, $y \notin l(K)$ implies $d_L(x) \neq d_L(y)$, and analogously for d_R* [1].

Each faithful transition can be regarded as a transformation step plus a change-of-context [11]. This is modelled by additional deletion and creation of elements before and after the actual step.

4.3 Graph Transitions with Application Conditions

Using positive and negative application conditions [9], the embedding of the left-hand side of a rule in a given graph can be restricted, thus increasing the expressiveness of the formalism.

Definition 3. (rules with application conditions)
An application condition $A(p) = (AP(p), AN(p))$ for a graph transformation rule $p : s = (L \xleftarrow{l} K \xrightarrow{r} R)$ consists of two sets of typed graph morphisms $AP(p), AN(p) \subseteq \mathcal{MOR}(L)$ starting from L, that contain positive and negative constraints, respectively. $A(p)$ is called positive (negative) *if $AN(p)$ $(AP(p))$ is empty.*

Let $L \xrightarrow{\hat{l}} \hat{L}$ be a positive or negative constraint and $L \xrightarrow{d_L} G$ a typed graph morphism (cf.Fig. 5). Then d_L P-satisfies $\hat{l}$, if there exists a typed graph morphism $\hat{L} \xrightarrow{d_{\hat{L}}} G$ such that $d_{\hat{L}} \circ \hat{l} = d_L$. d_L N-satisfies $\hat{l}$, if it does not P-satisfy $\hat{l}$.

Let $A(p) = (AP(p), AN(p))$ be an application condition and $L \xrightarrow{d_L} G$ a typed graph morphism. Then d_L satisfies $A(p)$, if it P–satisfies all positive constraints and N-satisfies all negative constraints from $A(p)$.

A graph transformation rule with application condition is a pair $\hat{p} = (p, A(p))$ consisting of a graph transformation rule $p : s = (L \xleftarrow{l} K \xrightarrow{r} R)$ and an application condition $A(p)$ for p. It is applicable *to a graph G via $L \xrightarrow{d_L} G$ if d_L satisfies $A(p)$.*

---

[1] This condition means that d_L and d_R satisfy the identification condition of the DPO approach [4] with respect to l and r.

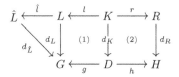

Fig. 5. DPB graph transition and rule with application condition.

Two examples of rules with positive and negative constraints are given in Fig. 3.

Definition 4. (graph transition with application condition) *Let $\hat{p} = (p, A(p))$ be a graph transformation rule with application condition, where $s = (L \xleftarrow{l} K \xrightarrow{r} R)$. A graph transition from G to H via the rule $\hat{p}$, denoted by $G \overset{\hat{p}/d}{\rightsquigarrow} H$, is a graph transition via a rule p, such that $d_L \in d$ satisfies the application condition of $\hat{p}$.*

Faithful transitions as introduced in Section 4.2 capture our intuition about a loose interpretation of contracts which can be specified by graph transformation rules with application conditions.

4.4   Semantic Compatibility and Syntactic Matching Relations

The concept of transition allows us to formalize semantically the desired notion of compatibility: Provider and requestor rules are semantically compatible if (1) every transition via the provider rule can be regarded as a transition via the requestor rule, and (2) applicability of the requestor rule implies applicability of the provider rule.

Definition 5. (semantic compatibility) *Let $\hat{p}_1 = (p_1, A(p_1))$ and $\hat{p}_2 = (p_2, A(p_2))$ be graph transformation rules with application conditions, where $s_1 = (L_1 \xleftarrow{l_1} K_1 \xrightarrow{r_1} R_1)$ and $s_2 = (L_2 \xleftarrow{l_2} K_2 \xrightarrow{r_2} R_2)$. We say that $\hat{p}_1$, is semantically compatible with $\hat{p}_2$, in symbols $\hat{p}_2 \models \hat{p}_1$, iff*

1. *for all spans $t = (G \xleftarrow{g} D \xrightarrow{h} H)$ and transitions $G \overset{p_2/d_2}{\rightsquigarrow} H$, there exists a transition $G \overset{p_1/d_1}{\rightsquigarrow} H$ using the same bottom span t, where $d_1 = (d_{L_1}, d_{K_1}, d_{R_1})$ and $d_2 = (d_{L_2}, d_{K_2}, d_{R_2})$ (cf. Fig. 6 on the right), and*
2. *if $d_{L_1} : L_1 \to G$ satisfies the application condition of $\hat{p}_1$, then $d_{L_2} : L_2 \to G$ satisfies the application condition of $\hat{p}_2$.*

This definition reflects the desired relation between contracts, but can hardly be applied for an algorithm determining contract compatibility. Therefore, we introduce a relation of *syntactic matching* that encompasses ideas presented in Section 3 and has more constructive character.

Definition 6. (syntactic matching) *Let $\hat{p}_1 = (p_1, A(p_1))$ and $\hat{p}_2 = (p_2, A(p_2))$ be graph transformation rules with application conditions, where*

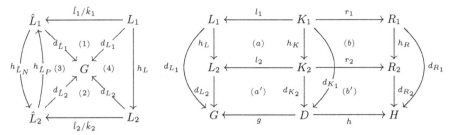

Fig. 6. Semantic compatibility and syntactic matching relations.

$s_1 = (L_1 \xleftarrow{l_1} K_1 \xrightarrow{r_1} R_1)$ and $s_2 = (L_2 \xleftarrow{l_2} K_2 \xrightarrow{r_2} R_2)$. We say that $\hat{p}_1$ syntactically matches with $\hat{p}_2$, in symbols $\hat{p}_2 \vdash \hat{p}_1$, iff

1. there exist graph homomorphisms $h_L : L_1 \to L_2$, $h_K : K_1 \to K_2$, and $h_R : R_1 \to R_2$ such that the diagrams (a) and (b) represent a faithful transition (cf. Fig. 6 on the right), and
2. (a) for all $\hat{l}_2 : L_2 \to \hat{L}_2 \in AP(p_2)$ there exist $\hat{l}_1 : L_1 \to \hat{L}_1 \in AP(p_1)$ and a graph homomorphism $h_{\hat{L}_P} : \hat{L}_2 \to \hat{L}_1$ such that the corresponding square in Fig. 6 on the left commutes, and
   (b) for all $\hat{k}_1 : L_1 \to \hat{L}_1 \in AN(p_1)$ there exist $\hat{k}_2 : L_2 \to \hat{L}_2 \in AN(p_2)$ and a graph homomorphism $h_{\hat{L}_N} : \hat{L}_1 \to \hat{L}_2$ such that the corresponding square in Fig. 6 on the left commutes.

An example of syntactic matching is given in Section 3 for the graph transformation rules specifying the contracts of the required operation bookHotel() and the provided operation reservHotel().

Next, we present a theorem ensuring the soundness and completeness of our approach.

Theorem 1. (soundness and completeness of matching) *Assume two graph transformation rules with application conditions $\hat{p}_1$ and $\hat{p}_2$. Then $\hat{p}_2 \vdash \hat{p}_1$ if and only if $\hat{p}_2 \models \hat{p}_1$.*

Proof Sketch. Soundness: Assume two graph transformation rules with application conditions $\hat{p}_1$ and $\hat{p}_2$. We show that $\hat{p}_2 \vdash \hat{p}_1$ *implies* $\hat{p}_2 \models \hat{p}_1$, i.e., Def. 6 entails Def. 5.1/2, respectively.

1. It is necessary to demonstrate that for each faithful transition via the second rule there is a faithful transition via the first rule. By assumption, there exist graph homomorphisms between the first and the second rule (h_L, h_K, h_R), forming a faithful transition (cf. Fig. 6 on the right). Now, both transitions can be vertically composed using the composition of the underlying pullback squares. The faithfulness of the composed transition follows from the preservation of the identification condition under the composition of pullback squares.

2. We have to demonstrate that if d_{L_1} satisfies the application condition of $\hat{p}_1$, then d_{L_2} satisfies the application condition of $\hat{p}_2$. This induces two problems (cf. Def. 3):

   (a) d_{L_2} (cf. Fig. 6 on the left) must *P-satisfy* all positive constraints of $\hat{p}_2$. Since $h_{\hat{L}_P}$ exists by assumption (Def. 6.2(a)), $d_{\hat{L}_2}$ can be constructed by $d_{\hat{L}_1} \circ h_{\hat{L}_P}$. It is not difficult to see that $d_{\hat{L}_2} \circ \hat{l}_2 = d_{L_2}$.

   (b) d_{L_2} (cf. Fig. 6 on the left) must *N-satisfy* all negative constraints of $\hat{p}_2$, i.e., there does not exist $d_{\hat{L}_2} : \hat{L}_2 \to G$. This can be proved by assuming existence of $d_{\hat{L}_2}$ and showing a contradiction.

   Combining (a) and (b), we obtain that d_{L_2} satisfies the application condition of $\hat{p}_2$.

Completeness: Assume two graph transformation rules with application conditions $\hat{p}_1$ and $\hat{p}_2$. To prove $\hat{p}_2 \models \hat{p}_1$ *implies* $\hat{p}_2 \vdash \hat{p}_1$, i.e. Def. 5 entails Def. 6.1/2, respectively.

1. To show that there exist graph homomorphisms between the first and the second rule, we can apply $\hat{p}_2$ to the graph L_2 at the identity mapping. If $\hat{p}_2$ is applicable, we can create a (faithful) transition $L_2 \overset{p_2/id}{\rightsquigarrow} R_2$ with the span $t = (L_2 \overset{l_2}{\leftarrow} K_2 \overset{r_2}{\to} R_2)$. Consequently (see Def. 5.1), there exists a faithful transition $L_2 \overset{p_1/h}{\rightsquigarrow} R_2$ via the first rule using the same bottom span t. If we can not apply $\hat{p}_2$ to the graph L_2, then the premise is false, and the conclusion is trivially true.

2. Two questions have to be considered for the second part of Def. 6:

   (a) Existence of a graph homomorphism $h_{\hat{L}_P} : \hat{L}_2 \to \hat{L}_1$ between the graphs representing the patterns for positive constraints of the rules. We can apply $\hat{p}_1$ to the graph $\hat{L}_1$ at $d_{L_1} := \hat{l}_1 \in AP(p_1)$. Since d_{L_1} satisfies $\hat{l}_1$, there exists d_{L_2} satisfying the constraint $\hat{l}_2 \in AP(p_2)$ of $\hat{p}_2$ (see Def. 5.2). That implies existence of a graph homomorphism $h_{\hat{L}_P} : \hat{L}_2 \to \hat{L}_1$. The commutativity of the corresponding square in Fig. 6 on the left comes out of the commutativity of the diagrams (1),(2),(3), and (4). The commutativity of the diagrams (1),(2), and (3) can easily be shown. To prove that the diagram (4) commutes one has to assume existence of a typed graph morphism $m : L_1 \to \hat{L}_1 := d_{L_2} \circ h_L \neq d_{L_1}(= \hat{l}_1)$ and find out a contradiction.

   (b) Existence of a graph homomorphism $h_{\hat{L}_N} : \hat{L}_1 \to \hat{L}_2$ between the graphs representing patterns for negative constraints of the rules. We can apply $\hat{p}_1$ to the graph $\hat{L}_2$ at some d_{L_1}. By Def. 5.2 if d_{L_1} *satisfies* $\hat{k}_1 \in AN(p_1)$, then d_{L_2} *satisfies* $\hat{k}_2 \in AN(p_2)$. We can reformulate this as: if d_{L_2} *does not satisfy* $\hat{k}_2$, *then* d_{L_1} *does not satisfy* $\hat{k}_1$.

   Now we try to apply $\hat{p}_2$ to the graph $\hat{L}_2$ at $d_{L_2} := \hat{k}_2$. It is possible to see that the premise of the statement above is true (d_{L_2} *does not satisfy* $\hat{k}_2$), so is the conclusion, i.e., d_{L_1} *does not satisfy* $\hat{k}_1$. This may happen only if there exists a graph homomorphism $h_{\hat{L}_N} : \hat{L}_1 \to \hat{L}_2$, which was required to be proved. The commutativity of the corresponding square in Fig. 6 on the left comes out of the commutativity of the diagrams (1),(2),(3), and (4). The only problem here is the commutativity of the diagram (4) which can be demonstrated in the same way as in (a).

Combining two parts of the proof we can conclude that $\hat{p}_2 \vdash \hat{p}_1$ if and only if $\hat{p}_2 \models \hat{p}_1$.

Two final sections discuss approaches related to our work and summarize the main results.

5   Related Work

The problem of discovering a component or service satisfying specific requirements is not a new one. A significant amount of work has been done in the area of Component-Based Software Engineering (CBSE) to increase the reliability and maintainability of software through reuse. Central here is the development of the techniques for reasoning about component descriptions and component matching. These techniques differ in the constituents involved in the matching procedure (e.g., operation signatures, behavioral specifications) and the way these constituents are specified (e.g., logic formulas, algebraic specification languages).

One of the most elaborate approaches, along with a thorough overview of related work, is presented by Zaremski and Wing in [19] and [20], who have developed matching procedures for two levels of specifications semantic information (signatures and specifications) and two levels of granularity (functions and modules). Structural and behavioral information about components is given using the algebraic specification language Larch/ML.

A pre/post-condition style of specification, like in [20], is also utilized by other authors. For example, in the work of Perry [17] operations are specified with pre- and post-conditions in first order logic. Order-sorted predicate logic (OSPL) is employed by Jeng and Cheng in [13] for component and function specifications. Basically, two features differentiate our approach from the works described above. The first one is the operational interpretation of graph transformation rules. Second, we have proposed a visual, model-based approach which provides better usability, because it can be more easily integrated into the standard model-driven techniques for software development.

Matching required and provided interfaces is also an issue present in modularization approaches for algebraic specification languages and typed graph transformation systems (TGTS), in particular for the composition of modules. An algebraic module specification MOD in [5] consists of four parts called import IMP, export EXP, parameter PAR, and body BOD. All components are given by algebraic specifications, which are combined through specification morphisms. TGTS-modules in [8] are composed of three TGTS, IMP, EXP, and BOD, the only difference being the absence of a parameter part. IMP and BOD are related by a simple inclusion morphism, whereas EXP and BOD are connected by a refinement morphism, allowing a temporal or spatial decomposition of rules.

Composition of modules MOD_1 and MOD_2 is based on the morphisms connecting the import interface of MOD_1 with the export interface of MOD_2. Relating required (import) and provided (export) services, this morphism has a similar role like the matching relation in this paper. A detailed comparison with [5] is hampered by the conceptual distance between the two formalisms, i.e., graph transformation and algebraic specifications.

The interconnection mechanism in [8] is incomparable to our notion: On the one hand, the cited paper allows an extension and renaming of types, an issue that we did not discuss at all in this paper. On the other hand, the relation between rules is more general in our case, i.e., it is the most general notion allowing the entailment of applicability from import to export as well as the entailment of effects in the opposite direction.

Finally, let us mention several approaches in the Semantic Web context. Paolucci et al. in [16] propose a solution for automation of Web service discovery based on DAML-S, a DAML-based language for service description. While required and provided service descriptions contain specifications of pre-conditions and effects, the matching procedure in [16] merely compares input and output parameters of services. Basically, such a kind of matching can be considered as an extended variant of the structural compatibility. Sivashanmugan et al. in [18] extend WSDL using DAML+OIL ontologies to support semantics-based discovery of Web Services. The authors emphasize the importance of matchmaking not only for input and output parameters, but also for functional specifications of operations. Since the work contains only conceptual descriptions of the matching procedure, we can not provide a more formal analysis.

Hausmann et al. in [10] use graph transformation rules defined over a domain ontology to represent service specifications and introduce a relation between them. The strength of this work is the implementation of the matching procedure in a prototypical tool chain. Informally, the basic ideas introduced in [10] are similar to those of this paper, but there is a number of technical differences. While the matching procedure has been precisely defined in a set-theoretic notation, the authors of [10] did not provide a formal operational semantics along with a semantic compatibility relation. As a consequence, the correctness of the proposed formalism has been justified only by means of examples. Besides, the lack of application conditions limits the expressiveness of contracts to basic positive statements.

Another approach is presented by Pahl in [15]. He proposed to use description logic for service specification and introduced a contravariant inference rule capturing service matching. This approach is closely related to ours because of the pre/post style of service specification and the contravariant character of the matching. But, as it was correctly stated by the author, the expressiveness of description logic has negative implications for the complexity of reasoning. Unlike our approach, the service matching in [15] has a problem with decidability that can be guaranteed only under certain restrictions (the set of predicates must be close under negation). Although, the sub-graph problem, in general, is not solvable in polynomial time, there exist a number of heuristic solutions which make it appear realistic.

6   Conclusion

In this paper we have developed formal concepts for a UML-based approach to service specification and matching based on conditional graph transformation rules for modeling operation contracts. We have used a loose interpretation of rules via DPB graph transitions to obtain an operational understanding of contracts and a corresponding semantic compatibility relation, and we have established a syntactic relation providing necessary and sufficient conditions for the semantic one.

Several issues remain for future work. First, the formal presentation needs to be extended to typed graphs with attributes [12] and sub-typing [2]. Second, the compatibility relation should be improved to allow the matching of one requestor rule against a spatial/temporal composition of several provider rules [8].

Third, the practical application of the concepts discussed in this paper requires a mapping to the Web service platform, consisting of XML-based standards like SOAP and WSDL. The first part of this mapping has to relate the type systems of both levels, i.e. a type graph of GTS and an XML-schema of a WSDL specification. The second part should associate operation signatures given by UML interfaces with the corresponding specifications of operations in a WSDL document. The last part should provide the mapping of contracts into an adequate XML-representation. This should be integrated with WSDL and should support the implementation of the corresponding matching procedures. While there are isolated examples for all three mappings in the literature, their integration yet remains an open issue.

References

1. The OpenTravel^TM Alliance. Message users guide, December 2003.
   http://www.opentravel.org/.
2. R. Bardohl, H. Ehrig, J. de Lara, O. Runge, G. Taentzer, and I. Weinhold. Node type inheritance concepts for typed graph transformation. Fachberichte Informatik 2003-19, Technical University Berlin, 2003.
3. A. Corradini, U. Montanari, and F. Rossi. Graph processes. *Fundamenta Informaticae*, 26(3,4):241–266, 1996.
4. A. Corradini, U. Montanari, F. Rossi, H. Ehrig, R. Heckel, and M. Löwe. Algebraic approaches to graph transformation, Part I: Basic concepts and double pushout approach. In G. Rozenberg, editor, *Handbook of Graph Grammars and Computing by Graph Transformation, Volume 1: Foundations*, pages 163–245. World Scientific, 1997. Preprint available as Tech. Rep. 96/17, Univ. of Pisa,
   http://www.di.unipi.it/TR/TRengl.html.
5. H. Ehrig and B. Mahr. *Fundamentals of algebraic specification 2: module specifications and constraints*. Springer-Verlag, 1990.
6. H. Ehrig, M. Pfender, and H.J. Schneider. Graph grammars: an algebraic approach. In *14th Annual IEEE Symposium on Switching and Automata Theory*, pages 167–180. IEEE, 1973.
7. D. Fensel and C. Bussler. The web service modeling framework WSFM. *Electronic Commerce Research and Applications*, 1(2):113–137, 2002.

8. M. Grosse-Rhode, F. Parisi-Presicce, and M. Simeoni. Refinements and modules for typed graph transformation systems. In J.L. Fiadeiro, editor, *Proc. Workshop on Algebraic Development Techniques (WADT'98), at ETAPS'98, Lisbon, April 1998*, volume 1589 of *LNCS*, pages 138–151. Springer-Verlag, 1999.
9. A. Habel, R. Heckel, and G. Taentzer. Graph grammars with negative application conditions. *Fundamenta Informaticae*, 26(3,4):287 – 313, 1996.
10. J.-H. Hausmann, R. Heckel, and M. Lohmann. Model-based discovery of web services. In *Proc. 2004 IEEE International Conference on Web Services (ICWS 2004) July 6-9, 2004, San Diego, California, USA.*, 2004. to appear.
11. R. Heckel, H. Ehrig, U. Wolter, and A. Corradini. Double-pullback transitions and coalgebraic loose semantics for graph transformation systems. *Applied Categorical Structures*, 9(1), January 2001. See also TR 97-07 at
    http://www.cs.tu-berlin.de/cs/ifb/TechnBerichteListe.html.
12. R. Heckel, J. Küster, and G. Taentzer. Confluence of typed attributed graph transformation systems. In A. Corradini and H.-J. Kreowski, editors, *Proc. 1st Int. Conference on Graph Transformation (ICGT 02), Barcelona, Spain*, Lecture Notes in Comp. Science. Springer-Verlag, October 2002.
13. J.-J. Jeng and B. H. C. Cheng. Specification matching for software reuse: A foundation. In *Proc. of the ACM SIGSOFT Symposium on Software Reusability (SSR'95)*, pages 97–105, April 1995.
14. H. Kreger. Web services conceptual architecture (WSCA 1.0), May 2001.
    http://www-3.ibm.com/software/solutions/webservices/pdf/WSCA.pdf.
15. C. Pahl. An ontology for software component matching. In M. Pezze, editor, *Fundamental approaches to software engineering: 6th international conference, FASE 2003*, volume 2621 of *LNCS*, pages 6–21. Springer, 2003.
16. M. Paolucci, T. Kawmura, T. Payne, and K. Sycara. Semantic matching of web services capabilities. In *Proc. First International Semantic Web Conference*, 2002.
17. D. E. Perry and S. S. Popovich. Inquire: Predicate-based use and reuse. In *Proc. of the 8th Knowledge-Based Software Engineering Conference*, pages 144–151, September 1993.
18. K. Sivashanmugam, K. Verma, A. Sheth, and J. Miller. Adding semantics to web services standards. In *Proc. First International Conference on Web Services (ICWS'03)*, 2003.
19. A.M. Zaremski and J.M. Wing. Signature matching: a tool for using software libraries. *ACM Transactions on Software Engineering and Methodology (TOSEM)*, 4(2):146 – 170, April 1995.
20. A.M. Zaremski and J.M. Wing. Specification matching of software components. In *Proc. SIGSOFT'95 Third ACM SIGSOFT Symposium on the Foundations of Software Engineering*, volume 20(4) of *ACM SIGSOFT Software Engineering Notes*, pages 6–17, October 1995. Also CMU-CS-95-127, March, 1995.

Representing First-Order Logic Using Graphs

Arend Rensink

Department of Computer Science, University of Twente
P.O.Box 217, 7500 AE, The Netherlands
rensink@cs.utwente.nl

Abstract. We show how edge-labelled graphs can be used to represent first-order logic formulae. This gives rise to recursively nested structures, in which each level of nesting corresponds to the negation of a set of existentials. The model is a direct generalisation of the *negative application conditions* used in graph rewriting, which count a single level of nesting and are thereby shown to correspond to the fragment $\exists\neg\exists$ of first-order logic. Vice versa, this generalisation may be used to strengthen the notion of application conditions. We then proceed to show how these nested models may be flattened to (sets of) plain graphs, by allowing some structure on the labels. The resulting formulae-as-graphs may form the basis of a unification of the theories of graph transformation and predicate transformation.

1 Introduction

Logic is about expressing and proving constraints on mathematical models. As we know, such constraints can for instance be denoted in a special language, such as First-Order Predicate Logic (FOL); formulae in that language can be interpreted through a satisfaction relation. In this paper we study a different, non-syntactic representation, based on graph theory, which we show to be equivalent to FOL. The advantage of this alternative representation is that it ties up notions from algebraic graph rewriting with logic, with potential benefits to the former.

We start out with the following general observation. A condition that states that a certain sub-structure should exist in a given model can often be naturally expressed by requiring the existence of a *matching* of (a model of) that sub-structure in the model in question. Illustrated on the *edge-labeled graphs* studied in this paper: The existence of entities x and y related by $a(x, y)$ and $b(y, x)$ (where a and b are binary relations) can be expressed by the requiring the existence of a matching of the following graph:

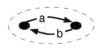

Note that this matching only implies that this sub-structure exists *somewhere* in the target graph; it does not say, for instance, that $a(x, y)$ and $b(y, x)$ always go together, or that the entities playing the roles of x and y are unrelated to other entities, or even that these entities are distinct from one another.

One particular context in which matchings of this kind play a prominent role is that of *algebraic graph rewriting* (see [3,8] for an overview). The basic building blocks of

H. Ehrig et al. (Eds.): ICGT 2004, LNCS 3256, pp. 319–335, 2004.

a given graph rewrite system are *production rules*, which (among other things) have so-called "left hand side" (LHS) graphs. A production rule *applies* to a given graph only if the rule's LHS can be matched by that graph;[1] moreover, the effect of the rule application is computed relative to the matching.

The class of conditions that matchings can express is fairly limited. For instance, no kind of *negative* information, such as the absence of relations or the distinctness of x and y in the example above, can be expressed in this manner. In the context of algebraic graph transformation, this observation has led to the definition of so-called *negative application conditions* (NACs) (see [10]). A NAC itself takes the form of a matching of the "base" graph into another. A LHS with NACs, interpreted as a logical constraint, is satisfied if a matching of the LHS exists which, however, cannot be *factored* through any of the NACs.[2] Consider, for instance, the following example of a graph with two NACs:

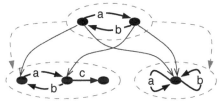

The base graph (here drawn on top) expresses, as before, that there are two entities, say x and y, such that $a(x, y)$ and $b(y, x)$; the first NAC adds the requirement that there is no z such that $c(y, z)$; and the second NAC adds the requirement that x and y are distinct. The combined structure thus represents the formula

$$\exists x, y: a(x, y) \wedge b(y, x) \wedge (\nexists z : c(y, z)) \wedge x \neq y \ .$$

Formally, a graph satisfies this condition if there is a matching of the base graph into it, that cannot be factored through a matching of either of the NAC target graphs.

Although, clearly, NACs strengthen the expressive power of graph conditions, it is equally clear that there are still many properties that can *not* be expressed in this way. For instance, any universally quantified positive formula is outside the scope of NACs. As a very simple example consider $\exists x: \forall y: x = y$ expressing that there exists precisely one entity. However, we can easily add more layers of "application conditions". For instance, the existence of a unique entity is represented by the following structure:

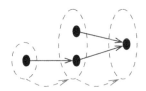

---

[1] In the single-pushout approach [8], the existence of a matching is also sufficient; in the double-pushout approach [3], there are some further conditions on the applicability of a rule.

[2] An important difference is that, in graph transformation, the issue is not whether a matching of a given LHS exists but to find *all* matchings. Seen in this light, the results of this paper concern the *applicability* of production rules.

At this point, readers familiar with the theory of *existential graphs* (see, e.g., [14,5]) may recognize the analogy between this stacking of graphs and the *cuts* used there to represent negation. See also the discussion of related work in Sect. 6. This paper is devoted to working out the ensuing theory in the category of edge-labeled graphs, providing a direct connection to the setting of algebraic graph transformation. We present the following results:

- Graph conditions with a stack of application conditions of height n, interpreted through matchings as sketched above, are expressively equivalent to a fragment of ($\forall$-free) FOL that may be denoted $\exists(\neg\exists)^n$ — for a precise statement see Th. 4. It is known that a higher value of n gives rise to a truly more expressive fragment of FOL; that is, for each n there is a property that can be expressed in $\exists(\neg\exists)^{n+1}$ and not in $\exists(\neg\exists)^n$. We prove equivalence through compositional translations back and forth.

  As a corollary, NACs carry the expressive power of $\exists\neg\exists$ — which indeed excludes universally quantified positive formulae, since those would translate to $\neg\exists\neg$ at the least. Another consequence is that more highly stacked graph conditions, providing the full power of FOL, can be integrated seamlessly into the theory of algebraic graph transformation. In the conclusions we briefly mention two such extensions that have been studied in the graph transformation literature.

- Graph conditions may be flattened, without loss of information, to simple, edge-labeled graphs, provided we add structure to the labels to reflect the stack height at which the nodes and edges were originally introduced. With hindsight this structure on the labels is indeed strongly reminiscent of the cuts in existential graphs, except that we avoid the need for representing cuts as an explicit structure on the graphs.

The remainder of this paper is structured as follows. In Sect. 2 we recall some definitions. In Sect. 3 we define graph predicates and provide a translation to FOL; in Sect. 4 we define the reverse translation. Sect. 5 shows how to flatten graph predicates. Finally, in Sect. 6 we summarize the results and discuss related work, variations and extensions.

2   Basic Definitions

We assume a countable universe of variables Var, ranged over by x, y, z, and a countable universe of relation (i.e., predicate) symbols Rel $\subseteq$ Lab (not including $=$), ranged over by a. The following grammar defines FOL, the language of first order logic with equality and binary relations:

$$\phi ::= x = x \mid a(x, x) \mid \neg\phi \mid \bigvee\Phi \mid \bigwedge\Phi \mid \exists X : \phi .$$

Here $\Phi \subseteq$ FOL and $X \subseteq$ Var are finite sets of formulae and variables, respectively. (So $\exists X : \phi$ is not second-order quantification but finitary first-order quantification.) We use $fv(\phi)$ [$fv(\Phi)$] for $\phi \in$ FOL [$\Phi \subseteq$ FOL] to denote the set of free variables in ϕ [Φ] (with the usual definition). Note that all sets of free variables are finite.

As models for FOL we use edge-labeled graphs without parallel edges. For this purpose we assume a countable universe of nodes Node, ranged over by v, w, and a countable universe of labels Lab $\supseteq$ Rel, ranged over by ℓ. Except in Sect. 5, we will have Lab $=$ Rel.

Definition 2.1 (graphs).

- A graph *is a tuple* $G = \langle N, E \rangle$ *where* $N \subseteq$ Node *is a set of* nodes *and* $E \subseteq N \times$ Lab $\times N$ *a set of* edges.
- *If* G *and* H *are graphs, a* graph morphism *from* G *to* H *is a tuple* $\mu = (G, H, f)$ *where* $f: N_G \to N_H$ *is such that* $(f(v), \ell, f(w)) \in E_H$ *for all* $(v, \ell, w) \in E_G$.
- *The category of graphs, denoted* Graph, *has graphs as objects and graph morphisms as arrows, with the obvious notions of identity and composition.*

We denote the components of a graph G by N_G and E_G (as already seen above), and we use $\mu: G \to H$ or $\mu \in$ Graph(G, H) to denote that μ is a morphism from G to H. For $\mu: G \to H$ we denote $src(\mu) = G$ and $tgt(\mu) = H$. The following result is standard.

Proposition 1. Graph *is a cartesian closed category with all limits and colimits.*

Every countable set A gives rise to a discrete graph $\langle A \rangle$, with $N = A$ and $E = \emptyset$. We also use $\langle v \xrightarrow{a} w \rangle$ to denote the one-edge graph with $N = \{v, w\}$ and $E = \{(v, a, w)\}$. Furthermore, for every $X \subseteq Y \subseteq$ Var (Y countable) we use $emb[X, Y] = (X, Y, id_X)$ for the morphism that embeds $\langle X \rangle$ in $\langle Y \rangle$, and for every $G \in$ Graph we write id_G for the identity morphism on G. The following rules define a modeling relation $\models$ between graph morphisms $\theta \in$ Graph$(\langle X \rangle, G)$ with $X \supseteq fv(\phi)$ (which combine the valuation of the logical variables in X with the algebraic structure of F) and FOL-formulae ϕ:

$$\begin{aligned}
\theta &\models x = y & &\text{if } \theta(x) = \theta(y) \\
\theta &\models a(x, y) & &\text{if } (\theta(x), a, \theta(y)) \in E_G \\
\theta &\models \neg\phi & &\text{if } \theta \not\models \phi \\
\theta &\models \bigvee \Phi & &\text{if } \theta \models \phi \text{ for some } \phi \in \Phi \\
\theta &\models \bigwedge \Phi & &\text{if } \theta \models \phi \text{ for all } \phi \in \Phi \\
\theta &\models \exists Y : \phi & &\text{if } \eta \models \phi \text{ such that } \theta = \eta \circ emb[X, X \cup Y] \ .
\end{aligned}$$

3   Graph Predicates and Conditions

For *finite* $G \in$ Graph we define Pred$[G]$, Cond$[G]$ as the smallest sets such that

- $p \subseteq$ Cond$[G]$ with p finite implies $p \in$ Pred$[G]$;
- $\alpha \in$ Graph(G, H) and $p \in$ Pred$[H]$ implies $(\alpha, p) \in$ Cond$[G]$.

The elements of Pred$[G]$ are called *(graph) predicates* on G and those of Cond$[G]$ *(graph) conditions* on G. Graph predicates can be thought of as finitely branching trees of connected graph morphisms, of finite depth; a graph condition is a single branch of such a tree. For $c \in$ Cond$[G]$ we write α_c for the morphism component, $T_c = tgt(\alpha_c)$ for the target of α_c and p_c for the predicate component; hence $c = (\alpha_c, p_c)$. The depth of predicates and conditions, which is quite useful as a basis for inductive proofs, is defined by:

$$\begin{aligned}
depth(p) &= \max\{depth(c) \mid c \in p\} \\
depth(c) &= 1 + depth(p_c) \ .
\end{aligned}$$

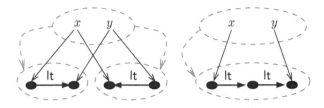

Fig. 1. Graph predicates for $lt(x, y) \lor lt(y, x)$ resp. $\exists z\colon lt(x, z) \land lt(z, y)$

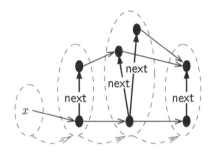

Fig. 2. Graph predicate for $\exists y\colon next(x, y) \land \forall z\colon (next(x, z) \Rightarrow z = y)$

It follows that the base case, $depth(p) = 0$, corresponds to $p = \emptyset$. Conditions have positive depth. We propose $\mathsf{Pred}[\langle X \rangle]$ as representations of FOL formulae over X. Note that in the introduction we discussed predicates consisting of a single condition only, and in the pictorial representation we omitted the source graph $\langle X \rangle$ (which anyway would be empty since the constraints discussed there are closed) and only displayed the structure from T_c onwards. Fig. 1 depicts two constraints with free variables accurately; Fig. 2 is another example, which shows multiple levels of conditions. The following defines satisfaction of a predicate $p \in \mathsf{Pred}[G]$, for arbitrary $\theta \in \mathsf{Graph}(G, H)$:

$$\theta \models p \text{ iff } \exists c \in p\colon \exists \mu\colon T_c \to H\colon \theta = \mu \circ \alpha_c, \mu \not\models p_c \,. \tag{1}$$

On the model side this generalizes $\models$ over FOL: here the source of θ can be an arbitrary graph, whereas there it was always discrete. An example is given in Fig. 3, which shows a model of the right-hand predicate of Fig. 1.

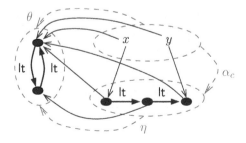

Fig. 3. Model satisfying the graph predicate for $\exists z\colon lt(x, z) \land lt(z, y)$

It follows that a predicate p should be seen as the *disjunction* of its conditions $c \in p$, and a condition c as the requirement that the structure encoded in T_c is present in the model, combined with the *negation* of p_c. This interpretation is formalized by recursively translating all $p \in \mathsf{Pred}[G]$ to formulae ϕ_p. For the translation, we assume that for all (sub-)conditions d occurring in the structure to be translated, the node sets of $src(\alpha_d)$ and $tgt(\alpha_d)$ are disjoint. (We show below that this assumption does not cause loss of generality, since we can always fulfill it by choosing isomorphic representatives of predicates and conditions.) Furthermore, we assume that for every node v occurring anywhere in the structure to be translated, there is a distinct associated variable x_v.

$$\phi_p = \bigvee \{\phi_c \mid c \in p\}$$
$$\phi_c = \exists\{x_v \mid v \in N_{T_c}\}\colon \bigwedge \{x_v = x_{\alpha_c(v)} \mid v \in N_G\} \wedge$$
$$\bigwedge \{a(x_v, x_w) \mid (v, a, w) \in E_{T_c}\} \wedge \neg\phi_{p_c} \ .$$

For every graph K occurring in p let $X_K = \{x_v \mid v \in N_K\}$ and let $\xi_K\colon \langle X_K\rangle \to K$ be given by $x_v \mapsto v$ for all $v \in N_K$. The following theorem is one of the main results of this paper.

Theorem 1. *Let $p \in \mathsf{Pred}[G]$ and $\theta \in \mathsf{Graph}(G, H)$; then $\theta \models p$ iff $\theta \circ \xi_G \models \phi_p$.*

Proof sketch. We prove the theorem together with an auxiliary result about conditions. First we extend the modeling relation to conditions, and we simplify the definition of $\models$ over Pred, as follows:

$$\theta \models c \text{ iff } \exists\mu\colon T_c \to H\colon \theta = \mu \circ \alpha_c, \mu \not\models p_c \tag{2}$$
$$\theta \models p \text{ iff } \exists c \in p\colon \theta \models c \ . \tag{3}$$

It follows that $\theta \models p$ iff $\theta \models c$ for some $c \in p$. The proof obligation is extended with:

If $c \in \mathsf{Cond}[G]$ then $\theta \models c$ iff $\theta \circ \xi_G \models \phi_c$.

The proof proceeds by mutual induction on these cases and on the depth of conditions. $\qquad\square$

4   Formulae as Graph Predicates

We now provide the inverse translation from formulae into graph predicates. For this purpose we need some constructions over predicates. First, for $\mu \in \mathsf{Graph}(H, G)$, $p \in \mathsf{Pred}[G]$ and $c \in \mathsf{Cond}[G]$, we define

$$p \circ \mu = \{c \circ \mu \mid c \in p\}$$
$$c \circ \mu = (\alpha_c \circ \mu, p_c) \ .$$

Clearly, $p \circ \mu \in \mathsf{Pred}[H]$ and $c \circ \mu \in \mathsf{Cond}[H]$. This construction satisfies:

Proposition 2. *Let $p \in \mathsf{Pred}[H]$, $\mu \in \mathsf{Graph}(G, H)$ and $\theta \in \mathsf{Graph}(G, K)$; then $\theta \models p \circ \mu$ iff there is an $\eta \in \mathsf{Graph}(H, K)$ such that $\eta \models p$ and $\theta = \eta \circ \mu$.*

Proof: if. Assume $\eta \models p$. It follows that, for some $c \in p$, there is a $\lambda: T_c \to K$ such that $\eta = \lambda \circ \alpha_c$ and $\lambda \not\models p_c$. But then $\theta = \eta \circ \mu = \lambda \circ \alpha_c \circ \mu = \lambda \circ \alpha_{c\circ\mu}$; since $c \circ \mu \in p \circ \mu$, it follows that $\theta \models p \circ \mu$.

Only if. Assume $\theta \models p \circ \mu$. It follows that, for some $c \in p$, there is a $\lambda: T_c \to K$ such that $\theta = \lambda \circ \alpha_{c\circ\mu} = \lambda \circ \alpha_c \circ \mu$ and $\lambda \not\models p_{c\circ\mu} \ (= p_c)$. But then $\lambda \circ \alpha_c \models p$, and hence $\eta = \lambda \circ \alpha_c$ fulfills the proof obligation. □

In the sequel we make heavy use of pushouts in Graph; therefore we introduce some auxiliary notation. Given $\alpha: H \to K$ and $\mu: G \to H$, we write $\alpha{\uparrow}\mu$ ("the remainder of α after μ") for the morphism opposite α in the pushout diagram; hence $\alpha{\uparrow}\mu: G \to L$ and $\mu{\uparrow}\alpha: K \to L$ are such that (among other properties) $(\alpha{\uparrow}\mu) \circ \mu = (\mu{\uparrow}\alpha) \circ \alpha$. We extend this notation to predicates $p \in \mathsf{Pred}[G]$ and conditions $c \in \mathsf{Cond}[G]$ as follows:

$$p{\uparrow}\mu = \{c{\uparrow}\mu \mid c \in p\}$$
$$c{\uparrow}\mu = (\alpha_c{\uparrow}\mu, p_c{\uparrow}(\mu{\uparrow}\alpha_c)) \ .$$

It follows that $p{\uparrow}\mu \in \mathsf{Pred}[H]$ and $c{\uparrow}\mu \in \mathsf{Cond}[H]$. By taking pushouts, essentially we merge the additional structure specified by μ with the existing conditions, in the "least obtrusive" way. These constructions clearly yield predicates, resp. conditions again. The following correspondence plays an important role in the sequel.

Proposition 3. *Let $p \in \mathsf{Pred}[G]$, $\mu \in \mathsf{Graph}(G, H)$ and $\theta \in \mathsf{Graph}(H, K)$; then $\theta \circ \mu \models p$ iff $\theta \models p{\uparrow}\mu$.*

Proof sketch. The proof strategy is similar to that of Th. 1, by mutual induction on the depth of predicates and conditions, alternating between "case Pred" in the proposition, and "case Cond" reading "If $c \in \mathsf{Cond}[G]$, then $\theta \circ \mu \models c$ iff $\theta \models c{\uparrow}\mu$." □

We now turn each $\mathsf{Pred}[G]$ and $\mathsf{Cond}[G]$ into a category. The arrows will essentially be proofs of implication. We define the hom-sets $\mathsf{Pred}[G](p, q)$ for $p, q \in \mathsf{Pred}[G]$ and $\mathsf{Cond}[G](c, d)$ for $c, d \in \mathsf{Cond}[G]$ as the families of smallest sets such that:

- $f: p \to q$ a function and $\gamma_c \in \mathsf{Cond}[G](c, f(c))$ a condition arrow for all $c \in p$ implies $(f, (\gamma_c)_{c \in p}) \in \mathsf{Pred}[G](p, q)$;
- $\mu: T_d \to T_c$ a function (in the reverse direction!) such that $\alpha_c = \mu \circ \alpha_d$ and $\pi \in \mathsf{Pred}[T_c](p_d{\uparrow}\mu, p_c)$ a compatible predicate arrow implies $(\mu, \pi) \in \mathsf{Cond}[G](c, d)$.

We let π range over sets of the form $\mathsf{Pred}[G](p, q)$, and use f_π and $\gamma_{\pi,c}$ for $c \in p$ to denote its components. Similarly γ ranges over sets of the form $\mathsf{Cond}[G](c, d)$, and μ_γ, π_γ denote its components. The following confirms the intuition that arrows between predicates are proofs of logical implication. The proof again goes by mutual induction (on the depth of p) of cases for Pred and Cond.

Proposition 4. *Let $\theta \in \mathsf{Graph}(G, H)$ and $p, q \in \mathsf{Pred}[G]$. If $\mathsf{Pred}[G](p, q)$ is non-empty then $\theta \models p$ implies $\theta \models q$.*

In preparation of the translation from FOL to graph predicates, we define the following operations over $c, d \in \mathsf{Cond}[G]$ and $p, q \in \mathsf{Pred}[G]$ (for arbitrary G):

$$c \times d = ((\alpha_c{\uparrow}\alpha_d) \circ \alpha_d, p_c{\uparrow}(\alpha_d{\uparrow}\alpha_c) \uplus p_d{\uparrow}(\alpha_c{\uparrow}\alpha_c)) \tag{4}$$

$$p + q = p \uplus q \tag{5}$$

$$p \times q = \{c \times d \mid c \in p, d \in q\} \tag{6}$$

$$!p = \{(id_G, p)\} \ . \tag{7}$$

In passing we note some facts about Pred and Cond.

Theorem 2. *For all $G \in$ Graph, $\mathrm{Cond}[G]$ is a category with products defined by (4) and initial object $(id_G, \emptyset)$; $\mathrm{Pred}[G]$ is a category with products defined by (6), coproducts defined by (5), initial object $\emptyset$ and terminal object $\{(id_G, \emptyset)\}$.*

Proof sketch. Concatenation in $\mathrm{Cond}[G]$ and $\mathrm{Pred}[G]$, for $\pi_i = (f_i, (\gamma_{i,c})_{c \in p_i}) \in \mathrm{Pred}[G](p_i, p_{i+1})$ and $\gamma_i = (\mu_i, \pi_i) \in \mathrm{Cond}[G](c_i, c_{i+1})$ $(i = 1, 2)$, is defined by

$$\pi_2 \circ \pi_1 = (f_2 \circ f_1, (\gamma_{2,f_1(c)} \circ \gamma_{1,c})_{c \in p_1}))$$
$$\gamma_2 \circ \gamma_1 = (\mu_1 \circ \mu_2, (\pi_2{\uparrow}\mu_1) \circ \pi_1) \ .$$

where for $\lambda \in \mathrm{Graph}(G, H)$ and $\pi \in \mathrm{Pred}[G](p, q)$, $\gamma \in \mathrm{Cond}[G](c, d)$, the remainders $\pi{\uparrow}\lambda \in \mathrm{Pred}[H](p{\uparrow}\lambda, q{\uparrow}\lambda)$ and $\gamma{\uparrow}\lambda \in \mathrm{Pred}[H](c{\uparrow}\lambda, d{\uparrow}\lambda)$ are defined by

$$\pi{\uparrow}\lambda = (\{(c{\uparrow}\lambda, f_\pi(c){\uparrow}\lambda) \mid c \in p\}, (\gamma_c{\uparrow}\lambda)_{c{\uparrow}\lambda \in q{\uparrow}\lambda})$$
$$\gamma{\uparrow}\lambda = (\mu{\uparrow}\lambda, \pi{\uparrow}\lambda) \ .$$

Note that, in order for $\pi{\uparrow}\lambda$ to be well-defined, we need to assume distinct $c{\uparrow}\lambda$. Since we are free to choose these objects up to isomorphism, this assumption causes no loss of generality. The projections pr_c and pr_d for the product in $\mathrm{Cond}[G]$ are given by

$$pr_c = (\alpha_d{\uparrow}\alpha_c, id_{p_c{\uparrow}(\alpha_d{\uparrow}\alpha_c)}) \quad (\in \mathrm{Cond}[G](c \times d, c))$$
$$pr_d = (\alpha_c{\uparrow}\alpha_d, id_{p_d{\uparrow}(\alpha_c{\uparrow}\alpha_d)}) \quad (\in \mathrm{Cond}[G](c \times d, d)) \ .$$

$\square$

The following proposition affirms that the operations defined above are appropriate for modeling FOL connectives. This partially follows from the characterization (in Th. 2) of $+$ and $\times$ as coproduct and product in Pred, plus Prop. 4 stating that arrows in Pred induce logical implication.

Proposition 5. *Let $\theta \in \mathrm{Graph}(G, H)$ and $p, q \in \mathrm{Pred}[G]$.*

1. *$\theta \models p + q$ if and only if $\theta \models p$ or $\theta \models q$.*
2. *$\theta \models p \times q$ if and only if $\theta \models p$ and $\theta \models q$.*
3. *$\theta \models !p$ if and only if $\theta \not\models p$.*

The final elements we need for the translation from FOL to graph predicates are representations for the base formulae, $x = y$ and $\mathsf{a}(x, y)$. These are given through graph morphisms $\alpha_{x=y}$ and $\alpha_{\mathsf{a}(x,y)}$, given pictorially in Fig. 4. The following table defines a function yielding for every $\phi \in$ FOL and finite $X \supseteq fv(\phi)$ a graph predicate $[\![\phi]\!]_X \in \mathrm{Graph}[\langle X \rangle]$.

Fig. 4. Graph morphisms for the base formulae $x = y$, resp. $\mathsf{a}(x, y)$

$$[\![x = y]\!]_X = \{(\alpha_{x=y}, \emptyset)\} \uparrow emb[\{x, y\}, X]$$
$$[\![\mathsf{a}(x, y)]\!]_X = \{(\alpha_{\mathsf{a}(x,y)}, \emptyset)\} \uparrow emb[\{x, y\}, X]$$
$$[\![\neg\phi]\!]_X = ![\![\phi]\!]_X$$
$$[\![\bigvee \Phi]\!]_X = \sum_{\phi \in \Phi} [\![\phi]\!]_X$$
$$[\![\bigwedge \Phi]\!]_X = \prod_{\phi \in \Phi} [\![\phi]\!]_X$$
$$[\![\exists Y.\phi]\!]_X = [\![\phi]\!]_{X \cup Y} \circ emb[X, X \cup Y] \ .$$

The following is the second half of the main correspondence result of this paper.

Theorem 3. *Let* $\phi \in$ FOL *and* $\theta \in$ Graph$(\langle X \rangle, G)$ *with* $X \supseteq fv(\phi)$; *then* $\theta \models \phi$ *iff* $\theta \models [\![\phi]\!]_X$.

Proof. By induction on the structure of ϕ. For the base formulae the result is immediate. For negation, disjunction and conjunction the result follows from Prop. 5, and for the existential from Prop. 2. ☐

It should be clear that there is a direct connection between the depth of graph predicates and the level of nesting of negation in the corresponding formula. We will make this connection precise. We use paths through the syntax tree of the formulae to isolate the relevant fragments of FOL. In the following theorem, a string of $\exists$ and $\neg$ indicates the set of all formulae for which, if we follow their syntax trees from the root to the leafs and ignore all operators except $\exists$ and $\neg$, all the resulting paths are prefixes of that string.

Theorem 4. *Let* $n \in$ Nat; *then the set of predicates* p *with* $depth(p) \leq n$ *is equivalent to the FOL-fragment* $\exists(\neg\exists)^n$.

Note that we could have formulated the same result while omitting $\exists$; we have included it to stress that $\forall$ is only allowed in its dual form, $\neg\exists\neg$.

Proof. Immediate from the two conversion mappings, ϕ_p for $p \in$ Pred and $[\![\phi]\!]$ for $\phi \in$ FOL; see Th. 1 and Th. 3. ☐

This result has consequences in the theory of algebraic graph transformations. The key insight is that the application conditions of [10] are exactly graph predicates of depth 0 (the positive conditions) and 1 (the negative conditions, or NACs) — where it should be noted that application conditions are closed formulae, so the corresponding graph predicates are in Pred$[\langle \emptyset \rangle]$; and indeed the presentation in [10] omits the base graph.

Corollary 1. *The application conditions proposed in [10] are equivalent to the FOL-fragment* $\exists\neg\exists$.

For instance, a useful property that can yet *not* be expressed through NACs is the uniqueness of nodes: the property $\exists y \colon \mathsf{next}(x, y) \wedge \forall z \colon (\mathsf{next}(x, z) \Rightarrow z = y)$, expressed in Fig. 2 by a graph predicate of depth 2, is outside the fragment $\exists \neg \exists$. More generally, as noted in the conclusion of [10], NACs cannot impose cardinality constraints, which in fact all have roughly the form $\exists X \colon \left(\bigwedge_{x, y \in X} x \neq y \right) \wedge \forall z \colon \bigvee_{x \in X} z = x$, and hence are in $\exists \neg \exists \neg$.[3]

5   Graph Predicates as Graphs

The way we defined them, graph predicates are highly structured objects. We now show that much of this structure can be done away with: there is an equivalent representation of graph conditions as sets of simple, flat graphs, in which the nesting structure is transferred to the *edge labels*. Graph predicates thus correspond to *sets* of such graphs. In this section we define the conversions back and forth. All proofs, being rather technical and uninteresting, are omitted. In this section, Lab is as follows:

$$\mathsf{Lab} = \mathsf{Rel} \cup (\mathsf{Rel}_= \times \mathsf{Nat}) \ .$$

- Rel is the set of relation symbols, as before. Plain relation symbols (i.e., without depth indicators, see below) are called *base labels*; they correspond to the base graph G of a predicate $p \in \mathsf{Pred}[G]$.
- $\mathsf{Rel}_= \times \mathsf{Nat}$, where $\mathsf{Rel}_= = \mathsf{Rel} \cup \{=\}$, consists of pairs of a relation symbol (this time including equality $=$, see below), together with a natural number indicating the *depth* of the edge. For Rel this will be the depth in p at which the edge is introduced; for $=$ it will be the depth at which the nodes are introduced or equated.

We use b to range over $\mathsf{Rel}_=$ and $^i\mathsf{b}$ as shorthand for (b, i); we use ℓ to range over Lab. Furthermore, we use $\bot$ to denote the *base depth* and regard it as an element smaller than any $i \in \mathsf{Nat}$, and such that $i - 1 = \bot$ if $i \in \{\bot, 0\}$. We use δ, ϵ to range over $\mathsf{Nat} \cup \{\bot\}$ and sometimes use $^\bot\mathsf{a}$ as equivalent notation for a ($\in \mathsf{Rel}$). We define the *depth* of nodes and edges as

- $depth(v) = i$ if $(v, \overset{i}{=}, v) \in E$, and $\bot$ otherwise;
- $depth(e) = i$ if $e = (v, {}^i\mathsf{a}, w)$.

(Node depth is well-defined due to the first well-formedness constraint below.) We call $a \in N \cup E$ *base* if $depth(a) = \bot$. In the remainder of this section we use RelGraph to denote the set of graphs over Rel used in the previous sections, and CondGraph to denote the set of *condition graphs*, which are graphs over Lab that satisfy the following well-formedness constraints, in addition to those already mentioned above:

---

[3] This is no longer true when matchings are required to be injective; see Sect. 6 for a conjecture about the increased expressiveness of that setting. Also, as remarked in the introduction, the double-pushout approach imposes further application conditions to ensure the existence of pushout complements, which we ignore here.

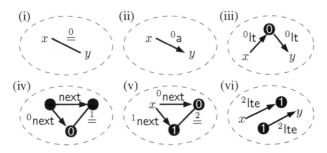

Fig. 5. Example condition graphs.

1. For any $v, w \in N$ and $b \in \mathsf{Rel}_=$ there is at most one δ such that $(v, {}^\delta b, w) \in E$.
2. If $(v, {}^\delta b, w) \in E$ then $depth(v), depth(w) \le \delta$.
3. If $(v, \overset{i}{=}, w) \in E$ then $(w, \overset{i}{=}, v) \in E$.

Note that $\mathsf{RelGraph} \subseteq \mathsf{CondGraph}$. Fig. 5 contains some example condition graphs. We indicate node depths by inscribing the depth inside nodes, and edge depths by appending the depth to the label. $\overset{i}{=}$-labeled edges, which are always bidirectional due to well-formedness condition 3, are drawn undirected. Graphs (i) and (ii) are flattenings of the morphisms $\alpha_{x=y}$ and $\alpha_{\mathsf{a}(x,y)}$ displayed in Fig. 4. Graph (iii) is the flattening of the right hand condition of Fig. 1; (iv) is the condition of Fig. 2 without the base level and (v) the complete condition. Graph (vi) represents the (right hand, connected) condition of Fig. 7 below.

Any graph morphism $\alpha \colon G \to C$ can be flattened to a condition graph, $flat(\alpha)$, by enriching G with the structure provided by α while keeping it distinguishable from the structure already present in G, so that α can be fully reconstructed (up to isomorphism of C). There are essentially three kinds of additional structure: fresh nodes of C, fresh edges of C, and node mergings, i.e., nodes on which α is non-injective. Node mergings and fresh nodes will be indicated using $\overset{0}{=}$-labeled edges, and fresh edges by a (non-base) depth indication. In general we allow $G \in \mathsf{RelGraph}$ and $C \in \mathsf{CondGraph}$. W.l.o.g. assume $N_G \cap N_C = \emptyset$; then $flat(\alpha) = (N, E) \in \mathsf{CondGraph}$ such that

$$N = N_G \cup (N_C \setminus \alpha(N_G))$$
$$E = E_G \cup \{(u, {}^0 a, v) \mid (u, a, v) \notin E_G, (\bar\alpha(u), a, \bar\alpha(v)) \in E_C\}$$
$$\cup \{(u, \overset{0}{=}, v) \mid u \ne_G v, \bar\alpha(u) = \bar\alpha(v)\}$$
$$\cup \{(u, {}^{i+1} b, v) \mid (\bar\alpha(v), {}^i b, \bar\alpha(v)) \in E_C\} \ .$$

Here $\bar\alpha = \alpha \cup id_{N \setminus N_G}$ extends α with identity mappings for the fresh nodes, and $=_G$ is the identity relation over N_G. Note that we do not only add depth indicators to the additional structural elements, but we also increment the depth indicators already present in C, so as to keep them distinct. For instance, note that, as expected, graphs (i) and (ii) in Fig. 5 indeed equal $flat(\alpha_{x=y})$ and $flat(\alpha_{\mathsf{a}(x,y)})$ (see Fig. 4).

For the inverse construction, we need to resurrect the original source and target graphs from the flattened morphism; or more generally, we construct morphisms from conditional graphs. In principle, the source graph is the sub-graph with base depth, whereas the target graph is the the entire condition graph, where, however, the depths are decremented and the nodes connected by $\overset{0}{=}$-edges are collected into single nodes.

In the following, given a binary relation ρ over a set A, for any $a \in A$ we define $[a]_\rho$ as the smallest set such that (i) $a \in [a]_\rho$, and (ii) if $b \in [a]_\rho$ and $\rho(b, c)$ or $\rho(c, b)$ then $c \in [a]_\rho$. Likewise, $A/\rho = \{[a]_\rho \mid a \in A\}$ is the partitioning of A according to ρ.

Let $G \in \mathsf{RelGraph}$, $C \in \mathsf{CondGraph}$ and $\mu = (G, C, f)$. We define $C|_\perp$ as the base part of C, and C^- as C considered "one level up", i.e., with all depth indicators decremented and all $\overset{0}{=}$-related nodes collected (considering $v \overset{0}{=} w$ iff $(v, \overset{0}{=}, w) \in E$). Using these, we construct $\phi_C \colon C|_\perp \to C^-$ mapping each $v \in N_C|_\perp$ to $[v]_{\overset{0}{=}}$.

$$C|_\perp = (N_C|_\perp, E_C|_\perp) \text{ where } A|_\perp = \{a \in A \mid depth(a) = \perp\} \tag{8}$$

$$C^- = (N/\overset{0}{=}, \{([u]_{\overset{0}{=}}, {}^{\delta-1}\mathsf{b}, [v]_{\overset{0}{=}}) \mid (u, {}^{\delta}\mathsf{b}, v) \in E_C \wedge (\mathsf{b} \in \mathsf{Rel} \vee \delta > 0)\}) \tag{9}$$

$$\mu|_\perp = (G, C|_\perp, f) \tag{10}$$

$$\phi_C = (C|_\perp, C^-, \{(v, [v]_{\overset{0}{=}}) \mid v \in N_C|_\perp\}) . \tag{11}$$

Resurrecting a morphism is left inverse to flattening, but not for all condition graphs right inverse. The latter is due to the fact that we have not bothered to give an exact characterization of those condition graphs that may be constructed by flattening. Such a characterization would be cumbersome and not add much to the current paper: the main purpose here is to show that graph conditions may be flattened without loss of information.

Proposition 6. *Let $G \in \mathsf{RelGraph}$ and $C \in \mathsf{CondGraph}$ with $\alpha \colon G \to C$.*

1. There exists an isomorphism μ such that $\phi_{flat(\alpha)} = \alpha \circ \mu$.
2. There exists an epimorphism $\mu \colon C \to flat(\phi_C)$.

We now extend these principles to graph conditions c, which are essentially nested morphisms. Here the depth indicators really come into play. The construction proceeds by flattening the sub-conditions in p_c, taking the union of the resulting graph as an extended target graph for α_c and then flattening (the extended) α_c.

$$flat(c) = flat(\beta) \text{ where } \beta = emb[T_c, \textstyle\bigcup_{d \in p_c} flat(d)] \circ \alpha_c \tag{12}$$

where $emb[G, H] \in \mathsf{Graph}(G, H)$ is given by (G, H, id_{N_G}) if $N_G \subseteq N_H$, $E_G \subseteq E_H$.

For the inverse construction, we need to reconstruct the $d \in p_c$ from $flat(c)$. For this purpose, we use the *connectedness* of $flat(c)$. A fragment of a condition graph will be taken as part of the same sub-condition if it is connected at depth > 0. For instance, graph (v) in Fig. 5 has one connected sub-condition, whereas graph (vi) has two.

The required notion of connectedness can be captured through the decomposition of morphisms into *primes*, as follows. For $\lambda \in \mathsf{Graph}(G, H)$ and $\mu \in \mathsf{Graph}(G, K)$ we define $\lambda +^G \mu = (\lambda \uparrow \mu) \circ \mu$; the superscript G stands for the source graph of the morphisms considered — more formally, this is the coproduct operation in a category of morphisms with source G (the slice category of Graph under G). Connectedness, in the above sense, is related to the ability to decompose a morphism into summands. We call μ *prime* if it has no non-trivial decomposition under $+^G$; that is, if $\mu = \sum^G M$ for a set of morphisms M (where $\sum^G \emptyset = id_G$) implies that M contains some isomorphic representative of μ. The following characterizes prime morphisms.

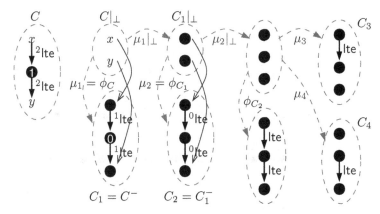

Fig. 6. Steps in the construction of a graph condition

Proposition 7. $\mu \in \mathsf{Graph}(G, H)$ *is prime iff one of the following holds:*

1. *μ is an epimorphism, and non-injective on exactly one pair of nodes;*
2. *μ is a monomorphism, and the set A of nodes and edges in H that do not occur as images of μ is connected by $\leftrightarrow \subseteq A \times A$, defined as the least relation such that $(v, \ell, w) \leftrightarrow v$ and $(v, \ell, w) \leftrightarrow w$ whenever $(v, \ell, w), v, w \in A$.*

The key property in the use of prime morphisms, stated in the following proposition, is that every morphism $\mu \colon G \to H$ can be decomposed into a finite number of them.

Proposition 8. *For all $\mu \in \mathsf{Graph}(G, H)$, there is a finite set of prime morphisms P, such that $\mu = \alpha \circ \sum^{G} P$ for some isomorphism α.*

Note that the decomposition P is not unique, even up to isomorphism; however, if μ is an isomorphism then $P = \emptyset$ is the only possibility. For the developments in this section it does not matter which decomposition we take; rather, we assume that $primes(\mu)$ is *some* (fixed) prime decomposition. Let $G \in \mathsf{RelGraph}$, $C \in \mathsf{CondGraph}$ and $\mu \colon G \to C$.

$$cond(\mu) = (\mu|_{\perp}, \{cond(\eta) \mid \eta \in primes(\phi_{tgt(\mu)})\}) \tag{13}$$
$$cond(C) = cond(\phi_C) . \tag{14}$$

Thus, $cond$ constructs a graph condition from a morphism by turning its target graph into a new morphism, and recursively calling itself on its prime decomposition. This terminates because in $\eta \in primes(\phi_D)$ with $D = tgt(\mu)$, all depth indicators of $tgt(\eta)$ have been decreased w.r.t. D; and if D is base then ϕ_D is an iso, hence $primes(\phi_D) = \emptyset$. For example, Fig. 6 shows several stages of constructing $cond(C)$, with C the graph on the left hand side. ϕ_C and ϕ_{C_1} are themselves prime, but $primes(\phi_{C_2}) = \{\mu_3, \mu_4\}$. From the figure we can see $cond(C) = (\mu_1|_{\perp}, \{(\mu_2|_{\perp}, \{(\mu_3|_{\perp}, \emptyset), (\mu_4|_{\perp}, \emptyset)\})\})$.

This construction gives us half of the desired correspondence (compare Prop. 6.2):

Proposition 9. *All $C \in \mathsf{CondGraph}$ have an epimorphism $\mu \colon C \to flat(cond(C))$.*

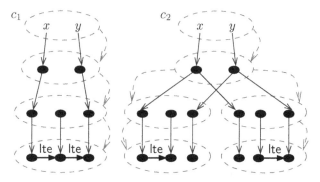

Fig. 7. An unconnected graph condition and its connected normal form

Still, this does not yet solve the problem, since graph conditions do not generally have the connectedness required to reconstruct them from their flattenings. For instance, $flat(c_1)$ for c_1 as in Fig. 7 yields C of Fig. 6, but $cond(C)$ is not isomorphic to c_1, and indeed the two are also inequivalent as properties over graphs. On the other hand, c_2 in Fig. 7 is equivalent to c_1 in this sense; $flat(c_2)$ yields graph (vi) of Fig. 5, from which $cond$ does construct a condition isomorphic to c_2.

To formulate the required connectedness property, we *enrich* the target graph T_c of conditions c with information about the connections made deeper in the tree of morphisms underlying c. For arbitrary $p \in \mathsf{Pred}[G]$ and $c \in \mathsf{Cond}[G]$ we define:

$$G_p^{\leftrightarrow} = (N_G, E_G \cup \{(v, \leftrightarrow, w) \mid \exists c \in p : \alpha_c(v) \text{ is connected to } \alpha_c(w) \text{ in } (T_c)_{p_c}^{\leftrightarrow}\})$$
$$\alpha_c^{\leftrightarrow} = (G, (T_c)_{p_c}^{\leftrightarrow}, f_{\alpha_c}) \ .$$

This brings us to the following definition of connectedness.

Definition 5.1. *Graph condition $c \in \mathsf{Cond}[G]$ is called* connected *if for all $d \in p_c$, $\alpha_d^{\leftrightarrow}$ is prime and d is connected.*

The following proposition lists the important facts about connected graph conditions: the graph conditions constructed by $cond$ (14) are always connected, connected graph conditions can be flattened without loss of information (compare Prop. 6.1), and for every graph condition there is an equivalence connected one.

Proposition 10. *Let $c \in \mathsf{Cond}[G]$.*

1. *If $c = cond(C)$ for some $C \in \mathsf{CondGraph}$, then c is connected.*
2. *If c is connected, then there is an isomorphism $\gamma : c \to cond(flat(c))$.*
3. *There is a connected $\bar{c} \in \mathsf{Cond}[G]$ such that $\theta \models c$ if and only if $\theta \models \bar{c}$.*

This brings us to the main result of this section, which states that every graph condition can be flattened to a condition graph expressing the same property. In order to represent a graph predicate $p \in \mathsf{Pred}[G]$, we use the *set* of condition graphs $\{flat(c) \mid c \in p\}$.

Theorem 5. *Let $c \in \mathsf{Cond}[G]$; then there is a $C \in \mathsf{CondGraph}$ such that $\theta \models cond(C)$ if and only if $\theta \models c$.*

6   Conclusions

We have presented an equivalent representation of first-order logic, as a recursively nested set of graph morphisms. We have defined compositional translations back and forth, and given an expressiveness result relating the recursive depth of our graph predicates to the corresponding fragment of FOL.

Subsequently, we have shown how the nested graph predicates can be translated, without loss of information, to flat graphs. We see as the main advantage of this that we can now use graph transformations to transform predicate graphs. This points the way to a potential connection between graph transformation and predicate transformation.

Graph constraints and conditional application conditions. At several points during the paper we already mentioned the work on (negative) application conditions within the theory of algebraic graph transformation, originally by [6] and later worked out in more detail in [10]. This paper is the result of an attempt to generalize their work.

Other related work in the context of graph transformations may be found in the *conditional application conditions* of [11] and in the *graph constraints* as proposed in, e.g., [12] and implemented in the AGG tool (cf. [9]). In fact it is not difficult to see, using the results of this paper, that conditional application conditions are expressively equivalent to the $\exists\neg\exists\neg\exists$-fragment of FOL, and graph constraints to the $\neg\exists\neg\exists$-fragment.

We conjecture that the requirement that all matches be injective increases the expressive power precisely by allowing *inequality* (but no other forms of negation) to occur in the context of the inner $\exists$.

Existential graphs. A large body of related work within (mainly) *artificial intelligence* centers around Peirce's *existential graphs* [14,15] and Sowa's more elaborate *conceptual graphs* [17]. The former were introduced as pragmatic philosophical models for reasoning over a century ago; the latter are primarily intended as models for knowledge representation. There are obvious similarities between those models and ours, especially concerning the use of nesting to represent negation and existential quantification. On the other hand, the thrust of the developments in existential and conceptual graphs is quite different from ours, making a detailed comparison difficult. New in the current paper seems to be the use of edge labels to encode the nesting structure; throughout the work on existential and conceptual graphs we have only seen this represented using so-called *cuts*, which are essentially sub-graphs explicitly describing the hierarchical structure. For us, the advantage of our approach is that our models can be treated as simple graphs, and as such submitted to existing graph transformation approaches.

More or less successful approaches to define a connection between existential, resp. conceptual, graphs and FOL can be found in [2,18,5]. In particular, in [18,5]) a complete theory of FOL is developed on the basis of the graph representations.

Variations. We have chosen a particular category of graphs and a particular encoding of FOL that turn out to work well together. However, it is interesting also to consider variations in either choice. For instance, our graphs do not have parallel edges, and our encoding does not allow reasoning about edge labels. It is likely that similar results can be obtained in an extended setting by moving to graph logics as in [1,4].

As an example of the use of such extensions, consider the so-called *dangling edge condition* that partially governs the applicability of a double-pushout rule (see also Footnote 1). This condition (under certain circumstances) forbids the existence of any edges outside an explicitly given set. In the setting of this paper, there is no uniform way to express such a constraint, since it requires the ability to refer to edges explicitly while abstracting from their labels.

Open issues. We list some questions raised by the work reported here.

- A direct semantics for condition graphs. The flat graph representations of Sect. 5 are shown to be equivalent by a translation back and forth to the nested graph predicate structures. Currently we do not have a modeling relation directly over condition graphs.
- The connection to predicate transformations. Traditional approaches to predicate transformation have to go to impressive lengths to represent pointer structures (e.g., [13]), whereas on the other hand graphs are especially suitable for this. Using the condition graph representation of predicates presented here, one can use graph transformations to construct or transform predicates.
- The extension of existing theory on graph transformation systems to support full graph predicates instead of (conditional) application conditions; for instance, rule independency in the context of negative application conditions developed in [10], and the translation of postconditions to application conditions in [11,12,7].

References

1. M. Bauderon and B. Courcelle. Graph expressions and graph rewritings. *Mathematical Systems Theory*, 20(2–3):83–127, 1987.
2. M. Chein, M.-L. Mugnier, and G. Simonet. Nested graphs: A graph-based knowledge representation model with FOL semantics. In Cohn, Schubert, and Shapiro, eds., *Principles of Knowledge Representation and Reasoning*, pp. 524–535. Morgan Kaufmann, 1998.
3. A. Corradini, U. Montanari, F. Rossi, H. Ehrig, R. Heckel, and M. Löwe. Algebraic approaches to graph transformation, part I: Basic concepts and double pushout approach. In Rozenberg [16], chapter 3, pp. 163–246.
4. B. Courcelle. The monadic second-order logic of graphs XI: Hierarchical decompositions of connected graphs. *Theoretical Comput. Sci.*, 224(1–2):35–58, 1999.
5. F. Dau. *The Logic System of Concept Graphs with Negation*, vol. 2892 of *LNCS*. Springer, 2003.
6. H. Ehrig and A. Habel. Graph grammars with application conditions. In Rozenberg and Salomaa, eds., *The Book of L*, pp. 87–100. Springer, 1986.
7. H. Ehrig, K. Ehrig, A. Habel, and K.-H. Pennemann. Constraints and application conditions: From graphs to high-level structures. In *International Conference on Graph Transformations*, LNCS. Springer, 2004. This volume.
8. H. Ehrig, R. Heckel, M. Korff, M. Löwe, L. Ribeiro, A. Wagner, and A. Corradini. Algebraic approaches to graph transformation, part II: Single pushout approach and comparison with double pushout approach. In G. Rozenberg [16], pp. 247–312.
9. C. Ermel, M. Rudolf, and G. Taentzer. The AGG approach: language and environment. In Ehrig et al., eds., *Handbook of Graph Grammars and Computing by Graph Transformation*, vol. II: Applications, Languages and Tools. World Scientific, Singapore, 1999.

10. A. Habel, R. Heckel, and G. Taentzer. Graph grammars with negative application conditions. *Fundamenta Informaticae*, 26(3/4):287–313, 1996.
11. R. Heckel and A. Wagner. Ensuring consistency of conditional graph grammars — a constructive approach. *ENTCS*, 2, 1995.
12. M. Koch, L. V. Mancini, and F. Parisi-Presicce. Conflict detection and resolution in access control policy specifications. In Nielsen and Engberg, eds., *Foundations of Software Science and Computation Structures*, vol. 2303 of *LNCS*, pp. 223–238. Springer, 2002.
13. C. Pierik and F. S. de Boer. A syntax-directed hoare logic for object-oriented programming concepts. In Najm, Nestmann, and Stevens, eds., *Formal Methods for Open Object-based Distributed Systems*, vol. 2884 of *LNCS*, pp. 64–78. Springer, 2003.
14. D. D. Roberts. *The Existential Graphs of Charles S. Peirce*. Mouton and Co., 1973.
15. D. D. Roberts. The existential graphs. *Computers and Mathematics with Applications*, 6:639–663, 1992.
16. G. Rozenberg, ed. *Handbook of Graph Grammars and Computing by Graph Transformation*, vol. I: Foundations. World Scientific, Singapore, 1997.
17. J. F. Sowa. *Conceptual Structures: Information Processing in Mind and Machine*. Addison-Wesley, 1984.
18. M. Wermelinger. Conceptual graphs and first-order logic. In *International Confence on Conceptual Structures*, vol. 954 of *LNAI*, pp. 323–337. Springer, 1995.

Coproduct Transformations
on Lattices of Closed Partial Orders[*]

Gemma Casas-Garriga and José L. Balcázar

Departament de Llenguatges i Sistemes Informàtics
Universitat Politècnica de Catalunya
{gcasas,balqui}@lsi.upc.es

Abstract. In the field of Knowledge Discovery, graphs of concepts are an expressive and versatile modeling technique that provides ways to reason about information implicit in the data. Typically, nodes of these graphs represent unstructured closed patterns, such as sets of items, and edges represent the relationships of specificity among them. In this paper we want to consider the case where data keeps an order, and nodes of the concept graph represent complex structured patterns. We contribute by first characterizing a lattice of closed partial orders that precisely summarizes the original ordered data; and second, we show that this lattice can be obtained via coproduct transformations on a simpler graph of so-called stable sequences. In the practice, this graph transformation implies that algorithms for mining plain sequences can efficiently transform the discovered patterns into a lattice of closed partial orders, and so, avoiding the complexity of the mining operation for the partial orders directly from the data.

1 Introduction

Formal Concept Analysis, mainly developed by [6], is based on the mathematical theory of complete lattices (e.g. [5]). This area has been used in a large variety of fields in computer science, such as in Knowledge Discovery where graphs of concepts are an expressive modeling technique to show structural relations implicit in the given set of data ([9, 10, 13]).

The two basic notions of Formal Concept Analysis are those of *formal context* and *formal concept*. In the main case of interest for Knowledge Discovery, a formal context consists of a binary relation R that can be regarded as a set of objects associated with a set of items (attributes holding in each object), that is, $R \subseteq \mathcal{O} \times \mathcal{I}$. On the other hand, a formal concept is a pair of a closed set of objects and a closed set of items linked by a Galois connection. To characterize a formal concept we need to define appropiate closure operators on the universe of items and objects respectively.

A closure operator Γ on a lattice, such as the one formed by the subsets of any fixed universe, satisfies the three basic closure axioms: monotonicity, extensivity

[*] This work is supported in part by MCYT TIC 2002-04019-C03-01 (MOISES).

H. Ehrig et al. (Eds.): ICGT 2004, LNCS 3256, pp. 336–351, 2004.

and idempotency. It follows from these properties that the intersection of closed sets, which are those sets that coincide with their closure, is also another closed set. One way of constructing closure operators is by composition of two derivation operators forming a Galois connection [6]. The standard Galois connection for a binary database R maps each family of objects to the set of the items that hold in all of them, and each set of items to the set of objects in which they hold. Then, the resulting closure operator Γ acts as follows: given R, the closure $\Gamma(Z)$ of a set of items $Z \subseteq \mathcal{I}$ includes all items that are present in all objects having all items in Z. The closed sets obtained are exactly the closed sets employed in closed set mining (see e.g. [9]), which in certain applications presents many advantages (see e.g. [3,9,10,13]). Similarly, we can consider the dual operator Γ^{-1} operating on the universe of the set of objects $\mathcal{O}$ and giving rise to closed set of objects, that is, $\Gamma^{-1}(O) = O$ where $O \subseteq \mathcal{O}$. For any binary database R, the closure systems of Γ and Γ^{-1} are isomorphic.

The closed sets found in the database R can be graphically organized in a hierarchical order, called the concept lattice, i.e. a graph where each node is a pair (O, Z) formed by the closed itemset $\Gamma(Z) = Z$ and the maximal set of objects O where Z is contained (symetrically, for $\Gamma^{-1}(O) = O$); so, each node is labelled by the pair of closed sets that is joined by the Galois connection. This concept lattice provides a comprehensive graphical representation that shows the structural relations between the concepts and summarizes, at the same time, all the characteristics of the binary data.

In this paper we want to analyze the case where R is *not* a binary relation, but the items keep an order in each one of the objects where they hold. Here we study the construction of a lattice where formal concepts model more complex structures, such as partial orders. Our contribution is to prove that this closure system of partial orders can be obtained by coproduct transformations on a simpler graph of so-called stable sequences studied recently ([4,11,12]). Algorithmically, this transformation avoids the computation of partial orders directly from the data.

2 Preliminaries

Let $\mathcal{I} = \{i_1, \ldots, i_n\}$ be a finite set of items. *Sequences* are ordered lists of itemsets where we assume that no item occurs more than once in a sequence. The input data we are considering consists of a database of ordered transactions that we model as a set of sequences, $\mathcal{D} = \{s_1, s_2, \ldots s_n\}$. Our notation for the component itemsets of a given sequence will be $s = \langle (I_1)(I_2) \ldots (I_n) \rangle$, where each $I_i \subseteq \mathcal{I}$ and I_i occurs before itemset I_j if $i < j$. Note that each I_i may contain several items that occur simultaneously; e.g. $\langle (AC)(B) \rangle$, where items A and C are given at the same time but always before item B. However, in order to simplify the definitions of the next theoretical discussion we consider for the moment that each I_i contains only one single item. Later in the paper, we will show how the simultaneity condition can be incorporated again into the theoretical framework. The set of *all* the possible sequences will be noted by $\mathcal{S}$.

From the point of view of Formal Concept Analysis, we can consider each input sequence as an object. Formally, data $\mathcal{D}$ can be transformed into an ordered context (see [4]) where *objects* of the context are sequences, *attributes* of the context are items, and the database becomes a ternary relation, subset of $\mathcal{O} \times \mathcal{I} \times \mathbb{N}$, in which each tuple $\langle o, i, t \rangle$ indicates that item i appears in the t-th element of the object o representing an input sequence s. A simple example of the described data and the associated context can be found in figure 1, where each object o_i of the formal context represents the corresponding input sequence, $s_i \in \mathcal{D}$. The context for a set of data $\mathcal{D}$ is only relevant to this paper to see objects $o_i \in \mathcal{O}$ and input sequences $s_i \in \mathcal{D}$ as equivalent, which eases the definition of the subsequent closure operator on ordered data.

Seq id	Sequence
s_1	$\langle (A)(B)(C)(D) \rangle$
s_2	$\langle (B)(C)(D)(A) \rangle$
s_3	$\langle (B)(C)(A)(D) \rangle$

(a) Collection of data $\mathcal{D}$

	A	B	C	D
o_1	1	2	3	4
o_2	4	1	2	3
o_3	3	1	2	4

(b)   Context for $\mathcal{D}$

Fig. 1. Example of ordered data $\mathcal{D}$ and its context.

Sequence $s = \langle (I_1) \ldots (I_n) \rangle$ is a *subsequence* of another $s' = \langle (I'_1) \ldots (I'_m) \rangle$ if there exist integers $j_1 < j_2 \cdots < j_n$ such that $I_1 \subseteq I'_{j_1}, \ldots, I_n \subseteq I'_{j_n}$. We note it by $s \subseteq s'$. For example, the sequence $\langle (A)(D) \rangle$ is a subsequence of the first and third input sequences in figure 1.

The *intersection* of a set of sequences $s_1, \ldots, s_n \in \mathcal{S}$ is the set of *maximal* subsequences contained in all the s_i. Note that the intersection of a set of sequences, or even the intersection of two sequences, is not necessarily a single sequence. For example, the intersection of the two sequences $s = \langle (A)(C)(B) \rangle$ and $s' = \langle (A)(B)(C) \rangle$ is the set of sequences $\{ \langle (A)(C) \rangle, \langle (A)(B) \rangle \}$: both are contained in s and s', and among those having this property they are maximal; all other common subsequences are not maximal since they can be extended to one of these. The maximality condition discards redundant information since the presence of, e.g., $\langle (A)(B) \rangle$ in the intersection already informs of the presence of each of the itemsets (A) and (B).

2.1 Stable Sequences and Closure Operators

A sequence s is *stable* in input data $\mathcal{D}$ if s is maximal in the set of objects where it is contained, that is, it cannot be extended. More formally, we say that:

Definition 1. *A sequence $s \in \mathcal{S}$ is stable if there exists no sequence s' with $s \subset s'$ s.t. they are both subsequences of the same set of objects (equivalently, input sequences).*

For instance, taking data from figure 1, sequence $\langle (B)(D) \rangle$ is not stable since it can be extended to $\langle (B)(C)(D) \rangle$ in all the objects where it belongs. However, $\langle (B)(C)(D) \rangle$ or $\langle (A)(D) \rangle$ are stable sequences in $\mathcal{D}$. The set of all stable sequences of data from figure 1 are shown in figure 2. The most relevant existing contributions on the mining of stable sequences are given by two algorithms, CloSpan [11] and BIDE [12], which find in a resonably efficient way all the stable sequences for the input dataset $\mathcal{D}$. Stable sequences are called "closed" there; we prefer a different term to avoid confusion with the closure operator. Note however, that these mentioned algorithms do not impose our condition of avoiding repetition of items in the input sequences.

Objects	Stable Sequences
1,2,3	$\langle (B)(C)(D) \rangle$
1,2,3	$\langle (A) \rangle$
2,3	$\langle (B)(C)(A) \rangle$
1,3	$\langle (A)(D) \rangle$
1	$\langle (A)(B)(C)(D) \rangle$
2	$\langle (B)(C)(D)(A) \rangle$
3	$\langle (B)(C)(A)(D) \rangle$

Fig. 2. Set of stable sequences and objects where they belong.

As it is formalized in [4], the set of stable sequences can be characterized in terms of a closure operator, named Δ, operating on the universe of sets of sequences. Briefly, the defined Galois connection maps each family of objects to the set of the maximal sequences that hold in all of them, and each set of sequences to the set of all objects in which they hold. It is proved there that these mappings indeed enjoy the properties of a Galois connection so that their composition provides the necessary closure operator. Again, a closed set of sequences are those coinciding with their closure, that is, $\Delta(\{s_1, \ldots s_n\}) = \{s_1, \ldots s_n\}$ where $\{s_1, \ldots s_n\} \subseteq \mathcal{S}$; similarly, a closed set of objects in this ordered context is defined by the dual closure operator, i.e. $\Delta^{-1}(O) = O$. A main result in [4] is that stable sequences are exactly those that belong to a closed set of sequences.

As in any other Galois connections (see [6]), this characterization gives immediately a lattice of formal concepts, that is, a graph where each node is a pair $(O, \{s_1, \ldots s_n\})$ where $\{s_1, \ldots s_n\}$ are stable sequences belonging to the closed set of objects O, and viceversa. For instance, in data of figure 1 we have that the set of sequences $\{\langle (B)(C)(D) \rangle, \langle (A)(D) \rangle\}$ is stable for the first and third objects; reciprocally, the set of objects formed by first and third transaction is closed for the set of stable sequences $\{\langle (B)(C)(D) \rangle, \langle (A)(D) \rangle\}$. So, $(\{o_1, o_3\}, \{\langle (B)(C)(D) \rangle, \langle (A)(D) \rangle\})$ will be a formal concept of the lattice. It is proved also in [4] that all the stable sequences mined by CloSpan or BIDE can be organized in different formal concepts of the same lattice, and this graph of stable sequences characterizes the non-redundant sequential patterns of the ordered data.

Interestingly enough for our subsequent contribution, the work in [4] also proves that if we rather deal with input sequences more than with objects, then the set of stable sequences of a formal concept is exactly the intersection of those input sequences renamed after the objects of the same concept. That is, renaming each object by its input sequence in $\mathcal{D}$, then a formal concept $(O, \{s_1, \ldots s_n\})$ becomes $(S, \{s_1, \ldots s_n\})$ where S are the input sequences of O, and we always have the following property: $\{s_1, \ldots s_n\} = \bigcap_{s \in S} s$.

3   Partial Orders on Sequences

Formally, we will model partial orders as a full subcategory of the set of directed graphs; so, as a starting point, we recall the most basic among the numerous definitions of graphs given in the literature.

Definition 2. *A directed graph is modeled as a triple* $G = (V, E, l)$ *where V is the set of vertices; $E \subseteq V \times V$ is the set of edges; and l is the injective labelling function mapping each vertex to an item, i.e. $l : V \to \mathcal{I}$.*

The set $\mathcal{I}$ in the labelling function is exactly the finite set of items defined in the preliminaries; this will be a fixed set in all the graphs belonging to our category. For our present work we consider that the labelling function l of a graph is *injective* (and not necessarily surjective). An edge between two vertices u and v will be denoted by $(u, v) \in E$, implying a direction on the edge from vertice u to v.

Definition 3. *A graph morphism $h : G \to G'$ between two graphs $G = (V, E, l)$ and $G' = (V', E', l')$ consists of an* injective *function $h_V : V \to V'$ that preserves labels (that is, $l' \circ h_V = l$), and $(u, v) \in E \Rightarrow (h(u), h(v)) \in E'$.*

Note here that the injectivity of the morphism h between any two graphs, whose labelling function must be also injective, forces the morphism h to be unique. So, if there are $h : G \to G'$ and $g : G' \to G$, it implies $h = g$ and $G = G'$. The composition of $h : G \to G'$ with a morphism $g : G' \to G''$ is the morphism $g \circ h : G \to G''$ consisting of the composed function $g_V \circ h_V$. It is well known that the good properties of graph morphisms turn the set of graphs into a category. From the category of the set of graphs, we will be specially interested in the following constructor operator.

Definition 4 (Coproduct). *The coproduct of a family of graphs $\{G_i\}$ is a graph $\widehat{G} = \coprod G_i$ in the same category and a set of morphisms $\{h_i : G_i \to \widehat{G}\}$ such that, for every graph G' and every family of morphisms $\{h'_i : G_i \to G'\}$, there is an unique morphism $g : \widehat{G} \to G'$ such that $g \circ h_i = h'_i$.*

In category-theoretic terms, the result $\widehat{G}$ of a coproduct is the initial object among all those candidates G'. Moreover, we know that this $\widehat{G}$ is unique in any coproduct since the morphisms $\{h_i\}$, $\{h'_i\}$ and g are injective and the family

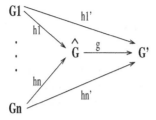

Fig. 3. Coproduct diagram.

of graphs considered here have an injective labelling function. Therefore, the coproduct of two graphs G_1 and G_2 in our category defines exactly a union of G_1 and G_2 where vertices in G_1 and G_2 with the same label are joined, and where the injectivity of morphisms ensures that all edges from both graphs are preserved.

From the set of all directed graphs, we will be interested in the full subcategory that models partial orders. A *partial order* (also called poset) is an acyclic directed graph $G_p = (V, E, l)$ such that the relation on V stablished by edges in E is reflexive, antisymmetric and transitive. The *sources* of a poset are those vertices that do not have any predecessor; similarly, the *sinks* are those vertices not preceded by any other vertex in the poset. Note that a poset may have different unconnected components, and so, it may have several source nodes.

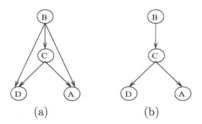

Fig. 4. Example of a partial order and its transitive reduction.

The graphical representation of partial orders is particularly useful for displaying results: we will display a poset by using arrows between the connected labelled vertices, and the symbol ∥ (parallel) to indicate trivial order among the different components of a poset. The *transitive reduction* of $G_p = (V, E, l)$ is the smallest relation resulting from deleting those edges in E that come from transitivity. Posets will be graphically depicted here by means of its transitive reduction to make them more understandable (as in figure 4(b)), but of course, all edges of the transitive closure are indeed included in E (figure 4(a)).

Some specific definitions we need on posets are the following ones.

Definition 5. *We define that a poset G_p is more general than another poset $G_{p'}$, noted by $G_p \unlhd G_{p'}$, if there exists a morphism from G_p to $G_{p'}$, i.e. $h : G_p \to G_{p'}$. Then, we also say that $G_{p'}$ is more specific than G_p.*

For instance, the partial order represented in figure 4 is more specific than the trivial order $\| A, B, C, D \|$ (parallelization of all items), but more general than (the transitive closure of) the total order $B \to C \to D \to A$.

Definition 6. *A partial order $G_p = (V, E, l)$ is compatible with a sequence s if: $\forall u \in V$ we have that item $l(u)$ is in s; and, $\forall (u, v) \in E$ we have that $\langle (l(u))(l(v)) \rangle \subseteq s$.*

In other words, a poset G_p is compatible with a sequence s if there exists a morphism from G_p to the poset obtained as the transitive closure of the sequence s. We will see a sequence as a partial order containing all the orders added by transitive closure, so we can express it for our convenience as $G_p \unlhd s$. For instance, the partial order of figure 4 is compatible with the second and third input sequences of the data in example 1. The trivial order is compatible with any sequence having all the items of the poset; so, at least there will be one poset (the trivial one) compatible with a given set of sequences. In case that sequences in S do not have any item in common, then we assume the existence of an empty poset compatible with them.

Definition 7. *We define a path from a poset $G_p = (V, E, l)$ as a sequence of items $\langle (i_1)(i_2) \ldots (i_n) \rangle$ such that for all consecutive i_j and i_{j+1} in the sequence, there exists $(u, v) \in E$ s.t. $l(u) = i_j$ and $l(v) = i_{j+1}$.*

For instance, sequences $\langle (B)(C)(D) \rangle$, or $\langle (B)(A) \rangle$, or $\langle (B)(C)(A) \rangle$ define paths of the poset shown in figure 4(a). We define a path to be *maximal* with respect to the inclusion of sequences. E.g., path $\langle (B)(A) \rangle$ is not maximal since it is a subsequence of the path $\langle (B)(C)(A) \rangle$. Note that posets are acyclic, so, the maximal paths in a poset G_p will always be defined between sources and sinks of G_p (of course, avoiding the shortcuts of the transitive closure). Note also that since a poset is actually a graph, we are still able to operate coproducts on them; although this does not necesarily imply that the coproduct of two partial orders is another partial order.

Next, we will consider the goal of summarizing the input sequences as the most specific partial orders compatible with them.

3.1   A Closure System of Partial Orders

This section presents our first contribution: the construction and visual display of a concept lattice where nodes contain partial orders, and the relationships among them will be representative of the relationships in the input ordered data. We will construct this lattice by avoiding the formalization of a closure operator; however, we will show that the constructed family of formal concepts is indeed a closure system.

We say that partial order G_p is *closed for a set of sequences* S if G_p is the most specific poset from all those posets compatible with all $s \in S$. For instance, given the set of sequences $S = \{\langle(B)(C)(A)\rangle, \langle(B)(C)(D)\rangle\}$ we have two maximal posets compatible with them: the trivial order $G_{p_1} = \| B, C \|$ and the total order $G_{p_2} = B \rightarrow C$; but only G_{p_2} is closed for S since $G_{p_1} \trianglelefteq G_{p_2}$. Next proposition 1 ensures unicity of closed posets for a set of sequences S.

Proposition 1. *Given a set of sequences S there is exactly one single closed poset for S.*

Proof. As mentioned after definition 6, there is at least one compatible poset with all the sequences in S, which is the trivial order with the shared items of all sequences in S. Now, assume to the contrary that we had two most specific posets compatible with all $s \in S$, named G_{p_1} and G_{p_2} with corresponding morphisms into s. If so, we can construct a third poset $\widehat{G}$ from the union of G_{p_1} and G_{p_2}, i.e. their coproduct $\widehat{G} = \coprod G_{p_i}$ for $i = 1, 2$ (as we mentioned, the union of two graphs in this category can be formalized as their coproduct). The properties of the coproduct ensure that $\widehat{G}$ is unique, compatible with all $s \in S$, and more specific than G_{p_1} and G_{p_2}.   □

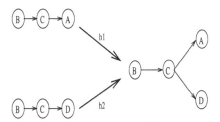

Fig. 5. Example of a coproduct.

We must point out an important remark after this proof: the coproduct is an operator defined on graphs, so, the coproduct of two partial orders may give a graph that is not another partial order. For instance, the coproduct of $G_{p_1} = A \rightarrow B$ and $G_{p_1} = B \rightarrow A$ leads to a graph with a cycle (then, not antisymmetric). However, in the proof of proposition 1 we are operating the coproduct on two posets simultaneously compatible with the same set of sequences S. Given that the sequences in S do not have repeated items by definition, it is not possible to get a cycle from the coproduct of two posets compatible with S; in other words, two partial orders whose union leads to a cycle cannot be both compatible with the same set S.

We say that *set of sequences S is closed* for a partial order G_p if S contains all the input sequences $s_i \in \mathcal{D}$ with which G_p is compatible. For example, the set of input sequences from figure 1 that are closed with respect to poset in figure 4 is $S = \{s_2, s_3\} = \{\langle(B)(C)(D)(A)\rangle, \langle(B)(C)(A)(D)\rangle\}$.

Now we are ready to define the notion of formal concept.

Definition 8. *A formal concept is a pair* (S, G_p) *where* G_p *is a closed partial order for the set of sequences* S, *and the set of sequences* S *is closed for the partial order* G_p.

Formal concepts (S, G_p) will be nodes of the concept lattice of partial orders. In practice, we will visualize these nodes principally by the closed poset G_p of the concept, and the dual S will be added as a list of object identifiers (thus, as it happens in general in Galois connections, these lists form a dual view of the same lattice that, in our case, is ordered by set-theoretic inclusion downwards); proposition 1 ensures that each node of the lattice can be represented by one single closed partial order. Edges in the lattice will be the specificity relationships among the concepts, named $\leq$, such that $(S_1, G_{p_1}) \leq (S_2, G_{p_2})$ if $G_{p_1} \trianglelefteq G_{p_2}$. The set of all concepts ordered by $\leq$ is called the concept lattice of partial orders of our context. Eventually, an artificial top representing a poset not belonging to any sequence is also added to the lattice and we note it by the unsatisfiable boolean constant $\square$. In figure 6 we show the lattice of closed concepts for the data of figure 1.

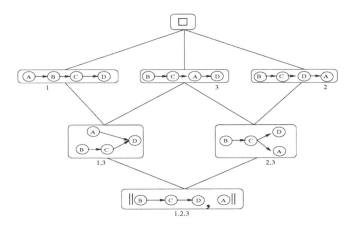

Fig. 6. Concept Lattice of partial orders for data in figure 1.

Although the lattice of closed partial orders has been characterized without defining any specific closure operator that ensures $\cap$-stability on concepts, we can indeed prove that our system is closed under intersection. Semantically speaking, intersection must preserve the maximal common substructure from the intersected objects. Again, because of the injectivity of the labelling function along with the uniqueness of morphisms, the intersection of two posets can be easily formalized by the product operator of category theory (dual to the coproduct). However, we introduce now a completely different characterization of intersection of posets in terms of the coproduct of maximal paths.

Definition 9. *The intersection of two partial orders $G_p \cap G_{p'}$ is $G_p \cap G_{p'} = \coprod t_i$ where $\{t_i\}$ is the family of maximal paths contained in both G_p and $G_{p'}$.*

Next lemma proves that the intersection as it is defined here preserves indeed the maximal common substructure from the intersected partial orders.

Lemma 1. *The result of $G_{p_1} \cap G_{p_2}$ is the most specific poset G_p from those where $G_p \trianglelefteq G_{p_1}$ and $G_p \trianglelefteq G_{p_2}$.*

Proof sketch. If $G_p = \coprod t_i$ where $\{t_i\}$ is the family of maximal paths contained in both G_{p_1} and G_{p_2}, then obviously there exists by construction a morphism $G_p \trianglelefteq G_{p_1}$ and $G_p \trianglelefteq G_{p_2}$. Any other poset $G_{p'}$ having as well morphisms to G_{p_1} and G_{p_2}, must have a morphism to G_p, i.e. $G_{p'} \trianglelefteq G_p$, otherwise we would contradict the fact that some path t_i used in the coproduct is not a maximal path. Thus, we exactly get a product diagram; so, G_p defines the result of a product operation on partial orders from category theory. □

Note that since intersection of posets is then associative. Next proposition 2 finally shows that our lattice is a system closed under intersection.

Proposition 2. *The intersection of closed posets is another closed poset.*

Proof sketch. We will reject the following false hypothesis: let (G_{p_1}, S_1) and (G_{p_2}, S_2) be two different closed partial orders for S_1 and S_2 respectively, and we suppose that $G_{p_1} \cap G_{p_2} = G_p$ is not another closed poset, i.e. G_p is not a part of any concept. In this case, let $G_{p'}$ be the closed poset immediately over G_p; that is, $G_p \trianglelefteq G_{p'}$ and it does not exist another closed $G_{p''}$ s.t. $G_{p''} \trianglelefteq G_{p'}$ and $G_p \trianglelefteq G_{p''}$. Then, it must be true that $G_{p'} \trianglelefteq G_{p_1}$ and $G_{p'} \trianglelefteq G_{p_2}$ (because both G_{p_1} and G_{p_2} are closed and different from G_p). But then, by lemma 1 we get a contradiction since G_p is not the most specific poset s.t. $G_p \trianglelefteq G_{p_1}$ and $G_p \trianglelefteq G_{p_2}$. □

This section shows that the input sequential data can be summarized by means of a concept lattice that presents a balance between generality and specificity for all the input sequences. As an example, observe lattice of figure 6 that gets an overview of the ordering relationships in the data. We also see that partial orders such as $\| A, B, C, D \|$ or $B \to D$ or $\| A, B \to D \|$ do not create a node in the concept lattice, i.e. they are not closed: this is because these posets turn to be redundant in describing our data, they are compatible with all the input sequences, but they are not specific enough to be closed.

The problem is now how to construct this useful concept lattice of closed partial orders. Currently, it is still a challenge how to deal with poset structures in the field of Knowledge Discovery: mining the partial orders directly from the data is a complex task incurring a substantial runtime overhead. For example, the work in [7] presents a method based on viewing a partial order as a generative model for a set of sequences and it applies different mixture model techniques. The final posets are not necessarily closed and so, they could be redundant; besides, they restrict the attention to a subset of partial orders called series-parallel partial orders (such as series-parallel digraphs) to avoid computational problems.

Note that here we do not restrict in any sense the form of the final closed posets. Another work worth mentioning here is [8], where the authors present the mining of episodes (whose hybrid structures are indeed partial orders). However, again these structures are not closed in any lattice and computational problems still persist.

4   Coproduct Transformations to Closed Posets

In this section we show that our lattice of closed posets can be indeed obtained via coproduct transformations on the simpler lattice of stable sequences; thus, providing an efficient way to derive those closed posets. We present, next, one of the main results that will lead to our characterization.

Theorem 1. *Let G_p be the most specific partial order compatible with a set of sequences S; then the set of maximal paths of poset G_p defines exactly the intersection of all sequences in S, that is $\bigcap\limits_{\forall s \in S} s$.*

Proof. We will prove both directions: maximal paths from G_p come from the intersection of all $s \in S$; and that intersections of all $s \in S$ define the maximal paths of G_p.

$\Rightarrow$) Let $t = \langle (i) \dots (j) \rangle$ define a maximal path between a source node, labelled with item i, and a sink node, labelled with item j, of the poset G_p, and suppose that t does not belong to the intersection of sequences in S. This implies that either $1/ t$ does not belong to some $s \in S$; or $2/ t$ belongs to all $s \in S$ but t is not maximal (i.e, there exists another t' belonging to the intersection of S s.t. $t \subset t'$). In the first case, it would mean that we started from a poset G_p not compatible with the set of sequences S; in the second case, it would mean that we started from a poset G_p which is not the most specific, since with $G_p \cup t'$ (formalized as the coproduct) we would get a more specific poset. In any case we are contradicting the original formulation of G_p in the theorem, so it is true that all maximal paths in G_p come from the intersection of sequences in S.

$\Leftarrow$) Let t be a sequence belonging to the intersection of all $s \in S$, such that it does not define a maximal path between a source and a sink in G_p. This implies that G_p is not the most specific for S, since we could add the path defined by t to the poset G_p and get a more specific poset still compatible with all S. Again, we reach a contradiction that makes the original statement true. Again, we must insist on the fact that sequences in S do not have repeated items by definition, so the sequence t considered here always leads to a path for G_p with no cycles. $\qquad\square$

To illustrate this theorem, let us consider any formal concept (S, G_p) of our lattice of partial orders: the maximal paths of the closed poset G_p are defined exactly by the intersection of all sequences in the closed set of sequences S. For instance, taking the lattice in figure 6: the intersection of the closed set $S = \{s_2, s_3\} = \{\langle (B)(C)(D)(A)\rangle, \langle (B)(C)(A)(D)\rangle\}$, is the set of sequences $\{\langle (B)(C)(D)\rangle, \langle (B)(C)(A)\rangle\}$, which coincides exactly with the maximal paths of the closed poset G_p in the same node. The next corollary follows immediately from the main theorem.

Corollary 1. *Let G_p be the most specific partial order compatible with a set of sequences S; then the set of all paths from G_p is defined exactly by subsequences of some sequence in the intersection of S, that is $t' \subseteq \bigcap_{\forall s \in S} s$.*

Proof. It immediately follows from theorem 1: if the maximal paths from G_p are defined by the intersection of all $s \in S$, then any subsequence of a maximal path defines a shorter path in G_p. □

Clearly after corollary 1, it is possible to reconstruct the transitive closure of a closed poset G_p with the intersections of the closed set of sequences S where G_p is maximally contained. The coproduct transformation will help in this procedure, as it is shown in next theorem 2. Again, for our notational purposes we consider that a sequence is indeed a poset with all the proper edges added by transitive closure.

Theorem 2. *Let G_p be the most specific partial order compatible with a set of sequences S, then G_p is the result of the coproduct transformation on $\bigcap_{\forall s \in S} s$.*

Proof. We know by theorem 1 that the intersection of sequences in S are exactly those maximal paths of the most specific partial order G_p compatible with S. Then, it is possible to prove that a poset G_p comes from the coproduct transformations of its maximal paths. This can be easily proved by induction on the number of paths of the poset G_p, and taking into account that the coproduct operator is associative. Moreover, because these paths come from intersections of S and S is restricted to not having repetition of items, then the coproduct transformations do not lead to any cycle here. □

This last theorem concludes that the closed poset G_p from formal concept, (S, G_p), can be generated by the coproduct transformations on the intersections of S. Besides, for the same reason given after proposition 1, we can be sure that coproduct transformations always return a valid poset in theorem 2: two posets G_{p_1} and G_{p_2} whose coproduct transformation has a cycle cannot be both compatible with the same set of sequences S at the same time. To exemplify theorem 2, let us take the closed set of sequences $S = \{s_2, s_3\} = \{\langle (B)(C)(D)(A) \rangle, \langle (B)(C)(A)(D) \rangle\}$, whose intersection gives the set of sequences $\{\langle (B)(C)(D) \rangle, \langle (B)(C)(A) \rangle\}$. The coproduct transformation on these intersections is given in figure 5, and it exactly returns the closed poset in that formal concept.

Next section will explain the relation of these intersections of S with the stable sequences; thus, reaching a final lattice transformation of stable sequences to closed posets by means of coproduct operations on its nodes.

4.1   Lattice Transformation and Algorithmic Consequences

As introduced in the preliminaries, the work in [4] shows that stable sequences can be characterized by the closure operator Δ. Therefore, the set of all stable sequences from a database $\mathcal{D}$ can be organized in formal concepts $(O, \{s_1, \ldots, s_n\})$, where $\{s_1, \ldots, s_n\}$ are a set of stable sequences for the closed set of objects O. Since each object in the ordered context represents indeed an input sequence from $\mathcal{D}$, we can rewrite the formal concepts as $(S, \{s_1, \ldots, s_n\})$ where S is the set of input sequences renamed after O. From this point of view it is possible to prove as in [4] that stable sequences are the intersection of S, that is, $\{s_1, \ldots, s_n\} = \bigcap\limits_{s \in S} s$.

Actually, this last observation leads naturally to the following theorem.

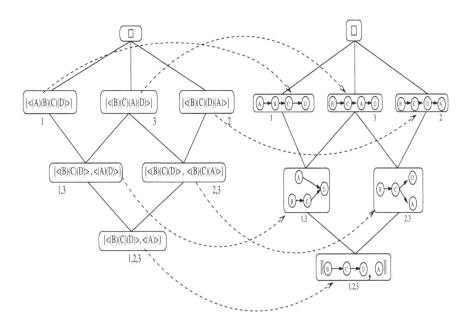

Fig. 7. Lattice transformation to closed partial orders.

Theorem 3. *A lattice of stable sequences can be transformed into a lattice of closed partial orders by rewriting each node via coproduct transformations.*

Proof. Let $(O, \{s_1, \ldots, s_n\})$ be a formal concept of stable sequences, that we can rewrite as $(S, \{s_1, \ldots, s_n\})$. We have that $\{s_1, \ldots, s_n\}$ are stable sequences (maximal) in S, and S is the set of *all* the input sequences in which they appear. Let us construct a partial order G_p via coproduct transformations on these the stable sequences $\{s_1, \ldots, s_n\}$. We have that G_p is the most specific partial order for S: because $\{s_1, \ldots, s_n\}$ is equivalent to $\bigcap\{s | s \in S\}$ (as shown in [4]), and then, it immediately follows from theorem 1 and theorem 2 that G_p is the most specific poset for S. Thus, G_p is a closed poset for S. At the same time, S must be the maximal set of input sequences for the poset G_p to not contradict the fact of being O a maximal set of objects for the stable sequences $\{s_1, \ldots, s_n\}$; so, S is a closed set of input sequences for G_p. Therefore, (S, G_p) can be derived from from $(S, \{s_1, \ldots, s_n\})$ via coproduct transformations.   □

Figure 7 shows the set of all stable sequences from data in figure 1 organized in formal concepts. Each node of this lattice represents a set of stable sequences together with the list of object identifiers (input sequences) where they are maximally contained. We partially order sets of stable sequences in the lattice as in [4]: $\{s_1, \ldots, s_n\} \preceq \{s'_1, \ldots, s'_m\}$ if and only if $\forall s_i \exists s'_j, \ s_i \subseteq s'_j$. It can be graphically seen that each node of the lattice of stable sequences can be transformed into a node of the lattice of partial orders by means of a coproduct. Dashed lines of figure 7 indicate a node rewriting by means of a coproduct transformation on the stable sequences.

In practice, this transformation carries several consequences. Mining complex structured patterns directly from the data, such as partial orders, graphs, molecules, etc, is a complex task due to the overhead incurred by the algorithms (see [7, 8]). However, the presented lattice transformation allows the construction of closed posets in the data by transforming the graph of stable sequences. So, the algorithmic contributions to the mining of stable sequences, such as CloSpan ([11]) and still more efficient, BIDE ([12]), can be used to mine partial orders by performing coproduct transformations on the discovered stable sequences.

4.2   Simultaneity Condition Revisited

To consider input sequences $s = \langle (I_1)(I_2) \ldots (I_n) \rangle$ where each I_i may contain several simultaneous items, we must redefine the kind of posets we deal with. In particular, the labelling function l must be a function mapping each vertex to a set of items, i.e. $l : V \to 2^\mathcal{I}$. Again, this function is required to avoid repetitions of items in the nodes of the partial order (since we are not allowing repetitions in the input sequences either); so, the injectivity condition of l implies here that the items in each node are unique in the poset. With the new labelling function, the path from a poset is a sequence of itemsets and not single items. All the presented results hold with the new definitions.

5   Conclusions and Further Work

This paper shows that a simple lattice of stable sequences can be transformed into a lattice of closed partial orders: the first step proves that the maximal paths of each closed poset G_p are characterized by the intersections of those input sequences where G_p is contained. Then, the convenient coproduct operator is used to naturally allow the formalization of G_p as the transformation on those intersections. The last step is to prove that the intersections of input sequences are indeed stable sequences, so that it is possible to rewrite each node of the lattice of stable sequences into a closed partial order.

This transformation performed on lattices is of great importance in the field of Knowledge Discovery, where the mining of partial orders directly from the data is a complex task. In particular, it implies that algorithms for mining plain stable sequences can efficiently transform the discovered patterns into a lattice of closed partial orders that best summarizes the data.

The next step to complete our transformation framework would be to consider repetition of items in the input sequences, and so, to allow a non-injective labelling function in graphs. In this case, the coproduct operator does not work because the result of the transformation might not be unique. However, the pushout operator seems to fit in the new conditions: the pushout constructor can be used to formalize exactly the union and intersection of partial orders, and to allow the transformation process on the intersections of sequences. The problem with the pushout operator is to define a convenient target graph so that the pushout diagram commutes; it is still unclear at the moment how to achieve this. We also plan to study as a future work other complex structures, such as concept lattices of graphs or molecules: there we hope to find out which substitution mechanisms are required to obtain closed structures for more complex data.

Acknowledgements

The authors thank Elvira Pino for her useful explanations on category theory.

References

1. Adamek J., Herrlich H., and Strecker G. Abstract and Concrete Categories. John Wiley, New York, 1990.
2. Agrawal R., and Srikant S. *Mining Sequential Patterns.* In Proc. of the Int'l Conf. on Data Engineering, pages 3-14. 1995.
3. Balcázar J.L., and Baixeries J. *Discrete Deterministic Datamining as Knowledge Compilation.* Workshop on Discrete Mathematics and Data Mining, in SIAM Int. Conf, 2003.
4. Casas-Garriga G. *Towards a Formal Framework for Mining General Patterns from Structured Data.* Proc. of the 2nd. Multi-relational Datamining, in KDD Int. Conf, 2003.
5. Davey B.A., and Priestly H.A. *Introduction to Lattices and Order.* Cambridge, 2002.
6. Ganter B., and Wille R. *Formal Concept Analysis. Mathematical Foundations.* Springer, 1998.
7. Mannila H., and Meek C. *Global Partial Orders from Sequential Data.* Int. Conf. on Knowledge Discovery in Databases. 2000.
8. Mannila H., Toivonen H., and Verkamo A.I. *Discovery of frequent episodes in event sequences.* Data Mining and Knowledge Discovery, 1(3):259-289. 1997.

9. Pasquier N., Bastide Y., Taouil R., and Lakhal L. *Closed set based discovery of small covers for association rules.* In Proc. of the 15th. Conference on Advanced Databases, pages 361-381. 1999.
10. Pfaltz J.L., and Taylor C.M. *Scientific Knowledge Discovery through Iterative transformations of Concept Lattices.* Workshop on Discrete Mathematics and Data Mining, in SIAM Int. Conf., pages 65-74. 2002.
11. Yan X., Han J., and Afshar R. *CloSpan: Mining Closed Sequential Patterns in Large Datasets.* Int. Conference SIAM Data Mining. 2003.
12. Wang J., and Han J. *BIDE: Efficient Mining of Frequent Closed Sequences.* Int. Conf. on Data Engineering. 2004.
13. Zaki M. *Generating non-redundant Association Rules.* In Proc. of the Sixth Int. Conference on Knowledge Discovery and Data Mining, pages 34-43. 2000.

Parsing String Generating Hypergraph Grammars

Sebastian Seifert and Ingrid Fischer

Lehrstuhl für Informatik 2, Universität Erlangen–Nürnberg
Martensstr. 3, 91058 Erlangen, Germany
Sebastian.T.Seifert@stud.informatik.uni-erlangen.de
Ingrid.Fischer@informatik.uni-erlangen.de

Abstract. A string generating hypergraph grammar is a hyperedge replacement grammar where the resulting language consists of string graphs i.e. hypergraphs modeling strings. With the help of these grammars, string languages like $a^n b^n c^n$ can be modeled that can not be generated by context-free grammars for strings. They are well suited to model discontinuous constituents in natural languages, i.e. constituents that are interrupted by other constituents. For parsing context-free Chomsky grammars, the Earley parser is well known. In this paper, an Earley parser for string generating hypergraph grammars is presented, leading to a parser for natural languages that is able to handle discontinuities.

1 Discontinuous Constituents in German

One (of many) problems when parsing German are **discontinuous constituents** [1]. Discontinuous constituents are constituents which are separated by one or more other constituents and still belong together on a semantic or syntactic level. An example[1] for a discontinuous constituent is

(1) *Er hat schnell gearbeitet.*
    He has fast   worked.
    He (has) worked fast.

The verb phrase *hat gearbeitet ((has) worked)*[2] is distributed; the finite verb part, the auxiliary verb *hat (has)*, is always in the second position in a German declarative sentence. The infinite verb part, the past participle *gearbeitet (worked)*, is usually in the last position of a declarative sentence, only a few exceptions like relative clauses or appositions can be put after the infinite verb part. Another (more complicated) German example of discontinuous constituents is

---

[1] The German examples are first translated word by word into English to explain the German sentence structure and then reordered into a correct English sentence.
[2] The present perfect in German can be translated either in present perfect or in past tense in English.

H. Ehrig et al. (Eds.): ICGT 2004, LNCS 3256, pp. 352–367, 2004.

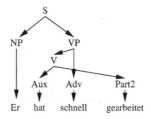

Fig. 1. The phrase structure tree for the German sentence *Er hat schnell gearbeitet.* (*He (has) worked fast*).

(2) Kleine grüne Autos habe ich keine gesehen die   mir gefallen hätten.
  Small green cars   have I   none seen   that me pleased would have.
  I did not see any small green cars that would have pleased me.

Here, the noun phrase *keine kleinen grünen Autos, die mir gefallen hätten* (*any small green cars that would have pleased me*) is distributed in three parts[3]. There are also discontinuous conjunctions as in

(3) Weder   Max noch Lisa haben die Aufgabe verstanden.
  Neither Max nor Lisa have the task   understood.
  Neither Max nor Lisa understood the task.

In this example, the discontinuity of *weder . . . noch* is also present in the English translation *neither . . . nor*.

It is typical for all the above examples that the syntactical or semantical connection between the two parts of the discontinuous constituent cannot be expressed with general context–free Chomsky grammars. One would wish for a phrase structure tree as shown in Figure 1 for example (1). Of course it is possible to construct a weakly equivalent context–free Chomsky grammar to parse such a sentence, but it must contain some work-around for the discontinuous constituent like attaching one part of the discontinuous constituent in another production than the other. The main advantage of context-free string generating hypergraph grammars lies in their possibility to describe discontinuous constituents in a context–free formalism. A more detailed description can be found in [3]. It is desireable to have a parser based on this representation formalism for discontinuous constituents. With the help of a parser, larger grammar descriptions can be developed and tested. An often used parser for context–free Chomsky grammars is the Earley parser [4]. The main goal of this paper is to describe an Earley parser for context–free string generating hypergraph grammars. Therefore, first a short introduction into this type of grammars is given and their application in natural language modeling is shown. Related work is shortly described. Finally the modified Earley parser is described in detail.

---

[3] This syntactic phenomenon is also called *split topicalization* [2].

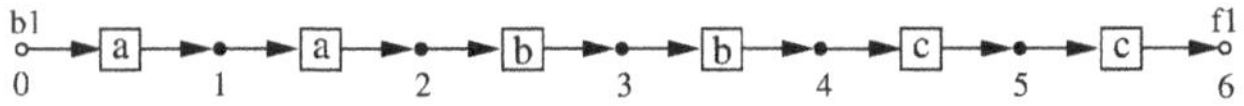

Fig. 2. The hypergraph representing $a^2b^2c^2$.

2   String Generating Hypergraph Grammars

Hypergraph grammars have been studied extensively in the last decades. Introductions and applications can be found in [5], [6]. A subset of hypergraph grammars, context–free string generating hypergraph grammars, are described in detail in [7], [8], [5]. In this section, only an overview of the most important definitions is given.

A **labeled directed hyperedge** is a labeled edge that is not only connected to one source and one target node as an ordinary graph edge, but to

- a sequence of arbitrary length of source nodes $\langle s_1, s_2, \ldots, s_n \rangle$ and
- a sequence of arbitrary length of target nodes $\langle t_1, t_2, \ldots, t_m \rangle$ with $n, m \geq 0$.

The points where the hyperedge is connected to hypergraph nodes are called **tentacles**. The **type** (m, n) of a hyperedge consists of the number of source tentacles m and the number of target tentacles n.

A **hypergraph** (E, V, s, t, l, b, f) consists of

- a finite set of hyperedges E
- a finite set of nodes V
- a source function $s : E \to V^*$ assigning a sequence of source nodes to each edge
- a target function $t : E \to V^*$ assigning a sequence of target nodes to each edge
- a labeling function $l : E \to A$ assigning a label from a given alphabet A to each edge; the label of an edge determines its type
- a sequence of external source nodes b
- a sequence of external target nodes f

A hypergraph has the **type** (m, n) if it has m external source nodes and n external target nodes. A hypergraph H with nodes $\{v_0, v_1, \ldots, v_n\}$ and edges $\{e_1, \ldots, e_n\}$ is called a **string graph**, if $s(e_i) = v_{i-1}$ and $t(e_i) = v_i$, $i = 1 \ldots n$. The node v_0 is the only external source node and v_n is the only external target node. This string hypergraph is a hypergraph based representation of a normal string. The letters (or words) of the string are the labels of the hyperedges. The letters labeling the hyperedges must be ordered in the string graph in the same way as in the underlying string. The hypergraph representation of the string $a^2b^2c^2$ is show in Figure 2.

A **hyperedge replacement rule** consists of a left hand side edge that is replaced by the hypergraph on the right hand side of the rule. Hyperedge and hypergraph must have the same type. In Figure 3 the hyperedge replacement rules generating the string language $a^n b^n c^n$ are shown.

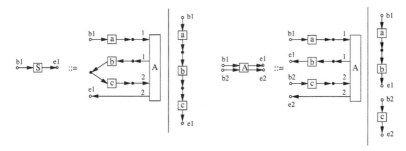

Fig. 3. Hyperedge replacement rules to generate the string language $a^n b^n c^n$.

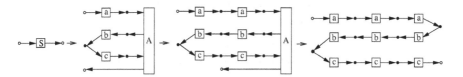

Fig. 4. The derivation of $a^3 b^3 c^3$ using the grammar given in Fig. 3.

When replacing a hyperedge with a graph as given in a production, the edge is first removed from its original host graph. In the resulting hole, the right hand side graph of the production is inserted except for its external nodes. The attaching nodes of the tentacles of the removed hyperedge are used instead: the ith source or target node of the removed edge replaces the ith external source or target node of the right hand side hypergraph. The derivation for $a^3 b^3 c^3$ is given in Figure 4 using the productions given in Figure 3. In Figure 5 the corresponding derivation tree is shown.

A **context–free hyperedge replacement grammar** $G = (T, N, P, S)$ consists of

- a finite set of terminal edge labels T
- a finite set of nonterminal edge labels N
- a finite set of productions P where each production has one hyperedge labeled with a nonterminal label on its left hand side
- a starting hyperedge labeled S.

The productions given in Figure 3 are part of a hypergraph grammar with terminal symbols $\{a, b, c\}$, nonterminal symbols $\{A, S\}$ and start symbol S.

A context–free hyperedge replacement grammar is a **string generating grammar** if the language generated by the grammar consists only of string graphs. This is the case in Figure 3. $a^n b^n c^n$ cannot be generated by a context–free Chomsky grammar.

Please note that for the rest of this paper we assume that the string generating hypergraph grammars are reduced, cycle–free and ϵ–free [9]. There are no unconnected nodes. If a hyperedge replacement grammar generates a string language, the start symbol must have one source and target tentacle. Each node

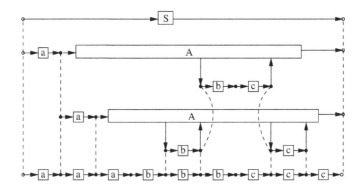

Fig. 5. The derivation tree for the derivation given in Fig. 4. The dotted lines mark the nodes must be matched.

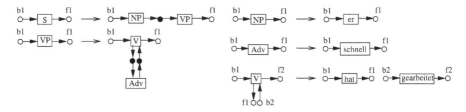

Fig. 6. Hyperedge replacement productions for *Er hat schnell gearbeitet. (He (has) worked fast).*

is connected to at most one source tentacle and one target tentacle; otherwise no string language will be generated [10].

In the next section, a natural language example for context–free string generating hyperedge replacement grammars is provided.

3   A Natural Language Example

The productions shown in Figure 6 are necessary to generate example (1) *Er hat schnell gearbeitet. (He (has) worked fast.).* Most of them resemble the usual grammar productions used in Chomsky grammars for natural languages. $S \to NP\ VP$ splits a sentence S into a noun phrase (NP) and a verbal phrase (VP). In our example, the noun phrase becomes the personal pronoun *er (he).* The verbal phrase has two parts, an adverb *Adv* and the verb V. This is the most interesting rule, since the use of hypergraphs becomes evident. With their help, it is possible to split the verb V into two parts. The two parts are separated by the adverb. On the right hand side of the corresponding rule, the hyperedge labeled V modeling the verb has two source nodes and two target nodes. Figuratively spoken, one enters the verbal phrase through $b1$, leaves it for the adverb through $f1$, reenters for the second half of the verbal phrase through $b2$, and leaves again through $f2$. Finally, the verb has to be derived into its two halves, the auxiliary *hat* (*has*) and the past participle *gearbeitet* (*worked*). *Adv* is derived into *schnell* (*fast*).

4   Related Work

There are a lot of occurences of discontinuous constituents in natural languages like parenthical placement, right node raising, relative clause extraposition, scrampling and heavy NP shift. It is not possible to ignore these phenomena or to always use a work-around. A formalism is needed that can handle these problems. The most famous German treebank, *the Negra corpus* [11] and the largest English treebank, the Penstate treebank [12], contain notations for discontinuous constituents. Several proposals have been made on how to extend context-free grammars. Chomsky himself suggested transformations that move constituents around in addition to grammars [13]. Transformations should help to eliminate the need for discontinuous constituents. Other formalisms like *Lexical Functional Grammar* [14] have a context-free backbone extended with feature structures. Unification of feature structures after parsing ensures that discontinuous constituents that belong together are found. This is called *functional uncertainty*. Nevertheless, discontinuous constituents have to be split into seperate parts within the context-free backbone. An overview on discontinuous constituents in HPSG (*Head-Driven Phrase Structure Grammar*) using a similar mechanism can be found in [15]. Other approaches based on phrase structure grammars separate word order from the necessary constituents [16]. These ideas have been extended in [17] where the *discontinuous phrase structure grammar* is formally defined. As an extension for *Definite Clause Grammars*, *Static Discontinuity Grammars* are presented in [18]. In this approach, several rules can be applied in parallel. There may be gaps between the phrases generated by the rules. This way, several rules can handle one discontinuous constituent in parallel. Compared to all these approaches, string generating hypergraph grammars have one advantage: the discontinuous constituent is modelled with one symbol! The only implemented parser for general hypergraph grammars is described in [19]. This parser is similar to the Cocke–Kasami–Younger parser [16] for general context-free Chomsky grammars and embedded into *Diagen*, a diagram editor generator.

5   Earley-Parsing

The Earley parser for general context-free string grammars was first presented in [4] and is now widely used and has been extended in several ways [9] within natural language processing. It implements a top-down search, avoiding backtracking by storing all intermediate results only once.

When parsing with string grammars, positions at the beginning of the string, between the letters and at the end of the string to be parsed are numbered. When parsing *aabbcc* seven positions are necessary, ranging from 0 to 6. From position

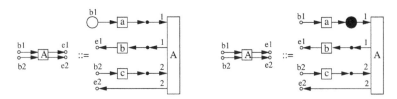

Fig. 7. Two examples for "dotted" rules using a grammar rule shown in Fig. 3.

0 to position 1 the first a can be found, from 1 to 2 the second a, etc. The last letter c lies between 5 and 6. This numbering scheme is easily transferred onto string hypergraphs; the nodes in the hypergraph are numbered from 0 to 6 as done at the bottom of Fig. 2. These position numbers are necessary for the main data structure of the Earley algorithm, the **chart**. When parsing a string $s_0 s_1 \ldots s_{n-1}$ consisting of n letters, the chart is a $(n+1) \times (n+1)$ table. In our running example we have a 7×7 table.

In this table, sets of **chart entries** are stored. Entries are never removed from the chart and are immutable after creation. A chart entry at position (i, j) in the chart contains information about the partial derivation trees for parsing the substring $s_i \ldots s_{j-1}$. This information consists of the currently used grammar production and information about the progress made in completing the subtree given by this production. For string grammars, chart entries are visualized by so called **dotted rules**, where the dot marks the parsing progress. If the dot is at the end of the rule's right hand side, the chart entry is finished or **inactive**, else, it is **active**, i.e. ready to accept a terminal or an inactive chart entry. This concept can easily be transferred to hypergraphs. Two dotted hypergraph rules are shown in Fig. 7. The node larger than the others symbolizes the **dot**. It marks what parts of the right hand side of the rule have already been found. On the left side of Fig. 7, nothing has been found yet, as the dot is at a external source node of the graph. On the right side, a has already been found, the next symbol that has to be found is A. If the dot, also called the current node, is one of the external target nodes, the edge is **inactive**, otherwise it is **active**. Both rules shown in Fig. 7 are active.

The classical Earley algorithm consists of three steps that alternate until the possibilities to apply one of them are exhausted. These steps are called `shift`, `predict` and `complete`. The algorithm processes one terminal of the input string from left to right at a time by applying the `shift` operation. `shift` extends all active chart entries ending at the position of the new terminal symbol and expect this terminal symbol (it follows their dot). The extended chart entries are added to the chart. `complete` performs the analogous operation for inactive chart entries. Since an inactive chart entry represents a complete derivation subtree, active chart entries that expect the left hand side symbol of the inactive entry can be extended, advancing the dot by one position. `predict` is applied whenever an active entry is inserted into the chart. If the inserted entry expects a nonterminal symbol, `predict` inserts new chart entries starting at the active entry's ending position and carrying the expected symbol on their rule's left hand side. If these prediction entries become inactive during parsing, the prediction was correct and they will be used to `complete` the chart entry for which they were predicted. A successful parse has been found when a chart entry with the grammar's start symbol S at its rule's left hand side has been inserted into position $(0, n)$ (where n is the length of the input string).

Our variant of the Earley algorithm consists of modifications of these three main steps (procedure names differ slightly to avoid misunderstandings), and is thus very similar to the classical Earley algorithm. The main difference lies in the role of inactive chart entries. In the classical algorithm, an inactive chart entry is always **finished**, i.e. all right hand side symbols have been matched to a part of the input string. In our algorithm, a chart entry becomes inactive when the current node dot reaches an external target node of the right hand side hypergraph. While this inactive chart entry will already be used for the `complete` step, the chart entry is not necessarily finished, as there may be several external source and target nodes. `predict` will insert **continuation entries**, active entries that restart the parsing of an inactive entry through another external source node, if it encounters a nonterminal hyperedge over which the dot already has stepped before.

In our extension of the Earley algorithm, a **chart entry** e consists of the following information:

- **rule(e)**, the hypergraph rule that is used
- **currentnode(e)**, the dot inside the rule's right hand side
- **entrynode(e)**, one of the external source nodes of the rule
- **from(e)**, the index of the first symbol of the input string that is covered by this chart entry.
- **to(e)**, the index of the last covered symbol of the input string plus one
- **predecessor(e)**[4], the active chart entry extended to create e, or null
- **continuation-of(e)**[5], an inactive chart entry that has been continued by this edge or one of its predecessors, or null
- **parts(e,h)**, for any hyperedge h that is part of the rule's right hand side, either a chart entry describing a derivation of this edge, or null

The entry node is the external source node used to "enter" the hypergraph during parsing. In Fig. 7 the entry node is both times $b1$. Both *from* and *to* are integers between $0 \ldots n$ where n is the length of the input string. The substring ranging from s_{from} to s_{to-1} has been matched to a path in the hypergraph between the entry node and the current node. E.g. the chart entry from 3 to 5 encompasses $s_3 s_4$ of the string to be parsed.

parts(e,h) is defined for hyperedges h that are part of the right hand side hypergraph of *rule(e)* and for which some derivation has been found during the parsing process up to now. Its value is the terminal or the chart entry representing the derivation of this edge. It is set during the `complete` step.

---

[4] Actually, only either *parts* or *predecessor* is necessary for the algorithm. We use *parts* for algorithmic purposes, but include *predecessor* for the visualization of the chart.

[5] 'of' refers to the function's value, not its parameter.

We will now introduce an Earley-style algorithm for parsing strings generated by a hypergraph grammar with the properties mentioned at the end of section 2. While we explain the algorithm in detail, a running example will be provided in Figures 8 to 14. The example visualizes all chart entries created when parsing *aabbcc* with the grammar given in Fig. 3. The chart entries are numbered for convenience. For each chart entry *e*, first *from(e)* and *to(e)* is given, followed by *rule(e)* and *entrynode(e)*. *currentnode(e)* is given indirectly by drawing this node larger than the others. The current node is set to the entry node of the right hand side in the example. The brackets () following each terminal or nonterminal edge *h* represent *parts(e,h)*. These brackets are empty in Fig. 8 if no parts have been found yet. The last three columns are useful to visualize how a chart entry is created: *continuation-of* points to the inactive chart entry of which the current entry is a continuation entry. *predecessor* points to the active chart entry from which the current entry has been grown in the **complete** step. In the last column, the operation that created the chart entry is stated.

The main procedure, **parse**, returns all possible derivation trees for a given string of terminals and the start symbol S.[6]

```
procedure parse:
  returns: trees, a set of derivation trees
  parameters: S, a nonterminal label
              input, a string of terminal symbols
    trees := {}
    for all rules r where label(left-hand-side(r)) = S
        p := initial-prediction(r)
        if p is not in chart[0, 0]
          insert p into chart[0, 0]
          predict-for(p)
    for i = 0 .. length(input)-1
        shift(i, input[i])
    for all inactive entries e in chart[0, length(input)]
        if label(left-hand-side(rule(e))) = S
          insert generate-tree(e) into trees
    return trees
```

Since we are using a top-down parsing approach, it is necessary to recursively predict the leftmost parts of possible derivation trees, starting with the start symbol S, and using the procedure **predict-for** (detailed below).

initial-prediction(r) creates a chart entry from 0 to 0 using rule *r*. Its current node is set to the only possible entry node, since S is of the type $(1, 1)$. *parts* is null. In Fig. 8, two entries are inserted for the two S rules of the grammar. *predecessor* and *continuation-of* are null.

The main part of parsing happens in the **shift**-loop: one terminal symbol at a time is used to complete existing, active chart entries. Further predictions and completions are performed recursively, as detailed below in the procedure **shift**. Finally, each derivation tree we found during a successful parse is represented

---

[6] The indentation of the pseudo-code marks the end of loops and alternatives.

from to	rule with current node and parts	entry node	conti nuation of	prede cessor

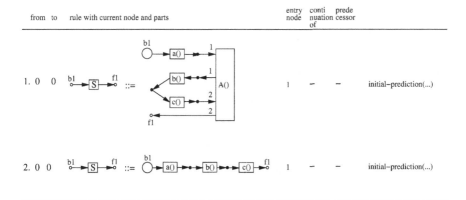

Fig. 8. Chart entries for strings ending at node 0 when parsing *aabbcc*.

by a chart entry with the label S from 0 to n. The corresponding derivation tree is trivially built by recursion on the entry's *parts*.

Next, the function `predict-for` is explained:

```
procedure predict-for:
  returns: nothing
  parameters: e, an active chart entry
    h := hyperedge following currentnode(e)
    if label(h) is terminal label, abort
    if parts(e,h) is defined
        // we have reached a hyperedge that has already been traversed
        c := generate-continuation(parts(e,h),e)
        if c is not in chart[to(e), to(e)]
            insert c into chart[to(e), to(e)]
            predict-for(c)
    else
        for all rules r where label(left-hand-side(r)) = label(h)
          c := generate-prediction(e,r)
          if c is not in chart[to(e), to(e)]
            insert c into chart[to(e), to(e)]
            predict-for(c)
```

`predict-for` inserts **prediction entries** for an active chart entry e that expects a nonterminal symbol. What e expects to parse next is determined by looking at the hyperedge h following the dot. In the case of a terminal, no prediction is necessary, since the entry will be completed if the matching terminal is shifted.

If h has not been used for parsing before *(parts(e,h) = null)*, we proceed analogous to the classical Earley algorithm:

The function **generate-prediction** returns a new active chart entry c that may become the root of a possible derivation tree of the hyperedge following

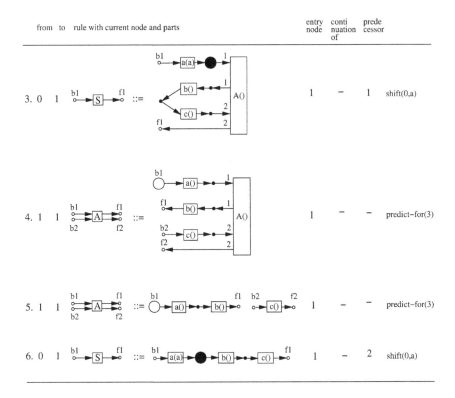

from	to	rule with current node and parts	entry node	continuation of	predecessor	
3. 0	1		1	–	1	shift(0,a)
4. 1	1		1	–	–	predict–for(3)
5. 1	1		1	–	–	predict–for(3)
6. 0	1		1	–	2	shift(0,a)

Fig. 9. Chart entries for strings ending at node 1 when parsing *aabbcc*.

currentnode(e), using the rule r. No part of it has yet been matched to the input string. c starts and ends at *to(e)*, *rule(c)* := r, *currentnode(c)* is set to the begin node that corresponds to *currentnode(e)*, using the unique mapping between a hyperedge's nodes and its replacement hypergraph's nodes. *parts(c,x)* is undefined for all hyperedges x. *predecessor* and *continuation-of* are null. This entry is inserted into the chart.

In Fig. 9, two such prediction entries are shown. In chart entry 3, after having shifted over the first a in the input string, the hyperedge following the current node is labeled A. Two chart entries must be predicted for the two rules with left hand side A.

A phenomenon that is new compared to parsing with context-free Chomsky grammars (e.g. the classical Earley algorithm) is that multiple, separate substrings may form a derivation of the same nonterminal symbol. In order to cope with this, we introduce the concept of a **continuation chart entry**. When, during parsing, the current node reaches a hyperedge of the right hand side hypergraph to which a portion of the input string has already been matched, we predict an active chart entry that is consistent with the last chart entry that represented this nonterminal. **generate-continuation** returns such a contin-

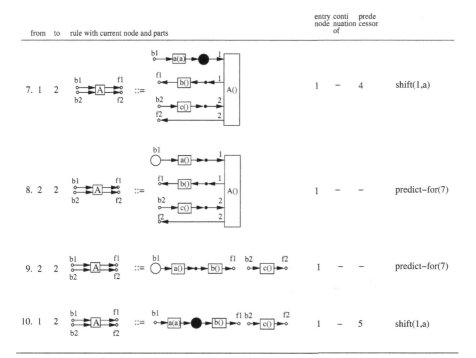

Fig. 10. Chart entries for strings ending at node 2 when parsing *aabbcc*.

uation entry that represents an already partially matched possible derivation of h. This nonterminal hyperedge has already been "entered" once through a source node and has been "left" again through a target node. Now this hyperedge is "reentered" through another source node. The chart entry c returned by *generate-continuation(p,e)* starts and ends at *to(e)*, *rule(c)* := *rule(p)*, $\forall x$ *parts(c,x)* := *parts(p,x)*, *predecessor(c)* := *null*, c is a continuation of p. *currentnode(c)* is determined the same way as above.

In Fig. 13, the chart entry 15 is predicted from the chart entry 14. In chart entry 14 the hyperedge A is reentered through its second source node. $A(11)$ states that this hyperedge has been handled before in chart entry 11, whose continuation will now be predicted.

Next, **shift** and **complete-with** are explained:

```
procedure shift:
  returns: nothing
  parameters: position, a non-negative integer
              t, a terminal symbol
    for all active chart entries where to(e) = position
      h := hyperedge following currentnode(e)
      if(label(h) = t)
          insert completion(e,t) into chart[from(e), to(e)+1]
```

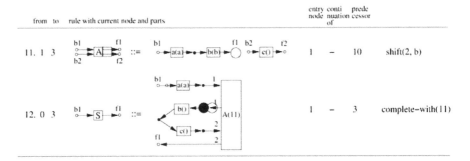

Fig. 11. Chart entries for strings ending at node 3 when parsing *aabbcc*.

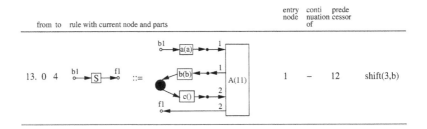

Fig. 12. Chart entries for strings ending at node 4 when parsing *aabbcc*.

```
if completion(e,t) is inactive
    complete-with(completion(e,t))
else
    predict-for(completion(e,t))


procedure complete-with:
 returns: nothing
 parameters: ia, an inactive chart entry
   for all active entries e where to(e) = from(ia)
       if expects(e,ia)
           insert completion(e,ia) into chart[from(ia), to(ia)]
           if completion(e,ia) is inactive
               complete-with(completion(e,ia))
           else
               predict-for(completion(e,ia))
```

shift and complete-with perform similar operations: they try to extend active chart entries with a new part. If this completion is possible (see below), the resulting chart entry represents a larger part of the input string and is inserted into the chart.

If the completed entry is inactive, other chart entries can be completed with it. If it is active (expecting more input), we insert prediction entries.

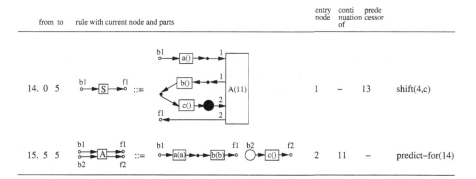

Fig. 13. Chart entries for strings ending at node 5 when parsing *aabbcc*.

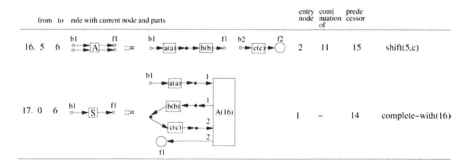

Fig. 14. Chart entries for strings ending at node 6 when parsing *aabbcc*.

The function `expects(e,ia)` is extended compared to the original Earley algorithm. `expects(e,ia)` determines if a given inactive edge *ia* will be accepted for completion of *e*. Please note that parsing of an inactive edge is not necessarily finished; an edge is inactive if the current node, the dot, has reached a target node of the rule. If the label or type of the left hand side of *ia*'s rule differs from *e*'s expected nonterminal edge label or type, *expects(e,ia)* is false. If the node used to enter *ia*, *entrynode(ia)*, does not correspond to *currentnode(e)*, *expects(e,ia)* is false. And if *ia* is a continuation chart entry, but the inactive entry that has been continued does not match *parts(e,h)*, *ia* represents a different derivation of the hyperedge than the one we assumed the last time it was traversed; therefore, *expects(e,ia)* is false. It is true otherwise.

The function `completion(e,x)` creates a new chart entry, either by accepting a terminal symbol *x*, or by accepting an inactive edge *x* into the partial derivation tree. Let *h* be the hyperedge following the current node inside *e*. The new chart entry *c := completion(e,x)* is identical to *e*, except for the following modifications: The substring of the input covered by *c* reaches from the start of the active entry *e*, *from(e)*, to the end of the inactive entry *x*, *to(x)*, or to *to(e)+1* for a terminal label *x*. *predecessor(c)* is set to *e*. *parts(c,h)* is set to *x*, i.e. *c*'s derivation consists of the same subtrees as e's except for the new part x. If x is a continuation edge, *parts(e,h)* is already defined; the change of definition is intended, since all information held by *parts(e,h)* is also held by x. Furthermore, the current node pointer is advanced over h, using *currentnode(x)* to determine the correct target node of h, unless x is a terminal symbol. This is possible because of the unique mapping between the source and target nodes of h and the source and target nodes of the replacement hypergraph.

In Fig. 9, the `shift` over the first a of the string to be parsed is shown. The dot moves over the hyperedge labeled a, the entrynode is still 1 and the predecessor is the first chart entry. In Fig. 11, a `complete-with` is shown. In chart entry 12, the hyperedge labeled A is completed the first time. The inactive chart entry 11 is taken by the active chart entry 3.

The complexity of our algorithm is comparable to the underlying, classical Earley algorithm, $O(n^3)$. The usage of more complicated data structures introduces a constant penalty factor on space and time complexity in several places (or a factor of $O(k)$ where k is the maximum number of right hand side hyperedges in a grammar; but we regard k as a constant). Since the only point in our algorithm where it distinctively differs from classical Earley, aside from transferring its concepts onto string generating hypergraphs grammars, is the prediction of continuation entries, and only one such continuation entry is inserted during prediction (instead of several prediction entries), the complexity class of the algorithm itself does not increase [10].

6   Conclusion

String generating hypergraph grammars are a theoretical concept introduced in [5]. Context-free hyperedge replacement can model string languages that are not context-free in the usual Chomskian sense. In this paper an Earley-based parser was presented for string generating hypergraph grammars. The major extensions compared to the original Earley algorithm are the introduction of inactive chart entries that can be activated again. This is the case when a hypergraph was "entered" through one external source node, "left" through an external target node and "reentered" again through a new external source node. The parser described has been implemented in Java [10]. A German grammar and lexicon is currently developed.

This parser can be extended in may ways as shown in [9]. First, it is interesting to add an agenda and implement bottom-up parsing or parsing with probabilities. For linguistic applications, it is useful to attribute hyperedges and hypergraphs with feature structures that are combined with unification.

References

1. Trask, R.: A dictionary of Grammatical Terms in Linguistics. Roudledge, New York, London (1993)

2. Nolda, A.: Gespaltene Topikalisierung im Deutschen. In: Bericht des III. Ost-West-Kolloquiums für Sprachwissenschaft, Berlin, Germany (2000)
3. Fischer, I.: Modelling discontinuous constituents with hypergraph grammars. In . L. Pfaltz, M. Nagl, B.B., ed.: Applications of Graph Transformation with Industrial Relevance Proc. 2nd Intl. Workshop AGTIVE'03. Volume 3062 of Lecture Notes in Computer Science., Charlottesville, USA,, Springer Verlag (to appear)
4. Earley, J.: An efficient context-free parsing algorithm. Communications of the ACM **13** (1970) 94–102
5. Habel, A.: Hyperedge Replacement: Grammars and Languages. Volume 643 of Lecture Notes in Computer Science. Springer-Verlag, Berlin (1992)
6. Drewes, F., Habel, A., Kreowski, H.J.: Hyperedge replacement graph grammars. In Rozenberg, G., ed.: Handbook of Graph Grammars and Computing by Graph Transformation. Vol. I: Foundations. World Scientific (1997) 95–162
7. Engelfriet, J., Heyker, L.: The string generating power of context-free hypergraph grammars. Journal of Computer and System Sciences **43** (1991) 328–360
8. Engelfriet, J., Heyker, L.: Context–free hypergraph grammars have the same term-generating power as attribute grammars. Acta Informatica **29** (1992) 161–210
9. Jurafsky, D., Martin, J.H.: Speech and Language Processing: An Introduction to Natural Language Processing, Computational Linguistics, and Speech Recognition. Prentice Hall (2000)
10. Seifert, S.: Ein Earley–Parser für Zeichenketten generierende Hypergraphgrammatiken. Studienarbeit, Lehrstuhl für Informatik 2, Universität Erlangen–Nürnberg (2004)
11. Brants, T., Skut, W.: Automation of treebank annotation. In Powers, D.M.W., ed.: Proceedings of the Joint Conference on New Methods in Language Processing and Computational Natural Language Learning: NeMLaP3/CoNLL98. Association for Computational Linguistics, Somerset, New Jersey (1998) 49–57
12. Marcus, M.P., Santorini, B., Marcinkiewicz, M.A.: Building a large annotated corpus of english: The penn treebank. Computational Linguistics **19** (1994) 313–330
13. Chomsky, N.: Syntactic Structures. The Hague: Mouton (1957)
14. Kaplan, R.M., Maxwell, III, J.T.: An algorithm for functional uncertainty. COLING-88 (1988) 297–302
15. Müller, S.: Continuous or discontinuous constituents? a comparison between syntactic analyses for constituent order and their processing systems. Research on Language and Computation, Special Issue on Linguistic Theory and Grammar Implementation **2** (2004) 209–257
16. Naumann, S., Langer, H.: Parsing — Eine Einführung in die maschinelle Analyse natürlicher Sprache. Leifäden und Monographen der Informatik. B.G. Teubner, Stuttgart (1994)
17. Bunt, H.: Formal tools for the description and processing of discontinuous constituents. In: Discontinuous Constituency. Mouton De Gruyter (1996)
18. Dahl, V., Popowich, F.: Parsing and generation with static discontinuity grammars. New Generation Computers **8** (1990) 245–274
19. Minas, M.: Spezifikation und Generierung graphischer Diagrammeditoren. Shaker-Verlag, Aachen (2001)

Composition of Path Transductions

Tanguy Urvoy

Université Paul Sabatier, IRIT, Batiment 1R1,
118 route de Narbonne - 31062 Toulouse cedex 4
urvoy@irit.fr
http://www.irit.fr/recherches/TYPES/SVF/urvoy/

Abstract. We propose to study two infinite graph transformations that we respectively call *bounded* and *unbounded path transduction*. These graph transformations are based on path substitutions and graph products. When graphs are considered as automata, path transductions correspond to rational word transductions on the accepted languages. They define strict subclasses of monadic transductions and preserve the decidability of monadic second order theory.
We give a generalization of the Elgot and Mezei composition theorem from rational word transductions to path transductions.

Introduction

As a theoretical model for systems verification, the study of infinite transitions graphs has become an active research area. The most fundamental dynamic behavior of these (oriented, labelled and simple) graphs is the language of their path labels: a class of automata corresponds to each class of graphs.

To be interesting for formal verification, an infinite graph is also expected to have some decidable properties like reachability or a decidable monadic second order theory. These decidability results for graphs are often strongly connected with the classes of languages recognized by the corresponding automata: graphs hierarchies share many properties with classical languages hierarchies [16].

A standard way to show a property on a given infinite structure is to define this structure by transformations from a known generator. If the transformations used are well suited, the property will be inherited from the generator. A transformation which preserve a given property is said to be *compatible* with this property.

For example, *monadic transductions* [7] and *unfolding* [8] are compatible with monadic second order logic: any graph which is built by these transformations from a finite graph will have a decidable monadic theory [5]. Monadic second order logic is very expressive and thus undecidable for many graphs. For these graphs one must resign oneself to weaker decidability results like reachability or reachability under rational control. To study such properties, one must consider graph transformations which are strictly weaker than monadic transductions.

This article is the continuation of [17] which extended to infinite graphs the classical language theoretic notion of *abstract family* [11, 12]. Many families of

H. Ehrig et al. (Eds.): ICGT 2004, LNCS 3256, pp. 368–382, 2004.
© Springer-Verlag Berlin Heidelberg 2004

graphs, like higher order pushdown graphs [5], Petri nets transitions graphs or automatic graphs [3] can be characterized by using a common and restricted set of simple graph transformations from specific generating systems [18].

Abstract families are a way to build generic proofs: we consider *families of graphs* without specifying what the graphs are. We only require that the families are closed under certain operations. In language theory, rational word transductions [12, 2] are one of these fundamental operations.

Rational transductions have been naturally extended from words to trees [15] but no convenient automaton based definition of graph transduction have been given. In this article we study two graph transformations that we respectively call *bounded path transduction* and *unbounded path transduction*. These transformations are not defined in terms of finite automaton nor in terms of logic but rather in terms of path substitution, path morphism and graph product. They define strict subclasses of *monadic transductions* [7] which preserve the decidability of monadic second order theory but their principal characteristic is to respect the orientation of edges and thus being *compatible* with language transformations.

Given a path transduction T, we can build a word rational transduction $\hat{T}$ such that for any automaton $\mathcal{G}$ (finite or not) we have:

$$L(T(\mathcal{G})) = \hat{T}(L(\mathcal{G})) \ .$$

Conversely, given a rational word transduction R, we easily build a path transduction T verifying $R = \hat{T}$.

The most important and new result is the generalization of the Elgot and Mezei composition closure theorem from rational word transductions to path transductions.

This article is divided into four parts : the first part recalls some basic definitions about graphs, automata and rational word transductions; the second one introduces the notion of *(language) compatible* transformation and gives a formal definition of path transductions; the third one is devoted to bounded path transductions and the last to unbounded path transductions.

1   Preliminaries

1.1   Graphs and Automata

We use the standard definition of graphs and automata but here the state space is not supposed to be finite.

Definition 1. *A Σ-Γ-graph structure (or more simply a Σ-graph or a graph) is a tuple $\langle \Sigma, \Gamma, Q, (R_a)_{a \in \Sigma}, (P_b)_{b \in \Gamma} \rangle$ where:*

- *Σ is an alphabet of edge labels;*
- *Γ is an alphabet of states labels;*
- *Q is a countable set of states (or vertices);*
- *for each label $a \in \Sigma$, $R_a \subseteq Q \times Q$ is the relation labelled by a;*
- *for each label $b \in \Gamma$, $P_b \subseteq Q$ is the unary predicate (the set) labelled by b;*

Recall that the composition of two relations R and S on Q is the relation

$$R \cdot S = \{(p, q) \in Q \times Q \mid \exists r \in Q, \ pRr, rSq\} \ .$$

The mapping which associates to each symbol $a \in \Sigma$ its relation R_a is extended to words according to the following rules:

$$
\begin{aligned}
R_\varepsilon &= \{(p, p) \mid p \in Q\} && (\varepsilon \text{ denotes the empty word}) \\
R_{au} &= R_a \cdot R_u && a \in \Sigma, u \in \Sigma^* \ .
\end{aligned}
\tag{1}
$$

We will also consider the relation labelled by a set L of words:

$$R_L = \{(p, q) \mid \exists w \in L, \ pR_w q\} \ .$$

The *path label language* of $\mathcal{G}$ between two sets of vertices A and B is the set of finite words

$$L(\mathcal{G}, A, B) = \{w \in \Sigma^* \mid R_w \cap A \times B \neq \emptyset\} \ .$$

An *automaton* is a graph structure with two unary predicates : P_i for the initial states and P_f for the final ones. The language $L(\mathcal{A})$ of an automaton $+$ is the path label language $L(\mathcal{A}, P_i, P_f)$.

1.2   Word Transductions

Rational word transductions are fundamental tools for the study of formal languages. We give here some basic definitions of this concept. See [1, 2, 12] for further details.

Definition 2. *A rational transduction between Σ_1^* and Σ_2^* is a rational subset of the monoid $\Sigma_1^* \times \Sigma_2^*$ for the canonical concatenation $(u, v) \cdot (w, x) = (uw, vx)$.*

A rational transduction is usually defined, as in Figure 1, by a finite automaton, called a transducer, which is labelled by pairs of words. An equivalent way to define such a relation is to give its morphism decomposition:

Property 1. Any rational transduction R is the composition of an inverse morphism f^{-1}, an intersection with a rational language K, and a direct morphism g:

$$R = f^{-1} \cdot (\cap K) \cdot g = \{(f(w), g(w)) \mid w \in K\} \ .$$

A relation R between Σ_1^* and Σ_2^* is *faithful* when for any $w \in \Sigma_2^*$, $R^{-1}(w)$ is finite ; it is *continuous* when $R^{-1}(\{\varepsilon\}) = \{\varepsilon\}$. These two properties are important for the study of *quasi-real-time acceptors* [10] because a transduction which is both faithful and continuous is unable to erase more than a bounded number of letters.

We give here a representation of faithful and continuous rational transductions which was given in [4]. Recall that a morphism α from Σ_1^* to Σ_2^* is *strictly alphabetic* when $\alpha(a) \in \Sigma_2$ for any $a \in \Sigma_1$.

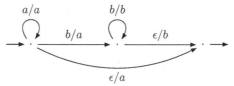

Fig. 1. A transducer recognizing the function $a^n b^m \mapsto a^{n+1} b^m$.

Theorem 1. *(Boasson and Nivat) A rational transduction R between Σ^* and Σ_2^* is faithful and continuous if and only if there exists an alphabet Π, a rational set $K \in Rat(\Pi^*)$, a morphism f from Π^* to Σ^* and a strictly alphabetic morphism α from Π^* to Σ_2^* such that*

$$R = \{(f(w), \alpha(w)) \mid w \in K\} \ .$$

The Elgot and Mezei composition theorem is an important property of rational transductions.

Theorem 2. *(Elgot and Mezei)*
The rational transductions are closed under composition.

A proof of this result can be found in [1] or [2].

Because faithfulness and continuity are preserved by composition we have the following corollary.

Corollary 1. *Faithful and continuous rational transductions are closed by composition.*

Rational transductions and their closure by composition are fundamentals in the theory of *abstract families of languages* (AFL). Figure 2 gives some of the most usual closure operators and the names of the corresponding abstract families. See [12, 2] or [10] for further details on this theory.

2   Language Compatible Graph Transformations: Path Transductions

We study graph transformations which are independent of the vertices naming convention or, in other words, invariants with respect to graph isomorphism. Another fundamental restriction is that we only consider graph transformations that are "compatible" with language transformations.

Definition 3. *A graph transformation T of arity n is compatible (with language transformations) if there exists a language transformation $\hat{T}$ of arity n such that for any sequence $(\mathcal{A}_i)_{1 \leq i \leq n}$ of automata, we have*

$$L(T(\mathcal{A}_1, \ldots, \mathcal{A}_n)) = \hat{T}(L(\mathcal{A}_1), \ldots, L(\mathcal{A}_n)) \ .$$

Many graph transformations are impossible to translate into language transformations. Consider, for example, the simple transformation B consisting of adding a backward edge labelled by $\bar{a}$ for each edge labelled by an a. The non compatibility of this transformation is illustrated by Figure 3.

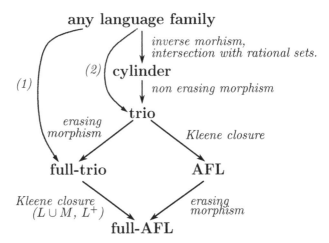

Fig. 2. Different abstract families of languages from the less constrained to the most constrained. (1) stands for rational transductions and (2) for faithful and continuous rational transductions.

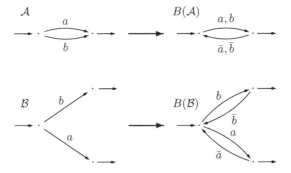

Fig. 3. The two finite automata $\mathcal{A}$ and $\mathcal{B}$ are both accepting the language $\{a, b\}$, but their respective images by B do not accept the same languages.

2.1   Path Morphisms and Path Substitutions

A morphism is a mapping which replace letters by words. More generally, a substitution is a relation which replaces letters by languages. Morphisms, inverse morphism, substitutions and inverse substitutions are well-known language transformations but these transformations can be generalized as graph/automata transformations.

Definition 4. *Let Σ_1 and Σ_2 be alphabets and $h \subseteq \Sigma_1 \times \Sigma_2^*$ be a relation.*

The substitution generated by h is the relation $\bigcup_{n \geq 0} h_n$ defined inductively by:

$$h_0 = \{(\varepsilon, \varepsilon)\}$$
$$h_{n+1} = \{(au, vw) \in \Sigma_1^{n+1} \times \Sigma_2^* \mid a \in \Sigma_1, v \in h(a), w \in h_n(u)\} .$$

We use the same symbol to denote the substitution $\bigcup_{n \geq 0} h_n$ and its generator h.

Example 1. If h is the rational substitution defined by $h(c) = a$ and $h(d) = (ab)^*$, the image of the word $cdcd$ by h is the rational language $a(ab)^*a(ab)^*$. By inverse we have $h^{-1}(abab) = \{w \mid abab \in h(w)\}$ hence $h^{-1}(abab) = d^+$.

When the relation is a function, which associates a unique word to each letter, the generated substitution is a morphism, when this image is a rational (resp. finite) language, the substitution is called rational (resp. finite).

Inverse rational substitution is a simple and well defined operation on graphs. Each path labelled by a word in the language $h(a)$ is replaced by an arc labelled by a. This path transformation do not add vertices to the graph.

Definition 5. *(Caucal [6])*
If h is a substitution from Σ_1^* to Σ_2^* and $\mathcal{G}$ is a graph, then $h^{-1}(\mathcal{G})$ is the graph $\left\langle \Sigma_1, \Gamma^{\mathcal{G}}, Q^{\mathcal{G}}, (R_a^{h^{-1}(\mathcal{G})})_{a \in \Sigma_1}, (P_b)_{b \in \Gamma^{\mathcal{G}}} \right\rangle$ where

$$R_a^{h^{-1}(\mathcal{G})} = R_{h(a)}^{\mathcal{G}} \text{ for each } a \in \Sigma_1 .$$

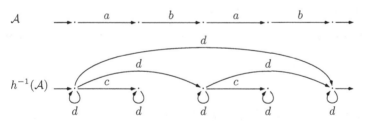

Fig. 4. An automaton $\mathcal{A}$ and its inverse rational substitution according to $h(c) = a$ and $h(d) = (ab)^*$. In other words $R_c^{h^{-1}(\mathcal{A})} = R_a^{\mathcal{A}}$ and $R_d^{h^{-1}(\mathcal{A})} = (R_a^{\mathcal{A}} \cdot R_b^{\mathcal{A}})^*$.

From the algebraic point of view this transformation is natural: the set of relations $\{R_w^{\mathcal{G}} \mid w \in \Sigma_2^*\}$ is a monoid for the composition operator. When h is rational (resp. finite), each relation $R_a^{h^{-1}(\mathcal{G})}$ is a rational (resp. finite) subset of this monoid. Figure 4 gives an example of inverse rational substitution. Note that the languages $L(h^{-1}(\mathcal{A}))$ and $h^{-1}(L(\mathcal{A}))$ coincide: inverse substitution is a compatible transformation.

Lemma 1. *Let $\mathcal{G} = \langle \Sigma_1, \Gamma, Q, (R_a)_{a \in \Sigma_1}, (P_b)_{b \in \Gamma} \rangle$ be a graph, let i, f be two letters of Γ and let g, h be substitutions between Σ_2^* and Σ_1^*.*

1. $\forall w \in \Sigma_2^*,\ R_w^{h^{-1}(\mathcal{G})} = R_{h(w)}^{\mathcal{G}}$;
2. $L(h^{-1}(\mathcal{G}), P_i, P_f) = h^{-1}(L(\mathcal{G}, P_i, P_f))$;
3. $g^{-1}(h^{-1}(\mathcal{G})) = (g \cdot h)^{-1}(\mathcal{G})$ (☞ *relational composition.*).

Recall that a monoid morphism α from Σ_1^* to Σ_2^* is *strictly alphabetic* when $\alpha(a) \in \Sigma_2$ for any $a \in \Sigma_1$. If α is strictly alphabetic, we write $\alpha(\mathcal{G})$ for the graph obtained by replacing each R_a by $R_{\alpha(a)}$. Direct strictly alphabetic mapping is, like inverse morphism, a special case of inverse finite substitution.

It is harder to give for graphs a definition of direct rational substitution that extend cleanly the language transformation. The most natural way, given a continuous[1] substitution h, is to use an *edge replacement* [9] to replace each edge of the graph labelled by a symbol a by the automaton $\mathcal{G}_a$ recognizing the language $h(a)$.

As proved in [17], such an (oriented) edge replacement may be simulated by using only finite inverse substitution and product.

2.2   Graph Products

The notion of graph product is well known in automata theory for its correspondence with the intersection of languages.

Definition 6. *Let $\mathcal{G}$ and $\mathcal{H}$ be two graphs. The product $\mathcal{G} \times \mathcal{H}$ is the graph*

$$\langle \Sigma^{\mathcal{G}} \cap \Sigma^{\mathcal{H}}, Q^{\mathcal{G}} \times Q^{\mathcal{H}}, (R_a)_{a \in \Sigma^{\mathcal{G}} \cap \Sigma^{\mathcal{H}}}, (P_b)_{b \in \Gamma^{\mathcal{G}} \cap \Gamma^{\mathcal{H}}} \rangle$$

where

$$R_a = \{((p,p'),(q,q')) \mid pR_a^{\mathcal{G}}q, p'R_a^{\mathcal{H}}q'\} \quad \text{for each } a \in \Sigma^{\mathcal{G}} \cap \Sigma^{\mathcal{H}} \text{ and}$$
$$P_b = P_b^{\mathcal{G}} \times P_b^{\mathcal{H}} \quad \text{for each } b \in \Gamma^{\mathcal{G}} \cap \Gamma^{\mathcal{H}} .$$

Lemma 2. *Let $\mathcal{G}$ and $\mathcal{H}$ be two graphs.*

1. *For any word w, we have:*

$$R_w^{\mathcal{G} \times \mathcal{H}} = \{((p,p'),(q,q')) \mid pR_w^{\mathcal{G}}q,\ p'R_w^{\mathcal{H}}q'\} \ ;$$

2. *for any couple of letters i, f:*

$$L(\mathcal{G} \times \mathcal{H}, P_i^{\mathcal{G}} \times P_i^{\mathcal{H}}, P_f^{\mathcal{G}} \times P_f^{\mathcal{H}}) = L(\mathcal{G}, P_i^{\mathcal{G}}, P_f^{\mathcal{G}}) \cap L(\mathcal{H}, P_i^{\mathcal{H}}, P_f^{\mathcal{H}}) \ ;$$

3. *For any morphism f, we have $f^{-1}(\mathcal{G} \times \mathcal{H}) = f^{-1}(\mathcal{G}) \times f^{-1}(\mathcal{H})$.*

---

[1] if $\varepsilon \in h(a)$ we need ε-transitions or vertex fusion.

2.3   Definition of Path Transductions

In [17], an *abstract family of graphs (AFG)* is defined as a graph family closed by inverse finite substitution and product by finite graphs. A *full-AFG* is an *AFG* which is also closed by inverse rational substitution. These two definitions only ask the considered family to be closed by these operations but a question was still remaining for *principal AFG* (resp. *principal full-AFG*). An *AFG* (respectively a *full-AFG*) is *principal* when all its elements are derived from a single graph (the generator) by a finite sequence of *AFG* (resp. *full-AFG*) operations.

Does the AFG transformations generate strictly increasing (for inclusion) chains of graph families ? If not, how many successive transformations are needed to obtain the whole family from its generator ?

The answer to the second question is three and the transformations obtained by composition of the three basic AFG transformations used are, as proved in the next sections, what we call *path transductions*. We distinguish two classes of *path transductions* : *bounded path transduction* which are only local transformations and *unbounded path transduction* which may act more globally on the graph.

Definition 7.

1. *A bounded path transduction is the composition of an inverse morphism, a product by a finite graph and a direct strictly alphabetic morphism;*
2. *an unbounded path transduction is the composition of an inverse rational substitution, a product by a finite graph and an inverse rational substitution.*

It is important to notice that the definition of *bounded path transductions* is a direct generalization to graphs of Boasson Nivat's characterization of faithful and continuous rational word transductions (Theorem 1). The only difference is that instead of using word morphisms, we use *path morphisms* and instead of using intersection with rational sets we use product with finite graphs.

The characterization of general rational word transductions with inverse substitution is less usual but easy to deduce from Property 1.

Property 2. Any rational word transduction is the composition of an inverse rational word substitution, an intersection by a rational set and an inverse rational substitution.

Theorem 3. *1. Given a word transduction t we can construct an unbounded path transduction T such that $\hat{T} = t$.*

2. *given a faithful and continuous word transduction t we can construct a bounded path transduction T such that $\hat{T} = t$.*

Proof:
The main proof is already given in [17]. For (2) we can use theorem 1.   □

2.4   Examples

Here is the example of a bounded transduction T, defined by a morphism f, a finite automaton $\mathcal{H}$ and a strictly alphabetic morphism α.

$$f : \begin{cases} A \mapsto a \\ B \mapsto \bar{a} \\ C \mapsto aa \\ D \mapsto \varepsilon \\ E \mapsto \bar{a} \end{cases}, \quad \mathcal{H} : \quad 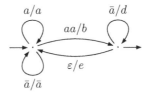 \quad , \text{ and } \quad \alpha : \begin{cases} A \mapsto a \\ B \mapsto \bar{a} \\ C \mapsto b \\ D \mapsto c \\ E \mapsto d \end{cases}.$$

This path transduction may also be represented directly, as in figure 5, by a finite transducer. The application of the path transduction is a kind of "left synchronized graph product".

Fig. 5. Another representation of T.

The one letter *Dyck graph*, illustrated on Figure 6 is a representation of the positive integers with $+1$ and -1 operations respectively encoded by the labels a and $\bar{a}$. This graph is a generator for the one counter automata transition graphs.

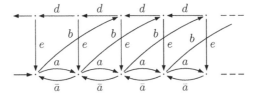

Fig. 6. The one letter Dyck graph Δ_1.

By applying T to Δ_1, we obtain the graph of figure 7.

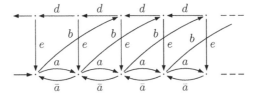

Fig. 7. The graph $T(\Delta_1)$.

Figure 8, give an example of unbounded path transduction and its application to Δ_1. The $\xrightarrow{\bar{a}^*/c}$ edge of the transducer leads to infinite degrees in the final graph.

The two letters *Dyck graph*, illustrated on Figure 9 is a binary tree with backward transitions. Its path label language from and to the "root" is the semi-Dyck language a generator of context-free languages but its "geometry" is also

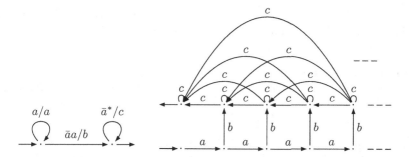

Fig. 8. An unbounded path transduction and its application to Δ_1.

generating, by path transductions, the pushdown graphs and prefix recognizable graphs which are the prefix transitions graphs of finite (resp. recognizable) word rewriting systems[2] [13, 6].

We can deduce the decidability of monadic second order theory for these graphs from the fundamental result of Rabin [14] about the binary tree. More examples like Petri nets transitions graphs or automatic graphs can be found in [18].

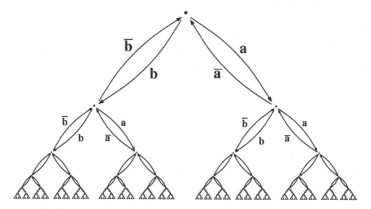

Fig. 9. A generator of pushdown AFG and prefix recognizable full-AFG. This graph has a decidable second order monadic theory [14].

3   Composition of Bounded Path Transductions

In this section we show that the classical Elgot and Mezei theorem naturally extends from faithful and continuous rational word transductions to bounded path transductions.

---

[2] The definition of [13] use a restriction to a reachable component and the one of [6] use a restriction to rational sets of vertices. We consider here the prefix rewriting graphs without these restrictions.

To establish this result, we use the basic properties of inverse morphism and product (Lemma 1 and Lemma 2) but the keys of the proof are the two following "delayed forgetting" lemmas.

Lemma 3. *If f is a morphism and α is a strictly alphabetic morphism, then we can construct a morphism g and a strictly alphabetic morphism β such that for any graph $\mathcal{G}$, we have*

$$f^{-1}(\alpha(\mathcal{G})) = \beta(g^{-1}(\mathcal{G})) .$$

Proof: We first remark that the relational composition $h = f \cdot \alpha^{-1}$, which verifies $h^{-1}(\mathcal{G}) = f^{-1}(\alpha(\mathcal{G}))$, is a finite substitution. For each letter $a \in Dom(h)$, we define the (finite) alphabet $\Sigma_a = \{[a, w] \mid w \in h(a)\}$ and take $\beta([a, w]) = a$ and $g([a, w]) = w$ to get $g(\beta^{-1}(a)) = h(a)$ □

In other words, the direct strictly alphabetic morphism, which forgets informations, may be applied after the inverse morphism.

Lemma 4. *If $\mathcal{H}$ is a graph and α is a strictly alphabetic morphism, then for any graph $\mathcal{G}$, we have*

$$\alpha(\mathcal{G}) \times \mathcal{H} = \alpha\left(\mathcal{G} \times \alpha^{-1}(\mathcal{H})\right) .$$

Proof: Let a be a label of $\mathcal{H}$.

$$
\begin{aligned}
R_a^{\alpha(\mathcal{G}) \times \mathcal{H}} &= \Big\{ ((p, p'), (q, q')) \mid p R_a^{\alpha(\mathcal{G})} q, \ p' R_a^{\mathcal{H}} q' \Big\} \quad \text{(by product def.)} \\
&= \Big\{ ((p, p'), (q, q')) \mid \exists b, \ a = \alpha(b), \ p R_b^{\mathcal{G}} q, \ p' R_a^{\mathcal{H}} q' \Big\} \quad \text{(by $\alpha(\mathcal{G})$ def.)} \\
&= \Big\{ ((p, p'), (q, q')) \mid \exists b, \ a = \alpha(b), \ p R_b^{\mathcal{G}} q, \ \exists c, \ c = \alpha(b), \ p' R_c^{\alpha^{-1}(\mathcal{H})} q' \Big\} \\
&= \Big\{ ((p, p'), (q, q')) \mid \exists b, \ a = \alpha(b), \ p R_b^{\mathcal{G}} q, \ p' R_a^{\alpha^{-1}(\mathcal{H})} q' \Big\} \quad \text{(because $c = a$)} \\
&= R_a^{\alpha\left(\mathcal{G} \times \alpha^{-1}(\mathcal{H})\right)}
\end{aligned}
$$

□

Theorem 4. *Bounded path transductions are closed by composition.*

Proof: Let T_1 and T_2 be two bounded path transductions. Our aim is to show that the relation $T = T_1 \cdot T_2$ remains a bounded path transduction. Let $\mathcal{G}$ be a graph and let $\mathcal{H} = T_2(T_1(\mathcal{G}))$ hence

$$\mathcal{H} = \alpha_2(f_2^{-1}(\alpha_1(f_1^{-1}(\mathcal{G}) \times \mathcal{F}_1)) \times \mathcal{F}_2)$$

where f_i are morphisms, $\mathcal{F}_i$ are finite graphs and α_i are strictly alphabetic morphisms. By Lemma 3, there exists a morphism g and a strictly alphabetic morphism β such that $\alpha_1 \cdot f_2^{-1} = g^{-1} \cdot \beta$, hence

$$\mathcal{H} = \alpha_2(\beta(g^{-1}(f_1^{-1}(\mathcal{G}) \times \mathcal{F}_1)) \times \mathcal{F}_2) .$$

By Lemma 2, g being a morphism, we have:

$$g^{-1}(f_1^{-1}(\mathcal{G}) \times \mathcal{F}_1) = g^{-1}(f_1^{-1}(\mathcal{G})) \times g^{-1}(\mathcal{F}_1) ,$$

hence

$$\mathcal{H} = \alpha_2(\beta(g^{-1}(f_1^{-1}(\mathcal{G})) \times g^{-1}(\mathcal{F}_1)) \times \mathcal{F}_2) \ .$$

By Lemma 1, the composition of the two inverse morphisms f_1^{-1} and g^{-1} is an inverse morphism: if we set $h = g \cdot f_1$ and $\mathcal{F} = g^{-1}(\mathcal{F}_1)$, then we have:

$$\mathcal{H} = \alpha_2(\beta(h^{-1}(\mathcal{G}) \times \mathcal{F}) \times \mathcal{F}_2) \ .$$

By Lemma 4, we may substitute $\beta(h^{-1}(\mathcal{G}) \times \mathcal{F}) \times \mathcal{F}_2$ by $\beta\left(h^{-1}(\mathcal{G}) \times \mathcal{F} \times \beta^{-1}(\mathcal{F}_2)\right)$. With $\mathcal{K} = \mathcal{F} \times \beta^{-1}(\mathcal{F}_2)$ and $\gamma = \alpha_2 \circ \beta$: for any graph $\mathcal{G}$

$$\alpha_2(f_2^{-1}(\alpha_1(f_1^{-1}(\mathcal{G}) \times \mathcal{F}_1)) \times \mathcal{F}_2) = \gamma(h^{-1}(\mathcal{G}) \times \mathcal{K}) \ \square$$

4   Composition of Unbounded Path Transductions

In this section we show that the classical Elgot and Mezei theorem naturally extends from rational word transductions to unbounded path transductions. The construction is similar as the one of Section 3 but, unlike inverse morphism, inverse rational substitution is not compatible with graph product. In general situations we have

$$h^{-1}(\mathcal{G} \times \mathcal{H}) \neq h^{-1}(\mathcal{G}) \times h^{-1}(\mathcal{H}) \ .$$

So we need a stronger "delay" property than Lemma 4. This property is true up to isolated vertices.

A vertex/state is *isolated* if it is not connected to an edge nor labelled by a predicate. More formally recall that the image of a relation $R \subseteq Q \times Q$ by a function f of domain Q is $f(R) = \{(f(p), f(q)) \mid pRq\}$. We say that two Σ-Γ-graphs $\mathcal{G}$ and $\mathcal{H}$ are *isomorphic up to isolated vertices*, and we write $\mathcal{G} \approx \mathcal{H}$, if there is an injective function f from $Q^{\mathcal{G}}$ to $Q^{\mathcal{H}}$ such that

$$\forall a \in \Sigma, \ f(R_a^{\mathcal{G}}) = R_a^{\mathcal{H}} \text{ and }, \ \forall b \in \Gamma, \ f(P_b^{\mathcal{G}}) = P_b^{\mathcal{H}} \ .$$

Lemma 5. *If $\mathcal{H}$ is a finite graph and h is a rational word substitution then there exist a rational substitution k, a morphism f and a finite graph $\mathcal{K}$ such that for any graph $\mathcal{G}$, we have*

$$h^{-1}(\mathcal{G}) \times \mathcal{H} \approx k^{-1}\left(f^{-1}(\mathcal{G}) \times \mathcal{K}\right) \ .$$

Proof: For any letter a appearing in $ran(h)$[3], we define a fresh symbol τ_a (used to denote an ε-transition). We define the alphabetic morphism f, by

$$\begin{cases} f(a) = a \\ f(\tau_a) = \varepsilon \end{cases} \text{ for any letter } a \text{ appearing in } ran(h).$$

We build the graph $\mathcal{K}$ from the graph $\mathcal{H}$ by edge replacement: each edge labelled by a is replaced by an automaton accepting the rational language $\tau_a h(a) \tau_a$.

---

[3] Recall that $ran(h) = \{v \mid \exists u, \ v \in h(u)\}$.

When $\varepsilon \notin h(a)$:

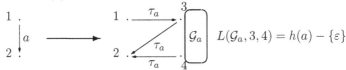

When $\varepsilon \in h(a)$:

In other words, we have a direct substitution $\mathcal{K} = k(\mathcal{H})$ with k defined by $k(a) = \tau_a h(a) \tau_a$ for all $a \in ran(h)$ (k is continuous). Note that for any couple of states $p, q \in Q^{\mathcal{K}}$, any label a and any word $u \in h(a)$, we have

$$pR_a^{\mathcal{H}}q \iff pR_{\tau_a u \tau_a}^{\mathcal{K}}q \tag{2}$$

and

$$pR_u^{\mathcal{G}}q \iff pR_{\tau_a u \tau_a}^{f^{-1}(\mathcal{G})}q \tag{3}$$

Let a be a label of $\mathcal{H}$:

$$
\begin{aligned}
R_a^{h^{-1}(\mathcal{G}) \times \mathcal{H}} &= \left\{ ((p,p'),(q,q')) \mid \exists u \in h(a),\ pR_u^{\mathcal{G}}q,\ p'R_a^{\mathcal{H}}q' \right\} \\
&= \left\{ ((p,p'),(q,q')) \mid \exists u \in h(a),\ pR_u^{\mathcal{G}}q,\ p'R_{\tau_a u \tau_a}^{\mathcal{K}}q' \right\} && \text{by (2)} \\
&= \left\{ ((p,p'),(q,q')) \mid \exists u \in h(a),\ pR_{\tau_a u \tau_a}^{f^{-1}(\mathcal{G})}q,\ p'R_{\tau_a u \tau_a}^{\mathcal{K}}q' \right\} && \text{by (3)} \\
&= \left\{ ((p,p'),(q,q')) \mid \exists u \in h(a),\ (p,p')R_{\tau_a u \tau_a}^{f^{-1}(\mathcal{G}) \times \mathcal{K}}(q,q') \right\} && \text{(Lemma 2)} \\
&= \left\{ ((p,p'),(q,q')) \mid \exists v \in k(a),\ (p,p')R_v^{f^{-1}(\mathcal{G}) \times \mathcal{K}}(q,q') \right\} && k(a) = \tau_a h(a) \tau_a .
\end{aligned}
$$

Hence, for any $a \in ran(h)$, we have

$$R_a^{h^{-1}(\mathcal{G}) \times \mathcal{H}} = R_a^{k^{-1}(f^{-1}(\mathcal{G}) \times \mathcal{K})} .$$

The two graphs are isomorphic up to the vertices of $Q^{\mathcal{K}} - Q^{\mathcal{H}}$ which are isolated in $k^{-1}(f^{-1}(\mathcal{G}) \times \mathcal{K})$. □

Theorem 5. *Up to isolated vertices, unbounded path transductions are closed by composition*

Proof: Let T_1 and T_2 be two unbounded path transductions. Our aim is to show that the relation $T = T_1 \cdot T_2$ remains an unbounded path transduction.

Let $\mathcal{G}$ be a graph and let $\mathcal{H} = T_2(T_1(\mathcal{G}))$ hence

$$\mathcal{H} = h_2^{-1}(g_2^{-1}(h_1^{-1}(g_1^{-1}(\mathcal{G}) \times \mathcal{F}_1)) \times \mathcal{F}_2)$$

where g_i and h_i are rational substitutions and $\mathcal{F}_i$ are finite graphs.

By composition of rational substitutions (Lemma 1) we deduce a rational substitution $g_3 = g_2 \cdot h_1$ such that

$$\mathcal{H} = h_2^{-1}(g_3^{-1}(g_1^{-1}(\mathcal{G}) \times \mathcal{F}_1) \times \mathcal{F}_2) \ .$$

From Lemma 5 we deduce a rational substitution k, a morphism f and a finite graph $\mathcal{K}$ such that

$$g_3^{-1}(g_1^{-1}(\mathcal{G}) \times \mathcal{F}_1) \times \mathcal{F}_2 \approx k^{-1}\left(f^{-1}(g_1^{-1}(\mathcal{G}) \times \mathcal{F}_1) \times \mathcal{K}\right) \ .$$

The function f being a morphism, by Lemma 2 we get

$$g_3^{-1}(g_1^{-1}(\mathcal{G}) \times \mathcal{F}_1) \times \mathcal{F}_2 \approx k^{-1}\left(f^{-1}(g_1^{-1}(\mathcal{G})) \times f^{-1}(\mathcal{F}_1)) \times \mathcal{K}\right) \ .$$

With $g = f \cdot g_1$, $h = h_2 \cdot k$, and $\mathcal{F} = f^{-1}(\mathcal{F}_1) \times \mathcal{K}$ for any graph $\mathcal{G}$ we have

$$h_2^{-1}(g_2^{-1}(h_1^{-1}(g_1^{-1}(\mathcal{G}) \times \mathcal{F}_1)) \times \mathcal{F}_2) \approx h^{-1}(g^{-1}(\mathcal{G}) \times \mathcal{F}) \ \square$$

5   Conclusion

In the theory of *languages and automata*, there was an asymmetry between an algebraic characterization for word languages and a "mechanical" one for their acceptors. Much work has been done to exhibit useful algebraic language transformations but the corresponding acceptors transformations remained scattered and hidden in some technical lemma.

Infinite labelled graphs/automata gives a natural framework to study the behavior of acceptors. We proposed the notion of path transduction which is a graph transformation. With this definition, the correspondence between the graph transformation and the language transformation becomes clearer than it is in [17]. The *AFG* and *full-AFG* defined in [17] may be respectively defined by closure under bounded and unbounded path transductions.

We also showed that path transductions are closed by composition, generalizing the classical Elgot and Mezei theorem from words to graphs. This property gives a canonical form for graphs in principal *AFG* like pushdown graphs [13] or prefix recognizable graphs [6].

To deepen the knowledge about graphs as language acceptors, a systematic study of language-compatible graph transformations and their application on infinite automata is required. The generalization of rational word transductions to path transductions is one step in this systematic study.

Path transductions are strictly weaker than monadic transductions, but their expressive power is important: for example the transitions graphs of Petri nets are the images of hypergrids by path transductions. Another interesting aspect that is not developed here is that, being weaker than monadic transductions, path transductions preserve also weaker decidability properties like reachability under rational control.

Acknowledgement

I would like to thank the referees (especially one) for their numberous remarks.

References

1. J.-M. Autebert and L. Boasson. *Transductions rationnelles – application aux langages algébriques.* Masson, Paris, 1988.
2. J. Berstel. *Transductions and context-free languages.* B.G. Teubner, Stuttgart, 1979.
3. A. Blumensath and E. Grädel. Automatic structures. In *Proceedings of 15th IEEE Symposium on Logic in Computer Science LICS 2000,* pages 51–62, 2000.
4. L. Boasson and M. Nivat. Sur diverses familles de langages fermèes par transductions rationelle. *Acta Informatica,* 2:180–188, 1973.
5. D. Caucal. On infinite terms having a decidable monadic theory. In *27th MFCS,* volume 2420 of *LNCS,* pages 165–176, Warsaw, 2002.
6. D. Caucal. On infinite transition graphs having a decidable monadic theory. *Theoretical Computer Science,* 290:79–115, 2003.
7. B. Courcelle. Monadic-second order definable graph transductions: a survey. *Theoretical Computer Science,* 126:53–75, 1994.
8. B. Courcelle and I. Walukiewicz. Monadic second-order logic, graph coverings and unfoldings of transition systems. *Annals of Pure and Applied Logic,* 1998.
9. J. Engelfriet. Context-free graphs grammars. In G. Rozenberg and A. Salomaa, editors, *Handbook of Formal Languages,* volume 3, chapter 3, pages 125–213. Springer-Verlag, 1997.
10. S. Ginsburg. *Algebraic and automata-theoretic properties of formal languages.* North-Holland, Amsterdam, 1975.
11. S. Ginsburg and S. Greibach. Abstract families of languages. *Mem. Am. Math. Soc.,* 87, 1969.
12. A. Mateescu and A. Salomaa. *Handbook of Formal Languages,* volume 1, chapter Aspects of classical language theory, pages 175–252. Springer-Verlag, 1997.
13. D. Muller and P. Schupp. The theory of ends, pushdown automata, and second-order logic. *Theoretical Computer Science,* 37:51–75, 1985.
14. M.O. Rabin. Decidability of second-order theories and automata on infinite trees. *Trans. Amer. Math. soc.,* 141:1–35, 1969.
15. J.-C. Raoult. A survey of tree transductions. In M. Nivat and A. Podelski, editors, *Tree Automata and Languages,* pages 311–325. North-Holland, Amsterdam, 1992.
16. W. Thomas. A short introduction to infinite automata. In *Proceedings of the 5th international conference Developments in Language Theory,* volume 2295, pages 130–144. LNCS, 2001.
17. T. Urvoy. Abstract families of graphs. In M. Ito and M. Toyama, editors, *Proceedings of the 6th international conference Developments in Language Theory,* volume 2450 of *LNCS,* pages 381–392, 2002.
18. T. Urvoy. *Familles abstraites de graphes.* PhD thesis, Université de Rennes I, 2003.

Translating Java Code
to Graph Transformation Systems[*]

Andrea Corradini[1], Fernando Luís Dotti[2], Luciana Foss[3], and Leila Ribeiro[3]

[1] Dipartimento di Informatica, Università di Pisa,
Pisa, Italy
andrea@di.unipi.it
[2] Faculdade de Informática, Pontifícia Universidade Católica do Rio Grande do Sul,
Porto Alegre, Brazil
fldotti@inf.pucrs.br
[3] Instituto de Informática, Universidade Federal do Rio Grande do Sul,
Porto Alegre, Brazil
{lfoss,leila}@inf.ufrgs.br

Abstract. We propose a faithful encoding of Java programs (written in a suitable fragment of the language) to Graph Transformation Systems. Every program is translated to a set of rules including some *basic rules*, common to all programs and providing the operational semantics of Java (data and control) operators, and the *program specific rules*, namely one rule for each method or constructor declared in the program.
Besides sketching some potential applications of the proposed translation, we discuss some desing choices that ensure its correctness, and we report on how do we intend to extend it in order to handle several other features of the Java language.

1 Introduction

Graph Transformation Systems (GTSs) were introduced about four decades ago as a generalization of Chomski Grammars to non-linear structures. Since then, the theory evolved quickly [24], and a growing community has applied GTSs to diverse application fields [7]. Along the years, GTSs have been used to provide an operational semantics to programming languages [22], to systems made of agents (or actors) interacting via message passing [15, 16, 6], to object-oriented specification formalisms [26], to process calculi [10], and to several other languages and formalisms.

By encoding a computational formalism into GTSs, usually one enjoys several benefits. Firstly, the representation of states as graphs is often quite abstract and intuitive, as it allows to ignore irrelevant details: in fact, graphs are considered usually "up to isomorphism", and this corresponds to considering states up to the renaming of bound variables or names. Secondly, the rich theory of GTSs

---

[*] Work partially supported by projects IQ-MOBILE (CNPq and CNR), PLATUS (CNPq), AGILE (EU FET – Global Computing) and SEGRAVIS (EU Reserach Training Network).

provides interesting results that may be applied to the original formalism via the encoding. As an example, the concurrent behavior of GTSs has been thoroughly studied and a consolidated theory of concurrency is now available, including event structure, process and unfolding semantics [23, 3]; the encoding of process calculi as GTSs equips such calculi with a concurrent operational semantics [10], to which the mentioned results can be applied.

Along these lines, in this paper we propose a translation of Java programs, written in a suitable fragment of the language, into Graph Trasformation Systems. The fragment of Java for which we present such an encoding is small, but still significant. It includes operators on primitive data types, class declarations, assignments, conditional statements, (static or instance) method invocations with parameter passing and value returning, and construction of new objects. A *Java declaration*, i.e., a closed set of class definitions, is translated into a *graph of types* and a set of *typed rules* that specify the (potential) behavior of the program. The states of execution of a Java program are represented by graphs which encode aspects related to both the data structures and the control of the program. As expected, information which are not relevant at run-time, like names of local variables or of formal parameters of methods, do not appear explicitly in the graphical representation of states. The rules resulting from the translation of a Java declaration are divided in two kinds: the *basic rules*, that are common to all Java programs and specify the operational meaning of the various (functional and control) operators of the language; and the *program specific rules*, one for each method or constructor, which replace a method/constructor call with the corresponding body. After presenting the translation, we discuss some design choices we made, and also how do we intend to extend the translation in order to handle other features of Java, including arrays, inheritance and multi-threading.

There are several potential applications of the kind of translation we propose: we sketch here two of them. Recently, various analysis techniques for GTSs have been proposed, some of which are based on the unfolding semantics (see, e.g., [2] for the finite-state case, and [4] for the general case). We intend to investigate how far such techniques can be applied to the systems obtained from the translation of Java programs, and thus, indirectly, to the Java programs themselves: the leading intuition is that relevant properties of Java programs can be formulated, in a GTSs setting, as structural properties of the graphs reachable in the system. Having this application in mind, we restricted the format of the rules obtained from the translation in order to match the constraints imposed by [2, 4], the most relevant of which is that rules must be injective.

Another application we have in mind is to provide a pure GTS-based semantics to *Java-attributed* GTSs, the graph transformation model adopted in tools like AGG [1]. In this model, the items of a graph can be associated with arbitray Java expressions (the *attributes*), and in a rewriting step such expressions are evaluated to compute the attributes of the newly created items. By exploiting the proposed translation of Java to GTS, we may translate a Java-attributed GTS to a plain typed GTS by first encoding the attributes as graphs, and then simulating an attributed graph rewriting step as a sequence of typed graph rewriting steps, where the evaluation of the attributes is performed graphically.

On the one hand, the translation just sketched would allow us to explore the applicability of the above mentioned analysis techniques to Java-attributed GTSs. On the other hand, we are confident that it will provide a formal ground on which to address a well-known limitation of the expressive power of attributed GTSs, namely the fact that graphical items may refer to attributes, but attributes cannot refer back to graphical items. For example, according to the standard approaches [17], it is not possible to define a graph where a vertex has as attribute a list of vertices of the graph itself.

The paper is organized as follows. Section 2 introduces the basics of typed (hyper)graph transformation systems according to the Single-Pushout Approach. Section 3 presents the fragment of Java that we shall consider, and Section 4 shows how to translate programs written in this fragment into typed GTSs. Section 5 discusses some design choices underlying the proposed translation and how do we intend to handle other features of Java, and Section 6 concludes sketching some subjects of future work.

2   Typed Graph Transformation Systems

In this section we recall the definition of graph transformation systems (GTSs) according to the Single-Pushout (SPO) approach [18,8]. However, it is worth stressing that we could have used equivalently other approaches, like for example the Double-Pushout approach: in fact, for the kind of graphs and rules generated by the translation of Java to GTSs, the two approaches are equivalent.

We shall define GTSs over (typed) *hyper*graphs, i.e., graphs where edges can be connected to any (finite) number of vertices. Graphically, an edge is depicted as a box (whose shape may vary), and the connections to the vertices are drawn as thin lines, called *tentacles*. Usually the tentacles of an edge are labeled by natural numbers: here we propose a slightly more general definition, allowing us to label tentacles with *labels* taken from a fixed set (see Figure 2).

The definition of the SPO approach is based on a category of graphs and *partial* morphisms.

Definition 1 (weak commutativity). *Given two partial functions $f, f' : A \to B$, we say that f is **less defined** than f' (and we write $f \leq f'$) if $dom(f) \subseteq dom(f')$ and $f(x) = f'(x)$ for all $x \in dom(f)$. Given two partial functions $f : A \to B$ and $f' : A' \to B'$, and two total functions $a : A \to A'$ and $b : B \to B'$, we say that the resulting diagram **commutes weakly** if $b \circ f \leq f' \circ a$.*

Now we introduce hypergraphs and partial morphisms. Each hyperedge is associated to a *finite, labeled set of vertices* (formally, to a partial function from a fixed set of labels to vertices). This definition makes use of a *labeling functor*, defined as follows: Let L be a set of labels. The labeling functor $\mathcal{I}_L : Set^P \to Set^P$ maps each set $V \in |Set^P|$ to the set of L-labeled sets over V (i.e., to the set of partial functions $L \to V$), and each partial function $f : V \to V' \in$

$Mor_{Set^P}(V, V')$ to the function $\mathcal{I}_L(f)$, defined for all $l : L \to V \in \mathcal{I}_L(V)$ as $\mathcal{I}_L(f)(l) = f \circ l$.

Definition 2 ((hyper)graph, (hyper)graph morphism). *A **(hyper) graph** $G = (V_G, E_G, c^G)$ over a set of labels L consists of a set of vertices V_G, a set of (hyper)edges E_G, and a total connection function $c^G : E_G \to \mathcal{I}_L(V_G)$, assigning a finite L-labeled set of vertices to each edge[1].*

*A **(partial) graph morphism** $g : G \to H$ is a pair of partial functions $g_V : V_G \to V_H$ and $g_E : E_G \to E_H$ which are weakly homomorphic, i.e., $\mathcal{I}_L(g_V) \circ c^G \geq c^H \circ g_E$ (if an edge is mapped, the corresponding vertices, if mapped, must have the same labels). A morphism is called total if both components are total. The category of hypergraphs and partial hypergraph morphisms is denoted by* **HGraphP** *(identities and composition are defined componentwise).*

$$
\begin{array}{ccc}
E_G & \xrightarrow{\ g_E\ } & E_H \\
{\scriptstyle c^G}\downarrow & \geq & \downarrow{\scriptstyle c^H} \\
\mathcal{I}_L(V_G) & \xrightarrow[\ \mathcal{I}_L(g_V)\ ]{} & \mathcal{I}_L(V_H)
\end{array}
$$

To distinguish different kinds of vertices and edges, we will use the notion of *typed hypergraphs*, analogous to typed graphs [5, 16]: every hypergraph is equipped with a morphism *type* to a fixed graph of types[2].

Definition 3 (typed hypergraphs). *Let TG be a fixed graph called the **graph of types**. A **typed hypergraph over** TG is a pair $HG^{TG} = (HG, type^{HG})$ where HG is a hypergraph called **instance graph** and $type^{HG} : HG \to TG$ is a total hypergraph morphism, called the **typing morphism**.*

A morphism between typed hypergraphs HG_1^{TG} and HG_2^{TG} is a partial graph morphism $f : HG_1 \to HG_2$ such that $type^{HG_1} \geq type^{HG_2} \circ f$. The category of hypergraphs typed over a graph of types TG, denoted by **THGraphP**(TG), *has hypergraphs over TG as objects and morphisms between typed hypergraphs as arrows (identities and composition are the identities and composition of partial graph morphisms).*

Definition 4 (graph transformation systems). *Given a graph of types TG, a **rule** $r : L \to R$ is a partial injective graph morphism in* **THGraphP**(TG) *such that 1) L and R are finite, 2) L has no isolated vertices, 3) r is total on the vertices of L (i.e., vertices are preserved), 4) r is not defined on at least one edge in L (i.e., the rule is consuming). A **graph transformation system** is a pair $\mathcal{G} = (TG, Rules)$ where TG is the graph of types and Rules is a set of rules over TG.*

Conditions 3) and 4), as well as the injectivity of rules, are required by the unfolding constructions or by the analysis techniques based on them (see [4, 2])

---

[1] Intuitively, for each edge $e \in E_G$, the partial function $c^G(e) : L \to V_G$ is defined on a label $a \in L$ iff e has a tentacle labeled a pointing to vertex $c^G(e)(a)$.

[2] Note that, due to the use of partial morphisms, this is not just a comma category construction: the morphism *type* is total whereas morphisms among graphs are partial, and we need weak commutativity instead of commutativity.

that we intend to apply to our systems. Condition 2) guarantees that vertices that become isolated have no influence on further rewriting. Thus one can safely assume that isolated vertices are removed by some kind of garbage collection, mitigating condition 3).

Definition 5 (derivation step, derivation). *Let $\mathcal{G} = (TG, Rules)$ be a GTS, $r : L \to R \in Rules$ be a rule, and G_1 be a graph typed over TG. A **match** for r in G_1 is a total morphism $m : L \to G_1$ in* **THGraphP**(TG). *A derivation step $G_1 \overset{r,m}{\Longrightarrow} G_2$ using rule r and match m is a pushout in the category* **THGraphP**(TG).

$$
\begin{array}{ccc}
L & \overset{r}{\longrightarrow} & R \\
{\scriptstyle m}\downarrow & (PO) & \downarrow{\scriptstyle m'} \\
G_1 & \underset{r'}{\longrightarrow} & G_2
\end{array}
$$

A derivation sequence of $\mathcal{G}$ is a sequence of derivation steps $G_i \overset{r_i,m_i}{\Longrightarrow} G_{i+1}$, $i \in \{0,\dots,n\}$, $n \in \mathbb{N}$, where $r_i \in Rules$ for all $i \in \{0,\dots,n\}$.

3 Java

Java [11] is a general-purpose object-oriented programming language designed to be highly portable and to ease the deployment of programs in distributed settings. Therefore the choice of a run-time environment based on interpretation (the Java Virtual Machine) and allowing for dynamic class loading from remote code bases. Garbage collection is also supported by the run-time environment. Java supports multi-threading as well as a synchronization mechanism close to Monitors. Altogether, the characteristics of Java led many developers to adhere to the language and currently a rich set of Java libraries can be found for the most different application areas.

In the following we present the BNF syntax of the fragment of Java considered in this paper and next a corresponding example of Java code. The fragment includes operators on the primitive data types **char**, **int** and **boolean**, class declarations, assignments, conditional statements, (static or instance) method invocations with parameter passing and value returning, and construction of new objects.

Declaration ::= Modifier **class** Id { ClassBody } | ;
Modifier ::= **static** | **public** | **private** | ε
ClassBody ::= Modifier ClassMember ClassBody | ; | ε
ClassMember ::= MethodDeclaration | ConstructorDeclaration | VarDeclaration
MethodDeclaration ::= MType Id (ParameterList) { Block }
MType ::= **void** | Type
ConstructorDeclaration ::= Id (ParameterList) { Block }
VarDeclaration ::= Type VarDeclarationList ;
VarDeclarationList ::= Id VarInitializer | VarDeclarationList , Id VarInitializer
VarInitializer ::= **=** Expression | ε
ParameterList ::= Type Id ParameterRest | ε

ParameterRest ::= , Type Id ParameterRest | ε
Block ::= BlockStatement Block | ε
BlockStatement ::= VarDeclaration | Statement
Statement ::= StatementExp | { Block } | if (Expression) Statement |
    if (Expression) Statement else Statement | return Expression ; | ;
StatementExp ::= Assignment | MethodInvocation | NewExpression |
    this (ArgumentList)
Assignment ::= QualifiedId = Expression | FieldAccess = Expression
FieldAccess ::= Primary . Id
Primary ::= FieldAccess | MethodInvocation | newExpression | (Expression)
    | this | Literal | null
Expression ::= Assignment | Expression InfixOp Expression | PrefixOp Expression
    | QualifiedId | Primary
MethodInvocation ::= Id (ArgumentList) | Primary . Id (ArgumentList)
NewExpression ::= new Id (ArgumentList)
InfixOp ::= || | && | == | != | < | > | + | - | * | /
PrefixOp ::= ! | + | -
ArgumentList ::= Expression ArgumentRest | ε
ArgumentRest ::= , Expression ArgumentRest | ε
Literal ::= Integer | Character | true | false
Type ::= QualifiedId | BasicType
BasicType ::= char | int | boolean
QualifiedId ::= Id Qualification
Qualification ::= . Id Qualification | ε

We call a **Java declaration** a set *JSpec* of Java class declarations written in the above fragment of the language, closed under definitions (i.e., every class referred to by a class in *JSpec* is in *JSpec* as well), correct with respect to syntax and static semantics (i.e., the definitions compile safely), and satisfying some additional requirements described next.

Without loss of generality, we assume that all (static or instance) variables of the classes in *JSpec* are declared as `private`, (read/write access to a variable x can be provided by the standard accessor methods `getX`/`setX`); that no variable is accessed using the dot-notation "`class.var`" or "`obj.var`"[3], unless `obj` is `this`; and that all local variables of methods are initialized (possibly with the standard default value) at declaration time.

Figure 1 shows a sample Java declaration satisfying the above constraints.

4   Translating Java to GTSs

We present here the translation of Java declarations, as defined in the previous section, into graph transformation rules. A Java declaration *JSpec* is translated into a graph of types and a set of typed rules that specify the (potential) behavior of the program. The graphs which are rewritten represent aspects related to both the data structures and the control of the program under execution.

---

[3] This last assumption is necessary even if all variables are `private`.

```
public class LinkedList List  {              public class ListNode {
    private ListNode first;                      private Person element;
    private ListNode last;                       private ListNode next;

    public ListNode getFirst() {                 public Person getElement() {
        return first;}                               return element;}

    public ListNode getLast() {                  public ListNode getNext() {
        return last;}                                return next;}

    public void setFirst(ListNode f) {           public void setElement(Person e) {
        first=f;}                                    element = e;}

    public void setLast(ListNode l) {            public void setNext(ListNode n) {
        last=l;}                                     next = n;}

    public LinkedList( ) {                       public ListNode(Person element) {
        first = last = null; }                       this( element, null ); }

    public boolean isEmpty( ) {                  public ListNode(Person element, ListNode next ) {
        return first == null; }                      this.element  = element;
                                                     this.next     = next; }
    public void clear( ) {
        first = last = null; }               }
                                             public class Person {
    public boolean add( Person x ) {             private int identifier;
        if ( isEmpty() ) first = last = new ListNode( x );
        else { last.setNext(new ListNode( x )); public int getIdentifier() {
            last = last.getNext();  }                return identifier;}
        return true; }
    public boolean remove( ) {                   public void setIdentifier(int n) {
        if ( isEmpty() ) return false;               identifier = n;}
        else { first = first.getNext();
            return true; } }                     public Person(int identifier) {
}                                                    this.identifier = identifier; }
                                             }
```

Fig. 1. Example of Java Declaration.

In the graph modeling the current state of a computation every available data element is represented explicitly, including objects, variables, constant values of basic data types, and also the classes of *JSpec*, which are needed to handle static variables and methods. Each such data element is represented graphically by a vertex (its *identity*, typed with the corresponding Java data type) and by an edge carrying the relevant information, connected to the identity with a tentacle labeled by **val** (for constants) or **id** (for all other cases). Such edges may have additional tentacles: for example, the edge representing a class (depicted as a double box) has one tentacle for each static variable of the class; an edge representing a class instance has one tentacle for each *instance* variable of the class; an edge representing a variable has a **val** tentacle pointing to the (identity node of the) current value of the variable[4].

The control operators are also represented as hyperedges, and they are connected through suitable tentacles to all the data they act upon. Their operational behavior is specified by rules. The rules are divided in two kinds:

- **Basic rules:** These rules are common to all Java programs, and specify the operational meaning of the various (functional and control) operators of the language, including operators on primitive data types, like arithmetic and relational ones; assignment and **new** operators; **if** and **if-else** control struc-

---

[4] Technically, when restricted to the data part, the graphs we are considering are *term graphs* [20].

tures; `return` statement; `get` and `set` operators for accessing local variables; and rules for handling value indirections (`val` edges).
- **Program specific rules:** These rules encode the methods and constructors of the classes in $JSpec$. For each method and constructor declaration in $JSpec$ there is one rule, where the left-hand side encodes the method or constructor call and the right-hand side encodes the whole body of the method.

Conceptually, the execution of a method call is modeled by a derivation sequence where the first rule (a program specific one) replaces the method invocation edge by a graph that encodes the body of the method, and next the control edges in the body are evaluated sequentially using the corresponding rules. In order to guarantee that statements are evaluated in the same order as in the Java Virtual Machine, we supply every control edge with two additional tentacles (labeled `in` and `out`) and we use a unique `GO` hyperedge system-wide, acting as a token, to enforce sequential execution. In fact, every rule consumes the `GO` token connected to the `in`-vertex of the control edge (thus the rule cannot be applied if the token is not present), and generates the token again to the `out`-vertex.

For a given Java declaration $JSpec$, the following definition describes the graph of types $TG(JSpec)$ obtained from the translation.

Definition 6 (graph of types of a Java declaration). *Let $JSpec$ be a Java declaration. The graph of types $TG(JSpec) = (V, E, c)$ associated to $JSpec$ is the hypergraph defined as follows:*

- $V = BDT \cup CT \cup \{\circ\}$, *where $BDT = \{t \mid t$ is a basic data type used in $JSpec\}$, $CT = \{c \mid c$ is a class name in $JSpec\}$. Thus there is one vertex for each data type in $JSpec$, and an additional vertex "$\circ$" which is the type of all control vertices.*

- $E = \{\boxed{GO}, \boxed{if}, \boxed{+_b}, \boxed{-_b}, \boxed{*}, \boxed{/}, \boxed{<}, \boxed{>}, \boxed{||}, \boxed{\&\&}, \boxed{+_u}, \boxed{-_u}, \boxed{!}\} \cup CH \cup BH \cup CTE$, *where*
  - *subscripts b and u stand for "binary" and "unary";*
  - $CH = \{\boxed{C}, \boxed{C}, \boxed{null^C}, \boxed{new^C}, \boxed{\langle init \rangle_p^C}, \boxed{this^C}$
    $\mid C$ *is a class in $JSpec\} \cup \{\boxed{m_p^C} \mid m$ is a (static or instance) method in C with parameter type list $p\} \cup \{\boxed{\langle init \rangle_p^C} \mid$ there is a constructor with parameter type list p in C, or C has no constructors and p is the empty list\};*
  - $BH = \{\boxed{var^t}, \boxed{get^t}, \boxed{set^t}, \boxed{return^t}, \boxed{val^t}, \boxed{==^t}, \boxed{!=^t} \mid t$ *is a basic data type or a class in $JSpec\}$;*
  - $CTE = \{\boxed{cte_t} \mid cte$ *is a constant of a basic data type t in $JSpec\}$.*

- *The set of tentacle labels L is defined as $L = \{$g, exp, in, out, out_T, out_F, left, right, result, val, id, target, return, value$\} \cup NAT \cup PAR$, where $NAT = \{$at $\mid$ at is a (static or instance) variable name of a class $C \in CH\}$ and $PAR = \{i \mid i \in \mathbb{N} \wedge 0 < i \leq n$, where n is the maximum number of formal parameters of any method or constructor in $JSpec\}$.*

- *The connection function $c : E \rightarrow (L \rightarrow V)$ is defined as follows, where for each edge $e \in E$ the graph of the partial function $c(e) : L \rightarrow V$ is shown:*
  - $c(\boxed{GO}) = \{(\mathbf{g}, \circ)\};$
  - $c(\boxed{if}) = \{(\mathbf{exp}, boolean), (\mathbf{in}, \circ), (\mathbf{out\_T}, \circ), (\mathbf{out\_F}, \circ)\};$
  - $\forall op \in \{\boxed{+_b}, \boxed{-_b}, \boxed{*}, \boxed{/}\} . c(op) = \{(\mathbf{left}, int), (\mathbf{right}, int),$
    $(\mathbf{result}, int)\};$
  - $\forall op \in \{\boxed{>}, \boxed{<}\} . c(op) = \{(\mathbf{left}, int), (\mathbf{right}, int), (\mathbf{result}, boolean)\};$
  - $\forall op \in \{\boxed{||}, \boxed{\&\&}\} . c(op) = \{(\mathbf{left}, boolean), (\mathbf{right}, boolean),$
    $(\mathbf{result}, boolean)\};$
  - $c(\boxed{!}) = \{(\mathbf{right}, boolean), (\mathbf{result}, boolean)\};$
  - $c(\boxed{C}) = \{(\mathbf{at_1}, t_1), \ldots, (\mathbf{at_n}, t_n), (\mathbf{id}, C)\},$ *where n is the total number of* static *variables in C, and, for each $0 < i \leq n$, $\mathbf{at_i}$ is a static variable name in C and $t_i \in BDT \cup CT$ is its type;*
  - $c(\boxed{C}) = \{(\mathbf{at_1}, t_1), \ldots, (\mathbf{at_n}, t_n), (\mathbf{id}, C)\},$ *where n is the total number of* instance *variables in C, and, for each $0 < i \leq n$, $\mathbf{at_i}$ is a instance variable name in C and $t_i \in BDT \cup CT$ is its type;*
  - $c(\boxed{null^C}) = \{(\mathbf{val}, C)\};$
  - $c(\boxed{m_p^C}) = \{(\mathbf{1}, t_1), \ldots, (\mathbf{n}, t_n), (\mathbf{target}, C), (\mathbf{in}, \circ), (\mathbf{out}, \circ)\} \cup RET_m,$
    where $[t_1, \ldots, t_n]$ is exactly the parameter type list p of method m, and if the type t of method m is different from void, *then $RET_m = \{(\mathbf{return}, t)\}$ else $RET_m = \emptyset;$*
  - $c(\boxed{new^C}) = \{(\mathbf{target}, C), (\mathbf{return}, C), (\mathbf{in}, \circ), (\mathbf{out}, \circ)\};$
  - $c(\boxed{\langle init \rangle_p^C}) = \{(\mathbf{1}, t_1), \ldots, (\mathbf{n}, t_n), (\mathbf{target}, C), (\mathbf{in}, \circ), (\mathbf{out}, \circ)\},$
    where $[t_1, \ldots, t_n]$ is the parameter type list p of the constructor;
  - $c(\boxed{this^C}) = \{(\mathbf{target}, C), (\mathbf{result}, C), (\mathbf{in}, \circ), (\mathbf{out}, \circ)\};$
  - $c(\boxed{var^t}) = \{(\mathbf{id}, t), (\mathbf{val}, t)\};$
  - $c(\boxed{get^t}) = \{(\mathbf{target}, t), (\mathbf{result}, t), (\mathbf{in}, \circ), (\mathbf{out}, \circ)\};$
  - $c(\boxed{set^t}) = \{(\mathbf{target}, t), (\mathbf{value}, t), (\mathbf{result}, t), (in, \circ), (\mathbf{out}, \circ)\};$
  - $c(\boxed{return^t}) = \{(\mathbf{value}, t), (\mathbf{result}, t), (\mathbf{in}, \circ), (\mathbf{out}, \circ)\};$
  - $c(\boxed{val^t}) = \{(\mathbf{return}, t), (\mathbf{value}, t)\};$
  - $c(\boxed{==^t}) = \{(\mathbf{right}, t), (\mathbf{left}, t), (\mathbf{result}, boolean), (\mathbf{in}, \circ), (\mathbf{out}, \circ)\};$
  - $c(\boxed{! =^t}) = \{(\mathbf{right}, t), (\mathbf{left}, t), (\mathbf{result}, boolean), (\mathbf{in}, \circ), (\mathbf{out}, \circ)\};$
  - $c(\boxed{cte_t}) = \{(\mathbf{id}, t)\}.$

For each class $C \in JSpec$, the edge labeled $\boxed{C}$ represents the class itself, providing access to its static variables: it is assumed to be unique in any legal graph representing an execution state of a Java program, and it will be the target of all *static* method invocation, as well as of the *new* operator. The edges labeled

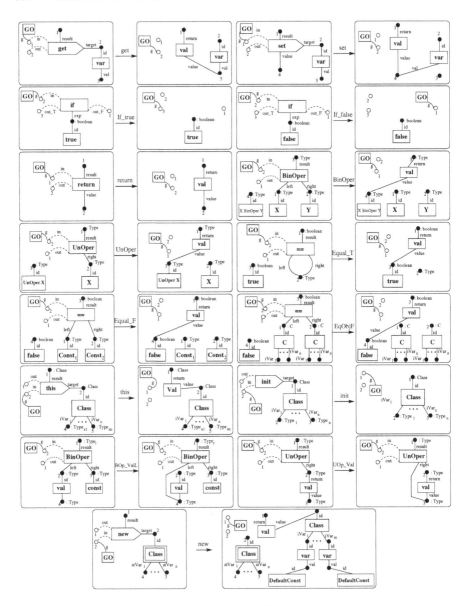

Fig. 2. Basic Rules.

$\boxed{C}$ represent instead instances of C, each connected to its private copy of the instance variables of C.

Concerning the rules obtained by the translation, many of the **basic rules** for the Java operators in the sub-language we consider are shown in Figure 2. Several of such rules actually are rule schemata, and represent the (finite, even if large) collection of all their instances. Some of the rules, including those handling the equality operator and the **val** edges, are discussed in Section 5.

As far as the **program specific rule** $r : L \to R$ encoding a method (or constructor) of the Java declaration is concerned, it is obtained as follows:

- the left-hand side L is the translation of the method invocation (see Figure 3(d) for a static method invocation);
- the right-hand side R is obtained by translating the body of the method; this translation is obtained by replacing every Java statement in the method body by a corresponding graph – as shown informally in Figure 3 – and connecting such (sub)graphs sequentially using the control tentacles;
- the partial morphism r preserves all the items of its source, but for the edge representing the method being called.

Figure 4 shows some of the rules resulting from the translation of the Java declaration of Figure 1. Currently, such translation is defined in a precise but informal way. We intend to formalize it using, for example, *pair grammars* [21].

5 Some Considerations About the Proposed Translation

In this section we first discuss the design choices underlying some of the rules presented in the previous section. Next we describe the way we intend to extend our translation to a larger fragment of Java.

Handling of **val** *edges.* To model the fact that an expression returns a value, we included a special *indirection* edge, labeled with **val**, linking the vertex where the result should be placed to the identity of the result value (see for example, in Figure 2, the rules **get**, which makes the value of the target variable accessible at vertex 1 of the left-hand side, and **set**, which implements an assignment by changing the value of the target variable and returning the assigned value). In order to get rid of the **val** edges, which is necessary because otherwise the execution cannot proceed, we need to add for each operator edge and for each tentacle through which the operator may access a data value, an auxiliary rule that allows to bypass any **val** edge connected to that tentacle. Rule schemata **BOp_ValL** and **UOp_Val** of Figure 2 show the shape of such auxiliary rules.

It is worth noting that the simpler solution of adding a single non-injective rule with a **val** edge in the left-hand side and a single node in the right-hand side (thus the rule would consume the edge, "merging" the two nodes connected to it), would not be legal according to Definition 4.

Handling of the equality operator "**==**". Rule **Equal_T** of Figure 2 shows that "**==**" is interpreted as *reference identity*, and this both for reference types (classes) and for basic data types. The correctness of this rule is ensured by assuming (as in [14]) that all constants of basic data types are present in the start graph, and that they are preserved by all rules. Such an assumption allows to deal with values of all data types uniformly at the graphical level (that is, using non-attributed graphs), at the price of making the size of the graphs practically unmanageable.

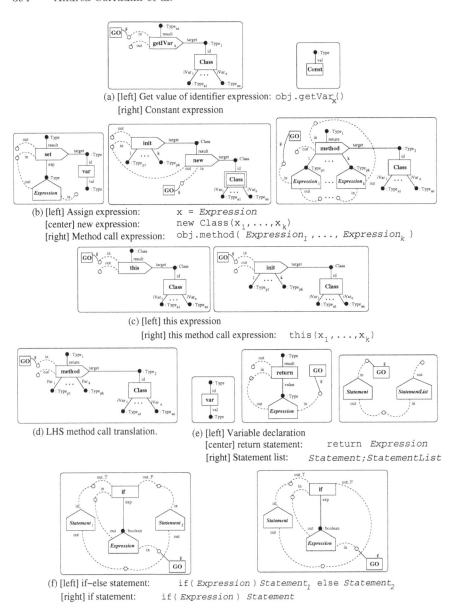

(a) [left] Get value of identifier expression: obj.getVar$_x$()
[right] Constant expression

(b) [left] Assign expression: x = Expression
[center] new expression: new Class(x$_1$,...,x$_k$)
[right] Method call expression: obj.method(Expression$_1$,..., Expression$_k$)

(c) [left] this expression
[right] this method call expression: this(x$_1$,...,x$_k$)

(d) LHS method call translation.   (e) [left] Variable declaration
[center] return statement: return Expression
[right] Statement list: Statement;StatementList

(f) [left] if–else statement: if(Expression) Statement$_1$ else Statement$_2$
[right] if statement: if(Expression) Statement

Fig. 3. Translation schema for Java statements.

Notice also that **Equal_F** is a schema that represents one such rule for each pair of distinct constants **Const$_1$** and **Const$_2$** of basic data types.

As a topic of further research, we intend to explore a different encoding making use of attributed graphs, where only values of basic data types would

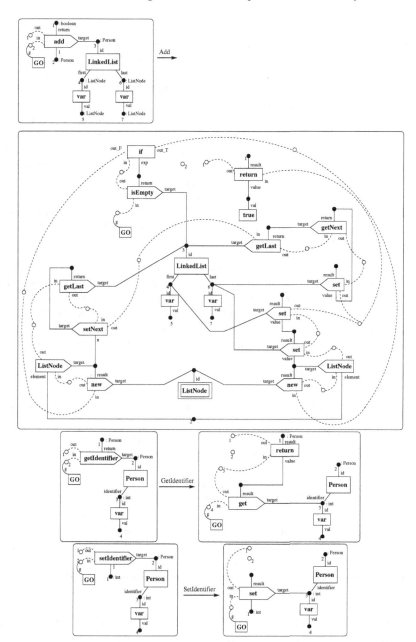

Fig. 4. The rules obtained from the translation of some instance methods of the Java declaration of Figure 1, namely add(Person x) of class LinkedList, and getIdentifier() and setIdentifier(int n) of class Person.

be allowed as attributes, and where equality on basic data types (as well as the other predefined operators) would be evaluated in a suitable algebra. This would reduce drastically the size of graphs as well as the number of rules, still having at our disposal a rich theory, thanks to the recent results presented in [9] which extend to *attributed* GTSs many classical results of *typed* GTSs.

Rule EqualObjF, states that "==" returns false if applied to two distinct instances of a given class. But such a rule could be applied, using a non-injective match, also to a match of rule Equal_T, leading to an incorrect result. There are several ways to fix this problem and to ensure correctness: (1) we could require that all matches are *injective* (actually, *edge-injectivity* would suffice); (2) we could establish a priority between rules Equal_T and EqualObjF; or, (3) we could use GTSs with negative application conditions. Both solutions (1) and (3) have a solid thoretical background (see [13, 12]), but we still have to study in depth how the various solutions affect the applicability of the analysis techniques we are interested in.

Iterative statements. The fragment of Java we considered does not include any kind of loops (while, for or do-while statements), which nevertheless can be simulated by recursive method invocations: these are in general computationally more expensive, but efficiency considerations go beyond the scope of this paper. It is quite evident that iterative statements cannot be handled like the other operators with a fixed set of basic rules. In fact, with the proposed rules, the edge representing an operator is deleted when the operator is evaluated, and thus it cannot be evaluated more than once. Instead, each iterative statement could be represented by a distinct edge, with a corresponding rule replacing that edge with the body of the statement: a solution basically equivalent to a recursive method invocation.

We started investigating how to extend the proposed translation to several other features of Java. **Arrays** could be modeled by adding ARRAY edges with the desired arity(ies), and corresponding rules for getting and setting the array element values. However, the most naive solution is not satisfactory, as it requires one copy of such rules for each legal index. **Multi-threading** could be modelled within our setting by adding multiple GO edges to create different execution threads (synchronization mechanisms would also map to corresponding manipulations of GO edges – for instance, the notion of "join" could be mapped to the use of counters and elimination of some GO edges until a next command should be executed). Concerning **inheritance**, we are investigating the use of var edges to model sub-class relations (a related approach has been proposed in [19]). The idea consists of adding to the graph of types one var edge for each pair of classes related by inheritance: this would allow to model the fact that a variable of type $Class_1$ can store objects of class $Class_2$ if $Class_2$ extends $Class_1$.

6   Conclusion and Future Work

In this paper we presented a translation of a fragment of the Java language to typed Graph Transformation Systems. The idea was to encode values of data types as well as Java control structures graphically, and to use rules to simulate the execution of Java programs.

The proposed translation is to be considered as a first contribution of a long-term research activity, and there are several topics for further work.

Besides addressing the translation of other features of Java, as sketched in the previous section, we intend to investigate the encoding of a relevant part of the Java Virtual Machine, for example following the approach taken in [25], and to explore how far a graphical specification method based on GTSs could compete with the approach based on Abstract State Machines. Modelling the JVM would be the natural way to address certain aspects of the language that we deliberately ignored here, including, for example, dynamic class loading and garbage collection.

In particular, the handling of garbage has a direct impact on the applicability of the analysis techniques developed for GTSs. Notice that besides the usual *data garbage* including the objects that are not reachable anymore, since we also represent graphically the control structures, our graphs may include *control garbage*, for example the control structures belonging to branches of computations which are not executed. A direction we are investigating is to get rid of garbage in the analysis phase at an abstract level. More precisely, let us mention that the analysis technique proposed in [2] applies to GTSs that are finite-state *up to isolated nodes*. In the same vein, we would like to deal with graphs encoding Java computational states *up to garbage*.

Acknowledgement

We would like to thank the anonymous referees for their detailed comments and constructive suggestions.

References

1. The AGG website. http://tfs.cs.tu-berlin.de/agg.
2. P. Baldan, A. Corradini, and B. König. Verifying Finite-State Graph Grammars: an Unfolding-Based Approach. In *Proc. of CONCUR'04*. Springer, 2004. To appear.
3. P. Baldan, A. Corradini, and U. Montanari. Unfolding and Event Structure Semantics for Graph Grammars. In *Proc. of FoSSaCS'99*, volume 1578 of *LNCS*, pages 73–89. Springer, 1999.
4. P. Baldan and B. König. Approximating the behaviour of graph transformation systems. In *Proc. of ICGT'02*, volume 2505 of *LNCS*, pages 14–29. Springer, 2002.
5. A. Corradini, U. Montanari, and F. Rossi. Graph processes. *Fundamenta Informaticae*, 26:241–265, 1996.
6. F. Dotti and L. Ribeiro. Specification of mobile code systems using graph grammars. In *Formal Methods for Open Object-based Systems IV*, pages 45–64. Kluwer Academic Publishers, 2000.

7. H. Ehrig, G. Engels, H.-J. Kreowski, and G. Rozenberg, editors. *Handbook of Graph Grammars and Computing by Graph Transformation. Vol. 2: Applications, Languages, and Tools.* World Scientific, 1999.

8. H. Ehrig, R. Heckel, M. Korff, M. Löwe, L. Ribeiro, A. Wagner, and A. Corradini. Algebraic Approaches to Graph Transformation II: Single Pushout Approach and comparison with Double Pushout Approach. In Rozenberg [24].

9. H. Ehrig, U. Prange, and G. Taentzer. Fundamental Theory for Typed Attributed Graph Transformation. In *Proc. of ICGT'04*, LNCS. Springer, 2004. This volume.

10. F. Gadducci and U. Montanari. A Concurrent Graph Semantics for Mobile Ambients. In *Proc. of MFPS'01*, volume 45 of *ENTCS*. Elsevier Science, 2001.

11. J. Gosling, B. Joy, Steele G., and G. Bracha. *The Java Language Specification. 2nd Edition.* Sun Microsystems, Inc., 2000. http://java.sun.com/docs/books/jls/.

12. A. Habel, R. Heckel, and G. Taentzer. Graph grammars with negative application conditions. *Fundamenta Informaticae*, 26(3/4):287–313, 1996.

13. A. Habel, J. Müller, and D. Plump. Double-Pushout Graph Transformation Revisited. *Mathematical Structures in Computer Science*, 11(5):637–688, 2001.

14. R. Heckel, J.M. Küster, and G. Taentzer. Confluence of Typed Attributed Graph Transformation Systems. In *Proc. of ICGT'02*, volume 2505 of *LNCS*, pages 161–176. Springer, 2002.

15. D. Janssens and G. Rozenberg. Actor grammars. *Mathematical Systems Theory*, 22:75–107, 1989.

16. Martin Korff. True Concurrency Semantics for Single Pushout Graph Transformations with Applications to Actor Systems. In *Proc. of IS-CORE'94*, pages 33–50. World Scientific, 1995.

17. M. Löwe, M. Korff, and A. Wagner. An algebraic framework for the transformation of attributed graphs. In *Term Graph Rewriting: Theory and Practice*, pages 185–199. John Wiley, 1993.

18. Michael Löwe. Algebraic approach to single-pushout graph transformation. *Theoretical Computer Science*, 109:181–224, 1993.

19. A.P. Lüdtke Ferreira and L. Ribeiro. Towards Object-Oriented Graphs and Grammars. In *Procs. FMOODS'03*, volume 2884 of *LNCS*, pages 16–31. Springer, 2003.

20. Detlef Plump. Term graph rewriting. In Ehrig et al. [7], chapter 1, pages 3–62.

21. Terrence W. Pratt. Pair Grammars, Graph Languages and String-to-Graph Translations. *Journal of Computer and System Sciences*, 6:560–59, 1971.

22. Terrence W. Pratt. Definition of Programming Language Semantics Using Grammars for Hierarchical Graphs. In *Proc. 1st Int. Workshop on Graph-Grammars and Their Application to Computer Science and Biology*, volume 73 of *LNCS*, pages 389–400. Springer, 1979.

23. Leila Ribeiro. *Parallel Composition and Unfolding Semantics of Graph Grammars.* PhD thesis, Technische Universität Berlin, 1996.

24. Grzegorz Rozenberg, editor. *Handbook of Graph Grammars and Computing by Graph Transformation. Vol. 1: Foundations.* World Scientific, 1997.

25. R. Stärk, J. Schmid, and E. Börger. *Java and the Java Virtual Machine: Definition, Verification, Validation.* Springer, 2001.

26. A. Wagner and M. Gogolla. Defining Operational Behaviour of Object Specifications by Attributed Graph Transformations. *Fundamenta Informaticae*, 26:407–431, 1996.

Extending Graph Rewriting for Refactoring⋆

Niels Van Eetvelde and Dirk Janssens

University of Antwerp
Department of Computer science
Middelheimlaan 1
2020 Antwerpen
{niels.vaneetvelde,dirk.janssens}@ua.ac.be

Abstract. Refactorings are transformations that change the structure of a program, while preserving the behavior. The topic has attracted a lot of attention recently, since it is a promising approach towards the problem of program erosion. Nevertheless a solid theoretical foundation is still lacking. In this paper we focus on improving the expressive power of graph rewriting rules, so that more refactorings can be expressed by single rules. Two new mechanisms are introduced: a notion of refinement of graphs, enabling one to treat specific substructures (e.g. syntax trees of expressions) as a whole, and a notion of duplication, enabling one to use parts of rewriting rules as prototypes that can be instantiated several times. Both mechanisms can be viewed as ways to specify large or infinite sets of rules in a concise way. It is shown that the refactorings *PushdownMethod*, *ExtractMethod* and *InlineMethod* can be expressed using the proposed techniques.

1 Introduction

Refactorings are software transformations that restructure object-oriented programs while preserving their behavior [1–3]. The key idea is to redistribute instance variables and methods across the class hierarchy in order to prepare the software for future extensions. If applied well, refactorings improve the design of software, make software easier to understand, help to find bugs, and help to program faster [1]. Refactoring is already supported by a lot of tools (see e.g. [4]) and is even one of the cornerstones of the *Extreme Programming* [5] methodology, but a formal basis for it is still lacking. This paper continues the exploration of graph transformation as such a formal basis. In [6] and [7] the basic idea is to represent the program source code as a simple, typed graph, and model these refactorings as graph transformations on this metamodel. Unfortunately, many refactorings cannot be expressed directly as rules, mainly because it is difficult to express common operations like the copying and moving of program trees. One possible solution is to break down the refactorings into smaller steps, which then correspond to single rule applications. This approach has been used in [7],

---

⋆ Work partially supported through FWO - project "A Formal Foundation for Software Refactoring" and EU-TMR network SegraVis through Universiteit Antwerpen.

H. Ehrig et al. (Eds.): ICGT 2004, LNCS 3256, pp. 399–415, 2004.

where refactorings correspond to transformation units [8]. A disadvantage of this approach is that such description may be less transparent than one where each refactoring corresponds to a single rule: one may end up with a description where most of the complexity is in the control structure of the transformation unit, and where the rules are relatively trivial. Thus in this paper we aim at extending the notion of a graph rewriting rule in an existing embedding-based approach in such a way that the effect of each refactoring is described by just one rule.

Two mechanisms are presented: the first one is a notion of refinement (for labels, graphs and rules) loosely based on the notion of a hierarchical graph [9, 10]. In contrast to the approach in [10], however, both nodes and edges are refined, and the refinement is based on labels. The second mechanism enables one to duplicate certain parts of a rule. Both mechanisms can be viewed as ways to specify large or infinite sets of rules in a concise way. There are two main approaches to graph rewriting: on the one hand the approach based on double pushouts, and on the other hand the one based on embedding. This paper uses graph transformation with embedding, where applying a rule to a graph G happens in three stages: a subgraph of G is removed, it is replaced by a new graph, and this new graph is connected to the remaining part of G according to an embedding mechanism that is provided as part of the rule. It is well known that a rule with embedding may be viewed as a specification of a potentially infinite set of (simpler) rules in the double pushout approach. Since, in this paper, it is our aim to make single rules more expressive, it is natural to choose for the embedding-based approach. For more information about the various approaches to graph rewriting we refer to [11].

In Section 2 some basic notions and notations concerning graphs and graph rewriting are recalled. In Section 3 the two main constructions, refinement and duplication, are motivated in terms of rules describing refactorings. In Sections 5 and 4 the two constructions are introduced, and illustrated using the *Push Down Method* refactoring. Finally, in Section 6 it is demonstrated that the constructions provide sufficient expressive power to allow single rule descriptions of the refactorings *Extract Method* and *Inline Method*.

2   Graphs and Graph Transformation

In this section we recall some basic notions and notation concerning graphs and embedding-based graph rewriting. We consider graphs over an alphabet $\Sigma = \Sigma_v \cup \Sigma_e$, where Σ_v and Σ_e are the sets of node and edge labels, respectively.

2.1   Graphs

Definition 1 (graph). *A graph is a 3 - tuple (V, E, lab) where V is a finite set, $E \subseteq (V \times \Sigma_e \times V)$, and $lab : V \to \Sigma_v$.*

The *atoms* of G is the set $A_G = \{V_G \cup E_G\}$. For an edge $e = (v, a, w)$, we denote $lab(e) = a$. Hence lab is viewed as a labeling function defined on A_G. The following notations will be used throughout this paper. Let G be a graph, and let x be an atom of G. Then G_x is the graph defined as follows:

- if $x \in V_G$, then G_x is the discrete graph consisting of the node x, labeled by $lab_G(x)$.
- if $x \in E_G, x = (u, a, w)$, then G_x is the graph consisting of the edge x and its source u and target w, where u and w are labeled by $lab_G(u)$ and $lab_G(w)$, respectively.

The *adjacency set* $Adj(G)$ of a graph G is defined by $Adj(G) = \{(e, v) | e \in E_G, v \in V_G, v = source(e)\} \cup \{(v, e) | e \in E_G, v \in V_G, v = target(e)\}$.

The *induced subgraph* of a graph $G = (V, E, lab)$ on a set of nodes $S \subseteq V$ is the graph $G' = (S, E', lab')$ such that $E' = E \cap (S \times \Sigma_e \times S)$ and lab' is the restriction of lab to S.

2.2   Graph Rewriting

The basic idea of graph rewriting is that local transformations of graphs are described as applications of rules.

The approach used in this paper is a special case of Local Action Systems [12], and a variant of ESM systems [13]. The presentation used here is somewhat more basic, since the emphasis is on the form of individual rules, rather than on the description of concurrent processes composed from them. A rule consists of four components: the first two are graphs, called the left-hand side and the right-hand side of the rule, respectively. The other two components specify the embedding mechanism, as in [14]. A rule r is applied to a graph G by matching the left-hand side with a subgraph of G, replacing this subgraph by a copy of the right-hand side, and finally connecting this copy to the remaining part of G using the embedding mechanism. Formally, the fact that a graph G is transformed into a graph H, using a specific matching morphism i_L and specific morphisms i_R and i_{rest}, may be expressed using the notion of a *step*.

Definition 2 (rule). *A* rule *is a 4-tuple* (L, R, in, out) *where* L, R *are graphs and* $in, out \subseteq V_L \times V_R \times \Sigma_e \times \Sigma_e$.

Definition 3 (step). *A* step *is a 5-tuple* $(G, H, P, i_L, i_R, i_{rest})$ *such that the following holds:*

1. *G and H are graphs, $P = (L, R, in, out)$ is a rule, $i_L : L \to G_L$ is an isomorphism of L onto a subgraph G_L of G, $i_R : R \to H_R$ is an isomorphism of R onto a subgraph of H, and $i_{rest} : G_{rest} \to H_{rest}$ is an isomorphism, where G_{rest} and H_{rest} are the induced subgraphs of G and H on $(V_G \backslash i_L(V_L))$ and $(V_H \backslash i_R(V_R))$, respectively*
2. *For each $x \in i_R(V_R)$, $y \in V_H \backslash i_R(V_R)$, $a \in \Sigma_e$, $(x, a, y) \in E_H$ if and only if there exists an edge $(x', a', y') \in E_G$ such that $x' \in i_L(V_L), y' \in V_G \backslash i_L(V_L), y = i_{rest}(y')$ and $(i_L^{-1}(x'), i_R^{-1}(x), a', a) \in out$*
3. *For each $x \in i_R(V_R)$, $y \in V_H \backslash i_R(V_R)$, $a \in \Sigma_e$, $(y, a, x) \in E_H$ if and only if there exists an edge $(y', a', x') \in E_G$ such that $x' \in i_L(V_L), y' \in V_G \backslash i_L(V_L), y = i_{rest}(y')$ and $(i_L^{-1}(x'), i_R^{-1}(x), a', a) \in in$.*

Thus, informally, an occurrence G_L of L, is replaced by an occurrence H_R of R. An item $(v, w, \lambda, \delta) \in in$ is used to redirect incoming edges from $i_L(v)$ to $i_R(w)$, at the same time changing its label from λ to δ. The interpretation for *out* is similar, but for outgoing edges.

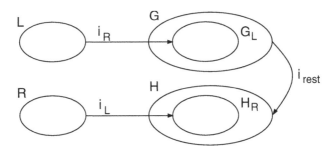

Fig. 1. A step.

3   Refactoring as Graph Transformation

To describe refactorings as graph transformations, a graph representation of programs is needed. The representation used in this paper is a simplified version of the one in [6], and is similar to a syntax tree. It contains only four types of nodes: C, M, V and E. The first three node types model the basic entities in object-oriented programs: classes, methods and variables. No distinction is made between variables, attributes and formal parameters of methods because this information can be easily extracted from the context (e.g. a V-node connected to a class is an attribute). The fourth type of node is E, which can be any other syntax tree element, like an assignment, a call, or a block structure node. To express the relations between the program entities, a set of edge labels is introduced. A summary of the types is given in Table 1. To make the representation of the refactorings in Section 6 easier to understand, a few subtypes of the E type are introduced. They are *Ca, As, B* for method calls, assignments and block structure nodes.

3.1   Example

In [15], a java program for the simulation of a token ring Local Area Network is considered. Such a network consists of a set of nodes, which are connected to each other. Packet forwarding is done on a hop by hop basis. Different types of nodes can exist on the network, like nodes offering printing services, or user workstations. In Figure 2 a simplified program graph representing this simulation is given. The class *Node* has two methods *accept* and *send* for forwarding packets (modeled with the *Packet* class), and an *originate* method for creating

Table 1. Types of nodes and edges in the graph representation.

Edge	description	Node	description
l	method lookup	*C*	Class
m	entity membership	*M*	Method Signature
i	inheritance	*V*	Variable
a	read/write access	*E*	Syntax Tree Node
p	actual/formal parameter		
c	dynamic call		
e	cascaded expression		
lhs	left hand side of assignment		
rhs	right hand side of assignment		
lv	local variable		

packets. Two subtypes of *Node* exist: *PrintServer* and *Workstation*, which provide printing and security services. To keep the example clear, the parameter relations between the different methods and the *Packet* class are omitted. It is

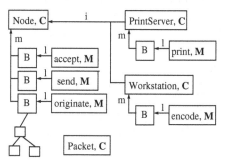

Fig. 2. Part of a program graph for a LAN simulation program.

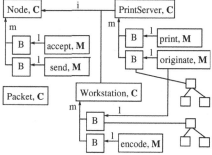

Fig. 3. The LAN program after the Push Down Method refactoring.

obvious that this design is not optimal, because a printing server should not originate new packets although it inherits this method from its parent class. It turns out that the *originate* method is too specific for the *Node* class. The *Push Down Method* refactoring can solve this problem by moving this behaviour from the *Node* class to its subclasses. The resulting program is given in Figure 3. Note that, for guaranteeing the preservation of behaviour, the refactoring duplicates the method to all of the subclasses. A further refactoring step should then remove the *originate* implementation from *PrintServer* when there are no references to the method from other classes. The required graph rewriting rule is depicted in Figure 4, and the embedding relation for it is given in Table 2. Note that in this case no edges need to be relabeled: for each item of the table, the 3rd and 4th components are equal. It is however clear that this rule needs to be adapted whenever the method body of the *originate* method changes. Hence it would be

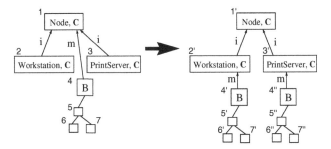

Fig. 4. Concrete rule for the push down method.

useful to introduce a more abstract representation of the rule, as depicted in Figure 5, where the label α of node 4 serves as a variable for trees representing method bodies. Such a rule could then be refined by replacing this node by the desired syntax tree. A construction allowing this is presented in Section 5.

Table 2. Embedding relation for the *Push Down Method* rule.

Incoming	Outgoing
$(1, 1', x, x)$	$(1, 1', x, x)$
$(2, 2', x, x)$	$(2, 2', x, x)$
$(3, 3', x, x)$	$(3, 3', x, x)$
$(4, 4', x, x), (4, 4'', x, x)$	$(4, 4', x, x), (4, 4'', x, x)$
$(5, 5', x, x), (5, 5'', x, x)$	$(5, 5', x, x), (5, 5'', x, x)$
$(6, 6', x, x), (6, 6'', x, x)$	$(6, 6', x, x), (6, 6'', x, x)$
$(7, 7', x, x), (7, 7'', x, x)$	$(7, 7', x, x), (7, 7'', x, x)$

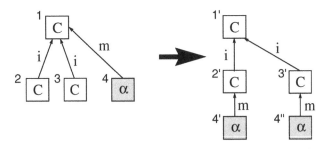

Fig. 5. Metarule for the push down method.

A second problem becomes apparent when one realizes that the number of subclasses of *Node* is not necessarily fixed. If a new subclass *DHCPServer* is added, a new refactoring rule has to be created. Rather than to specify a new rule for each number of subclasses, one would like to have a mechanism to designate a part of a rule that may be duplicated an arbitrary number of times, yielding more, similar rules. Hence the designated part serves as a prototype for the duplicated versions. In the case of *Push Down Method* the rule would specify explicitly how *one* subclass is handled, as in Figure 6, and then the part of the rule consisting of the nodes 2, 2' and 3' could be used as the part to be duplicated.

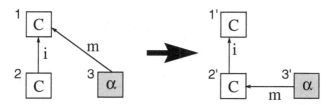

Fig. 6. Prototype rule for the push down method refactoring.

The example suggests that two mechanisms are needed to improve the expressivity of embedding based rules: one allowing for the refinement of items, an one allowing duplication. The latter, which is the simpler one, is introduced in the next section, whereas refinement is introduced in Section 5.

4   Duplication in Rewrite Rules

The aim of this section is to introduce a notion of duplication in productions. The part of a rule $P = (L, R, in, out)$ that is to be duplicated is designated by giving its set of nodes: a subset D of $V_L \cup V_R$, called the *duplication set*. For each $n \in \mathbb{N}^+$, a new version P_n of P is constructed by creating n new copies G_n of the subgraph G of $L \oplus R$ induced by D. ($L \oplus R$ is the graph obtained by putting together L and R, without overlap, and considering them as one graph.) Hence in general each new copy G_i contains a part of the left-hand side as well as a part of the right-hand side of P_n. The new copies G_i of G are connected by edges to the nodes of L and R outside D in the same way as G is in $L \oplus R$, and they are not connected to each other. The notion of duplication is formally defined as follows:

Definition 4 (duplication). *Let G be the subgraph of $L \oplus R$ induced by D, and let $G_1, G_2, \ldots, G_n$ be pairwise disjoint copies of G. For each $v \in D$, let v_i be its image in G_i and let $G_i = (V_i, E_i, lab_i)$. Let G_{rest} be the induced subgraph of $L \oplus R$ on $V_{(L \oplus R)} \backslash D$. Then the rule P_n is constructed as follows:*

- $V_{P_n} = (V_{(L \oplus R)} \backslash D) \cup (\bigcup_{i=1}^{n} V_i)$
- $lab_{P_n}(w) = \begin{cases} lab_{L \oplus R}(w) \ \textit{if } w \in V_{(L \oplus R)} \backslash D \\ lab_{L \oplus R}(v) \ \textit{if } w = v_i \textit{ for some } i \end{cases}$
- $E_{P_n} = E_{G_{rest}} \cup (\bigcup_{i=1}^{n} E_i) \cup \{(w, a, v_i) | \ w \in V_{(L \oplus R)} \backslash D, v \in D, (w, a, v) \in E_{L \oplus R}\} \cup \{(v_i, a, w) | \ w \in V_{(L \oplus R)} \backslash D, (v, a, w), v \in D, \in E_{L \oplus R}\}$

The embedding relation in_n of P_n contains, for each item (v, w, a, b) in the original embedding relation in, the following items. For a node v of D, let v_i denote the corresponding node of G_i.

- If $v, w \notin D$, then $(v, w, a, b) \in in_n$
- If $v \in D, w \notin D$, then $(v_i, w, a, b) \in in_n$, for each $i : 1 \le i \le n$
- If $v \notin D, w \in D$, then $((v, w_i, a, b) \in in_n$, for each $i : 1 \le i \le n$
- If $v \in D, w \in D$, then $(v_i, w_i, a, b) \in in_n$, for each $i : 1 \le i \le n$

The relation out_n is obtained in a similar way. A schematic representation of duplication in a rule is depicted in Figure 7.

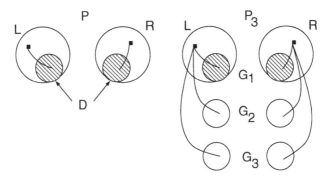

Fig. 7. Schematic representation of the duplication.

5   Refinement

The notion of refinement allows one to view a graph in a hierarchical way: some nodes and edges on the higher level represent graphs, and these graphs serve as building blocks to construct the lower level graph. The notion is inspired by the work of [10] on hierarchical graphs. However there are several differences: not only nodes, but also edges are refined, and there are only two levels in the hierarchy. Moreover the refinement of graphs is based on refinement of labels. Thus it is assumed that the information about the relationship between high-level labels and the corresponding sets of building blocks, as well as about the way these building blocks should be connected to each other, is provided together with the alphabets. In this way the problem of specifying the refinement is decoupled from the problem of designing rules. In this section, first the information associated to the alphabets is considered. Then it is described how this is used to obtain a notion of refinement for graphs. Finally, the latter notion is combined with that of graph rewriting, leading to a mechanism for the application of so-called metarules (rules containing labels that can be refined).

5.1   Alphabets

The notion of refinement is based on the introduction of new "high level" labels. These labels are variables in the sense that they represent an arbitrary element of a set of graphs (e.g. syntax trees of method bodies). Thus this set of graphs serves as the type or set of possible values of the variable. Thus, formally, the set $\Sigma = \Sigma_v \cup \Sigma_e$ of labels is extended by a set Δ_{var}. The high-level representation mentioned above is a graph over the alphabet $\Delta = \Delta_{var} \cup \Sigma$. Since there are variables for both nodes and edges, $\Delta_{var} = \Delta_{var,v} \cup \Delta_{var,e}$ and $\Delta_v = \Delta_{var,v} \cup \Sigma_v$, $\Delta_e = \Delta_{var,e} \cup \Sigma_e$. For a graph G over Δ_{var}, an atom x of G is a *variable* atom if $lab_G(x) \in \Delta$, and it is a *constant* atom otherwise.

Since the graphs considered are representations of programs, the way variable labels may be used in them is subject to restrictions similar to the ones that describe the metamodel of Section 3. The set $Adj(\Delta)$ contains all pairs (α, β) of node and edge labels that may be adjacent in valid graphs, i.e. in graphs that are valid representations of programs in the chosen metamodel. Thus $Adj(\Delta) = \{(\alpha, \beta) | \alpha \in \Delta_v, \beta \in \Delta_e, \alpha$ may occur as label of the source of a β-labeled edge$\} \cup \{(\beta, \alpha) | \alpha \in \Delta_v, \beta \in \Delta_e, \alpha$ may occur as label of the target of a β-labeled edge$\}$.

5.2 Refinement of Labels

An important issue in the proposed construction is the description of the way the refined atoms of a graph over Δ are glued together. The well-known pushout construction, in the category of graphs with label-preserving graph morphisms, is used for this, and hence the way graphs are glued is specified by *spans*.

Definition 5 (span). *Let G, H be graphs. A (G, H)-span is a 5-tuple (G, h_1, S, h_2, H) where G, H and S are graphs over Σ and $h_1 : S \to G, h_2 : S \to H$ are injective graph morphisms.*

One needs spans of a restricted form to specify the way constant atoms are connected with refined versions of abstract atoms:

1. for $a \in \Sigma_v$ and a Σ-graph G, a (G, a)-span is a span of the form
$$G \xleftarrow{a} \bullet \xrightarrow{a} \bullet$$
2. for $b \in \Sigma_e$ and a Σ-graph G, a (G, b)-span is a span of the form
$$G \xleftarrow{x} \bullet \longrightarrow \boxed{\overset{x'}{\bullet} \overset{b}{\longrightarrow} \bullet} \quad (x \text{ is mapped to } x').$$
3. for $b \in \Sigma_e$ and a Σ-graph G, a (b, G)-span is a span of the form
$$G \xleftarrow{x} \bullet \longrightarrow \boxed{\overset{x'}{\bullet} \overset{b}{\longleftarrow} \bullet} \quad (x \text{ is mapped to } x').$$

The relationship between variable labels and the sets of graphs which are their possible refinements is determined by a *refinement set*: for each variable label $\alpha \in \Delta_{var}$, this refinement set is denoted $\mathcal{R}(\alpha)$ and contains a set of graphs over Σ. Similarly, one should also specify in which way the refinements of an edge label may be connected to (or overlap whith) the refinements of the node labels of its source and target. This information must be available for each $(\alpha, \beta) \in Adj(\Delta)$ and is given in the form of a set of spans. Furthermore, one needs to specify the information about the way refined versions of variable labels may be connected to constant nodes and edges. This can be expressed using the special spans. Formally, it is assumed that the following information is given:

– for each $\alpha \in \Delta_{var}$, a set $\mathcal{R}(\alpha)$ of graphs over Σ.
– for each pair $(\alpha, \beta) \in Adj(\Delta)$, a set $\mathcal{R}(\alpha, \beta)$ such that
  • if α, β are variables, then $\mathcal{R}(\alpha, \beta)$ is a set of (x, y)-spans where $x \in \mathcal{R}(\alpha)$ and $y \in \mathcal{R}(\beta)$.
  • if α is variable and β is constant, then $\mathcal{R}(\alpha, \beta)$ is a set of (x, β)-spans where $x \in \mathcal{R}(\alpha)$.
  • if α is constant and β is variable, then $\mathcal{R}(\alpha, \beta)$ is a set of (α, x)-spans where $x \in \mathcal{R}(\beta)$.

Moreover it is assumed that, for each variable α, each $G_\alpha \in \mathcal{R}(\alpha)$, and each constant β such that $(\alpha, \beta) \in Adj(\Delta), \mathcal{R}(\alpha, \beta)$ contains exactly one (G_α, β)-span, and similar for the symmetric case (α constant and β variable). Hence, informally, there is only one way to connect the refined version of a variable to a concrete atom. This restriction is not fundamental to the notion of a refinement; it just turns out that we do not need the more general case for the purpose of this paper.

The use of these sets can be illustrated by the *Push Down Method* refactoring. For example, a variable α in Figure 5 is used for representing a syntax tree, thus, $\mathcal{R}(\alpha)$ is the set of trees representing method bodies. A typical example for the special spans is the connection between the constant m-edge and the variable α-node. This is specified in the set $\mathcal{R}(\alpha, m)$.

5.3   Refinement of Graphs

A graph G containing variable atoms can be refined by replacing each variable atom with label α by a graph chosen in $\mathcal{R}(\alpha)$, and gluing the so obtained building blocks together according to the pairs (e, v) and (v, e) in $Adj(G)$. Formally we define a refinement to be an assignment of graphs and spans to the atoms and the elements of $Adj(G)$, respectively. Evidently, this assignment has to be consistent with the information about refinement given together with the alphabets.

Definition 6 (refinement). *Let G be a graph over Δ. A refinement of G is a mapping ref defined on $A_G \cup Adj(G)$, such that*

1. *for each atom $x \in A_G$: $ref(x)$ is a graph over Σ*
2. *for each $(x, y) \in Adj(G)$, $ref(x, y)$ is a $(ref(x), ref(y))$-span*
3. *for each variable atom x of G, $ref(x) \in \mathcal{R}(lab_G(x))$*
4. *for each constant atom x of G, $ref(x) = G_x$*
5. *for each $(x, y) \in Adj(G)$ such that at least one of x and y is a variable atom, $ref(x, y) \in \mathcal{R}(lab_G(x), lab_G(y))$*
6. *for each $(x, y) \in Adj(G)$ such that both x and y are constant, $ref(x, y)$ is the span $(G_x \hookleftarrow G_y \xrightarrow{id} G_y)$ if x is an edge and y is a node, and $(G_x \xleftarrow{id} G_x \hookrightarrow G_y)$ if x is a node and y is an edge. Here $\xrightarrow{id}$ denotes the identity mapping.*

The notion is illustrated by Figure 8. Observe that the refinement *ref* is completely determined by the elements $ref(x)$ and $ref(x, y)$ where both x and y

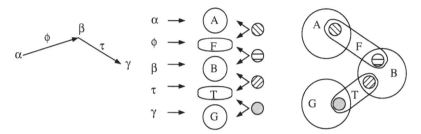

Fig. 8. refinement scheme for three connected nodes.

are variable atoms. Given a graph G over Δ and a refinement *ref* of G, one may now use the graphs *ref*(x) as building blocks which are glued together using the spans *ref*(x, y). It is well known that this construction can be defined formally using the notion of a pushout in the category of graphs.

Definition 7 (refined version of a graph). *Let G be a graph over Δ and let ref be a refinement. The refined version of G, denoted ref(G), is the common pushout object of the graphs ref(x), $x \in A_G$, and the spans ref(x, y), $(x, y) \in Adj(G)$.*

The pushout object *ref*(G) is unique up to isomorphism, Thus *ref*(G) is in fact a set of isomorphic graphs (i.e. an abstract graph). For our purpose it suffices to choose an arbitrary concrete representative. Also, observe that each constant atom of G corresponds to a unique item of *ref*(G). Thus it makes sense to distinguish between two kinds of atoms in *ref*(G):

– constant atoms, that correspond to constant atoms of G
– atoms that belong to the refined versions of variable items of G

5.4   Combining Refinement with Graph Rewriting

As explained earlier, our aim is to introduce graph rewriting rules over the extended alphabet Δ, i.e. including variable labels. Such a rule, called a *metarule*, can be refined, yielding a possibly large or infinite set of rules over Σ. These can be applied in the usual way to graphs representing programs. Although, in principle, one could generate all rules corresponding to a metarule $r = (L, R, in, out)$, and then apply one of them to a graph G, it is more realistic, from an implementation point of view, to do the instantiation on-the-fly and apply the metarule to G, going through the following phases.

a. Find an *occurrence*, i.e. determine the place in G where a rule, derived from the metarule, fits: this results in a refined version L_{low} of L as well as in a graph morphism m relating L_{low} to a subgraph of G.
b. Using this occurrence, and in particular the refinement determined by it, to construct a refined version R_{low} of the right-hand side R.
c. construct the embedding relations of the refined rule.
d. apply the so constructed rule to the occurrence chosen in a.

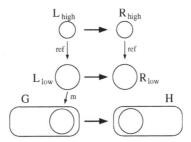

Fig. 9. Application of a metarule.

The situation is depicted in Figure 9. We will now explain in more detail how the steps a - c are carried out and lead to a formal definition of a metarule.

Determine an Occurrence. The usual situation in graph rewriting is that the place where a rule may be applied is determined by an injective graph morphism from the left-hand side of the rule into the graph that is rewritten. When applying a metarule to a graph G over Σ, the situation is slightly more complex: finding an occurrence now is a combination of finding a suitable refinement as well as a morphism.

Definition 8 (occurrence). *Let G be a graph over Δ and let H be a graph over Σ. An occurrence of G in H is a triple $(ref, \hat{G}, m)$ where ref is a refinement of G, $\hat{G}$ is a concrete representative of ref(G), and m is an injective morphism of $\hat{G}$ into H. The refinement should be compatible with the labeling of G:*

- *if x, x' are variable atoms of G and $lab_G(x) = lab_G(x')$, then ref$(x) = ref(x')$*
- *if $(x, y), (x', y') \in Adj(G)$, $(lab_G(x), lab_G(y)) = (lab_G(x'), lab_G(y'))$, and at least one of x, y is variable, then ref$(x, y) = ref(x', y')$*

Thus, the same labels and pairs of labels are refined in the same way throughout G. For the purpose of this paper, i.e. when the mechanism is used to describe refactoring, it is useful to consider only *maximal* occurrences, i.e. occurrences such that the image of $\hat{G}$ in H is maximal (formally, there exists no occurrence $(ref', \hat{G}', m')$ such that $\hat{G} \neq \hat{G}'$ and there is an injective morphism from $\hat{G}$ into $\hat{G}'$). This restriction will avoid a situation where the refined rule matches e.g., not a complete method body but only a part of it.

Construction of ref(R). To be able to construct a unique refinement of the right hand side of a metarule, given a specific occurrence of its left hand side L, it is essential that the relevant variable labels and pairs of labels of the rule have a unique refinement assigned to them by the occurrence. Therefore, for each variable item x in R, an atom $\tilde{x}$ of L, carrying the same label, should exist. And similarly, for each combination of variable atoms (x, y) in $Adj(R), (\tilde{x}, \tilde{y}) \in Adj(L)$. Then an occurrence of L determines a refinement $\widetilde{ref}$ of R:

- For each variable atom x of R : $\widetilde{ref}(x) = ref(\tilde{x})$
- For each $(x, y) \in Adj(R)$ with x, y variable: $\widetilde{ref}(x, y) = ref(\tilde{x}, \tilde{y})$

Refinement of the Embedding. The refinement of a metarule requires the refinement of its embedding relations as well. One distinguishes between two kinds of nodes in the refined versions of L and R: those that correspond to constant nodes of L or R, and the other nodes. For the nodes of the first kind, the embedding relations are explicitly given as part of the metarule. Nodes of the second kind belong to the refinement of a variable item. Because of the restrictions on metarules, each node x of this kind in R has a counterpart $\tilde{x}$ in L. When the refined rule is applied, then the embedding edges of $\tilde{x}$ are redirected to x. Hence all elements $(\tilde{x}, x, a, a)$, for each label a of Σ, must be added to the embedding relations of the metarule to obtain the embedding relations of the refined rule. From the previous paragraphs, the formal definition of a metarule can now be deduced.

Definition 9 (metarule). *A metarule is a rule $r = (L, R, in, out)$ over the extended alphabet Δ, but satisfying the following restrictions:*

1. *Only constant nodes of L and R occur in the embedding relations in and out.*
2. *For each variable item x of R, there exists an item $\tilde{x}$ of L such that $lab_R(x) = lab_L(\tilde{x})$.*
3. *If x, y are variable items of R, and $(x, y) \in Adj_R$, then $(\tilde{x}, \tilde{y}) \in Adj_L$*

5.5   Application to *Push Down Method*

The final version of the refactoring corresponds indeed to the one in Figure 6. This rule is called the prototype rule and contains only one superclass. The duplication set is $D = \{2, 2', 3'\}$ (i.e. on both sides of the rule, equally many copies of the subclass node have to be made, but the method body only needs to be duplicated on the right hand side).

5.6   Combining Refinement and Duplication

As duplication is intended to be applied to a metarule before refinement, it must be able to handle variable atom duplication. Therefore a few adaptations to the definition of duplication are useful to strengthen the expressivity of the rules. Consider a variable atom x with label α in the duplication set D. In the duplication operation, as defined in Section 4, the part of the production corresponding to D is duplicated, including the variable item. Thus, if duplicated i times, leads to i versions of x, all labeled α. In an occurrence this matches only a graph containing i identical substructures, each of which is a refinement of α. However, often one wants something else: one wants to allow different refinements for each version (for example, the γ variable in the Inline Method refactoring, described in Section 6.2). Therefore, some of the variable labels may be designated as *special*. For special labels, a new version is created for each duplicate. Formally, let $\Delta_{special} \subseteq \Delta_{var}$ be the set of special labels. For each special label $\alpha \in \Delta_{special}$, let $\alpha_1, \alpha_2, \ldots$ be new variable labels, let $\mathcal{R}(\alpha_i) = \mathcal{R}(\alpha)$ and let $\mathcal{R}(\alpha_i, \beta) = \mathcal{R}(\alpha, \beta)$ for each label β. Let for each $\alpha \in \Delta$, and $i \geq 1$,

$$\phi_i(\alpha) = \begin{cases} \alpha & \text{if } \alpha \text{ is not special} \\ \alpha_i & \text{if } \alpha \text{ is special} \end{cases}$$

The definition of duplication (Definition 4) is modified as follows:

- The $G_i = (V_i, E_i, lab_i)$ are pairwise disjoint and there exists a function $f_i :$ $V_G \to V_i$ such that
  - f_i is bijective
  - for each $x \in V_i$, such that $x = f_i(v)$: $lab_i(x) = \phi_i(v)$
  - $E_i = \{(f_i(x), \phi_i(a), f_i(y)) \mid (x, a, y) \in E_G\}$
- E_{P_n} contains the edges $E_{G_{rest}}$ $\cup$ $(\bigcup_{i=1}^{n} E_i)$ $\cup$ $\{(w, \phi_i(a), f_i(v)) | w$ $\in V_{(L \oplus R)} \backslash D, v \in D, (w, a, v) \in E_{L \oplus R}\} \cup \{(f_i(v), \phi_i(a), w) | w \in V_{(L \oplus R)} \backslash D, v$ $\in D, (v, a, w) \in E_{L \oplus R}\}$

6   Using Metarules for Refactoring

In this section, two other refactorings which can be expressed using the proposed extensions are presented: *Extract Method* and *Inline Method*. For each, an informal definition and a sketch of the metarule is given.

6.1   Extract Method

The *Extract Method* refactoring simplifies methods by identifying large parts of existing methods as stand-alone methods. The input for the refactoring is a block of code inside the method, specified by the programmer. A small code example illustrating the refactoring is given below. On the left side, the original program is listed. When the programmer decides that the method sequence a(m), b(x), c() is a coherent part of the method, this sequence can be grouped in a new method. Inside the original method, a call to the new method is added. The refactored code is shown in the right hand side:

```
public class c1 {
    void method1(int m) {
        int x;
        start();
        a(m);
        b(x);
        c();
        stop();
    }
}
```

```
public class c1 {
    void execute(int m, int x) {
        a(m);
        b(x);
        c();
    }
    void method1(int m)  {
        int x;
        start();
        execute(m, x);
        stop();
    }
}
```

When modelling this refactoring by a metarule, one needs to consider the different variable aspects of the refactoring:

1. The expression block selected by the programmer is variable (e.g. the programmer could have chosen to extract only the `a(m);b(x)` sequence).
2. The position of the expression inside the original method body is variable
3. The variables can be accessed from arbitrary points in the syntax tree (e.g. the variable `x` could be accessed by the method `b` instead of `a`)
4. The number of local variables and parameters accessed by the expression is unknown. This information is needed because these variables should be passed as parameters to the new method (e.g. there could be a second local variable `y`, accessed by the expression block). This variable should then be added to the parameter list of `execute()`.

The first problem is solved using a variable node label δ, similar to the label for the method body in the *Push Down Method* refactoring. The second problem is solved using an edge with variable label α, α can be refined into a path from the root of the method body to the start of the expression. The third problem can also be modeled using the refinement of edges. In this case, the refined version of the high level edge contains a set of access edges from the syntax tree to the parameters and local variables. Such an edge is needed for every local variable and parameter. The prototype refactoring, showing the metarule for a situation where only one method parameter and one local variable is accessed, is depicted in Figure 10. The fourth problem is solved by defining two duplication sets with the refactoring, the first one for the local variables: $D_1 = \{4, 4', 6'', 7''\}$, and the second one for the parameters: $D_2 = \{2, 2', 6', 7'\}$.

The embedding relation for this refactoring is very simple: all incoming and outgoing edges to or from nodes on the left hand are redirected to the corresponding nodes on the right hand side.

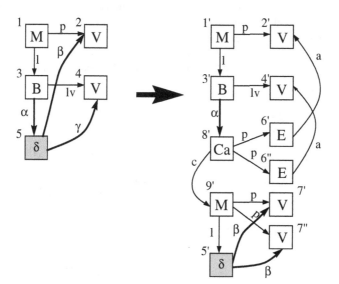

Fig. 10. The Extract Method Metarule.

6.2   Inline Method

The *Inline Method* is the opposite refactoring of Extract Method and replaces a call to a method by an inlined version of this method. Although this refactoring seems in general not a good idea, because it causes code duplication, it might be a good choice for a software engineer because it allows one to reduce the number of (short) class methods and, consequently, the complexity of the class interface. The refactoring is illustrated in the code example below. The method `print` of the class `C1` is very short, actually it does nothing more than calling the similar system function. In the refactored version, it is therefore replaced by an inlined version of the method.

```
public class C1 {                      public class C1 {
    void print(int i) {                    void method(int a, int b) {
        System.out.println(i);                 dosomethingwith(a, b);
    }                                          System.out.println(a);
    void method(int a, int b) {            }
        dosomethingwith(a, b);         }
        print(a);
    }
}
```

The metarule is depicted in Figure 11. Both the method containing the call (node 3) and the obsolete method (node 7) are involved in the rule. The method call can occur at any place inside the method. Thus, the syntax tree path from the method root to the call is modelled by the variable edge α. A special label γ is used to model the actual parameters of the old method, because the duplication of this variable node requires that each copy should be refined to a different actual parameter. The access edges from the syntax tree of the old method to its local variables need to be modeled as well. This is done using the variable edge with label δ. The general outline of the refactoring execution is simple: at the

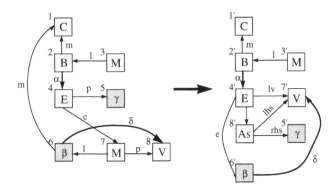

Fig. 11. The Inline Method Metarule.

place of the call a new codeblock is added. At the start of the block, a new local variable is created for every parameter, and the actual parameter expression is assigned to it. Finally, the syntax tree is copied and and the references to the parameters are updated to become references to the new local variables. One duplication set $D = \{5, 5', 8, 8', 7'\}$ is used, to take care of the unknown number of parameters of the method. Note that this version of Inline method is not very optimal, because of the creation of new variables. The code can still be made more efficient by applying subsequent refactorings such as *Inline Temporary Variable* [1] to remove redundant variables. The embedding is similar as in the previous case.

References

1. Fowler, M.: Refactoring - Improving the Design of Existing Code. Addison Wesley (1999)
2. Opdyke, W.F.: Refactoring Object-Oriented Frameworks. PhD thesis, University of Illinois, Urbana-Champaign, IL, USA (1992)
3. Roberts, D.: Practical Analysis for Refactoring. PhD thesis, University of Illinois at Urbana-Champaign (1999)
4. Roberts, D., Brant, J., Johnson, R.: A refactoring tool for Smalltalk. Theory and Practice of Object Systems **3** (1997) 253–263
5. Beck, K., Fowler, M.: Planning extreme programming. Addison-Wesley (2001)
6. Mens, T., Demeyer, S., Janssens, D.: Formalising behaviour preserving program transformations. In: Graph Transformation. Volume 2505 of Lecture Notes in Computer Science., Springer-Verlag (2002) 286–301 Proc. 1st Int'l Conf. Graph Transformation 2002, Barcelona, Spain.
7. Paolo Bottoni, Francesco Parisi Presicce, G.T.: Specifying integrated refactoring with distributed graph transformations. In: Proceedings of AGTIVE 2003. (2003) 227–242
8. Kreowski, H.J., Kuske, S.: Graph transformation units with interleaving semantics. Formal Aspects of Computing **11** (1999) 690–723
9. Engels, G., Schürr, A.: Encapsulated hierarchical graphs, graph types and meta types. Electronic Notes in Theoretical Computer Science **2** (1995)
10. Drewes, F., Hoffmann, B., Plump, D.: Hierarchical graph transformation. Lecture Notes in Computer Science **1784** (2000) 98–113
11. Engelfriet, J., Rozenberg, G.: Handbook of Graph Grammars and Computing by Graph Transformation. Volume 1. World Scientific (1997)
12. Janssens, D.: Actor grammars and Local Actions. In: Handbook of Graph Grammars and Computing by Graph Transformation. World Scientific (1999) 56–106
13. Janssens, D.: ESM systems and the composition of their computations. Graph Transformations in Computer Science **776** (1994) 203–217
14. Janssens, D., Mens, T.: Abstract semantics for ESM systems. Fundamenta Informaticae **26** (1996) 315–339
15. Janssens, D., Demeyer, S., Mens, T.: Case study: Simulation of a LAN. In: Electronic Notes in Theoretical Computer Science. Volume 72, issue 4., Elsevier (2002)

Derivations
in Object-Oriented Graph Grammars

Ana Paula Lüdtke Ferreira[1] and Leila Ribeiro[2]

[1] Universidade do Vale do Rio dos Sinos – UNISINOS
[2] Universidade Federal do Rio Grande do Sul – UFRGS
anapaula@exatas.unisinos.br, leila@inf.ufrgs.br

Abstract. This work aims to extend the algebraic single-pushout approach to graph transformation to model object-oriented systems structures and computations. Graphs whose sets of nodes and edges are partially ordered are used to model the class hierarchy of an object-oriented system, making the inheritance and overriding relations explicit. Graphs typed over such structures are then used to model systems of objects, whose behaviour is given by an object-oriented graph grammar. It will be shown how the derivations which arise from those grammars are compatible with the way object-oriented programs are defined and executed.

1 Introduction

The principles behind the object-oriented paradigm – *data and code encapsulation, information hiding, inheritance* and *polymorphism* – fit very well into the needs of modular system development, distributed testing, and reuse of software faced by system designers and engineers, making it perhaps the most popular paradigm of system development in use nowadays. The most distinguished features of object-oriented systems are inheritance and polymorphism, which make them considerably different from other systems in both their architecture and model of execution.

Inheritance is the construction which permits a class (in the class-based approach [5]) or an object (in the object-based approach [25]) to be specialized from an already existing one. This newly created object carries (or "inherits") all the data and the actions belonging to its primitive object, in addition to its own data and actions. If this new object is further extended using inheritance, then all the new information will also be carried along. The relation "inherits from" induces a hierarchical relationship among the defined classes of a system, which can be viewed as a set of trees (single inheritance) or as an acyclic graph (multiple inheritance). The subclass polymorphism [4] relation follows exactly the same structure.

Subclass polymorphism specifies that an instance of a subclass can appear wherever an instance of a superclass is required. Hence, in order to apply it, the inheritance relationship must be made explicit by any formalism aiming to model

H. Ehrig et al. (Eds.): ICGT 2004, LNCS 3256, pp. 416–430, 2004.
© Springer-Verlag Berlin Heidelberg 2004

object-oriented systems. One of the most useful ways inheritance and polymorphism can be used is through method redefinition. The purpose of method redefinition is to take advantage of the polymorphism concept through the mechanism known as *dynamic binding*, which is an execution time routine to determine what piece of code should be called when a message is sent to an object.

Object-oriented programs usually make use of inheritance (which is the most common way of code reuse) and method redefinition to achieve their goals. Therefore, it should be expected that formalisms for the specification of object-oriented architectures or programs reflect those concepts, otherwise the use of such formalisms will neglect concepts that have a major influence in their organization and model of execution.

Graph grammars have been used to specify various kinds of software systems, where the graphs correspond to the states and the graph productions to the operations or transformations of such systems [15]. System specifications through graphs often rely on labelled graphs or typed graphs to represent different system entities [1], [2], [8], [10], [18], [20], [22], [23]. However, neither labelling nor typing reflect the inheritance relation among objects, and polymorphism cannot be applied if it is not made explicit. To do so, a class, if represented by a node or edge in a graph, should have a multiplicity of labels or types assigned to it, which is not compatible with the usual way labelling or typing are portrayed (as functions).

This work aims to extend the algebraic single-pushout approach to graph grammars [12], [19] to model object-oriented systems structures and computations. Graphs whose sets of nodes and edges are partially ordered [9] are used to model the class hierarchy of an object-oriented specification, making the inheritance and overriding [6], [24] relations explicit. Typing morphisms and graph morphisms are compatible with the order structure, adequately reflecting inherited attributes and methods. Some preliminary results were presented in [16], and here we go further into the theory. The main result of this paper is to show that a direct derivation in an object-oriented graph grammar is a pushout on the category of object-oriented graphs and their morphisms. It is an important result in the sense that the algebraic approach to graph grammars rely on categorical constructs to express most of its results. Having proven that those constructs exist in the new setting, all results derived from them can be automatically used.

This paper is structured as follows: Section 2 presents class-model graphs, which are graphs whose node and edge sets are special partially ordered sets, and whose structure fits the way object-oriented programming specifications are built. Class-model graphs are meant to portray the structure of object-oriented systems, and they will provide a typing structure to the graphs and grammars presented in the rest of the text. Section 3 presents the graphs and grammars used to represent object-oriented systems and programs. It is also shown how the semantics of dynamic binding can be given in terms of pushouts in a suitable category. Finally, Section 4 presents some final remarks and directions for future work.

2 Class-Model Graphs

Inheritance, in the context of the object-orientation paradigm, is the construction which permits an object[1] to be specialized from an already existing one. When only single inheritance is allowed, the hierarchical relationship among the classes of a system can be formalized as a set of trees. A *strict relation*, defined below, formalizes what should be the fundamental object-oriented hierarchical structure of classes.

Definition 1 (Strict relation). *A finite binary relation* $R \subseteq A \times A$ *is said a strict relation if and only if it has the following properties: if* $(a, a') \in R$ *then* $a \neq a'$ *(R has no reflexive pairs); if* $(a, a_1), (a_1, a_2), \ldots, (a_{n-1}, a_n), (a_n, a') \in R$, $n \geqslant 0$, *then* $(a', a) \notin R$ *(R has no cycles); for any* $a, a', a'' \in A$, *if* $(a, a'), (a, a'') \in R$ *then* $a' = a''$ *(R is a function).*

Notice that the requirement concerning the absence of cycles and reflexive pairs on strict relations is consistent with both the creation of classes and redefinition of methods (overriding). A class is defined as a specialization of at most one class (single inheritance), which must exist prior to the creation of the new one. A method can only redefine another method (with the same signature) if it exists in an already existing primitive class. Hence, neither a class can ever be created nor a method can be redefined in a circular or reflexive way. It is easy to prove that the reflexive and transitive closure of a strict relation R is a partial order, so in the rest of this paper we will assume that result.

Definition 2 (Strict ordered set). *A* strict ordered set *is a pair* $\langle P, \sqsubseteq_P^* \rangle$ *where* P *is a set,* $\sqsubseteq_P$ *is a strict relation, and* $\sqsubseteq_P^*$ *is its reflexive and transitive closure.*

Object-oriented systems consist of instances of previously defined classes which have an internal structure defined by attributes and communicate among themselves solely through message passing. That approach underlies the structure of the graphs used to model those systems, as defined next.

Definition 3 (Class-model graph). *A class-model graph is a tuple* $\langle V_\sqsubseteq, E_\sqsubseteq,$ $L, src, tar, lab \rangle$ *where* $V_\sqsubseteq = \langle V, \sqsubseteq_V^* \rangle$ *is a finite strict ordered set of vertices,* $E_\sqsubseteq = \langle E, \sqsubseteq_E^* \rangle$ *is a finite strict ordered set of (hyper)edges,* $L = \{\text{attr}, \text{msg}\}$ *is an unordered set of two edge labels,* $src, tar : E \to V^*$ *are monotone order-preserving functions, called respectively* source *and* target *functions,* $lab : E \to L$ *is the edge* labelling *function, such that the following constraints hold:*

Structural constraints: *for all* $e \in E$, *the following holds:*
 – *if* $lab(e) = \text{attr}$ *then* $src(e) \in V$ *and* $tar(e) \in V^*$, *and*
 – *if* $lab(e) = \text{msg}$ *then* $src(e) \in V^*$ *and* $tar(e) \in V$.
 Sets $\{e \in E \mid lab(e) = \text{attr}\}$ *and* $\{e \in E \mid lab(e) = \text{msg}\}$ *will be denoted by* $E|_{\text{attr}}$ *and* $E|_{\text{msg}}$, *respectively.*

---

[1] Object-based and class-based models are actually equivalent in terms of what they can represent and compute [25], so we will sometimes use the terms *object* and *class* interchangeably. We hope this will not cause any confusion to the reader.

Order relations constraints: *for all $e \in E$, the following holds:*

1. *if $(e, e') \in \sqsubseteq_E$ then $lab(e) = lab(e') = $ msg,*
2. *if $(e, e') \in \sqsubseteq_E$ then $src(e) = src(e')$,*
3. *if $(e, e') \in \sqsubseteq_E$ then $(tar(e), tar(e')) \in \sqsubseteq_V^+$, and*
4. *if $(e', e) \in \sqsubseteq_E$ and $(e'', e) \in \sqsubseteq_E$, with $e' \neq e''$, then $(tar(e'), tar(e'')) \notin \sqsubseteq_V^*$ and $(tar(e''), tar(e')) \notin \sqsubseteq_V^*$.*

That approach underlies the structure of the graphs used to model those systems. Each graph node is a class identifier, hyperarcs departing from it correspond to its internal attributes, and messages addressed to it consist of the services it provides to the exterior (i.e., its methods). Notice that the restrictions put to the structure of the hyperarcs assure, as expected, that messages target and attributes belong to a single object.

Inheritance and overriding hierarchies are made explicit by imposing that graph nodes (i.e., the objects) and message edges (i.e., methods) are strict ordered sets (Definition 2). Notice that only single inheritance is allowed, since $\sqsubseteq_V$ is required to be a function. The relation between message arcs, $\sqsubseteq_E$, establishes which methods will be redefined within the derived object, by mapping them. The restrictions applied to $\sqsubseteq_E$ ensure that methods are redefined consistently, i.e., only two message arcs can be mapped (1), their parameters are the same (2), the method being redefined is located somewhere (strictly) above in the class-model graph (under $\sqsubseteq_V^+$) (3), and only the closest message with respect to relations $\sqsubseteq_V$ and $\sqsubseteq_E$ can be redefined (4).

Since so many structures in computer science can be represented as graphs, and a number of other structures in the same field are adequately represented by order relations, the idea of combining the two formalisms is appealing. However, this combination does not appear often in the literature. In [3], for instance, "partially ordered graphs" are defined, which consist of ordinary labelled graphs together with a tree structure on their nodes. Partially ordered graph grammars are also defined, which consist of graph productions and tree productions, which must assure that the rewriting process maintains the tree structure. They are applied on lowering the complexity of the membership problem of context sensitive string grammars. Graphs labelled with alphabets equipped with preorders (i.e., reflexive and transitive binary relations) appear in [21] to deal with variables within graph productions. Unification of terms can be achieved (by the rule morphism) if the terms are related by the order relation, which means that the ordering is actually a sort of variable typing mechanism. The concluding remarks of this work present some ideas on using the framework to describe inheritance, but this direction seems not have been pursuit.

It can be shown that class-model graphs and suitable morphisms between them form a cocomplete category, and the two most common operations of object-oriented system extension, namely *object extension by inheritance* and *object creation by aggregation*, can be expressed in terms of colimits in it [17]. Our focus, however, lies on object-oriented *systems*, i.e., a collection of instances from the classes defined in a class-model graph. Such structure will be characterized as a graph typed over a class-model graph.

Remark 1. For any partially ordered set $\langle P, \sqsubseteq_P \rangle$, an induced partially ordered set $\langle P^*, \sqsubseteq_{P^*} \rangle$ can be constructed, such that for any two strings $u = u_1 \ldots u_n, v = v_1 \ldots v_n \in P^*$, we have that $u \sqsubseteq_{P^*} v$ if and only if $|u| = |v|$ and $u_i \sqsubseteq_P v_i$, $i = 1, \ldots, |u|$. If $P_\sqsubseteq$ and $Q_\sqsubseteq$ are partially ordered sets, any monotone function $f : P \to Q$ can be extended to a monotone function $f^* : P^* \to Q^*$, with $f^*(p_1 \ldots p_n) = f(p_1)f(p_2) \ldots f(p_n) = q_1 \ldots q_n$ where $p_1 \ldots p_n \in P^*$ and $q_1 \ldots q_n \in Q^*$.

Definition 4 (Class typed graph). *A class typed graph $G^{\mathcal{C}}$ is a tuple $\langle G, t, \mathcal{C} \rangle$, where $\mathcal{C} = \langle V_\sqsubseteq, E_\sqsubseteq, L, src, tar, lab \rangle$ is a class-model graph, G is a hypergraph, and t is a pair of total functions $\langle t_V : V_G \to V, t_E : E_G \to E \rangle$ such that $(t_V \circ src_G) \sqsubseteq_{V^*} (src \circ t_E)$, and $(t_V \circ tar_G) \sqsubseteq_{V^*} (tar \circ t_E)$.*

Class typed graphs reflect the inheritance of attributes and methods within the object-oriented paradigm. Notice that they are hypergraphs typed over a class-model graph. However, the typing morphism is more flexible than the traditional one [8]: an edge can be incident to any node which is connected to its typing edge in the underlying order relation. This definition reflects the idea that an object can use any attribute one of its primitive classes have, since it was inherited when the class was specialized.

Remark 2. For all diagrams presented in the rest of this text, $\mapsto$-arrows denote total morphisms whereas $\to$-arrows denote arbitrary morphisms (possibly partial). For a partial function f, $dom(f)$ represents its domain of definition, $f?$ and $f!$ denote the corresponding domain inclusion and domain restriction. Each morphism f within category **SetP** can be factorized into components $f?$ and $f!$.

Definition 5 (Class typed graph morphism). *Let $G_1^{\mathcal{C}} = \langle G_1, t_1, \mathcal{C} \rangle$ and $G_2^{\mathcal{C}} = \langle G_2, t_2, \mathcal{C} \rangle$ be two class typed graphs typed over the same class-model graph $\mathcal{C}$. A class typed graph morphism $h : G_1^{\mathcal{C}} \to G_2^{\mathcal{C}}$ between $G_1^{\mathcal{C}}$ and $G_2^{\mathcal{C}}$, is a pair of partial functions $h = \langle h_V : V_{G_1} \to V_{G_2}, h_E : E_{G_1} \to E_{G_2} \rangle$ such that the diagram (in **SetP**)*

$$
\begin{array}{ccccc}
E_{G_1} & \xleftarrow{h_E?} dom(h_E) \xrightarrow{h_E!} & E_{G_2} \\
{\scriptstyle src_{G_1}, tar_{G_1}} \downarrow & & \downarrow {\scriptstyle src_{G_2}, tar_{G_2}} \\
V_{G_1}^* & \xrightarrow{\quad h_V^* \quad} & V_{G_2}^*
\end{array}
$$

commutes, for all elements $v \in dom(h_V)$, $(t_{2V} \circ h_V)(v) \sqsubseteq_{V^*}^* t_{1V}(v)$, and for all elements $e \in dom(h_E)$, $(t_{2E} \circ h_E)(e) \sqsubseteq_E^* t_{1E}(e)$. If $(t_{2E} \circ h_E)(e) = t_{1E}(e)$ for all elements $e \in dom(h_E)$, the morphism is said to be strict.

A graph morphism is a mapping which preserves hyperarcs origins and targets. Ordinary typed graph morphisms, however, cannot describe correctly morphisms on object-oriented systems because the existing inheritance relation among objects causes that manipulations defined for objects of a certain kind are valid to all objects derived from it. So, an object can be viewed as not being uniquely typed, but having a type *set* (the set of all types it is connected via the relation on nodes). Defining a graph morphism compatible with the underlying order relations assures that polymorphism can be applied consistently.

The following theorem has been proven in [16][2], and will be used to prove the remaining results of this paper:

Theorem 1 (Category GraphP($\mathcal{C}$)). *There is a category* **GraphP**($\mathcal{C}$) *which has class typed graphs over a class-model graph $\mathcal{C}$ as objects and class typed graph morphisms as arrows.*

Remark 3. Some usual notation from order theory is used in the rest of this paper. More specifically, given a partially ordered set $\langle P, \sqsubseteq_P \rangle$, a subset $A \subseteq P$ is an *upper set* if whenever $x \in A$, $y \in P$ and $x \sqsubseteq_P y$ we have $y \in A$. The upper set of $\{x\}$, with $x \in P$ (also called the set of all elements *above* x) is denoted by $\uparrow x$. A *lower set* and the set of all elements below some element $x \in A$, denoted by $\downarrow x$, is defined dually. An element $x \in P$ is called an *upper bound* for a subset $A \subseteq P$, written $A \sqsubseteq x$, if and only if $a \sqsubseteq_P x$ for all $a \in A$. The set of all upper bounds of A is denoted by $ub(A)$, and if is has a least element (i.e., an element which is below all others), that element is denoted by $lub(A)$ or $\sqcup A$. *Lower bounds*, the set of all lower bounds $lb(A)$, and the greatest lower bound $glb(A)$ or $\sqcap A$, can be defined dually.

Definition 6 (Attribute and message sets). *Let $\langle G, t, \mathcal{C} \rangle$ be a class typed graph. Then the following functions are defined:*

- *the* attribute set function $attr_G : V_G \to 2^{E_G}$ *return for each vertex $v \in V_G$ the set $\{e \in E_G \mid src_G(e) = v \wedge lab(t_E(e)) = \text{attr}\}$;*
- *the* message set function $msg_G : V_G \to 2^{E_G}$ *returns for each vertex $v \in V_G$ the set $\{e \in E_G \mid tar_G(e) = v \wedge lab(t_E(e)) = \text{msg}\}$;*
- *the* extended attribute set function, $attr_{\mathcal{C}}^* : V \to 2^E$, *where $attr^*(v) = \{e \in E \mid lab(e) = \text{attr} \wedge src(e) \in \uparrow v\}$;*
- *the* extended message set function, $msg_{\mathcal{C}}^* : V \to 2^E$, *where $msg^*(v) = \{e \in E|_{\text{msg}} \mid tar(e) \in \uparrow v \wedge \neg \exists e' \in E|_{\text{msg}} : tar(e') \in \uparrow v \wedge e' \sqsubseteq_E e\}$.*

For any class typed graph $\langle G, t, \mathcal{C} \rangle$ there is a total function $t_E^* : 2^{E_G} \to 2^{E_\mathcal{M}}$, which can be viewed as the extension of the typing function to edge (or node) sets. The function t_E^*, when applied to a set $E \in 2^{E_G}$, returns the set $\{t_E(e) \in E_\mathcal{M} | e \in E\} \in 2^{E_\mathcal{M}}$. Notation $t_E^*|_{\text{msg}}$ and $t_E^*|_{\text{attr}}$ will be used to denote the application of t_E^* to sets containing exclusively message and attribute (respectively) hyperarcs. Now, given the functions already defined, we can present a definition of the kind of graph which represents an object-oriented system.

---

[2] Class typed graphs used to be called "hierarchical graphs", we have changed this denomination to avoid unnecessary confusion with *hierarchic graphs*, which are graphs where a subgraph can be abstracted to one node and a number of edges between two abstracted subgraphs to one edge [1].

Definition 7 (Object-oriented graph). *Let $\mathcal{C}$ be a class-model graph. A class typed graph $\langle G, t, \mathcal{C} \rangle$ is an object-oriented graph if and only if the following diagram (in* **Set***)*

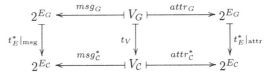

can be constructed, and all squares commute. If, for each $v \in V_G$, the function $t_E^*|_{\mathrm{attr}}(attr_G(v))$ is injective, $G^{\mathcal{C}}$ is said a strict object-oriented graph. If $t_E^*|_{\mathrm{attr}}(attr_G(v))$ is also surjective, $G^{\mathcal{C}}$ is called a complete object-oriented graph.

It is important to realize what sort of message is allowed to target a vertex on an object-oriented graph. The left square on the diagram presented in Definition 7 ensures that an object can only have a message edge targeting it if that message is typed over one of those returned by the extended message set function. It means that the only messages allowed are the least ones in the redefinition chain to which the typing message belongs. This is compatible with the notion of *dynamic binding*, since the method actually called by any object is determined by the actual object present at a certain point of the computation.

Object-oriented graphs can also be *strict* or *complete*. Strict graphs require that nodes do not possess two arcs typed as the same element on the underlying object-model graph. The requirement concerned the injectivity of t_E^* guarantees that there will be no such exceeding attribute connected to any vertex. For an object-oriented graph to be *complete*, however, it is also necessary that all attributes defined on all levels along the node hierarchy are present. The definition of a complete object-oriented graph is coherent with the notion of inheritance within object-oriented framework, since an object inherits all attributes, and exactly those, from its primitive classes.

Object-oriented systems are often composed by a large number of objects, which can receive messages from other objects (including themselves) and react to the messages received. Object-oriented graphs also may have as many objects as desired, since the number and type of attributes (arcs) in each object (vertex) is limited, but the number and type of vertices in the graph representing the system is restricted only by the typing morphism.

Theorem 2. *There is a category* **OOGraphP**$(\mathcal{C})$ *with object-oriented graphs as objects and class typed graph morphisms as arrows.*

Proof. (sketch) It is easy to see that object-oriented graphs are just a special kind of class typed graphs, and that **OOGraphP**$(\mathcal{C})$ is a subcategory of **GraphP**$(\mathcal{C})$(Theorem 1).

3   Object-Oriented Graphs Grammars

Complete object-oriented graphs (Definition 7) can model an object-oriented system. However, in order to capture the system computations, a graph grammar formalism should be introduced. The most fundamental notion in a graph grammar is the concept of graph production, or rule. When a rule is intended to represent some system evolution, it must reflect the way this evolution takes place. The rule restrictions presented in this text are object-oriented programming principles, as described next.

First, no object may have its type altered nor can any two different elements be identified by the rule morphism. This is accomplished by requiring the rule morphism to be injective on nodes and arcs (different elements cannot be merged by the rule application), and the mapping on nodes to be invertible (object types are not modified).

The left-hand side of a rule is required to contain exactly one element of type message, and this particular message must be deleted by the rule application, i.e., each rule represents an object reaction to a message which is consumed in the process. This demand poses no restriction, since systems may have many rules specifying reactions to the same type of message (non-determinism) and many rules can be applied in parallel if their triggers are present at an actual state and the referred rules are not in conflict [12], [22]. Systems' concurrent capabilities are so expressed by the grammar rules, which can be applied concurrently (accordingly to the graph grammar semantics), so one object can treat any number of messages at the same time.

Additionally, only one object having attributes will be allowed on the left-hand side of a rule, along with the requirement that this same object must be the target of the above cited message. This restriction implements the principle of *information hiding*, which states that the internal configuration (implementation) of an object can only be visible, and therefore accessed, by itself.

Finally, although message attributes can be deleted (so they can have their value altered[3]), a corresponding attribute must be added to the rule's right-hand side, in order to prevent an object from gaining or losing attributes along the computation. Notice that this is a *rule* restriction, for if a vertex is deleted, its incident edges will also be deleted [15]. This situation is interesting in modelling distributed systems, when pointers to objects could be made meaningless if the objects they point to are deleted. If a vertex is deleted, no rule that uses it can ever be applied, which is compatible with the idea that following a broken link is not meaningful. These consequences, however, will not be explored here.

Definition 8 (Object-oriented rule). *An object-oriented rule is a tuple* $\langle L^C, r, R^C \rangle$ *where* $L^C = \langle L, t_L, \mathcal{C} \rangle$ *and* $R^C = \langle R, t_R, \mathcal{C} \rangle$ *are strict object-oriented graphs and* $r = \langle r_V, r_E \rangle : L^C \to R^C$ *is a class typed graph morphism holding the following properties:*

- r_V *is injective and invertible,* r_E *is injective,*
- $\{v \in V_L \mid \exists e \in E_L : src_L(e) = v \land (lab \circ t_L)(e) = \text{attr}\}$ *is a singleton, whose unique element is called* attribute vertex,

---

[3] Graphs can be enriched with algebras in order to deal with sorts, values and operations. Although we do not develop these concepts here, they can easily be added to this framework.

- $\{e \in E_L \mid (lab \circ t_L)(e) = \mathrm{msg}\}$ *is a singleton, whose unique element is called* left-side message, *having as target object the attribute vertex,*
- *the left-side message does not belong to the domain of* r,
- *for all* $v \in V_L$ *there is a bijection* $b_v : \{e \in E_L \mid src_L(e) = v, lab(t_L(e)) = attr\} \leftrightarrow \{e \in E_R \mid src_R(e) = r_V(v), lab(t_R(e)) = attr\}$, *such that* $t_R \circ b_v = t_L$ *and* $t_L \circ b_v^{-1} = t_R$, *and*
- *for all* $v \in V_R$, *such that* $v \notin im(r_V)$ *the diagram*

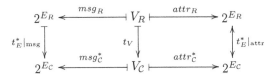

exists, commutes, and $t_E^*(attr_R(v))$ *is a bijection.*

The last condition is needed to assure that creation of objects does not generate an incomplete system (i.e., a system with an object lacking attributes). Notice that r_V is not required to be surjective, so new vertices can be added by the rule, as long as all created vertices must have exactly the attributes defined along its class-model graph.

Definition 9 (Object-oriented match). *An* object-oriented match *between a strict object-oriented graph* L^C *and a complete object-oriented graph* G^C *is a class typed graph morphism* $m = \langle m_V, m_E \rangle : L^C \to G^C$ *such that* m_V *is total,* m_E *is total and injective, and for any two elements* $a, b \in L$, *if* $m(a) = m(b)$ *then either* $a, b \in dom(r)$ *or* $a, b \notin dom(r)$.

Definition 10 (Object-oriented direct derivation). *Given a complete object-oriented graph* $G^C = \langle G, t_G, C \rangle$, *an object-oriented rule* $r : L^C \to R^C$, *and an object-oriented match* $m : L^C \to G^C$, *their* object-oriented direct derivation, *or* rule application, *can be computed in two steps:*

1. *Construct the pushout of* $\Phi(r) : \Phi(L^C) \to \Phi(R^C)$ *and* $\Phi(m) : \Phi(L^C) \to \Phi(G^C)$ *in* **GraphP**, *where* Φ *is the forgetful functor which sends an object-oriented graph to a hypergraph without any typing structure and a class typed graph morphism to a hypergraph morphism,* $\langle H, r' : G \to H, m' : R \to H \rangle$ *[12];*
2. *equip the result with the following typing structure on nodes and edges, resulting in the graph* $H^C = \langle H, t_H, C \rangle$ *where,*
   - *for each* $v \in V_H$, $t_H(v) = \sqcap \left(t_G(r'^{-1}(v)) \cup t_R(m'^{-1}(v)) \right)$,
   - *for each* $e \in E_H|_{attr}$, $t_H(e) = \sqcap \left(t_G(r'^{-1}(e)) \cup t_R(m'^{-1}(e)) \right)$,
   - *for each* $e \in E_H|_{msg}$, $t_H(e) = e'$, *where* $e' \in msg_C^*(t_H(tar_H(e)))$, *and* $e' \sqsubseteq_E^* \sqcap [t_G(r'^{-1}(e)) \cup t_R(m'^{-1}(e))]$.

The tuple $\langle H^C, r', m' \rangle$ *is then the resulting derivation of rule* r *at match* m.

An object-oriented derivation collapses the elements identified simultaneously by the rule and by the match, and copies the rest of the context and the added elements. Element typing, needed to transform the result into an object-oriented graph is achieved by getting the greatest lower bound (with respect the partial order relations on nodes and edges) of the elements mapped by morphisms m' and r' to the same element (the other elements have their types merely copied). The object-oriented rule restriction concerning object types (which cannot be altered by the rule) assures that it always exist, as shown in more detail below. Messages, however, need some extra care. Since graph L^C presents a single message, which is deleted by the rule application, a message on H^C comes either from G^C or from R^C. If it comes from G^C, which is an object-oriented graph itself, no retyping is needed. However, if it comes from R, in order to assure that H^C is also an object-oriented graph, it must be retyped according to the type of the element it is targeting on the graph H^C. Notice that this element can have a different type from the one in the rule, since the match can be done to any element belonging to the lower set of the mapped entity.

It is important to realize that an object-oriented direct derivation is well defined, as shown by the proposition below.

Proposition 1. *The structure proposed in Definition 10 always exists.*

Proof. V_H is a pushout in **SetP**, hence r'_V is injective (because r_V is injective), and r'_V and m'_V are jointly surjective. For any $h \in V_H$, let $m'^{-1}_V(h) = \{r \in V_R \mid m'_V(r) = h\}$ and $r'^{-1}_V(h) = \{g \in V_G \mid r'_V(g) = h\}$. Now, for any $h \in V_H$, $r'^{-1}_V(h)$ is a singleton (because r'_V is invertible) with unique element g, then, by definition, $\sqcap t_G(r'^{-1}_V(h)) = g$; for any $h \in V_H$ such that $m'^{-1}_V(h) \notin im(r_V)$, $m'^{-1}_V(h)$ is also a singleton, and the same reasoning applies; for any $h \in V_H$ such that $m'^{-1}_V(h) \subseteq im(r_V)$, we have that $(t_R \circ m'^{-1}_V)(h) =_{V_C^*} (t_L \circ r_V^{-1} \circ m'^{-1}_V)(h)$; but, for the same element h we have that $(t_G \circ r'^{-1}_V)(h) \sqsubseteq^*_{V_C^*} (t_L \circ m_V^{-1} \circ r'^{-1}_V)(h)$. But, as mentioned before, V_H is a pushout, then $(r_V^{-1} \circ m'^{-1}_V)(h) = (m_V^{-1} \circ r'^{-1}_V)(h)$ and therefore $(t_G \circ r'^{-1}_V)(h) \sqsubseteq^*_{V_C^*} (t_R \circ m'^{-1}_V)(h)$. Since $\sqcap t_G(r'^{-1}_V(h))$ is defined, $\sqcap (t_G(r'^{-1}_V(h)) \cup t_R(m'^{-1}_V)(h))$ always exists.

For all attributes arcs in E_L, E_R and E_G, the morphism can be reduced to an ordinary partial graph morphism (since they are only connected to themselves via the order relation), and then $\sqcap (t_G(r'^{-1}_E(h)) \cup t_R(m'^{-1}_E(h)))$ is always well defined. For message arcs, however, notice that object-oriented rules require that the only message in E_L is deleted by the rule. It means that messages arcs of E_H either come from E_G or from E_R. A message e is typed over the least element from the overriding relation with respect to its actual target vertex's type. This is assured by the fact that the typing edge must belong to the set $msg^*_C(t_H(tar_H(e)))$, which, by definition, assures that there is only one edge of choice (there are no related message arcs in the set $msg^*_C(v)$, for any vertex type v. □

Lemma 1. *The object $H^C = \langle H, t_H, C \rangle$ built according to Definition 10 is a complete object-oriented graph.*

Proof. (sketch) $L^\mathcal{C}$, $R^\mathcal{C}$ are strict object-oriented graphs, and $G^\mathcal{C}$ is a complete one. If $H^\mathcal{C}$ is an object-oriented graph, the following diagram can be constructed:

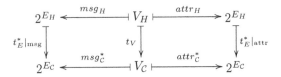

Notice that the set of added vertices $V_R \setminus r_V(V_L)$ can be viewed as a complete object-oriented graph, for it has all necessary attributes. Now, restricting the reasoning to the set of mapped vertices and attributes, one has the following: for each $v \in V_L$, let b_v be the bijection existing between the attribute edges from $A_L \subseteq E_L|_{attr}$ and $A_R \subseteq E_R|_{attr}$ defined as the last object-oriented rule restriction in Definition 8. Match m between the rule's left-side and graph $G^\mathcal{C}$ is total on vertices and arcs, and injective on arcs, and by the characteristics of the pushout construction, function m'_E is also total and injective on arcs. Notice that all edges from $G^\mathcal{C}$ are either belonging to the image of m_E (the mapped edges) or not (the context edges). Since the context edges are mapped unchanged to the graph $H^\mathcal{C}$ (and so there is a natural bijection between them), it must exist a bijection $B : E_G \leftrightarrow E_H$ which implies the existence of the trivial bijection $2^B : 2^{E_G} \to 2^{E_H}$, and since the sets V_G and V_H are isomorphic if we disregard the added vertices (note the existence of an implicit property of an object-oriented rule that prevents it from deleting vertices, since a deletion of a vertex implies the deletion of an edge, which cannot occur, otherwise there would be no bijection b_v), it can be concluded that the right square on the diagram can be constructed. The same reasoning applies to the left square of the diagram, since the rule application assures that messages are typed right. Hence, one can conclude that $H^\mathcal{C}$ is a complete object-oriented graph.   □

Lemma 2. *The morphisms $r' : G^\mathcal{C} \to H^\mathcal{C}$ and $m' : R^\mathcal{C} \to H^\mathcal{C}$, built according to Definition 10 are hierachical graph morphisms.*

Proof. (sketch) The morphisms r' and m' preserve the order structure, which is a sufficient condition to assure that they are actually class typed graph morphisms.   □

Theorem 3. *Given an object-oriented graph $G^\mathcal{C}$, an object-oriented rule $r : L^\mathcal{C} \to R^\mathcal{C}$, and an object-oriented match $m : L^\mathcal{C} \to G^\mathcal{C}$, the resulting derivation of rule r at match m, $\langle H^\mathcal{C}, r', m' \rangle$ is a pushout of the arrows r and m in the category $\mathbf{OOGraphP}(\mathcal{C})$.*

Proof. Proposition 1 assures that a direct derivation can always be constructed, and Lemmas 1 and 2 show that the resulting object and morphisms belong to the category $\mathbf{OOGraphP}(\mathcal{C})$. Then, let $\langle H^\mathcal{C}, r' : G^\mathcal{C} \to H^\mathcal{C}, m' : R^\mathcal{C} \to H^\mathcal{C} \rangle$ be the result obtained by application of rule r under match m. Now, let $H'^\mathcal{C}$ be an object-oriented graph and $h_R : R^\mathcal{C} \to H'^\mathcal{C}$, $h_G : G^\mathcal{C} \to H'^\mathcal{C}$ be two class typed graph morphisms such that $h_R \circ r = h_G \circ m$. Then let $h : H^\mathcal{C} \to H'^\mathcal{C}$ be the class typed graph morphism built as follows: for all graph element (node or edge) $e \in dom(h_R) \cap dom(m')$, $h(m'(e)) = h_R(e)$; for all $e \in dom(h_G) \cap dom(r')$, $h(r'(e)) = h_G(e)$; it is easy to see that, by construction, $h_R = h \circ m'$ and $h_G = h \circ r'$. Notice that h is a class typed graph morphism, since all elements $e \in H$ are typed over the greatest lower bound (respecting the concerning order relation) of the elements mapped to them. It means that if exists an element $e' \in H'$ such that there are elements $e_g \in G$ and $e_r \in R$ with $h_G(e_g) = e' = h_R(e_r)$, then (h_G and h_R are class typed graph morphisms) $t_{H'}(e') \sqsubseteq^*_{V_\mathcal{C}} t_G(e_g)$ and $t_{H'}(e') \sqsubseteq^*_{V_\mathcal{C}} t_R(e_r)$; since $t_H(e_h) = \sqcap (t_G(r'^{-1}(e_h)) \cup t_R(m'^{-1}(e_h)))$ if e_h is a vertex or an attribute, and $t_H(e_h) \in msg^*_\mathcal{C}(t_H(tar_H(e_h)))$ if e_h is a message, then $t_{H'}(e') \sqsubseteq^*_{V_\mathcal{C}} t_H(e_h)$ for any $e' = h(e_h)$. Hence, h is a class typed graph morphism.

Suppose there is another class typed graph morphism $h' : H^\mathcal{C} \to H'^\mathcal{C}$ such that $h_R = h' \circ m'$ and $h_G = h' \circ r'$ but $h' \neq h$. Then there must be at least one graph element $e \in H$ such that $h(e) \neq h'(e)$. But r' and m' are jointly surjective, so all elements of H belong to the domain of h, so if there is an element $e \in H$ such that $h(e) \neq h'(e)$, the equalities $h_R = h' \circ m'$ and $h_G = h' \circ r'$ would not hold.   □

The purpose of method redefinition is to take advantage of the polymorphism concept through the mechanism known as *dynamic binding*. Dynamic binding is usually implemented in object-oriented languages by a function pointer virtual table, from which it is decided which method should be called at that execution point. This decision is often presented as an algorithm, which inspects the runtime type of the message receiving object to determine what is the closest redefinition (if any) of the method being sent. This decision is modelled in our work by the direct derivation construction, which assures that the type chosen for methods being sent (i.e., messages created by rule application) is the least possible regarding the redefinition chain given by the order relation on message edges.

The algebraic approach to graph grammars rely on categorical constructs to express most of its results. Semantics of computations, for instance, are usually given as categories of concrete or abstract derivations [7], which rely on the fact that direct derivations are pushouts on a suitable category of graphs and graph morphisms. It means that if the constructs used within the classical theory of graph grammars can be proven to exist in the new setting, the conclusions drawn from the former could be automatically transferred to the latter. Having a direct derivation characterized as a pushout construction in the category **OOGraphP**$(\mathcal{C})$ is hence fundamental to inherit all previously achieved theoretical results.

4   Conclusions and Future Work

This paper presented a graph-based formal framework to model object-oriented specifications and computations. More specifically, an extension of the algebraic single-pushout approach to (typed) graph grammars was developed, where typing morphisms are compatible with the order relations defined over nodes and edges. The typing graphs, called class-model graphs, have partially ordered sets of nodes and edges, whose base structure follows the possible hierarchy of types when single inheritance in used.

The four core characteristics of the object-oriented paradigm for system development are contemplated in this work. *Encapsulation* of data and methods is achieved by structural constraints on the class-model graphs, which assure that an attribute belong to exactly one class, and a method can be received by exactly one object. *Information hiding* is obtained through restrictions on the rules used to model computations: an object which is receiving a message has access only to its attributes, although it can send messages to any object it has knowledge of (again, through its own attributes). *Inheritance* appears through the typing morphism, which allows an object to make use of any attribute or method defined along the hierarchy of types provided by the order relation on the class-model graph nodes. *Polymorphism* is implemented by the morphisms between object-oriented graphs, assuring that an object belonging to a class can be mapped to any object belonging to one of its derived classes. Additionally, dynamic binding occurs in the model of computation provided by those grammars, where the application of a rule which sends a message to an object has as its direct derivation a message typed according to the actual type of the object mapped by the morphism, which is compatible with the way it is implemented in object-oriented programming languages.

A significant advantage to the use of a formal framework for object-oriented system specification is in the ability to apply rigorous inference rules so as to allow reasoning with the specifications and deriving conclusions on their properties. Fixing the sort of rules to be used within a graph grammar, properties regarding the computational model can be derived. Being this a formal framework, the semantics of operations (such as system and grammar composition) can also be derived.

Graph grammars are well suited for system specification, and object-oriented graph grammars, as presented in this text, fill the need for the key features of object-oriented systems be incorporated into a formal framework.

References

1. ANDRIES, M., ENGELS, G., HABEL, A., HOFFMANN, B., KREOWSKI, H.-J., KUSKE, S., PLUMP, D., SCHURR, A., AND TAENTZER, G. Graph transformation for specification and programming. *Science of Computer Programming 34* (1999), 1–54.
2. BLOSTEIN, D., FAHMY, H., AND GRBAVEC, A. Practical use of graph rewriting. Tech. Rep. 95-373, Queen's University, Kingston, Ontario, Canada, 1995.
3. BRANDENBURG, F. J. On partially ordered graph grammars. In *3rd International Workshop on Graph Grammars and their Application to Computer Science* (Warrenton, Virginia, USA, 1986), H. Ehrig, M. Nagl, G. Rozenberg, and A. Rosenfeld, Eds., Lecture Notes in Computer Science 291, Springer-Verlag, pp. 99–111.

4. CARDELLI, L., AND WEGNER, P. On understanding types, data abstraction and polymorphism. *ACM Computing Surveys 17*, 4 (1985), 471–522.

5. COOK, W. R. *Object-oriented programming versus abstract data type*, vol. 489 of *Lecture Notes in Computer Science.* Springer, Berlin, 1990, pp. 151–178.

6. COOK, W. R., HILL, W. L., AND CANNING, P. S. Inheritance is not subtyping. In *POPL'90 - 17th Annual ACM Symposium on Principles of Programming Languages* (January 1990), Kluwer Academic Publishers.

7. CORRADINI, A., EHRIG, H., LÖWE, M., MONTANARI, U., AND PADBERG, J. The category of typed graph grammars and its adjunctions with categories of derivations. In *[11]* (1994), pp. 56–74.

8. CORRADINI, A., MONTANARI, U., AND ROSSI, F. Graph processes. *Fundamentae Informatica 26*, 3-4 (1996), 241–265.

9. DAVEY, B. A., AND PRIESTLEY, H. A. *Introduction to Lattices and Order*, 2 ed. Cambridge University Press, Cambridge, 2002. 298p.

10. DOTTI, F. L., AND RIBEIRO, L. Specification of mobile code using graph grammars. In *Formal Methods for Open Object-Based Distributed Systems IV* (2000), Kluwer Academic Publishers, pp. 45–64.

11. EHRIG, H., ENGELS, G., AND ROZENBERG, G., Eds. *5th International Workshop on Graph Grammars and their Application to Computer Science* (Williamsburg, 1994), Lecture Notes in Computer Science 1073, Springer-Verlag.

12. EHRIG, H., HECKEL, R., KORFF, M., LÖWE, M., RIBEIRO, L., WAGNER, A., AND CORRADINI, A. Algebraic approaches to graph transformation. Part II: single-pushout approach and comparison with double pushout approach. In *[13]*. ch. 4, pp. 247–312.

13. EHRIG, H., KREOWSKI, H.-J., MONTANARI, U., AND ROZEMBERG, G. *Handbook of Graph Grammars and Computing by Graph Transformation*, vol. 1 (Foundations). World Scientific, Singapore, 1996.

14. EHRIG, H., KREOWSKI, H.-J., MONTANARI, U., AND ROZEMBERG, G. *Handbook of Graph Grammars and Computing by Graph Transformation*, vol. 3 (Concurrency, Parallelism, and Distribution). World Scientific, Singapore, 1999.

15. EHRIG, H., AND LÖWE, M. Parallel and distributed derivations in the single-pushout approach. *Theoretical Computer Science 109* (1993), 123–143.

16. FERREIRA, A. P. L., AND RIBEIRO, L. Towards object-oriented graphs and grammars. In *Proceedings of the 6th IFIP TC6/WG6.1 International Conference on Formal Methods for Open Object-Based Distributed Systems (FMOODS 2003)* (Paris, November 19-21 2003), E. Najm, U. Nestmann, and P. Stevens, Eds., vol. 2884 of *Lecture Notes in Computer Science*, Springer-Verlag, pp. 16–31.

17. FERREIRA, A. P. L., AND RIBEIRO, L. A graph-based semantics for object-oriented programming constructs. In *Proceedings of the International Conference on Category Theory and Computer Science (CTCS2004)* (Copenhagen, DK, August 12-14 2004), To appear as a special issue of Eletronic Notes in Theoretical Computer Science. 15p.

18. KORFF, M. *Generalized Graph Structure Grammars with Applications to Concurrent Object-Oriented Systems.* PhD Thesis, Technische Universität Berlin, Berlin, 1995.

19. LÖWE, M. *Extended Algebraic Graph Transformation.* PhD thesis, Technischen Universität Berlin, Berlin, Feb 1991.

20. MONTANARI, U., PISTORE, M., AND ROSSI, F. Modeling concurrent, mobile and coordinated systems via graph transformations. In *[14]*. ch. 4, pp. 189–268.

21. PARISI-PRESICCE, F., EHRIG, H., AND MONTANARI, U. Graph rewriting with unification and composition. In *3rd International Workshop on Graph Grammars and their Application to Computer Science* (Warrenton, Virginia, USA, 1986), H. Ehrig, M. Nagl, G. Rozenberg, and A. Rosenfeld, Eds., Lecture Notes in Computer Science 291, Springer-Verlag, pp. 496–514.

22. RIBEIRO, L. *Parallel Composition and Unfolding Semantics of Graph Grammars.* PhD Thesis, Technische Universität Berlin, Berlin, June 1996. 202p.

23. TAENTZER, G. *Parallel and Distributed Graph Transformation Formal Description and Application to Communication-Based Systems.* PhD Thesis, TU Berlin, Berlin, 1996.

24. TROYER, O. D., AND JANSSEN, R. On modularity for conceptual data models and the consequences for subtyping, inheritance and overriding. In *Proceedings of the 9th IEEE Conference on Data Engineering (ICDE 93)* (1993), IEEE CS Press, pp. 678–685.

25. UNGAR, D., CHAMBERS, C., CHANG, B.-W., AND HÖLZLE, U. Organizing programs without classes. *Lisp and Symbolic Computation 3*, 4 (1991).

Tutorial Introduction to Graph Transformation: A Software Engineering Perspective

Luciano Baresi[1] and Reiko Heckel[2]

[1] Politecnico di Milano, Italy
baresi@elet.polimi.it
[2] University of Paderborn, Germany
reiko@upb.de

Abstract. As a basic introduction to graph transformation, this tutorial is not only intended for software engineers. But applications typical to this domain, like the modeling of component-based, distributed, and mobile systems, model-based testing, and diagram languages provide well-known examples and are therefore used to give a survey of the motivations, concepts, applications, and tools of graph transformation.

Introduction. Graphs and diagrams provide a simple and powerful approach to a variety of problems that are typical to computer science in general, and software engineering in particular. In fact, for most activities in the software process, a variety of visual notations have been proposed, including state diagrams, Structured Analysis, control flow graphs, architectural description languages, function block diagrams, and the UML family of languages. These notations produce models that can be easily seen as graphs and thus graph transformations are involved, either explicitly or behind the scenes, when specifying how these models should be built and interpreted, and how they evolve over time and are mapped to implementations. At the same time, graphs provide a universally adopted data structure, as well as a model for the topology of object-oriented, component-based and distributed systems. Computations in such systems are therefore naturally modeled as graph transformations, too.

Graph transformation has originally evolved in reaction to shortcomings in the expressiveness of classical approaches to rewriting, like Chomsky grammars and term rewriting, to deal with non-linear structures. The first proposals, appearing in the late sixties and early seventies [10, 11], are concerned with rule-based image recognition, translation of diagram languages, etc.

In this tutorial, we will introduce the basic concepts and approaches to graph transformation, demonstrate their application to software engineering problems, and provide a high-level survey of graph transformation tools.

Concepts. Fundamental approaches that are still popular today [12] include the algebraic or double-pushout (DPO) approach [6], the node-label controlled (NLC) [9] approach, the monadic second-order (MSO) logic of graphs [4], and the PROGRES approach [13] which represents the first major application of graph transformation to software engineering [7].

H. Ehrig et al. (Eds.): ICGT 2004, LNCS 3256, pp. 431–433, 2004.

In this tutorial we first introduce a simple core model (a set-theoretic presentation of the double-pushout approach [6]) whose features are common to most graph transformation approaches and which provides the foundation for further elaboration. Then, we discuss alternatives and extensions, addressing in particular the two fundamentally different views of graph transformation referred to as the *gluing* and the *connecting* approach. They differ for the mechanism used to embed the right-hand side of the rule in the context (the structure left over from the given graph after deletion): In a gluing approach like DPO, the new graph is formed by gluing the right-hand side with the context along common vertices. In a connecting approach like NLC, the embedding is realized by a disjoint union, with as many new edges as needed to connect the right-hand side with the rest of the graph.

This feature, which provides one possible answer to a fundamental problem, the replacement of substructures in an unknown context, is known in software engineering-oriented approaches like PROGRES and FUJABA by the name of *set nodes* or *multi objects*. Other extensions of the basic approach include *programmed* transformations, concerned with controlling the (otherwise non-deterministic) rewrite process, as well as *application conditions*, restricting the applicability of individual rules.

Structural constraints over graphs, comparable to invariants or integrity constraints in data bases, deal with the (non-) existence of certain patterns, including paths, cycles, etc. They are expressed as cardinalities, in terms of first- or higher-order logic, or as graphical constraints.

Applications. After introducing concepts and approaches, we present by way of examples typical applications to software engineering problems, like

− model and program transformation,
− syntax and semantics definition of visual languages, and
− visual behavior modeling and programming.

In particular, we distinguish between the use of graph transformation as a *modeling notation* (and semantic model) to reason on particular problems, like functional requirements or architectural reconfigurations of individual applications, and its use as a *meta language* to specify the syntax, semantics, and manipulation of visual modeling languages, like the UML. Halfway between applications and languages lies the *specification of application families*, including architectural styles and policies.

Tools. The last part of the tutorial surveys tool support for, and by means of, graph transformation, i.e., general-purpose environments for specifying, executing, and analyzing graph transformation systems, and environments for dedicated application areas.

Universal graph transformation tools are represented, e.g., by PROGRES (*PROgrammed Graph REwriting Systems* [13, 14]), one of the first fully developed programming languages based on graph transformation, and AGG [8], which implements the algebraic approach by means of a rule interpreter and associated analysis techniques. Quite different is the support offered by FU-JABA [1], an environment for round trip engineering between UML and Java based on graph transformation as the operational model underlying UML collaboration diagrams.

As examples for more application-specific tools, DIAGEN and GENGED [2] provide support for the generation of graphical editors based on the definition of visual languages through graph grammars.

In conclusion, the aim of this tutorial is to provide a sound introduction to graph transformation, its concepts, applications, and tools in the context of software engineering. This shall enable attendees to work their way through the relevant literature and to benefit from the presentations at the conference. A previous version of this tutorial together with an accompanying paper has been presented in [3].

References

1. From UML to Java and Back Again: The Fujaba homepage. `www.upb.de/cs/isileit`.
2. R. Bardohl and H. Ehrig. Conceptual model of the graphical editor GenGed for the visual definition of visual languages. In H. Ehrig, G. Engels, H.-J. Kreowski, and G. Rozenberg, editors, *Proc. 6th Int. Workshop on Theory and Application of Graph Transformation (TAGT'98), Paderborn, November 1998*, volume 1764 of *LNCS*, pages 252 – 266. Springer-Verlag, 2000.
3. L. Baresi and R. Heckel. Tutorial introduction to graph transformation: A software engineering perspective. In *Proc. of the First International Conference on Graph Transformation (ICGT 2002), Barcellona, Spain*, pages 402–429, oct 2002.
4. B. Courcelle. The monadic second-order logic of graphs i, recognizable sets of finite graphs. *Information and Computation*, 8521:12–75, 1990.
5. H. Ehrig, G. Engels, H.-J. Kreowski, and G. Rozenberg, editors. *Handbook of Graph Grammars and Computing by Graph Transformation, Volume 2: Applications, Languages, and Tools*. World Scientific, 1999.
6. H. Ehrig, M. Pfender, and H.J. Schneider. Graph grammars: an algebraic approach. In *14th Annual IEEE Symposium on Switching and Automata Theory*, pages 167–180. IEEE, 1973.
7. G. Engels, R. Gall, M. Nagl, and W. Schäfer. Software specification using graph grammars. *Computing*, 31:317–346, 1983.
8. C. Ermel, M. Rudolf, and G. Taentzer. The AGG approach: Language and tool environment. In Engels et al. [5], pages 551 – 601.
9. D. Janssens and G. Rozenberg. On the structure of node-label controlled graph grammars. *Information Science*, 20:191–216, 1980.
10. J. L. Pfaltz and A. Rosenfeld. Web grammars. *Int. Joint Conference on Artificial Intelligence*, pages 609–619, 1969.
11. T. W. Pratt. Pair grammars, graph languages and string-to-graph translations. *Journal of Computer and System Sciences*, 5:560–595, 1971.
12. G. Rozenberg, editor. *Handbook of Graph Grammars and Computing by Graph Transformation, Volume 1: Foundations*. World Scientific, 1997.
13. A. Schürr. Programmed graph replacement systems. In Rozenberg [12], pages 479 – 546.
14. A. Schürr, A.J. Winter, and A. Zündorf. The PROGRES approach: Language and environment. In Engels et al. [5], pages 487–550.

Tutorial on DNA Computing and Graph Transformation

Tero Harju[1], Ion Petre[2], and Grzegorz Rozenberg[3]

[1] Department of Mathematics, University of Turku,
FIN-20014 Turku, Finland
harju@utu.fi
[2] Department of Computer Science, Åbo Academi University, FIN-20520 Turku,
Finland
ion.petre@abo.fi
[3] Leiden Institute for Advanced Computer Science, Leiden University,
Niels Bohrweg 1, 2333 CA Leiden, The Netherlands
rozenber@liacs.nl

DNA computing, or more generally molecular computing, is an exciting research area at the intersection of mathematics, computer science and molecular biology. Research in DNA computing can be roughly divided in two streams: DNA computing *in vitro* and *in vivo*. The former is concerned with building (specialized) DNA-based computers in test tubes, while the latter is concerned with implementing some computational components in living cells, as well as with studying the computational processes taking place in the living cells.

In this tutorial, we shall discuss the computational nature of an intricate DNA processing taking place in single cell organisms called *ciliates*, in the process of *gene assembly*. We shall discuss the role of graph transformations in modelling and studying gene assembly process. In particular, we demonstrate that graph transformations provide a suitable level of abstraction, and useful technical tools for studying gene assembly. On the other hand, the gene assembly process inspires a new computing paradigm, computing by folding and recombination, which induces novel questions and challenges for research on graph transformation. For details, we refer to the recent monograph [1].

1 Nuclear Dualism of Ciliates

Ciliates are an ancient group of single cell organisms that have a unique nuclear dualism: they have two kinds of nuclei, the germline *micronucleus* and the somatic *macronucleus*. The genetic information is encoded in different ways in the two types of nuclei. In the micronucleus, the genes are placed in long continuous DNA molecules interrupted by noncoding spacer DNA. In the macronucleus, the DNA is present in short, gene-size molecules. During sexual reproduction ciliates convert the micronuclear genome into the macronuclear genome, eliminating all noncoding DNA, and assembling the micronuclear form of each gene into its macronuclear form. This assembly process is one of the most involved DNA processing known in Nature, producing the shortest known DNA molecules. Moreover, gene assembly is fascinating also from computational point of view.

H. Ehrig et al. (Eds.): ICGT 2004, LNCS 3256, pp. 434–436, 2004.

Each micronuclear gene is divided in several segments called MDSs (*Macronuclear Destined Segments*), which are interrupted by noncoding segments called *IES*s (*Internal Eliminated Segments*), see Fig. 1(a), where MDSs are represented by rectangles and the IESs are represented by line segments. The MDSs can occur in the micronuclear DNA sequences in a scrambled order, and moreover, some of them may even be inverted. The macronuclear version of the same gene, illustrated in Fig. 1(b), consists of MDSs from its micronuclear form that are spliced together in the *orthodox* order $M_1 M_2 \ldots M_k$. The central role in assembling the MDSs in the orthodox order is played by the so-called *pointers*, short nucleotide sequences bordering each MDS. As it turns out, the pointer in the end of the i-th MDS (the *outgoing* pointer of M_i) is identical to the pointer occurring in the beginning of the i+1-st MDS (the *incoming* pointer of $M_i + 1$). This facilitates the DNA recombination and the assembly of the two MDSs in a bigger coding sequence, see Fig. 1(b).

(a) M_1 M_3 (b) M_k

Fig. 1. The structure of (a) micronuclear and (b) macronuclear gene.

Formally, each MDS M_i, for $i = 2, \ldots, k-1$, is of the form $M_i = (\pi_i, \mu_i, \pi_{i+1})$ while M_1 has the form $M_1 = (\beta, \mu_1, \pi_2)$, and M_k has the form $M_k = (\pi_k, \mu_k, \varepsilon)$. The segments μ_i are the *bodies*, π_i are the pointers, and the special symbols β in M_1 and ε in M_k are merely *markers* for the beginning and the end of the macronuclear gene. Note that each pointer has two occurrences in two different MDSs: one incoming and one outgoing.

During gene assembly the micronuclear gene is converted to the macronuclear gene. In this process, all IESs from the micronuclear gene are removed, while the MDSs are spliced together on their common pointers to form the macronuclear gene. Thus, a gene in a macronucleus is presented as a string $M_1 M_2 \ldots M_k$, while its micronuclear predecessor is described by a string $I_1 M'_{t_1} I_2 M'_{t_2} I_3 \ldots I_k M'_{t_k} I_{k+1}$, where each I_i represents an IES, and each M'_{t_i} represents either M_{t_i} or its inversion $\overline{M}_{t_i}$. Here $M_{t_1} \ldots M_{t_k}$ is a permutation of the string $M_1 M_2 \ldots M_k$.

2   Formalisms: Legal Strings and Overlap Graphs

Let $\Pi = \{2, 3, \ldots, k\}$ and $\overline{\Pi} = \{\overline{2}, \overline{3}, \ldots, \overline{k}\}$ be the alphabets for the pointers and their inversions, for some $k \geq 2$. For $p \in \Pi$, let $\bar{\bar{p}} = p$. Strings over $\Pi \cup \overline{\Pi}$ that have exactly two occurrences from $\{p, \overline{p}\}$, for each pointer $p \in \Pi \cup \overline{\Pi}$, are called *legal strings*. A pointer p in a legal string δ is *negative*, if $\delta = \delta_1 p \delta_2 p \delta_3$, and p is *positive*, if $\delta = \delta_1 p \delta_2 \overline{p} \delta_3$.

Since we are mostly interested in the assembly strategies, we ignore the IESs in the micronucleus, and we write $i(i + 1)$ for the MDS $M_i = (\pi_i, \mu, \pi_{i+1})$ and $\overline{(i + 1)i}$ for the inversion of M_i. (The beginning and the end markers are ignored in the coding.) This associates a unique legal string to each micronuclear gene.

Pointers $p, q \in \Pi$ *overlap* in a legal string δ, if δ can be written as $\delta = \delta_1 p_1 \delta_2 q_1 \delta_3 p_2 \delta_4 q_2 \delta_5$, where $p_1, p_2 \in \{p, \overline{p}\}$ and $q_1, q_2 \in \{q, \overline{q}\}$, for some strings δ_i. The *signed overlap graph* γ_δ of δ has Π as its set of vertices, and there is an (undirected) edge pq if p and q overlap in δ. Moreover, each vertex p is labelled by $+$ or $-$ depending whether the pointer p is positive or negative in δ. This associates a unique signed graph to each micronuclear gene.

It was postulated in [2, 3] that gene assembly is accomplished through the following three molecular operations: *ld-excision*, *hi-excision/reinsertion*, and *dlad-excision/reinsertion*, see Fig. 2, that first fold a molecule, cut is made (on some pointers), and then recombination takes place.

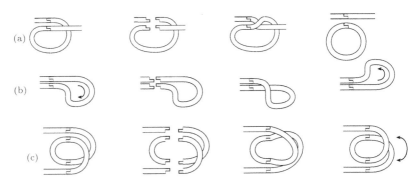

Fig. 2. The molecular operations (a) ld, (b) hi, and (c) dlad.

One can express these operations in terms of the overlap graphs (as well as on legal strings). In this approach: (1) *ld* corresponds to removing a negative isolated vertex p^-; (2) *hi* corresponds to removing a positive vertex p^+, and complementing the neighbourhood of p (also changing the signs of these vertices); (3) *dlad* corresponds to removing two adjacent negative vertices p^- and q^-, and complementing the edge set between the neighbourhoods. The universality result now states that every overlap graph reduces to the empty graph by a finite number of these three graph theoretic operations. Such a sequence of operations describes the assembly strategy of a micronuclear gene.

References

1. Ehrenfeucht, A., Harju, T., Petre, I., Prescott, D. M., and Rozenberg, G., *Computation in Living Cell. Gene Assembly in Ciliates*, Springer-Verlag, 2004, xviii + 201 pages.
2. Ehrenfeucht, A., Prescott, D. M., and Rozenberg, G., Computational aspects of gene (un)scrambling in ciliates. In *Evolution as Computation*, Landwerber, L., Winfree, E. (eds.), 45–86, Springer Verlag, Berlin, Heidelberg, 2000.
3. Prescott, D. M., Ehrenfeucht, A., and Rozenberg, G., Molecular operations for DNA processing in hypotrichous ciliates, *Europ. J. Protistology* **37** (2001) 241–260.

Workshop TERMGRAPH 2004

Maribel Fernández

Department of Computer Science,
King's College London,
Strand, London, WC2R 2LS, UK
maribel@dcs.kcl.ac.uk

1 Introduction

Term graph rewriting is concerned with the representation of expressions as graphs and their evaluation by rule-based graph transformation. The advantage of using graphs rather than strings or trees is that common subexpressions can be shared, which improves the efficiency of computations in space and time. Sharing is ubiquitous in implementations of programming languages: many implementations of functional, logic, object-oriented and concurrent calculi are based on term graphs. Term graphs are also used in symbolic computation systems and automated theorem proving.

Research in term graph rewriting ranges from theoretical questions to practical implementation issues. Many different research areas are included, for instance: the modelling of first- and higher-order term rewriting by (acyclic or cyclic) graph rewriting, the use of graphical frameworks such as interaction nets and sharing graphs to model strategies of evaluation (for instance, optimal reduction in the lambda calculus), rewrite calculi on cyclic higher-order term graphs for the semantics and analysis of functional programs, graph reduction implementations of programming languages, and automated reasoning and symbolic computation systems working on shared structures.

The Second International Workshop on Term Graph Rewriting (TERM-GRAPH 2004) takes place in Rome, on Saturday 2nd October 2004, as a satellite event of the Second International Conference on Graph Transformation (ICGT 2004). The previous TERMGRAPH Workshop took place in Barcelona, in 2002, also as a satellite event of ICGT.

2 Topics and Aims

Topics of interest include:

- Theory of first-order and higher-order term graph rewriting.
- Graph rewriting in lambda-calculus and related calculi.
- Applications in functional, logic, object-oriented and concurrent languages.
- Applications in automated reasoning and symbolic computation.
- Implementation issues.

H. Ehrig et al. (Eds.): ICGT 2004, LNCS 3256, pp. 437–438, 2004.

The aim of this workshop is to bring together researchers working in these different domains and to foster their interaction, to provide a forum for presenting new ideas and work in progress, and to enable newcomers to learn about current activities in term graph rewriting.

3 Organisation

The Workshop TERMGRAPH 2004 is organised by:

- Andrea Corradini, University of Pisa , Italy;
- Maribel Fernandez, King's College London, United Kingdom, (chair);
- Fabio Gaducci, University of Pisa, Italy;
- Detlef Plump, University of York, United Kingdom;
- Femke van Raamsdonk, Free University of Amsterdam, The Netherlands.

and it includes sessions on programming language design using graphical structures, graphical encodings of functional and object-oriented programming languages, and applications of term graphs in logic.

The Proceedings of TERMGRAPH will appear in Elsevier's *Electronic Notes in Theoretical Computer Science*.

More information about the workshop, including the programme and electronic versions of all the accepted papers, can be found on the workshop web page:

http://www.dcs.kcl.ac.uk/staff/maribel/TERMGRAPH.html

Acknowledgements

I would like to thank the authors of the papers, the members of the Programme Committee and organisers of TERMGRAPH, the additional external reviewers, and the organisers of ICGT 2004, for helping make TERMGRAPH 2004 a successful workshop!

Workshop on Graph-Based Tools

Tom Mens[1], Andy Schürr[2], and Gabriele Taentzer[3]

[1] Université de Mons-Hainaut, Belgium
tom.mens@umh.ac.be
[2] Technische Universität Darmstadt, Germany
andy.schuerr@es.tu-darmstadt.de
[3] Technische Universität Berlin, Germany
gabi@cs.tu-berlin.de

Abstract. Graphs are well-known, well-understood, and frequently used means to depict networks of related items. They are successfully used as the underlying mathematical concept in various application domains. In all these domains, tools are developed that store, retrieve, manipulate and display graphs. It is the purpose of this workshop to summarize the state of the art of graph-based tool development, bring together developers of graph-based tools in different application fields and to encourage new tool development cooperations.

1 Motivation

Graphs are an obvious means to describe structural aspects in various fields of computer science. They have been successfully used in application areas such as compiler compiler toolkits, constraint solving problems, generation of CASE tools, model-driven software development, pattern recognition techniques, program analysis, software engineering, software evolution, software visualization and animation, and visual languages. In all these areas, tools have been developed that use graphs as an important underlying data structure. Since graphs are a very general structure concept, still novel application areas emerge which are of interest.

It is a challenge to handle graphs in an effective way. Using graphs inside tools, the following topics play an important role: efficient graph algorithms, empirical and experimental results on the scalability of graphs, reusable graph-manipulating software components, software architectures and frameworks for graph-based tools, standard data exchange formats for graphs, more general graph-based tool integration techniques, and meta CASE tools or generators for graph-based tools. The aim of the Second Workshop on Graph-Based Tools (GraBaTs) is to bring together developers of all kinds of graph-based tools in order to exchange their experiences, problems, and solutions concerning the efficient handling of graphs. The first workshop on this topic [5] took place two years ago as a satellite event of the 1. Int. Conference of Graph Transformation (ICGT) [2].

The GraBaTs workshop is of special relevance for the conference on graph transformation: In many cases, the application of graph transformation technology requires the existence of reliable, user-friendly and efficiently working graph

H. Ehrig et al. (Eds.): ICGT 2004, LNCS 3256, pp. 439–441, 2004.

transformation tools. These tools in turn have to be built on top of basic services or frameworks for graphs which are the main topic of our workshop.

2   Workshop Issues

The workshop aims at bringing together tool developers from different fields, dealing with graphs from different perspectives. In the following, we give an overview on the most important perspectives.

2.1   Graph-Based Tools for Visual Modelling Techniques

Nowadays it is well accepted to use graphs or graph-like structures to describe the underlying structures of visual modelling techniques. The European Research Training Network SegraVis [8] is dedicated to develop new solutions for the definition and implementation of visual modelling languages. Mainly, graph transformation [3] and meta-modelling approaches [6] use graphs to define visual modelling languages. For a model driven development, especially model transformations are needed, either to validate models or to translate them into executable code.

Especially, graph-based tools are considered for generating a modelling environment for a sub-language of the UML [9] containing at least a reasonable variant of Statechart Diagrams. Although this modelling technique is still graph-like, it contains already a number of typical visual concepts and serves, thus, very well as reference application. Contributions to this topic are meant to continue the discussion and comparison of visual language definition started at the "Statechart contest" at VLFM'01 [10]. This time, the emphasis is laid on graph-based approaches and tool support on one hand, and the discussion is extended with regard to different semantics and transformations on the other hand.

2.2   Graph Transformation

Being a satellite event of the international conference on graph transformation, this workshop puts a special emphasis on graph transformation tools. The kind of graphs used by these tools and also their rules, the basic means to manipulate graphs, differ w.r.t. to the formal definition of their semantics, the way how occurrences (matches) are searched for, and how matching rules are applied eventually [7, 3, 2].

In tools, graph transformation is mainly applied to visual languages, model transformation, specification, code generation, verification, restructuring, evolution and programming of software systems. Developers of graph transformation tools may profit from other workshop participants concerning more efficient realizations of basic functionality, while developers of other graph-based tools might find the graph transformation paradigm attractive to implement certain graph manipulations. The workshop may also provide insights to apply these tools to other application domains.

2.3   Common Exchange Formats for Graphs

To support interoperability between various graph-based tools, several initiatives on the development of common exchange formats for graphs have been founded. These formats are all based on the extensible markup language XML developed to interchange documents of arbitrary types. While the graph drawing community favors a format called GraphML [4], preceding discussions at several workshops of the re-engineering and graph transformation community led to the format GXL [1]. Another topic of interest in this direction is an exchange format for graph transformation systems called GTXL which is under development and which will be built on top of GXL.

3   Workshop Organization

The program committee of this workshop consists of Luciano Baresi, Ulrik Brandes, Holger Giese, Gabor Karsai, Scott Marshall, Mark Minas, Tom Mens, Andy Schürr, Gabriele Taentzer, Daniel Varro, Pieter Van Gorp, Andreas Winter, Albert Zündorf. Altogether, 20 papers have been submitted for GraBaTs. The workshop is planned for one day. More information about the workshop including its program and an electronic version of all accepted papers appearing in the electronic notes of Theoretical Computer Science (ENTCS) can be found on the workshop web page: http://tfs.cs.tu-berlin.de/grabats

References

1. *GXL* http://www.gupro.de/GXL, 2002.
2. A. Corradini, H. Ehrig, H.-J. Kreowski, and G. Rozenberg, editors. *Graph Transformation, 1st Int. Conference*, volume 2505. Springer LNCS, 2002.
3. H. Ehrig, G. Engels, H.-J. Kreowski, and G. Rozenberg, editors. *Handbook on Graph Grammars and Computing by Graph Transformation: Applications, Languages, and Tools*, volume 2. World Scientific, Singapore, 1999.
4. *The GraphML File Format*, 2004. Available at
   http://graphml.graphdrawing.org.
5. T. Mens, A. Schürr, and G. Taentzer, editors. *Graph-Based Tools (GraBaTs'02)*, volume 72. Electronic Notes in Theoretical Computer Science (ENTCS), 2002.
6. *Meta Object Facilities – version 1.4*, 2002. Available at
   http://www.omg.org/technology/documents/formal/mof.htm.
7. G. Rozenberg, editor. *Handbook of Graph Grammars and Computing by Graph Transformations, Volume 1: Foundations*. World Scientific, 1997.
8. *Syntactic and Semantic Integration of Visual Modeling Techniques*, 2004. Available at http://www.segravis.org.
9. *Unified Modeling Language – version 1.4*, 2002. Available at
   http://www.omg.org/technology/documents/formal/uml.htm.
10. *Symposium on Visual Languages and Formal Methods - Statecharts Modeling Contest*, 2004. Available at
    http://www2-data.informatik.unibw-muenchen.de/VLFM01/Statecharts.

Workshop on Petri Nets
and Graph Transformations

Hartmut Ehrig[1], Julia Padberg[1], and Grzegorz Rozenberg[2]

[1] Institute for Software Technology and Theoretical Computer Science
Technical University Berlin, Germany
{ehrig,padberg}@cs.tu-berlin.de
[2] Leiden Institute of Advanced Computer Science
Universiteit Leiden, The Netherlands
rozenberg@liacs.nl

The relationship between graph grammars and Petri nets is the topic of this workshop with the emphasis on work in progress and especially the transfer of concepts between both areas. Both areas are prominent specification formalisms for concurrent and distributed systems. It is well-known that the token game of place-transition nets can be modeled by double pushouts of discrete labeled graphs. This allows to relate basic notions of place-transition nets like marking, enabling, firing, steps, and step sequences to corresponding notions of graph grammars and to transfer semantical concepts from Petri nets to graph grammars. Since a marking of a net on one hand and a graph of a graph grammar on the other hand correspond to the state of a system to be modeled, graph grammars can be seen to generalize place-transition nets by allowing more structured states. The concurrent semantics of graph transformations presented in [BCM+99] of the Handbook of Graph grammars volume 3 and [Bal00] is strongly influenced by corresponding semantical constructions for Petri nets in [Win87].

More recently, the modification of the structure of Petri nets using graph transformations has led to Petri transformations, hat extend the classical theory of Petri nets by a rule-based technique that allows studying the changes of the Petri net structure. This area of Petri nets has been introduced about 10 years ago in order to allow in addition to the token game of Petri nets, where the net structure is fixed, also the change of the net structure [REP93,PER95,EGP99]. This enables the stepwise development of Petri nets using a rule-based approach in the sense of graph transformations, where the net structure of a Petri net is considered as a graph. The mutual influences of Petri nets and graph transformation systems is given as well by graph transformations that are used for the development, the simulation, or animation of various types of Petri nets. Using Petri net analysis for graph transformation systems can be achieved either by transferring Petri net notions or by analysing the corresponding net. Moreover, Petri nets can be used for the control of graph transformation systems. Another important link between graph transformations and Petri nets is to define visual languages and environments using graph transformations and to study Petri nets concerning simulation and animation as instantiation of corresponding concepts is visual languages and environments [BE03,EBE03].

H. Ehrig et al. (Eds.): ICGT 2004, LNCS 3256, pp. 442–444, 2004.

This workshop brings together people working especially in the area of low-level and high-level Petri nets and in the area of graph transformation and high-level replacement systems. According to the main idea and in order to further stimulate the research in this important area, this workshop triggers discussion and transfer of concepts across the borders of these and related areas.

We have 8 interesting contributions for the workshop that cover a wide range of topics. Some of them deal with the transformation of Petri nets in various contexts, in networks [PR04], for mobile processes [PPH4] or in visual environments [EE04]. Another focus area is the stepwise development of specific Petri net classes using transformations [PPH4,Rai04]. There graph transformations are either used directly or in terms of net transformations, based on high-level replacement systems. The extension of Petri nets by notions from the area of graph transformations are investigated in two contributions [BCM04,LO04] The use of Petri Nets as semantic domain is the topic of [BM04] and an operational semantic theory of graph rewriting is discussed in [SS04].

References

[Bal00]   P. Baldan. *Modelling Concurrent Computations: From Contextual Petri Nets to Graph Grammars.* PhD thesis, University of Pisa, 2000.

[BCM04]   P. Baldan, A. Corradini, and U. Montanari. Relating algebraic graph rewriting and extensions of Petri nets. Workshop on Petri Nets and Graph transformations, Satellite of ICGT 2004.

[BCM+99] P. Baldan, A. Corradini, U. Montanari, F. Rossi, H. Ehrig, and M. Löwe. Concurrent Semantics of Algebraic Graph Transformations. In G. Rozenberg, editor, *The Handbook of Graph Grammars and Computing by Graph Transformations, Volume 3: Concurrency, Parallelism and Distribution.* World Scientific, 1999.

[BE03]   R. Bardohl and C. Ermel. Scenario Animation for Visual Behavior Models: A Generic Approach Applied to Petri Nets. In G. Juhas and J. Desel, editors, *Proc. 10th Workshop on Algorithms and Tools for Petri Nets (AWPN'03)*, 2003.

[BM04]   L. Baresi and M. Pezze The use of Petri Nets as Semantic Domain for Diagram Notations Workshop on Petri Nets and Graph transformations, Satellite of ICGT 2004.

[EE04]   C. Ermel and K. Ehrig. . View Transformation in Visual Environments applied to Petri Nets. Workshop on Petri Nets and Graph transformations, Satellite of ICGT 2004.

[EBE03]   C. Ermel, R. Bardohl, and H. Ehrig. Generation of Animation Views for Petri Nets in GENGED. In H. Ehrig, W. Reisig, G. Rozenberg, and H. Weber, editors, *Advances in Petri Nets: Petri Net Technologies for Modeling Communication Based Systems*, Lecture Notes in Computer Science 2472. Springer, 2003.

[EGP99]   H. Ehrig, M. Gajewsky, and F. Parisi-Presicce. *High-Level Replacement Systems with Applications to Algebraic Specifications and Petri Nets*, chapter 6, pages 341–400. Number 3: Concurrency, Parallelism, and Distribution in Handbook of Graph Grammars and Computing by Graph Transformations. World Scientific, 1999.

[LO04]   M. Llorens and J. Oliver. Marked-Controlled Reconfigurable Nets: Petri
           Nets with Structural Dynamic Changes. Workshop on Petri Nets and
           Graph transformations, Satellite of ICGT 2004.

[PER95]   J. Padberg, H. Ehrig, and L. Ribeiro. Algebraic high-level net transforma-
           tion systems. *Mathematical Structures in Computer Science*, 5:217–256,
           1995.

[PPH4]   F. Parisi-Presicce, K. Hoffmann Higher order nets for Mobile Policies.
           Workshop on Petri Nets and Graph transformations, Satellite of ICGT
           2004.

[PR04]   M.C. Pinto and L. Ribeiro. Modeling Metabolic and Regulatory Networks
           using Graph Transformations. Workshop on Petri Nets and Graph trans-
           formations, Satellite of ICGT 2004.

[REP93]   L. Ribeiro, H. Ehrig, and J. Padberg. Formal development of concurrent
           systems using algebraic high-level nets and transformations. In *Proc. VII
           Simpósio Brasileiro de Engenharia de Software*, pages 1–16, 1993.

[Rai04]   D.C. Raiteri. The Translation of Parametric Dynamic Fault Trees in
           Stochastic Well-formed Nets as a Case of Graph Transformation. Work-
           shop on Petri Nets and Graph transformations, Satellite of ICGT 2004.

[SS04]   V. Sassobwe, P. Sobocinski. Congruences for contextual graph rewriting.
           BRICS Report RS-04-11, June 2004.

[Win87]   G. Winskel. Petri nets, algebras, morphisms, and compositionality. *Infor-
           mation and Computation*, 72:197–238, 1987.

Workshop on Software Evolution Through Transformations: Model-Based vs. Implementation-Level Solutions

Reiko Heckel[1] and Tom Mens[2]

[1] Universität Paderborn, Germany
reiko@upb.de
[2] Université de Mons-Hainaut, Belgium
tommens@vub.ac.be

Abstract. Transformations provide a general view of incremental and generative software development, including forward and reverse engineering, refactoring, and architectural evolution. Graphs provide a representation for all kinds of development artifacts, like models, code, documentation, or data. Based on these unifying views, the workshop provides a forum to discuss both the properties of evolution and its support by techniques and tools, including the tradeoffs and interconnection of solutions at model and implementation level, consistency management, source code analysis, model and program transformation, software visualization and metrics, etc.

1 Motivation and Objectives

Businesses, organizations and society at large are increasingly reliant on software at all levels. An intrinsic characteristic of software addressing a real-world application is the need to evolve [2]. Such evolution is inevitable if the software is to remain satisfactory to its stakeholders.

Changes to software artifacts and related entities tend to be progressive and incremental, driven, for example, by feedback from users and other stakeholders, such as bug reports and requests for new features, or more generally, changes of functional or non-functional requirements. In general, evolutionary characteristics are inescapable when the problem to be solved or the application to be addressed belongs to the real world.

There are different strategies to address this issue. Model-based software development using the UML, as proposed by the OMG's MDA initiative, addresses evolution by automating (via several intermediate levels) the transformation of platform-independent design models into code. In this way, software can be evolved at the model level without regard for technological aspects, relying on automated transformations to keep in sync implementations on different target platforms.

Classical re-engineering technology, instead, starts at the level of programs which, due to the absence or poor quality of models, provide the only definite

H. Ehrig et al. (Eds.): ICGT 2004, LNCS 3256, pp. 445–447, 2004.
© Springer-Verlag Berlin Heidelberg 2004

representation of a legacy system. The abstractions derived from the source code of these systems are not typically UML models, but they may play a similar role in the subsequent forward engineering steps.

Which approach is preferable in which situation, and how both strategies could be combined, is an open question. To answer that question and to implement any solutions deriving from these answers we require

- a *uniform understanding* of software evolution phenomena at the level of both models and programs, as well as of their interrelation;
- a *common technology basis* that is able to realize the manipulation of artifacts at the different levels, and to build bridges between them.

It is the hypothesis of this workshop that *graphs*, seen as conceptual models as well as data structures, defined by meta models or other means, and *transformations*, given as program or model transformations on graph- or tree-based presentations, provide us with unifying models for both purposes [1].

Transformations provide a very general approach to the evolution of software systems. Literally all activities that lead to the creation or modification of documents have a transformational aspect, i.e., they change a given structure into a new one according to pre-defined rules. As a common representation of artifacts like models, schemata, data, program code, or software architectures, graphs have been used both for the integration of development tools and as a conceptual basis to reason on the development process and its products.

Based on this conceptual and technological unification, it is the objective of the workshop to provide a forum for studying software evolution phenomena and discussing their support at different levels of abstraction.

2 Workshop Topics and Program

The workshop provides a forum for the discussion of problems, solutions, and experience in software evolution research and practice including, but not limited to

- graph-based models for analysis, visualization, and re-engineering
- software refactoring and architectural reconfiguration
- model-driven architecture and model transformations
- consistency management and co-evolution
- relation and tradeoffs between model- and program-based evolution

The program will combine presentations of position and technical papers with discussions on selected topics. The nomination of papers for presentation is determined through a formal review process.

Accepted papers will appear in an issue of Elsevier's Electronic Notes in Theoretical Computer Science. A preliminary version of the issue will be available at the workshop.

3 Program Committee

The following program committee is responsible for the selection of papers.

- Giulio Antoniol (Università degli Studi del Sannio, Italy)
- Andrea Corradini (Università di Pisa, Italy)
- Stephane Ducasse (University of Bern, Switzerland)
- Jean-Marie Favre (Institut d'Informatique et Mathématiques Appliquées de Grenoble, France)
- José Fiadeiro (University of Leicester, United Kingdom)
- Harald Gall (Technical University of Vienna, Austria)
- Martin Große-Rhode (Fraunhofer ISST Berlin, Germany)
- Reiko Heckel (Universität Dortmund, Germany) [co-chair]
- Anton Jansen (University of Groningen, The Netherlands)
- Dirk Janssens (University of Antwerp, Belgium)
- Juan F. Ramil (Open University, United Kingdom)
- Ashley McNeile (Metamaxim Ltd., London, United Kingdom)
- Tom Mens (Vrije Universiteit Brussel, Belgium) [co-chair]

4 Support

Due to support by the *European Science Foundation (ESF)* through the Scientific Network RELEASE (Research Links to Explore and Advance Software Evolution) the workshop is free of participation fees. It is the official kick-off meeting of the Working Group on Software Evolution of the *European Research Consortium for Informatics and Mathematics (ERCIM)* and a meeting of the *European Research Training Network SegraVis* on Syntactic and Semantic Integration of Visual Modeling Techniques.

References

1. R. Heckel, T. Mens, and M. Wermelinger, editors. *Proc. ICGT 2002 Workshop on Software Evolution through Transformations*, volume 72.4 of *Electronic Notes in TCS*, Barcelona, Spain, October 2002. Elsevier Science Publ.
2. M.M. Lehman and J.F. Ramil. Software evolution. invited keynote paper. In *International Workshop on Principles of Software Evolution, Vienna, Austria*, September 2001. A revised and extended version on an article to appear in Marciniak J. (ed.), Encyclopedia of Software Engineering, 2nd. Ed., Wiley, 2002.

Workshop on Logic, Graph Transformations, Finite and Infinite Structures

Bruno Courcelle and David Janin

University of Bordeaux-1, France
{courcell,janin}@labri.fr

Like in Barcelona in 2002, a satellite workshop will be organized during one day and a half.

Its main topics are the *use of logic* for representing graph properties and graph transformations, and the definition and the study of *hierarchical structurings of graphs*, in order to obtain decidability results and polynomial time algorithms. The two main motivations for this type of study are the construction of *efficient algorithms for particular types of graphs* and the *verification of programs* based on infinite graphs representing all possible computations. In many cases, the properties to verify are specified in certain logical languages. It is thus important to understand the relationships between the *logical expression of properties* (graph properties or properties of programs) and the *complexity of their verification*. This is the aim of *Descriptive Complexity*. The possibility of constructing logical formulas in the powerful but nevertheless restricted language of *Monadic Second-Order Logic* is frequently based on deep combinatorial properties. The **Graph Minor Theorem** and its extensions to *Matroids* and related notions like *Isotropic Systems* have a prominent place among such properties. They are linked to *tree-decompositions* and similar hierarchical decompositions. Other notions of hierarchical structuring like *modular and split decompositions* are also recognized as important. Among the numerous notions of graph structuring proposed by graph theoreticians, some fit better with logic and algorithmics. It is thus interesting to revisit these notions in a logical and algorithmic perspective.

The term "graph" in the title is actually a generic and convenient wording, covering *hypergraphs, relational structures, and related combinatorial structures, like matroids and partial orders*.

The **Graph Minor Theorem** proved by Robertson and Seymour is certainly a major result in Graph Theory, with many consequences for the construction of graph algorithms and the understanding of graph structure in the large. The notion of *tree-width* emerged from this work, and is the paradigmatic parameter in the theory of *Fixed Parameter Tractability*. *Clique-width* is another important parameter which has arisen from the theory of *Context-Free Graph Grammars*. This parameter is important for the construction of polynomial time algorithms and for the understanding of the structure of graphs obtained from trees by certain *graph transductions* (graph transformations) formalized in *Monadic Second-Order Logic*.

H. Ehrig et al. (Eds.): ICGT 2004, LNCS 3256, pp. 448–450, 2004.
© Springer-Verlag Berlin Heidelberg 2004

A natural question is whether the notion of *minor inclusion* (closely linked to tree-width) has some counter-part for "bounded clique-width". The answer is **Yes**, as the two lectures by S. Oum of Princeton University (who is working with Paul Seymour) will show. Two key notions are that of a *vertex-minor*, the desired alternative to that of a *minor*, and the notion of *rank-width*. The same sets of graphs have bounded rank-width and bounded clique-width, but the set of graphs of rank-width at most any fixed k is characterized by *finitely many excluded vertex-minors*.

The 15 year long standing conjecture made by D. Seese that the *sets of graphs for which Monadic Second-Order Logic is decidable* have a "*tree struc-ture*" equivalent to *bounded clique-width* can be "almost" proved as a conse-quence of these new results. "Almost" because the proved result uses the exten-sion of Monadic Second-Order Logic with a set predicate expressing *even cardi-nality*, so that the hypothesis is stronger. This conjecture has been presented in detail in the previous workshop held in Barcelona in 2002.

These new results owe much to the notions *of matroid and of isotropic system*, two combinatorial and algebraic concepts associated with graphs. Together with their representations in Monadic Second-Order Logic, they will be presented in the talks by P. Hliněny, D. Seese, S. Oum and B. Courcelle.

Several talks will be devoted to the definitions of *hierarchical decompositions of graphs, hypergraphs and matroids*, from which one can define parameters like tree-width, branch-width, clique-width and now *rank-width*. Decompositions of hypergraphs can be characterized in terms of *games*, extending a characteriza-tion of tree-width in terms of the *Robber and Cops game* (P. Seymour *et al.*). Furthermore the notions of tree-width and clique-width can also be used for *log-ical formulas*, and yield polynomial algorithms in situations where the general case (like for **SAT**, the Satisfiability problem) is not (known to be) solvable in polynomial time. These topics will be presented by F. Scarcello and J. Makowsky. Other decompositions like the *modular* and the *split decompositions* have inter-esting algorithmic properties, to be presented by D. Kratsch, M. Rao and A. Brandstädt. E. Fisher will consider the complexity of testing a graph property satisfied by few graphs, by using tools from the theory of random graphs.

In order to be able to formulate algorithmic questions about *infinite struc-tures* one needs to describe them by *finite tools* like automata, equation systems, logical formulas, or logically described transformations from fundamental struc-tures like trees. Infinite graphs and trees can be seen as representations of the *sets of all possible computations of processes*. The notion of a *winning strategy* for games on the infinite objects that represent their behaviours is a convenient way to express their correctness in an interactive context. There are intimate relations between games, automata and certain logical languages. They will be presented in the talks by D. Janin and D. Caucal.

The following talks have been confirmed at the date of July 10th.

B. Courcelle, Bordeaux (10 min.): Wellcome and presentation of the program.

P. Hliněny, Ostrava (45 min.): Are matroids interesting combinatorial struc-tures?

D. Seese, Karlsruhe (30 min.): Decidability and computability for matroids of bounded branch-width. (Joint work with P. Hliněny).

S. Oum, Princeton (45 min.): Rank-width, clique-width and vertex-minors.

B. Courcelle (30 min.): Isotropic systems, monadic second-order definition of vertex-minors and Seese's Conjecture. (Joint work with S. Oum).

S. Oum (45 min.): An algorithm for recognizing rank-width $< k$ and the well-quasi-ordering of the vertex-minor relation.

D. Kratsch, Metz (30 min.): Algorithms on $(P_5$, gem)-free graphs.

M. Rao, Metz (30 min.): On 2-join decomposition.

A. Brandstädt, Rostock (30 min.): Efficient algorithms for the Maximum Weight Stable Set problem: Graph decompositions and clique-width.

D. Janin, Bordeaux (45 min.): Automata for distributed games.

D. Caucal, Rennes (30 min.): Games on equational graphs.

F. Scarcello, Calabria (30 min.): Hypertree decompositions of hypergraphs: game-theoretic and logical representations.

J. Makowsky, Haïfa (45 min.) SAT related problems for clause sets of bounded clique-width. (Joint work with E. Fischer, A. Magid and E. Ravve).

E. Fisher, Haïfa (30 min.): The difficulty of testing for isomorphism against a graph that is given in advance.

Original articles related to this presentations will be invited for publication (after the usual refereeing process) to a special issue a journal, most likely the new electronic free access journal: **Logical Methods in Computer Science** (Web site http://www.lmcs-online.org)

The workshop is organized by the **International Conference on Graph Transformations** and sponsored by the European network **GAMES**, coordinated by E. Graedel (Aachen, Germany). The program committee consists of B. Courcelle, D. Janin (Bordeaux, France), H. Ehrig (Berlin, Germany) and E. Graedel (Aachen, Germany).

Author Index

Lecture Notes in Computer Science

For information about Vols. 1–3130

please contact your bookseller or Springer